Approximation Theory, Spline Functions and Applications

NATO ASI Series

Advanced Science Institutes Series

A Series presenting the results of activities sponsored by the NATO Science Committee, which aims at the dissemination of advanced scientific and technological knowledge, with a view to strengthening links between scientific communities.

The Series is published by an international board of publishers in conjunction with the NATO Scientific Affairs Division

A Life Sciences
B Physics

Plenum Publishing Corporation
London and New York

C Mathematical and Physical Sciences
D Behavioural and Social Sciences
E Applied Sciences

Kluwer Academic Publishers
Dordrecht, Boston and London

F Computer and Systems Sciences
G Ecological Sciences
H Cell Biology
I Global Environmental Change

Springer-Verlag
Berlin, Heidelberg, New York, London, Paris and Tokyo

NATO-PCO-DATA BASE

The electronic index to the NATO ASI Series provides full bibliographical references (with keywords and/or abstracts) to more than 30000 contributions from international scientists published in all sections of the NATO ASI Series.
Access to the NATO-PCO-DATA BASE is possible in two ways:

– via online FILE 128 (NATO-PCO-DATA BASE) hosted by ESRIN, Via Galileo Galilei, I-00044 Frascati, Italy.

– via CD-ROM "NATO-PCO-DATA BASE" with user-friendly retrieval software in English, French and German (© WTV GmbH and DATAWARE Technologies Inc. 1989).

The CD-ROM can be ordered through any member of the Board of Publishers or through NATO-PCO, Overijse, Belgium.

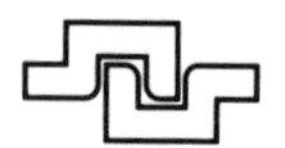

Series C: Mathematical and Physical Sciences - Vol. 356

Approximation Theory, Spline Functions and Applications

edited by

S. P. Singh

Memorial University of Newfoundland,
St. John's, NF, Canada

with the assistance of

Antonio Carbone

Università della Calabria,
Arcavacata di Rende (CS), Italy

R. Charron and B. Watson

Memorial University of Newfoundland,
St. John's, NF, Canada

Kluwer Academic Publishers

Dordrecht / Boston / London

Published in cooperation with NATO Scientific Affairs Division

Proceedings of the NATO Advanced Study Institute on
Approximation Theory, Spline Functions and Applications
Maratea, Italy
April 28–May 9, 1991

Library of Congress Cataloging-in-Publication Data

Approximation theory, spline functions, and applications / edited by
S.P. Singh with the assistance of Antonio Carbone ... [et al.].
p. cm. -- (NATO ASI series. Series C, Mathematical and
physical sciences ; 356)
"Published in cooperation with NATO Scientific Affairs Division."
"Proceedings of the NATO Advanced Study Institute on Approximation
Theory, Spline Functions, and Applications held in the Hotel villa
del Mare, Maratea, Italy between April 28, 1991 and May 9, 1991"--Pref.
Includes bibliographical references and index.
ISBN 0-7923-1574-X (acid free paper)
1. Approximation theory--Congresses. 2. Spline theory--Congresses. I. Singh, S. P. (Sankatha Prasad), 1937- .
II. North Atlantic Treaty Organization. Scientific Affairs
Division. III. Series: NATO ASI series. Series C, Mathematical and
physical sciences ; no. 356.
QA221.A68 1992
511'.42--dc20 91-43975

ISBN 0-7923-1574-X

Published by Kluwer Academic Publishers,
P.O. Box 17, 3300 AA Dordrecht, The Netherlands.

Kluwer Academic Publishers incorporates the publishing programmes of
D. Reidel, Martinus Nijhoff, Dr W. Junk and MTP Press.

Sold and distributed in the U.S.A. and Canada
by Kluwer Academic Publishers,
101 Philip Drive, Norwell, MA 02061, U.S.A.

In all other countries, sold and distributed
by Kluwer Academic Publishers Group,
P.O. Box 322, 3300 AH Dordrecht, The Netherlands.

Printed on acid-free paper

Printed in the Netherlands

This volume is dedicated to the memory of

Kanhaya Lal Singh

(February 15, 1944 - November 22, 1990)

Table of Contents

PREFACE

These are the Proceedings of the NATO Advanced Study Institute on Approximation Theory, Spline Functions and Applications held in the Hotel villa del Mare, Maratea, Italy between April 28,1991 and May 9, 1991.

The principal aim of the Advanced Study Institute, as reflected in these Proceedings, was to bring together recent and up-to-date developments of the subject, and to give directions for future research. Amongst the main topics covered during this Advanced Study Institute is the subject of univariate and multivariate wavelet decomposition over spline spaces. This is a relatively new area in approximation theory and an increasingly important subject. The work involves key techniques in approximation theory-cardinal splines, B-splines, Euler-Frobenius polynomials, spline spaces with non-uniform knot sequences. A number of scientific applications are also highlighted, most notably applications to signal processing and digital image processing. Developments in the area of approximation of functions examined in the course of our discussions include approximation of periodic phenomena over irregular node distributions, scattered data interpolation, Padé approximants in one and several variables, approximation properties of weighted Chebyshev polynomials, minimax approximations, and the Strang-Fix conditions and their relation to radial functions.

I express my sincere thanks to the members of the Advisory Committee, Professors B.Beauzamy, E.W. Cheney, J. Meinguet, D.Roux, and G.M. Phillips. My sincere appreciation and thanks go to A. Carbone, E. DePascale, R. Charron, and B. Watson for their excellent organization and smooth running of the Institute. I express my warm appreciation and thanks to Ivar Massabo for his help, cooperation and encouragement from the initial planning until the successful completion of the Institute.

I extend my thanks to my colleagues who helped me with the planning of the Institute.

I take this opportunity to extend my sincere thanks to the NATO Scientific Affairs Division for the generous support for the Institute, to the Universita della Calabria and Department of Tourism, Potenza, for the financial support, and to Memorial University of Newfoundland for secretarial help. Special thanks to Mrs. Philomena French for her excellent typing of parts of the manuscript.

Finally I thank the staff of Kluwer Academic Publishers for their understanding and cooperation.

This volume is dedicated to the memory of **Kanhaya Lal Singh** (February 15, 1944 - November 22, 1990) who always participated in and contributed to all our conferences.

S.P. Singh
St. John's. Newfoundland, Canada
September 26 , 91

List of Contributors

Aksoy, A. G., Department of Mathematics, Clarement McKenna College, Claremont, California, 91711, USA.

Bos, L., Department of Mathematics, University of Calgary, Calgary, Alberta, T2N 1N4, CANADA.

Campiti, M., Dipartimento di Matematica, Università di Bari, Traversa 200 Via Re David, 4, 70125 Bari, ITALY.

Casini, E., Department of Mathematics, University of Bologna, Piazza Porta S. Donato 5, Bologna, 40127, ITALY.

Cavaretta, A. S., Department of Mathematics and Computer Science, Kent State University, Kent, Ohio 44242, USA.

Cheney, W., Department of Mathematics, University of Texas, Austin, Texas 78712, USA.

Chui, C. K., Department of Mathematics, Texas A & M University, College Station, Texas 77843, USA.

Costabile, F., Dipartimento di Matematica, Università della Calabria, 87036 Arcavacata di Rende (Cosenza), ITALY.

Criscuolo, G., Istituto per le Applicazioni della Matematica - CNR Napoli, Via P. Castellino, 111, 80131 Napoli, ITALY.

Cuyt, A., Department of Mathematics & Computer Science, University of Antwerp (UIA), Universitaetsplein 1, B-2610 Wilrijk-Antwerp, BELGIUM.

De Blasi, F.S., Department of Mathematics, University of Roma II, 00133, Roma, ITALY.

Della Vecchia, B., Istituto per le Applicazioni della Matematica - CNR Napoli, Via P. Castellino, 111, 80131 Napoli, ITALY.

De Michelle, L., Dipartimento di Matematica, Università di Milano, Via C. Saldini, 50, 20133 Milano, ITALY.

De Pascale, E., Dipartimento di Matematica, Università della Calabria, 87036 Arcavacata di Rende (Cosenza), ITALY.

Deutsch, F., Department of Mathematics, Penn State University, University Park, PA 16802, USA.

Di Natale, M., Dipartimento di Matematica, Università di Milano, Via C. Saldini, 50, 20133 Milano, ITALY.

Gasca, M.V., Dep. de Matematica Aplicada, Universidad Zaragoza, Fac. Ciencias, 50009 Zaragoza, SPAIN.

Golitschek, M., Institut für Anglewandte Mathematik und Statistik, Universität Würzburg, 8700 Würzburg, GERMANY.

Gregory, J. A., Department of Mathematics, Brunel University, Uxbridge, Middlesex, UB8 3PH, U.K.

Guo, Shu Shen, Department of Mathematics, Hebei Normal University, Shijiazhuang, Hebei, 050016, P.R.C. (CHINA).

Isac, G., Department de Mathématiques, Collège Militaire Royal, St. Jean, Quebec, J0J 1R0, CANADA.

Lemarié-Rieusett, P. G., Université Paris-Sud, Mathematiques, Bat 425, 91405 Orsay Cedex, FRANCE.

Light, W., Department of Mathematics, University of Leicester, Leicester, LE1 7RH, U.K.

Lutterodt, C. H., Department of Mathematics, Howard University, Washington D. C. 20059, USA.

Marino, G., Dipartimento di Matematica, Università della Calabria, 87036 Arcavacata di Rende (Cosenza), ITALY.

Mastroianni, G., Dipartimento di Matematica, Università della Basilicata, Via N. Sauro, 85100 Potenza, ITALY.

Micchelli, C. A. , IBM T. J. Watson Research Center, P. O. Box 218, Yorktown Heights, NY 10598, USA.

Milman, P.D., Department of Mathematics, University of Toronto, Toronto, Ontario, M5S 1A1, CANADA.

Myjak, J, Dipartimento di Matematica Pura e Applicata, Università di L'Aquila, 67100 L'Aquila, ITALY.

Neamtu, M., University of Twente, Department of Applied Mathematics, P. O. Box 217, 7500 AE Enschede, THE NETHERLANDS.

Papini, P. L., Dipartimento di Matematica, Università di Bologna, Piazza di Porta S. Donato, 5, 40127 Bologna, ITALY.

Pená, J.M.F., Dep. Matematica Aplicada, Fac. Ciencias, Universidad de Zaragoza, 50009 Zaragoza, SPAIN.

Phillips, G. M., The Mathematical Institute, Unviersity of St. Andrews, St. Andrews, Fife KY16 9SS, SCOTLAND, U.K.

Pietramala, P., Dipartimento di Matematica, Università della Calabria, 87036 Arcavacata di Rende (Cosenza), ITALY.

Qu, Ruibin, Department of Mathematics and Statistics, Brunel University, Uxbridge, Middx, UB8 3PH, U.K.

Ricceri, B., Dipartimento di Matematica, Università di Catania, Città Universitaria, Viale A. Doria 6, 95125 Catania, ITALY.

Roux, D., Dipartimento di Matematica, Università di Milano, Via C. Saldini, 50, 20133 Milano, ITALY.

Schaback, R. A., Institut fur Numerische und Angewandte Math., Universität Göttingen, Lotzestrasse 16-18, 3400 Göttingen, GERMANY.

Sehgal, V.M. Department of Mathematics, University of Wyoming, Laramie, WY, 82071, USA.

Shekhtman, B., Department of Mathematics, University of South Florida, Tampa, FL 33620, USA.

Singh, S. P., Department of Mathematics and Statistics, Memorial University, St. John's, NF, A1C 5S7, CANADA.

Trombetta, G., Dipartimento di Matematica, Università della Calabria, 87036 Arcavacata di Rende (Cosenza), ITALY.

List of Participants

Aksoy, A. G., Department of Mathematics, Clarement McKenna College, Claremont, California, 91711, USA.

Arge, E., Inst. for Informatics, University of Oslo, P. O. Box 1080, Blindem, 0316 Oslo 3, NORWAY.

Bacopoulos, A. Department of Mathematics, National Technical University, Zografou, Athens, GREECE.

Baillon, J. B., Université Lyon I-IMI, 69622 Villeurbonne Cedex, FRANCE.

Behrends, E. Freie Universität Berlin, Math. Institut, 1000 Berlin 33, GERMANY.

Benda, N., Kappakokias 8, N. Pendeli, Athens, GREECE.

Bloom, T., Department of Mathematics, University of Toronto, Toronto, Ontario, M5S 1A1, CANADA.

Bos, L., Department of Mathematics, University of Calgary, Calgary, Alberta, T2N 1N4, CANADA.

Bozzini, M., Dipartimento di Matematica, Università di Milano, Via C. Saldini, 50, 20133 Milano, ITALY.

Buhman, M., Department of Mathematics, Magdalene College, Cambridge CB3 0AG, ENGLAND.

Campiti, M., Dipartimento di Matematica, Università di Bari, Traversa 200 Via Re David, 4, 70125 Bari, ITALY.

Canino, A., Dipartimento di Matematica, Università della Calabria, 87036 Arcavacate di Rende (Cosenza), ITALY.

Carbone, A., Dipartimento di Matematica, Università della Calabria, 87036 Arcavacata di Rende (Cosenza), ITALY.

Cavaretta, A. S., Department of Mathematics and Computer Science, Kent State University, Kent, Ohio 44242, USA.

Charron, R., Department of Mathematics and Statistics, Memorial University of Newfoundland, St. John's, NF, A1C 5S7, CANADA.

Cheney, W., Department of Mathematics, University of Texas, Austin, Texas 78712, USA.

Chiappinelli, R., Dipartimento di Matematica, Università della Calabria, 87036 Arcavacata di Rende (Cosenza), ITALY.

Chui, C. K., Department of Mathematics, Texas A & M University, College Station, Texas 77843, USA.

Criscuolo, G., Istituto per le Applicazioni della Matematica - CNR Napoli, Via P. Castellino, 111, 80131 Napoli, ITALY.

Cuyt, A., Department of Mathematics & Computer Science, University of Antwerp (UIA), Universitaetsplein 1, B-2610 Wilrijk-Antwerp, BELGIUM.

Della Vecchia, B., Istituto per le Applicazioni della Matematica - CNR Napoli, Via P. Castellino, 111, 80131 Napoli, ITALY.

De Michelle, L., Dipartimento di Matematica, Università di Milano, Via C. Saldini, 50, 20133 Milano, ITALY.

De Pascale, E., Dipartimento di Matematica, Università della Calabria, 87036 Arcavacata di Rende (Cosenza), ITALY.

Deutsch, F., Department of Mathematics, Penn State University, University Park, PA 16802, USA.

Di Natale, M., Dipartimento di Matematica, Università di Milano, Via C. Saldini, 50, 20133 Milano, ITALY.

Fidan Z., Ege Universitesi, Fen Fakultesi, Matematik Bolumu, Bornova-Izmir, TURKEY.

Fournier G., Department of Mathematics, University of Sherbrooke, Sherbrooke, Quebec, J1K 2R1, CANADA.

Franchetti, C., Dipartimento di Matematica Applicata, "G. Sansone", Università di Firenze, 50139 Firenze, ITALY.

Gasca, M.v., Dep. de Matematica Aplicada, Universidad Zaragoza, Fac. Ciencias, 50009 Zaragoza, SPAIN.

Golitschek, M., Institut für Anglewandte Mathematik und Statistik, Universität Würzburg, 8700 Würzburg, GERMANY.

Gonsor, D. E., Department of Mathematics, Kent State University, Kent, Ohio 44242, USA.

Gori, L., Dipartimento di Matematica, "Metodi e Modelli Matematici", Università "La Sapienza" di Roma, Via A. Scarpa, 10, 00161 Roma, ITALY.

Isac, G., Department de Mathématiques, Collège Militaire Royal, St. Jean, Quebec, J0J 1R0, CANADA.

Kurpinar E., Ege Universitesi, Fen Fakultesi, Matematik Bolumu, Bornova-Izmir, TURKEY.

Lassonde, M., Department Mathématiques, Université Blaise Pascal, 63177 Ausbière Cedex, FRANCE.

Lemarié-Rieusett, P. G., Université Paris-Sud, Mathematiques, Bat 425, 91405 Orsay Cedex, FRANCE.

Lewicki, G., Department of Mathematics, Jagiellonian University, 30-059 Krakow, Keymonta 4, POLAND.

Light, W., Department of Mathematics, University of Leicester, Leicester, LE1 7RH, U.K.

Lorentz, R. A., Gesellschaft fuer Mathematik und Datenverarbeitung, Schloss Birlinghoven, 5205 St. Augustin 1, GERMANY.

Lubuma, M., International Center for Theoretical Physics Trieste, P. O. Box 586, 34100 Trieste, ITALY.

Lutterodt, C. H., Department of Mathematics, Howard University, Washington D. C. 20059, USA.

Lyche, T., Institut for Informatics, University of Oslo, P. O. Box 1080, Blindem, 0316 Oslo 3, NORWAY.

Maddalena, L., Dipartimento di Mathematica, Università di Napoli, Via Mezzocannone, 16, 80134 Napoli, ITALY.

Marino, G., Dipartimento di Matematica, Università della Calabria, 87036 Arcavacata di Rende (Cosenza), ITALY.

Martinez-Legaz, J. E., Dept. Economia i Historia Economica, Universitat Autonoma de Barcelona, 08193 - Bellaterra, Barcelona, SPAIN.

Massabò, I., Dipartimento di Matematica, Università della Calabria, 87036 Arcavacata di Rende (Cosenza), ITALY.

Mastroianni, G., Dipartimento di Matematica, Università della Basilicata, Via N. Sauro, 85100 Potenza, ITALY.

Matos, A. C., Grupo di Mathematica Aplicada, Faculdade di Ciencias, Universidade do Porto, 4000 Porto, PORTUGAL.

Micchelli, C. A. , IBM T. J. Watson Research Center, P. O. Box 218, Yorktown Heights, NY 10598, USA.

Myjak, J, Dipartimento di Matematica Pura e Applicata, Università di L'Aquila, 67100 L'Aquila, ITALY.

Neamtu, M., University of Twente, Department of Applied Mathematics, P. O. Box 217, 7500 AE Enschede, THE NETHERLANDS.

Nugari, R., Dipartimento di Matematica, Università della Calabria, 87036 Arcavacata di Rende (Cosenza), ITALY.

Papini, P. L., Dipartimento di Matematica, Università di Bologna, Piazza di Porta S. Donato, 5, 40127 Bologna, ITALY.

Pedersen, H. L., Math. Inst., University of Copenhagen, 2100 Copenhagen, DENMARK.

Pená, J.M.F., Dep. Matematica Aplicada, Fac. Ciencias, Universidad de Zaragoza, 50009 Zaragoza, SPAIN.

Perri, U., Università di Reggo Calabria, Facoltà di Ingegneria, Viale E. Cuzzocrea, Reggio Calabria, ITALY.

Phillips, G. M., The Mathematical Institute, Unviersity of St. Andrews, St. Andrews, Fife KY16 9SS, SCOTLAND, U.K.

Pietramala, P., Dipartimento di Matematica, Università della Calabria, 87036 Arcavacata di Rende (Cosenza), ITALY.

Qu, Ruibin, Department of Mathematics and Statistics, Brunel University, Uxbridge, Middx, UB8 3PH, U.K.

Rassias, T.M., Department of Mathematics, University of Laverne, P. O. box 51105, Kifissia, 14510 Athens, GREECE.

Ricceri, B., Dipartimento di Matematica, Università di Catania, Città Universitaria, Viale A. Doria 6, 95125 Catania, ITALY.

Roux, D., Dipartimento di Matematica, Università di Milano, Via C. Saldini, 50, 20133 Milano, ITALY.

Schaback, R. A., Institut fur Numerische und Angewandte Math., Universität Göttingen, Lotzestrasse 16-18, 3400 Göttingen, GERMANY.

Schmets, J., Mathematics Inst., Avenue des Tilleuls, 15, Universitè de Liège, B-4000 Liege, BELGIUM.

Shekhtman, B., Department of Mathematics, University of South Florida, Tampa, FL 33620, USA.

Shraga Y., Department of Mathematical Physics, Soreq Nuclear Res. Center, 70600 Yavne, ISRAEL.

Singh, S. N., Department of Mathematics, Purvanchal University, Jaunpur, UP 222002,INDIA.

Singh, S. P., Department of Mathematics and Statistics, Memorial University, St. John's, NF, A1C 5S7, CANADA.

Sitharam, M., Nassestr. 2, Rsch. Inst. Dir. Math, University of Bonn, 5300 Bonn 1, GERMANY.

Stroem, K., Institut for Informatics, University of Oslo, P. O. Box 1080, Blindem, 0316 Oslo 3, NORWAY.

Thera, M., Department of Mathematiques, Université de Limoges, 87060 Limoges Cedex, FRANCE.

Trombetta, G., Dipartimento di Matematica, Università della Calabria, 87036 Arcavacata di Rende (Cosenza), ITALY.

Weber, H., Dipartimento di Matematica, Università della Basilicata, Via N. Sauro, 85100 Potenza, ITALY.

Zabrejko, P. P., Department of Mathematics, Belorussian State University, Minsk 220080, USSR.

APPROXIMATION BY FUNCTIONS OF NONCLASSICAL FORM

E. W. CHENEY
Mathematics Department
The University of Texas at Austin
Austin, Texas 78712, USA

ABSTRACT. This paper summarizes four expository lectures on these topics: (1) Interpolation on the circle, (2) Interpolation on higher-dimensional spheres, (3) Interpolation on Euclidean spaces, and (4) Ridge functions.

1. Introduction

This is the written account of four expository talks presented at the NATO conference. My purpose was to acquaint the audience with a number of relatively recent advances in the approximation of functions, especially *multivariate* functions. Emphasis was placed on new types of functions that have been found to be useful. Proofs of results (when given) were limited to simple cases, so that basic ideas could be easily conveyed. Full references to results quoted are given here.

Many cited results are taken from papers that I wrote with either Will Light, Xingping Sun, or Yuan Xu. They are to be fully credited with the results, of which this paper gives only a sampling.

2. Interpolation on the Circle

In this section, I draw upon work of Light, Xu, and myself in [24] and [42].

The unit circle in the plane is denoted by S^1. Its points are conveniently described by an angular variable θ, and so we can write

$$S^1 = \{(\cos\theta, \sin\theta) : -\infty < \theta < \infty\}$$

Interpolation on S^1 can be reduced to interpolation on the interval $[0, 2\pi)$, but for the moment we prefer to look upon S^1 as an interesting metric space, in which the distance between two points, say

$$x = (\cos\alpha, \sin\alpha) \qquad y = (\cos\beta, \sin\beta)$$

S. P. Singh (ed.), Approximation Theory, Spline Functions and Applications, 1–18.

is given by

$$d(x, y) = \min_{k \in \mathbb{Z}} |\alpha - \beta - 2k\pi| \tag{1}$$

This is the so-called **geodesic** distance; it is not to be confused with the Euclidean distance, which is

$$\|x - y\| = \sqrt{(\cos\alpha - \cos\beta)^2 + (\sin\alpha - \sin\beta)^2}$$

The space S^1 is the first (interesting) member in the sequence of spheres $S^1, S^2, \ldots, S^\infty$. Ultimately, we wish to develop methods for interpolating scattered data on these spaces. The definitions of these spheres are

$$S^m = \{x \in \mathbb{R}^{m+1} : \|x\| = 1\}$$
$$S^\infty = \{x \in \ell^2 : \|x\| = 1\}$$

In all cases, we can use this formula for the geodesic distance:

$$d(x, y) = \mathrm{Arccos}\langle x, y\rangle \tag{2}$$

Here $\langle x, y\rangle$ is the usual inner product, and the Arccosine function produces values in the interval $[0, \pi]$. A pair of points x, y on S^m is said to be an **antipodal** pair if $x + y = 0$, or equivalently, $d(x, y) = \pi$.

One might hope that S^1 would serve as a model for the higher dimensional spheres in interpolation theory. This hope is largely unrealized, however, because S^1 is not at all typical. For one thing, it has an elementary Abelian group structure that can be exploited in interpolation. This structure is obtained by "unrolling" the circle onto the real line and introducing an equivalence relation whereby a point is identified with any other point that differs from it by an integer multiple of 2π. In this way we see that

$$S^1 = \mathbb{R}/(2\pi\mathbb{Z}) = \mathbb{R}/\mathbb{Z}$$

This was already used in arriving at Equation (1).

On S^1, the most familiar method of interpolation is by means of trigonometric polynomials. It turns out that many other families of functions can be used. Locher [25] was apparently the first to make this observation in the case of equally-spaced nodes. We approach the topic by looking for **radial basis** functions that are suitable.

If (X, d) is a metric space and if $f : X \to \mathbb{R}$, we say that f is a **radial** function if it has the form

$$f(x) = \phi\big(d(x, \xi)\big) \tag{3}$$

for some fixed function ϕ and a fixed point $\xi \in X$. In a normed linear space, Equation (3) looks like this:

$$f(x) = \phi\big(\|x - \xi\|\big)$$

Such functions have been found to be very useful in the interpolation of scattered data in $\mathbb{R}^d$, when the Euclidean norm is employed. Two popular choices for ϕ are

$$\phi(t) = \sqrt{c^2 + t^2} \quad \text{and} \quad \phi(t) = 1/\sqrt{c^2 + t^2} \tag{4}$$

These lead respectively to **multiquadric** and **inverse multiquadric** surfaces. A history and bibliography of this topic is being prepared by the originator, Roland Hardy, and is to be presented at the Dublin IMACS Conference, July, 1991.

The standard interpolation problem involving the radial function in Equation (2) is as follows. Let n distinct points $y_1, y_2, \ldots, y_n$ (called **nodes**) be given in X. With each node there is associated a datum $r_i \in \mathbb{R}$. We seek to interpolate these data by a function of the form

$$F(x) = \sum_{j=1}^{n} c_j \phi\big(d(x, y_j)\big) \tag{5}$$

To accomplish this, one simply imposes the interpolation condition $F(y_i) = r_i$ $(1 \le i \le n)$. This leads to the **interpolation equations**

$$\sum_{j=1}^{n} c_j \phi\big(d(y_i, y_j)\big) = r_i \qquad (1 \le i \le n) \tag{6}$$

Whether this can be solved (for arbitrary data r_i) depends upon the **interpolation matrix**

$$A_{ij} = \phi\big(d(y_i, y_j)\big) \qquad (1 \le i, j \le n) \tag{7}$$

Open problem. For the particular spaces $\mathbb{R}^m$ and S^m find the most general family of functions ϕ for which the interpolation matrices in Equation (7) are nonsingular (for all n and for all sets of distinct nodes).

Now we specialize to S^1, and consider a basic case, corresponding to $c = 0$ in the first function in Equation (4). That is, $\phi(t) = t$. The interpolation matrix in this case is simply

$$A_{ij} = d(y_i, y_j) = \gamma(y_i - y_j)$$

in which we have employed the function

$$\gamma(t) = \min_{j \in \mathbb{Z}} |t + 2\pi j|$$

We observe these properties of γ:

$$\gamma \text{ is } 2\pi\text{-periodic} \tag{8}$$

$$d(x, y) = \gamma(x - y) \tag{9}$$

$$\gamma \text{ is even} \tag{10}$$

$$\gamma(x) + \gamma(x \pm \pi) = \pi \tag{11}$$

These can all be seen from the graph of γ. It looks like $|t|$ on the interval $|t| \leq \pi$ and is continued periodically.

We begin our analysis by showing that the interpolation matrix can be singular. Suppose that our set of nodes contains two antipodal pairs. By renumbering the nodes we can assume that $y_2 = y_1 + \pi$ and $y_4 = y_3 + \pi$. By Equation (11), we find that our basic functions $g_j(x) = d(x, y_j)$ are linearly dependent on each other. In fact,

$$\begin{aligned} & g_1(x) + g_2(x) - g_3(x) - g_4(x) \\ & \quad = \gamma(x - y_1) + \gamma(x - y_2) - \gamma(x - y_3) - \gamma(x - y_4) \\ & \quad = \gamma(x - y_1) + \gamma(x - y_1 - \pi) - \gamma(x - y_3) - \gamma(x - y_3 - \pi) \\ & \quad = \pi - \pi = 0 \end{aligned}$$

The full story is somewhat more complicated, and we refer to [24] for the following result.

Theorem. *For a set of n distinct nodes these properties are equivalent:*

(a) The set $\{g_1, \ldots, g_n\}$ is linearly independent.
(b) The node set contains at most one antipodal pair.
(c) the interpolation matrix is nonsingular.

We do not possess corresponding results for the radial function $\phi(t) = (c^2 + t^2)^{1/2}$ when $c \neq 0$.

Interpolation at uniformly spaced nodes on S^1 is quite a different matter, for we are able to identify a large class of useful radial functions. To fix the notation, let n be given, and set

$$y_j = 2\pi j/n \qquad (j \in \mathbb{Z})$$

We select a function ϕ and consider the $n \times n$ interpolation matrix

$$A_{ij} = \phi\big(d(y_i, y_j)\big) \qquad (1 \le i, j \le n)$$

If this matrix is nonsingular, then we can interpolate arbitrary data at the nodes $y_1, y_2, \ldots, y_n$ by a function having the form

$$x \mapsto \sum_{j=1}^{n} c_j \phi\big(d(x, y_j)\big)$$

For what functions ϕ is this possible? This question was answered by Locher in [25].

Since the values of $d(x, y)$ are in the interval $[0, \pi]$, we can assume without loss of generality that ϕ is even and 2π-periodic. We assume further that ϕ can be expanded in the cosine series whose coefficient sequence is in the space ℓ^1. Thus, we have

$$\phi(x) = \sum_{k=0}^{\infty} \alpha_k \cos kx \qquad \sum_{k=0}^{\infty} |\alpha_k| < \infty \tag{12}$$

These assumptions ensure that ϕ is continuous. An elementary calculation shows that $\phi(d(x, y)) = \phi(x - y)$ for all points x and y. It follows that the interpolation matrix A satisfies the equation

$$A_{ij} = \phi\big(d(y_i, y_j)\big) = \phi(y_i - y_j) = \phi(y_{i-j})$$

This indicates that A is a **circulant** matrix. We can refer to [9], or [20] for the eigenvalues of a circulant. For our special matrix, they are given by

$$\lambda_j = \sum_{\nu=0}^{n-1} \phi(y_\nu) e^{2\pi i j \nu / n} = \sum_{\nu=0}^{n-1} \phi(y_\nu) \cos(j y_\nu) \qquad (0 \le j \le n-1) \tag{13}$$

In this equation we recognize the discrete Fourier transform of the sequence $[\phi(y_\nu)]$. The discrete Fourier transform of order n can be denoted by $\mathcal{F}_n$ and defined by

$$\mathcal{F}_n u = w \qquad w_j = n^{-1} \sum_{\nu=0}^{n-1} u_\nu e^{-2\pi i \nu j / n} \qquad (j \in \mathbb{Z})$$

From this we have

$$\lambda_j = (\mathcal{F}_n u)_j \quad \text{where} \quad u_\nu = n\phi(y_{-\nu}) \tag{14}$$

This is how Locher interpreted the problem. His conclusion was, of course, that A is nonsingular if and only if the discrete Fourier transform of u (as above) had no zeros. By inserting the series that defines ϕ into Equation (13), one obtains another necessary and sufficient condition for the nonsingularity of A, namely

$$\sum_{k=0}^{\infty}(\alpha_{kn+j} + \alpha_{kn+n-j}) \neq 0 \qquad (0 \leq j \leq n-1)$$

This is Theorem 15 of [24]. A corollary of it is that if all the α_j in Equation (12) are positive, then for all n the interpolation matrix will be nonsingular. In fact, it will be symmetric and positive definite. Examples of such functions are

$$\begin{aligned}
&\phi(x) = (1 - 2a\cos x + a^2)^{-1} && (a \neq 0)\\
&\phi(x) = (1 + a\cos x)^{-1} && (|a| < 1)\\
&\phi(x) = \cos x(1 - 2a\cos x + a^2) && (a \neq 1)
\end{aligned}$$

Other conditions can be placed on ϕ to ensure the nonsingularity of A. As a sample, we cite a result from [24]:

Theorem. *If $\phi \in C^2[0,\pi]$, if $\int_0^\pi \phi(x)\,dx > 0$, and if ϕ' is nonnegative, nondecreasing, and nonconstant, then A is nonsingular for all even values of n.*

Examples of functions fulfilling the hypotheses in the preceding theorem are

$$\begin{aligned}
&\phi(x) = (c + x^2)^\beta && (\beta > 1/2\ ,\ c \geq 0)\\
&\phi(x) = (c + x^2)^{1/2} && (c > 0)\\
&\phi(x) = (c + x^2)^\beta && \left(0 < \beta < 1/2\ ,\ c \geq \pi^2/(1 - 2\beta)\right)\\
&\phi(x) = c - \cos\beta x && (c > 0\ ,\ 0 < |\beta| \leq 1/2)
\end{aligned}$$

Although we have used the nonsingularity of the interpolation matrix as the criterion for choosing ϕ, the interpolation itself can be accomplished without an inversion of the linear system. Instead, as discovered by Locher, a cardinal function can be given for the process. This is a function F in the linear span of the radial basis functions having the property $F(y_\nu) = \delta_{0\nu}$. By shifting this single function we obtain $F(y_i - y_j) = \delta_{ij}$. Hence the interpolant for data r_j at nodes y_j $(1 \leq j \leq n)$ is simply

$$\sum_{j=1}^{n} r_j F(x - y_j)$$

The cardinal function F in terms of ϕ is given by

$$F(x) = \frac{1}{n} \sum_{j=0}^{n-1} \sum_{\nu=0}^{n-1} \frac{1}{\lambda_\nu} e^{i\nu y_j} \phi(x - y_j) \tag{15}$$

Of course, the cardinal function exists precisely when the eigenvalues λ_j are all different from zero, for a fixed n.

In order to consider asymptotic questions, we label the nodes as y_{nj}, the eigenvalues as λ_{nj}, and the cardinal function as F_n. The accompanying interpolation operator is L_n. It provides an interpolant to any function g via the equation

$$(L_n g)(x) = \sum_{j=1}^{n} g(y_{nj}) F_n(x - y_{nj}) \qquad n = 1, 2, \ldots$$

Suppose that ϕ is chosen so that all L_n exist. Will it then be true that $\|L_n g - g\|_\infty \to 0$ for all $g \in C_{2\pi}$? The answer is "Yes" if and only if

(I) $\|L_n p - p\| \to 0$ for all trigonometric polynomials, and
(II) $\sup_n \|L_n\| < \infty$

Theorem. *If $\int_0^\pi \phi(x) \cos nx \, dx \neq 0$ and $\sum_{\nu=1}^{n} \phi(2\pi\nu/n) \cos(2\pi j\nu/n) \neq 0$ whenever $1 \leq j \leq n \in \mathbb{N}$ then $\|L_n p - p\|_\infty \to 0$ for all trigonometric polynomials p.* [42]

Corollary. *If $\phi(x) = \sum_{k=0}^{\infty} \alpha_k \cos kx$ with $\alpha_k > 0$ and $\sum \alpha_k < \infty$, then the translates of f form a fundamental set in $C_{2\pi}$. Indeed, the interpolation operators L_n discussed above have the property $L_n p \to p$ (uniformly) for each trigonometric polynomial p.*

This theorem and corollary address property (I) above. Property II is less easily verified. The norm of L_n as an operator on $C_{2\pi}$ is equal to the sup-norm of its Lebesgue function. The latter is

$$\Lambda_n(x) = \sum_{j=1}^{n} |F_n(x - y_{nj})| \tag{16}$$

So far, we have found no easy way to compute $\|\Lambda_n\|_\infty$ directly from the function ϕ. One can write $\Lambda_n(x)$ with the aid of Equations (16) and (15) and then substitute for λ_n the expression in Equation (13). Needless to say, the resulting formula for Λ_n is intractable. However in two cases, Xu and I were able to obtain some results:

(1) If $\phi(t) = |t|$ and if n is an odd integer at least 3, then L_n exists and $\|L_n\| \leq 3$.
(2) If $\phi(t) = t^2$ then L_n exists for all values of n, and for even n we have $\|L_n\| \leq 3$.

Many questions are left open. For example it would be desirable to know for what ϕ we have $\|L_n g - g\|_\infty \to 0$ for all $g \in C_{2\pi}$.

The papers [10] and [11] by Delvos are also concerned with this topic.

3. Interpolation on Spheres

In this section, I outline results recently obtained by Yuan Xu and myself [43] concerning interpolation on the sphere

$$S^m = \left\{x \in \mathbb{R}^{m+1} : \|x\| = 1\right\}$$

The starting point is a classical theorem of Schoenberg [34]. A function $\phi \in C[0,\pi]$ is said to be **positive definite** on S^m if the matrix $(\phi(d(x_i, x_j)))$ is nonnegative definite whenever $x_1, x_2, \ldots, x_n \in S^m$ and $n \in \mathbb{N}$. Here $d(x,y)$ is the geodesic distance on S^m as in Equation (2). Schoenberg proved that a function ϕ is positive definite on S^m if and only if it can be represented as

$$\phi(t) = \sum_{k=0}^{\infty} a_k p_k^{(\lambda)}(\cos t)\ , \qquad a_k \geq 0\ , \quad \sum a_k < \infty$$

In this equation, $p_k^{(\lambda)}$ is a Gegenbauer polynomial with normalization $p_k^{(\lambda)}(1) = 1$, and $\lambda = (m-1)/2$.

For interpolation on S^m at nodes $y_1, y_2, \ldots, y_n$ by a linear combination of radial basis functions $g_j(x) = \phi(d(x, y_j))$, we would like to have a positive definite matrix $\phi(d(y_i, y_j))$, and so require a variation of Schoenberg's theorem. We say that a continuous function ϕ is **strictly positive definite** on S^m if for any n and for any n distinct points $y_1, y_2, \ldots, y_n$ in S^m the matrix in question is positive definite. We did not discover a satisfactory characterization of such functions but obtained a simple sufficient condition.

Theorem. *If $\lambda = (m-1)/2$ and if*

$$\phi(t) = \sum_{k=0}^{\infty} a_k p_k^{(\lambda)}(\cos t)$$

with $a_k \geq 0$ for all k, $\sum a_k < \infty$, and $a_k > 0$ for $0 \leq k < n$, then for any n

distinct points y_i, the matrix $A_{ij} = \phi(d(y_i, y_j))$ is positive definite.

Here I shall give the proof for $m = 1$. In this case $\lambda = 0$ and $p_k^{(\lambda)}$ is the Chebyshev polynomial T_k. Assuming the hypotheses, we write $y_i = (\cos\theta_i, \sin\theta_i)$. Then

$$
\begin{aligned}
A_{ij} = \phi\big(d(y_i, y_j)\big) &= \sum_{k=0}^{\infty} a_k T_k\big(\cos(\theta_i - \theta_j)\big) \\
&= \sum_{k=0}^{\infty} a_k \cos\big(k(\theta_i - \theta_j)\big) \\
&= \sum_{k=0}^{\infty} a_k \cos k\theta_i \cos k\theta_j + \sum_{k=0}^{\infty} a_k \sin k\theta_i \sin k\theta_j \\
&= E_{ij} + F_{ij}
\end{aligned}
$$

Now any matrix of the form

$$(v\, v^T)_{ij} = (v_i\, v_j)$$

is a (rank 1) nonnegative definite matrix, since $c^T vv^T c = (v^T c)^T (v^T c) \geq 0$. Hence E and F, being sums of such matrices, are themselves nonnegative definite (as is A). Suppose now that $c^T Ac = 0$. Then $c^T Ec = 0$ and $c^T Fc = 0$. Hence, for F, we have

$$
\begin{aligned}
0 = \sum_{i=1}^{n}\sum_{j=1}^{n} c_i c_j F_{ij} &= \sum_{k=0}^{\infty} a_k \sum_{i=1}^{n}\sum_{j=1}^{n} c_i c_j \sin k\theta_i \sin k\theta_j \\
&= \sum_{k=0}^{\infty} a_k \left(\sum_{j=1}^{n} c_j \sin k\theta_j\right)^2
\end{aligned}
$$

Let $r = [n/2]$, and assume that $a_0, \ldots, a_r$ are positive. Then $\sum_{j=1}^{n} c_j \sin k\theta_j = 0$ for $1 \leq k \leq r$. Similarly $\sum_{j=1}^{n} c_j \cos k\theta_j = 0$ for $0 \leq k \leq r$, and consequently the functional $\mathcal{L}g = \sum_{j=1}^{n} c_j g(\theta_j)$ annihilates all trigonometric polynomials of degree at most r. Since this family of trigonometric polynomials is a Haar subspace of dimension $2r + 1 \geq n$, the coefficients c_i must all be zero. Thus the condition $c^T Ac = 0$ implies that $c = 0$, and A is positive definite. For $m = 1$, we have proved more than is required since $r \approx n/2$, but for the higher-dimensional spheres our proof requires $a_k > 0$ for $0 \leq k < n$. □

For the sphere S^∞, Schoenberg proved that if ϕ is positive definite then it has a representation

$$\phi(t) = \sum_{k=0}^{\infty} a_k \cos^k t \qquad a_k \geq 0 \,, \quad \sum_{k=0}^{\infty} a_k < \infty \tag{17}$$

The converse was given by Bingham [2]. Yuan Xu and I recently proved the following result in [44].

Theorem. *If $\phi(t)$ is as in Equation (17) with $a_m a_{m-1} > 0$ for some m, then for any $n \leq m+1$ and for any n distinct points $y_1, y_2, \ldots, y_n \in S^m$ the matrix $(\phi(d(y_i, y_j)))$ is positive definite.*

A simple consequence of this theorem is the following: In order for the function ϕ to be strictly positive definite on S^∞ it is sufficient that it have the form in Equation (17) and that infinitely many integers m have the property $a_m a_{m-1} > 0$.

4. Interpolation by Positive Definite Functions on $\mathbb{R}^d$

A function $\phi : \mathbb{R}^d \to \mathbb{C}$ is said to be **positive definite** if $\phi(x) = \overline{\phi(-x)}$ and if each matrix of the form

$$A_{kj} = \phi(x_k - x_j) \qquad (1 \leq k, j \leq n)$$

is nonnegative definite. That is,

$$c^* A c = \sum_k \sum_j \bar{c}_k c_j A_{kj} \geq 0$$

Here $x_1, \ldots, x_n$ are arbitrary points in $\mathbb{R}^d$, and n runs over all of $\mathbb{N}$.

A famous theorem of Bochner [3,4] characterizes the continuous positive definite functions on $\mathbb{R}^d$. They are precisely the functions that are Fourier transforms of nonnegative, finite, Borel measures:

$$\phi(x) = \int_{\mathbb{R}^d} e^{-i\langle x, y\rangle} \, d\mu(y)$$

In this setting, too, we require a refinement of this notion for use in interpolation on $\mathbb{R}^d$. We say that ϕ is **strictly positive definite** on $\mathbb{R}^d$ if for distinct points $x_1, \ldots, x_n$ in $\mathbb{R}^d$ the matrix $A_{kj} = \phi(x_k - x_j)$ is positive definite. When this circumstance arises, we can interpolate arbitrary data at the nodes x_j by a function of the form

$$F(x) = \sum_{j=1}^{n} c_j \phi(x - x_j)$$

In order to see the main principle at work here, let us prove this 1-dimensional result:

Theorem. *Let $f \in L^1(\mathbb{R})$, $f \geq 0$, $f \neq 0$. Then $\widehat{f}$ is strictly positive definite.*

Proof. Let $x_1, x_2, \ldots, x_n$ be distinct points in $\mathbb{R}$. Let $c = (c_1, c_2, \ldots, c_n) \in \mathbb{C}^n \setminus 0$. Put $A_{kj} = \widehat{f}(x_k - x_j)$. Then

$$\begin{aligned} c^* Ac &= \sum_k \sum_j \bar{c}_k c_j \int_{-\infty}^{\infty} f(y) e^{-i(x_k - x_j)y}\, dy \\ &= \int_{-\infty}^{\infty} f(y) \Big[\sum_k \bar{c}_k e^{-ix_k y}\Big] \Big[\sum_j c_j e^{ix_j y}\Big]\, dy \\ &= \int_{-\infty}^{\infty} f(y) \Big|\sum_j c_j e^{ix_j y}\Big|^2\, dy \\ &= \int_{-\infty}^{\infty} f(y) |g(y)|^2\, dy > 0 \qquad \square \end{aligned}$$

At the end we appeal to the following lemma.

Lemma. *Let $\lambda_1, \lambda_2, \ldots, \lambda_n$ be n distinct complex numbers, and let $g(z) = \sum_{j=1}^{n} c_j e^{\lambda_j z}$, where $c_j \in \mathbb{C}$ and $\sum_{j=1}^{n} |c_j| > 0$. Then g can have only a finite number of zeros in any bounded subset of $\mathbb{C}$.*

Proof. Use induction on n. If $n = 1$, then $g(z) = c_1 e^{\lambda_1 z}$ with $c_1 \neq 0$. This function has no zeros at all. Suppose now that the lemma has been established for $n < m$. Consider $g(z) = \sum_{j=1}^{m} c_j e^{\lambda_j z}$, with λ_j distinct and $\sum_{j=1}^{m} |c_j| > 0$. Suppose that g has an infinite number of zeros in a bounded set. These zeros have a point of accumulation. Since g is analytic, $g \equiv 0$. Consider

$$\begin{aligned} G(z) &= \frac{d}{dz}\left[e^{-\lambda_m z} g(z)\right] \\ &= \frac{d}{dz} \sum_{j=1}^{m} c_j e^{(\lambda_j - \lambda_m)z} = \sum_{j=1}^{m-1} c_j (\lambda_j - \lambda_m) e^{(\lambda_j - \lambda_m)z} \end{aligned}$$

Since $g \equiv 0$, the same is true of G. By the induction hypothesis, $0 = c_k(\lambda_j - \lambda_m)$ for $1 \leq j \leq m - 1$. It follows that $c_k = 0$ for $1 \leq j \leq m - 1$. Thus $g(z) =$

$c_m e^{\lambda_m z} = 0$. Hence c_m is also 0. □

Some strictly positive definite functions on $\mathbb{R}$ are listed here. These can be verified by consulting Oberhettinger [32].

$$f(x) = x^{-1} \sin ax \qquad f(x) = e^{-x^2/a}$$
$$f(x) = x^{-2} \cos x \sin^2(x/2) \qquad f(x) = x^{-1}(e^{-b|x|} - e^{-a|x|})$$
$$f(x) = e^{-a|x|} \qquad f(x) = \operatorname{sech}(ax)$$
$$f(x) = \cos(ax)e^{-b|x|} \qquad f(x) = \operatorname{Arctan}(a/|x|)$$
$$f(x) = (a^2 + x^2)^{-1} \qquad f(x) = \log(1 + e^{-\pi|x|})$$

There are many sources of information on positive definite functions. Some of them are [1], [13], [14], [15], [19], [22], [27], [31], [33], [36], [37].

5. Ridge Functions

In this section, I draw upon work of Xingping Sun and myself [39].

Let X be any normed linear space. A function $f : X \to \mathbb{R}$ is called a **ridge function** if it is of the form $f = g \circ \phi$ where $\phi \in X^*$ and $g \in C(\mathbb{R})$. (Thus ϕ is a continuous linear functional on X.)

When $X = \mathbb{R}^n$ (the most important case), we write

$$x = (\xi_1, \xi_2, \ldots, \xi_n)$$
$$a = (\alpha_1, \alpha_2, \ldots, \alpha_n)$$

$$\langle a, x \rangle = a \cdot x = \sum_{i=1}^{n} \alpha_i \xi_i$$

The latter formula expresses the most general linear functional on $\mathbb{R}$. Hence a ridge function in this setting is of the form

$$f(x) = g\big(\langle a, x \rangle\big) = g(\alpha_1 \xi_1 + \cdots + \alpha_n \xi_n)$$

Note that g is a function of one variable.

Ridge functions have a simple structure that permits them to be computed very efficiently. Moreover they can often be chosen to solve partial differential equations. For example, it is well known that the wave equation

$$\frac{\partial^2 u}{\partial \xi_1^2} = \frac{\partial^2 u}{\partial \xi_2^2}$$

becomes, under the change of variables $s = \xi_1 + \xi_2$, $t = \xi_1 - \xi_2$,

$$\frac{\partial^2 u}{\partial s \partial t} = 0$$

Its solution, obtained by integration, is

$$\begin{aligned} u &= g_1(s) + g_2(t) = g_1(\xi_1 + \xi_2) + g_2(\xi_1 - \xi_2) \\ &= g_1\big(\langle a, x\rangle\big) + g_2\big(\langle b, x\rangle\big) \end{aligned}$$

where $a = (1, 1)$ and $b = (1, -1)$.

A set G in a linear topological space E is said to be *fundamental* if its linear span is dense in E. Thus $\overline{\text{span } G} = E$. Equivalently: If $f \in E$ and if $\mathcal{N}$ is a neighborhood of 0, then there exist $g_1, \ldots, g_m \in G$ and $\lambda_1, \ldots, \lambda_m \in \mathbb{R}$ such that

$$f - \sum_{i=1}^{m} \lambda_i g_i \in \mathcal{N}$$

For a normed linear space, we use $\mathcal{N} = \{h : \|h\| < \varepsilon\}$ and then $\|f - \sum_{i=1}^{m} \lambda_i g_i\| < \varepsilon$.

Let X be any normed linear space. The space $C(X)$ consists of all continuous functions $f : X \to \mathbb{R}$. (They need not be bounded.) The usual algebraic structure is present:

$$\begin{aligned} (\alpha_1 f_1 + \alpha_2 f_2)(x) &:= \alpha_1 f_1(x) + \alpha_2 f_2(x) \\ (f_1 f_2)(x) &:= f_1(x) f_2(x) \end{aligned}$$

With these definitions, $C(X)$ becomes an *algebra*.

The usual function norms do not work on $C(X)$ since many elements are unbounded. However, $C(X)$ becomes a locally convex linear topological space if we define convergence as follows.

$$f_\nu \to 0 \;\text{ means }\; \|f_\nu\|_K \to 0 \;\text{ for all compact }\; K \subset X$$

Here we have introduced a family of seminorms by defining

$$\|f\|_K = \sup_{x \in K} |f(x)|$$

Now we can state the question we wish to answer in this lecture: For an arbitrary normed linear space X, do the ridge functions on X form a fundamental set in $C(X)$? Equivalently, we ask, If $f \in C(X)$, if K is compact in X, and if $\varepsilon > 0$, can we find $\phi_1, \ldots, \phi_m \in X^*$ and $g_1, \ldots, g_m \in C(\mathbb{R})$ such that

$$\left\| f - \sum_{i=1}^{m} g_i \circ \phi_i \right\|_K < \varepsilon$$

The most elementary case of interest is $X = \mathbb{R}^2$. Here the points are $x = (\xi_1, \xi_2)$ and the functionals are $\phi(x) = \langle v, x \rangle$ for $v \in \mathbb{R}^2$. Recall that the polynomials in two variables form a dense set in $C(\mathbb{R}^2)$. This is one case of the Stone-Weierstrass Theorem. Is it possible to approximate polynomials to arbitrary precision by linear combinations of ridge functions? If so, then the ridge functions will constitute a fundamental set. Since a polynomial in two variables has the form

$$p(x) = p(\xi_1, \xi_2) = \sum_{s=0}^{m} \sum_{r=0}^{m} a_{rs} \xi_1^r \xi_2^s$$

it is sufficient to prove that any monomial $\xi_1^r \xi_2^s$ can be approximated by a linear combination of ridge functions. More can be shown: that each monomial is a linear combination of "ridge polynomials." We try to solve this equation:

$$\sum_{j=0}^{k} c_j (\xi_1 + a_j \xi_2)^k = \xi_1^r \xi_2^s$$

By the binomial theorem, this becomes

$$\sum_{j=0}^{k} c_j \sum_{i=0}^{k} \binom{k}{i} \xi_1^{k-i} \xi_2^i a_j^i = \xi_1^r \xi_2^s$$

This can be rewritten as

$$\sum_{i=0}^{k} \xi_2^i \sum_{j=0}^{k} \binom{k}{i} c_j a_j^i \xi_1^{k-i} = \sum_{i=0}^{r+s} \delta_{is} \xi_1^{r+s-i} \xi_2^i$$

It suffices to put $k = r + s$ and to choose c_j so that

$$\sum_{j=0}^{k} \binom{k}{i} a_j^i c_j = \delta_{is} \qquad (0 \le i \le k)$$

Here s is fixed. The factor $\binom{k}{i}$ is a non-zero "row-factor" for the coefficient matrix and has no effect on the nonsingularity. The matrix that remains is (a_j^i), which is a Vandermonde matrix, known to be nonsingular if the points a_j are distinct. Thus the equations in question have a solution $(c_0, \ldots, c_k)$, provided that distinct a_j are chosen. This proves the next theorem.

Theorem. *If $[a_0, a_1, \ldots]$ is an infinite sequence of distinct real numbers, then the ridge functions of the form $(\xi_1 + a_i\xi_2)^k$, with $k, i \in \mathbb{N}$, form a fundamental set in $C(\mathbb{R}^2)$.*

Theorems of this type were first introduced by Hamaker and Solmon [18]. Extensions to $\mathbb{R}^d$ were given by Diaconis and Shahshahani [12]. In Sun and Cheney [39], it is proved that for a subset V of the unit sphere $\{x : \|x\| = 1\}$ in $\mathbb{R}^2$ these two properties are equivalent:

(1) $\{g \circ v : v \in V\ ,\ g \in C(\mathbb{R})\}$ is fundamental in $C(\mathbb{R}^2)$
(2) $\#V = \infty$.

In the same paper, we have given the next result.

Theorem. *For any normed linear space X, the set of all ridge functions on X is fundamental in $C(X)$.*

Proof. (This proof was suggested by an anonymous referee.) Let P be the linear subspace of $C(X)$ generated by the functions $x \mapsto [\phi(x)]^k$, where ϕ ranges over X^* and k ranges over $\mathbb{N}$. In order to prove that P is an algebra, it is enough to prove that $\phi^r\theta^s \in P$ whenever $\phi, \theta \in X^*$ and $r, s \in \mathbb{N}$. Recall from the preceding proof the equation

$$\sum_{j=0}^{r+s} c_j(\xi_1 + a_j\xi_2)^{r+s} = \xi_1^r\xi_2^s$$

Substitute $\xi_1 = \phi(x)$ and $\xi_2 = \theta(x)$, getting

$$\sum_{j=0}^{r+s} c_j\left[\phi(x) + a_j\theta(x)\right]^{r+s} = \left[\phi(x)\right]^r\left[\theta(x)\right]^s$$

Put $\theta_j = \phi + a_j\theta$. Then

$$\sum_{j=0}^{r+s} c_j\left[\theta_j(x)\right]^{r+s} = \left[\phi(x)\right]^r\left[\theta(x)\right]^s$$

Since $\sum c_j\theta_j^{r+s} \in P$, we conclude that $\phi^r\theta^s \in P$. This proves that P is a subalgebra of $C(X)$. Notice that P contains the constant functions and separates the points of X (by the Hahn-Banach Theorem). It follows (from a minor strengthening of the Stone-Weierstrass Theorem) that P is dense in $C(X)$. □

The form of the Stone-Weierstrass Theorem needed for non-compact spaces can be found as problem 44B, page 293 in Willard's book [41].

There is some interest in finding *subsets* of the family of all ridge functions that are still fundamental in $C(X)$. One can restrict the functionals ϕ and the continuous functions g in the expression $g \circ \phi$. Here is one such result [39].

Theorem. *Let $\mathcal{F}$ be any fundamental set in $C(\mathbb{R})$. Let X be any normed linear space. Let Φ be a subset of X^* such that $\{\phi/\|\phi\| : \phi \in X^* \setminus 0\}$ is dense in the unit sphere of X^*. Then*

$$\{f \circ \phi : f \in \mathcal{F}\ ,\ \phi \in \Phi\}$$

is fundamental in $C(X)$.

Example. In $\mathbb{R}^n$, let Φ consist of the integer lattice points in $\mathbb{R}^n$. (Thus $\Phi = \mathbb{Z}^n$.) Let $\mathcal{F}$ consist of the functions $t \mapsto 1, t, t^2, \ldots$. Then we see that the ridge functions $x \mapsto (i_1\xi_1 + \cdots + i_n\xi_n)^k$ form a fundamental set in $C(\mathbb{R}^n)$.

Other papers concerned with ridge functions, directly or tangentially, are [5], [6], [7], [8], [26], and [28].

References

1. R. Askey, "Orthogonal Polynomials and Special Functions," Regional Conference Series in Applied Mathematics, **21**, SIAM, Philadelphia, 1975.
2. N.H. Bingham, "Positive definite functions on spheres," Proc. Camb. Phil. Soc. **73** (1973), 145–156.
3. S. Bochner, "Vorlesungen über Fouriersche Integrale," Leipzig, 1932.
4. S. Bochner, "Monotone Funktionen...," Math. Ann. **108** (1933), 378–410.
5. D. Braess and A. Pinkus "Interpolation by ridge functions," preprint, 1991.
6. C.K. Chui and Xin Li, "Approximation by ridge functions and neural networks with one hidden layer," 1990.
7. G. Cybenko, "Approximation by superpositions of a sigmoidal function," Math. Control Signals Systems **2** (1989), 203–214.
8. W. Dahmen and C. Micchelli, "Some remarks on ridge functions," Approx. Theory and its Applications **3** (1987), 139–143.
9. P.J. Davis, "Circulant Matrices," Wiley-Interscience, New York, 1979.
10. F.-J. Delvos, "Convergence of interpolation by translation," Alfred Haar Memorial Volume, Colloquia Mathematica Societatis Janos Bolya **49** (1985), 273–287.
11. F.-J. Delvos, "Periodic interpolation on uniform meshes," J. Approximation Theory **51** (1987), 71–80.

12. P. Diaconis and M. Shahshahani, "On nonlinear functions of linear combinations," SIAM J. Sci. Stat. Comput. **5** (1984), 175–191.
13. W.F. Donoghue, "Distributions and Fourier Transforms," Academic Press, New York, 1969. (Volume 32 in the series Pure and Applied Mathematics.)
14. R.E. Edwards, "Functional Analysis," Holt, Rinehart and Winston, New York, 1965.
15. R.E. Edwards, "Integration and Harmonic Analysis on Groups," Cambridge University Press, 1972.
16. K. Guo and X. Sun, "Scattered data interpolation by linear combinations of translates of conditionally positive definite functions," to appear, J. Numer. Funct. Analysis and Optimization.
17. K. Guo, S. Hu, and X. Sun, "Conditionally positive definite functions and Laplace-Stieltjes integrals," preprint 1991.
18. C. Hamaker and D.C. Solmon, "The angles between the null spaces of X-rays," J. Math. Analysis and Appl. **62** (1978), 1–23.
19. S. Helgason, "Groups and Geometric Analysis," Academic Press, New York, 1984.
20. R.A. Horn and C.R. Johnson, "Topics in Matrix Analysis," Cambridge Univ. Press, 1991.
21. T. Husain, "Introduction to Topological Groups," W.B. Saunders, Philadelphia, 1966.
22. Y. Katznelson, "Harmonic Analysis," Dover Publ., New York, 1976.
24. W.A. Light and E.W. Cheney, "Interpolation by periodic radial basis functions," to appear, J. Math. Analysis and Appl.
25. F. Locher, "Interpolation on uniform meshes by translates of one function and related attenuation factors," Math. of Comp. **37** (1981), 403–416.
26. B.F. Logan and L.A. Shepp, "Optimal reconstruction of a function from its projections," Duke Math. J. **42** (1975), 645–659.
27. L. Loomis, "Abstract Harmonic Analysis," D. van Nostrand, New York, 1953.
28. W.R. Madych and S.A. Nelson, "Radial sums of ridge functions: a characterization," Math. Meth. Appl. Sci. **7** (1985), 90–100.
29. W.R. Madych and S.A. Nelson, "Multivariate interpolation and conditionally positive definite functions, II," Math. of Comp. **54** (1990), 211–230.
30. W.R. Madych and S.A. Nelson, "Multivariate interpolation and conditionally positive definite functions, I," Approx. Theory and its Appl. **4** (1988), no.4, 77–89.
31. M.A. Naimark, "Normed Rings," Wolters-Noordhoff, Groningen, 1964.
32. F. Oberhettinger, "Fourier Transforms of Distributions and Their Inverses," Academic Press, New York, 1973.
33. W. Rudin, "Fourier Analysis on Groups," Interscience, New York, 1963.

34. I.J. Schoenberg, "Positive definite functions on spheres," Duke Math. J. **9** (1942), 96–198.
35. I.J. Schoenberg, "Metric spaces and completely monotone functions," Annals of Math. **39** (1938), 811–841.
36. E.M. Stein and G. Weiss, "Introduction to Fourier Analysis on Euclidean Spaces," Princeton University Press, 1971.
37. J. Stewart, "Positive definite functions and generalizations, an historical survey," Rocky Mtn. J. Math. **6** (1976), 409–434.
38. X. Sun, "Conditionally positive definite functions and their application to multivariate interpolation," preprint, 1991.
39. X. Sun and E.W. Cheney, "The fundamentality of sets of ridge functions," preprint, 1990.
40. G. Szegö, "Orthogonal Polynomials," Amer. Math. Soc. Colloquium Publ., vol. XXIII, New York, 1959.
41. S. Willard, "General Topology," Addison-Wesley, Boston, 1970.
42. Y. Xu and E.W. Cheney, "Interpolation by periodic radial functions," preprint, 1991.
43. Y. Xu and E.W. Cheney, "Strictly positive definite functions on spheres," to appear, Proc. Amer. Math. Soc.
44. Y. Xu and E.W. Cheney, "A set of research problems in approximation theory," to appear in "Topics in Polynomials of One and Several Variables and Their Applications" edited by T.M. Rassias, H.M. Srivastava, and A. Yanushauskas, World Scientific Publishers, 1992.

WAVELETS — WITH EMPHASIS ON SPLINE-WAVELETS AND APPLICATIONS TO SIGNAL ANALYSIS

Charles K. Chui*
Center for Approximation Theory
Texas A&M University
College Station, TX 77843, U.S.A.

ABSTRACT. This is a tutorial on wavelets with special emphasis on spline-wavelets. It may be considered as a continuation of the tutorial article [2] written about a year ago for the Lancaster workshop. However, this present article is intended to be self-contained, as much as possible, with the danger of some overlapping. It is again written from the point of view of an approximation theorist with special interest in spline functions and applications to signal analysis.

1. Introduction

One of the most exciting recent development in modern mathematics is a very broad subject which may be called "Wavelet Analysis". Although there is still no unified notion, "wavelets", as it is usually called, has created a common link between pure and applied mathematics; between mathematics and information sciences; and in fact, among almost all subjects in physical science and technology. It enhances the physical interpretations and powerful techniques of Fourier Analysis by means of "adaptive localization" in both "time" and "frequency" domains.

Analogous to Fourier Analysis, in which the two entities of special importance are Fourier transform and Fourier series, the two most important issues in Wavelet Analysis are integral wavelet transform (IWT) and wavelet series. Yet, in contrast to Fourier Analysis, these two operations are intimately related. In fact, when a wavelet series is used to represent a function or analog signal f, the coefficients of this series are given by the IWT of f with respect to the "dual wavelet" and evaluated at binary scale (or frequency) levels and the corresponding dyadic positions. While the scale levels represent different frequency bands, the change of the window size governed by these levels automatically gives rise to the so-called "constant-Q filtering". Consequently, a wide frequency and narrow time window is used to study signals with high frequencies; and a narrow frequency and wide time window is used to study low-frequency signals. This is usually called the zoom-in and zoom-out effect of the IWT. The flexibility of the size of the time-frequency window makes the IWT the most natural method for studying time-frequency localizations, while the explicit display and separation of the IWT according to the time-frequency positions

* Research supported by NSF under Grant Number DMS-89-01345 and SDIO/IST managed by ARO under Contract Number DAAL 03-90-G-0091.

S. P. Singh (ed.), Approximation Theory, Spline Functions and Applications, 19–39.

in the wavelet series makes the wavelet series expansion the most desirable representation of the analog signal.

When spline functions are used to construct wavelets and their duals, both computation and implementation can be made extremely efficient. In particular, when a compactly supported spline-wavelet with minimal support, called a B-wavelet, is used to generate the B-wavelet series, the series has a special distinguished property. This property, which we call "complete oscillation" is an extreme contrast of the "total positivity" property of the B-spline series. It is one of the many ingredients of Wavelet Analysis that should be of special interest to the approximation theorist.

Other attractive aspects of Wavelet Analysis to the approximation theorist include the study of basis functions with minimum supports, approximation power versus vanishing moments, constructive aspects of wavelets, algorithms, recovery formulas, duals, Littlewood-Paley type inequalities and identities, and error estimations. We will only touch on a few of these topics. The interested reader is referred to the volumes [1,3,5,11,14,19].

2. Wavelet modeling of the cochlea

As is well known, the cochlea is the most essential organ in the human ear. It is a spiral-like device that lies behind the ear drum in the inner ear. Considered as a linear system, the audio reception (output) is given by the convolution of the audio signal (input) with the "system function". Let x denote the "position" in the cochlea and ω denote the frequency variable. Then if the cochlea system function is some real-valued function Θ, the corresponding transfer function $H_x(\omega)$ at the receptor position x is given by the Fourier transform of Θ with a delay due to the spiral geometry. So, by using the complex conjugate to facilitate our formulation, we may model the transfer function by

$$H_x(\omega) = \overline{\widehat{\Theta}(x - \ln \omega)}, \quad x > 0. \tag{2.1}$$

Therefore, if we consider a real-valued signal $f(t)$, then since

$$\hat{f}(\cdot - b)|_\omega = \hat{f}(\omega)e^{-ib\omega},$$

the reception $g(t-b, x)$ at the position x inside the cochlea at the time instant $t = b$ is given by the inverse Fourier transform of

$$H_x(\omega)\{\hat{f}(\omega)e^{-ib\omega}\}. \tag{2.2}$$

That is, from the assumption that both f and Θ are real-valued, we have

$$g(t-b, x) = \frac{1}{\pi} Re \int_0^\infty \{\hat{f}(\omega)e^{-ib\omega} H_x(\omega)\} e^{it\omega} d\omega;$$

or equivalently,

$$g(b, x) = \frac{1}{\pi} Re \int_0^\infty \hat{f}(\omega)e^{-ib\omega} H_x(\omega) d\omega.$$

This yields:

$$\begin{aligned} G(b,a) &:= g\left(b, \ln\frac{1}{a}\right) \\ &= \frac{1}{\pi} Re \int_0^\infty \hat{f}(\omega)\overline{\widehat{\Theta}(-\ln a - \ln\omega) e^{-ib\omega}} d\omega \\ &= \frac{1}{\pi} Re \int_0^\infty \hat{f}(\omega)\overline{\widehat{\Theta}\left(\ln\frac{1}{a\omega}\right) e^{-ib\omega}} d\omega. \end{aligned}$$

So, by setting

$$\widehat{\psi}(\omega) := \widehat{\Theta}\left(\ln\frac{1}{\omega}\right),$$

we have

$$\begin{aligned} G(b,a) &= \frac{1}{\pi} Re \int_0^\infty \hat{f}(\omega)\overline{\widehat{\psi}(a\omega) e^{ib\omega}} d\omega \\ &= \frac{1}{2\pi}\left\{\int_0^\infty \hat{f}(\omega)\overline{\widehat{\psi}(a\omega) e^{ib\omega}} d\omega \right. \\ &\quad \left. + \int_0^\infty \overline{\hat{f}(\omega)}\widehat{\psi}(a\omega) e^{ib\omega} d\omega \right\} \\ &= \frac{1}{a}\int_{-\infty}^\infty f(t)\overline{\psi\left(\frac{t-b}{a}\right)} dt; \end{aligned}$$

or

$$G(b,a) = \frac{1}{\sqrt{a}}(W_\psi f)(b,a), \tag{2.3}$$

where $W_\psi f$ is called the integral wavelet transform of f respect to the "mother wavelet" ψ, introduced by Grossmann and Morlet [15]. Observe that since $H_\infty(\omega) = 0$, we have

$$\int_{-\infty}^\infty \psi(t)dt = \widehat{\psi}(0) = \Theta(\infty) = 0.$$

Hence, if ψ has sufficiently fast decay at infinity, its graph behaves like a small wave.

3. The integral wavelet transform

Let ψ be a function in $L^2 := L^2(-\infty, \infty)$ with sufficiently fast decay at infinity. Then it can be used to localize (or window) an analog signal with finite energy represented by an L^2-function $f(t)$. More precisely, if we use two parameters, a and b, to adjust the width and position, respectively, of the window, then we have the so-called *integral wavelet transform* (IWT) of $f(t)$ with respect to the window function $\psi(t)$ introduced by Grossmann and Morlet [14], namely:

$$(W_\psi f)(b,a) := \frac{1}{\sqrt{a}}\int_{-\infty}^\infty f(t)\overline{\psi\left(\frac{t-b}{2}\right)} dt. \tag{3.1}$$

Suppose that $2\Delta_\psi$ and t^* are the width and center of ψ, respectively (cf. [2,4]). Then the function $f(t)$ in (3.1) is "localized" by a "time-window"

$$[b + at^* - a\Delta_\psi, b + at^* + a\Delta_\psi],$$

with center at $b+t^*$ and width $2a\Delta_\psi$. So, the positive parameter a is used to adjust the width, and the real parameter b that slides along the t-axis is to position the window.

The windowing concept described above is so simple that on the surface one does not expect any significant contribution from it to signal analysis. But on the contrary, this is really a major break-through in the field. The reason is that the time-windowing formula in (3.1) is identical to a frequency-windowing formula. To see this, let us consider the Fourier transform $\widehat{\psi}$ of ψ and assume that $\widehat{\psi}$ also has fast decay at infinity. Also, let $2\Delta_{\widehat{\psi}}$ and ω^* denote the width and center of $\widehat{\psi}$, respectively, and set

$$\eta(\omega) := \widehat{\psi}(\omega+\omega^*).$$

Furthermore, let us choose ψ so that $\omega^* > 0$. Then by the Plancherel identity, the IWT in (3.1) is also given by

$$(W_\psi f)(b,a) = \frac{\sqrt{a}}{2\pi}\int_{-\infty}^{\infty} \hat{f}(\omega)e^{ib\omega}\overline{\eta\Big(a\Big(\omega-\frac{\omega^*}{a}\Big)\Big)}d\omega. \tag{3.2}$$

Hence, with the exception of a linear-phase shift by b, the IWT also "localizes" the spectrum (or Fourier transform) $\hat{f}(\omega)$ by a "frequency-window"

$$\left[\frac{\omega^*}{a}-\frac{1}{a}\Delta_{\widehat{\psi}}, \frac{\omega^*}{a}+\frac{1}{a}\Delta_{\widehat{\psi}}\right],$$

with center at ω^*/a and width $2\Delta_{\widehat{\psi}}/a$. So, if a positive constant multiple of ω^*/a is used to represent the frequency variable, then the frequency-window widens while the time-frequency window narrows when high-frequency signals are observed. The opposite phenomenon occurs for low-frequency signals. This zoom-in and zoom-out capability of the IWT is what makes the IWT so important. Observe, in addition, that the ratio $\omega^*/a \div 2\Delta_{\widehat{\psi}}/a$ of the center frequency with the width of the "frequency band" is a constant. This is called *constant-Q* filtering.

To recover $f(t)$ from its IWT $W_\psi f$, Grossmann and Morlet [15] also introduced the formula:

$$f(t) = \frac{1}{C_\psi}\iint_{\mathbb{R}^2} (W_\psi f)(b,a)\frac{1}{|a|^{1/2}}\psi\left(\frac{b-t}{a}\right)\frac{dbda}{a^2}, \tag{3.3}$$

where

$$C_\psi := \int_{-\infty}^{\infty} \frac{|\widehat{\psi}(\omega)|^2}{|\omega|}d\omega. \tag{3.4}$$

For real-valued ψ and f, the recovery formula (3.3) can be reformulated so that the integral is taken over $(-\infty,\infty)\times(0,\infty)$ provided that the integral in the definition of C_ψ in (3.4) is also taken over $(0,\infty)$. This is more reasonable in applications to signal analysis.

Observe that for the formula (3.3) to be valid, the constant C_ψ in (3.4) must be finite. So, if $\widehat{\psi}$ is continuous at 0, it must be zero there; that is,

$$\int_{-\infty}^{\infty} \psi(t)dt = \widehat{\psi}(0) = 0,$$

and this is why ψ is called a wavelet, or small wave.

In the next section, we will be concerned with the reconstruction of $f(t)$ from only a discrete sample of its IWT, namely:

$$(W_\psi f)\left(\frac{k}{2^j}, \frac{1}{2^j}\right), \qquad j,k \in \mathbb{Z}.$$

4. Wavelet series and a classification of wavelets

In this paper, we will always use the notation:

$$\psi_{j,k}(x) = \psi(2^j x - k), \qquad j,k \in \mathbb{Z}, \tag{4.1}$$

where ψ is any function in L^2. Observe that we have dropped the normalization constant $2^{j/2}$ in (4.1). The reason for doing so is to facilitate the discussion and implementation of wavelet algorithms. Without this normalization constant, the "wavelet series" to be discussed in this section could be written as

$$f(x) = \sum_{j,k\in\mathbb{Z}} 2^{j/2} c_{j,k} \psi_{j,k}(x). \tag{4.2}$$

Suppose that this series represents an L^2 function f. Then how are the coefficients $c_{j,k}$ computed? The answer is clear if

$$\{2^{j/2}\psi_{j,k}(x)\}, \qquad j,k \in \mathbb{Z}, \tag{4.3}$$

is an orthonormal (o.n.) basis of L^2. In this case, we have

$$\begin{aligned} c_{j,k} &= \langle f, 2^{j/2}\psi_{j,k}\rangle \\ &= 2^{j/2}\int_{-\infty}^{\infty} f(x)\overline{\psi(2^j x - k)}dx \\ &= (W_\psi f)\left(\frac{k}{2^j}, \frac{1}{2^j}\right), \end{aligned} \tag{4.4}$$

where we have applied the definition of IWT in (3.1). Hence, indeed, the IWT and the orthogonal wavelet series are intimately related: on the jth octave (frequency band, or resolution level), the intensity (or magnitude) of the signal at the position $t = k/2^j$ is given by $(W_\psi f)(k/2^j, 2^{-j})$, which is the $(j,k)^{\text{th}}$ coefficient of the o.n. wavelet series.

DEFINITION 4.1. An L^2 function ψ is called an orthonormal (o.n.) wavelet, if the family in (4.3) is an o.n. basis of L^2.

The simplest o.n. wavelet is the Haar function, $\psi_H = \chi_{[0,1/2)} - \chi_{[1/2,1)}$, where χ_A denotes, as usual, the characteristic function of a set A. More recently, Daubechies [13] gave a very natural generalization of ψ_H by constructing o.n. wavelets with arbitrary smoothness and compact supports. Daubechies' fundamental results in [13] have greatly stimulated the current enthusiasm among both mathematicians and engineers in wavelet research and development. A very interesting result obtained in [13] is that any compactly supported o.n. wavelet which is either symmetric or antisymmetric relative to the center of its support must be the Haar function ψ_H. However, this restriction of the Daubechies wavelets limits their applications to many problems in signal analysis, since symmetry is equivalent to "linear phase" and antisymmetry is equivalent to "generalized linear phase" [8]. In image reconstruction from compressed data, for example, linear (or at least generalized linear) phase is needed to avoid distortion.

Hence, in order to retain the smoothness and compact support properties of ψ, the orthogonal property of $\{2^{j/2}\psi_{j,k}\}$ must be weakened. This leads to the second class of wavelets defined as follows.

DEFINITION 4.2. An L^2 function ψ is called a semi-orthogonal (s.o.) wavelet, if the family $\{2^{j/2}\psi_{j,k}\}$ as defined in (4.1) is a Riesz basis of L^2 and satisfies

$$\psi_{j,k} \perp \psi_{\ell,m}, \qquad j \neq \ell, \quad j,k,\ell,m \in \mathbb{Z}. \tag{4.5}$$

A general framework of compactly supported s.o. wavelets is given in our earlier work [8]. The most useful example is the class of all compactly supported s.o. wavelets ψ constructed as finite B-spline series [7]. The remaining sections of this paper will be devoted to the study of these so-called cardinal spline-wavelets. Before going into any details, we must go back to the wavelet series (4.2) and ask ourselves two important questions: How are the coefficients $c_{j,k}$ computed and what would make them related to the IWT?

First, it is clear that the dual basis $\{\psi^{\ell,m}\}$ relative to the Riesz basis $\{2^{j/2}\psi_{j,k}\}$ could be used to compute $c_{j,k}$ from f. Here, duality, of course, means that

$$\langle 2^{j/2}\psi_{j,k}, \psi^{\ell,m}\rangle = \delta_{j,\ell}\cdot\delta_{k,m}, \quad j,k,\ell,m \in \mathbb{Z}. \tag{4.6}$$

Hence, we have

$$c_{j,k} = \langle f, \psi^{j,k}\rangle. \tag{4.7}$$

However, to relate $c_{j,k}$ to the IWT, the basis functions $\psi^{j,k}$ must come from one single function $\widetilde{\psi}$, in the sense that

$$\psi^{j,k}(x) = 2^{j/2}\widetilde{\psi}(2^j x - k). \tag{4.8}$$

If this is the case, then we have

$$c_{j,k} = (W_{\widetilde{\psi}} f)\left(\frac{k}{2^j}, \frac{1}{2^j}\right) \tag{4.9}$$

as required, where the IWT is taken relative to $\widetilde{\psi}$, instead of ψ. We will call $\widetilde{\psi}$ the "*dual*" of ψ.

DEFINITION 4.3. An L^2 function ψ is called a wavelet, if the family $\{2^{j/2}\psi_{j,k}\}$ as defined in (4.1) is a Riesz basis of L^2, and if there exists some $\widetilde{\psi}$ in L^2 such that the dual basis $\{\psi^{j,k}\}$ relative to $\{2^{j/2}\psi_{j,k}\}$ is given by (4.8).

It is important to point out that $\widetilde{\psi}$ doesn't always exist. In other words, there are L^2 functions ψ each of which generates a Riesz basis $\{2^{j/2}\psi_{j,k}\}$ of L^2 but the dual basis $\psi^{j,k}$ is not generated by a $\widetilde{\psi}$ as in (4.8). A class of examples is given by

$$\eta(x) := \psi(x) - \bar{z}\sqrt{2}\psi(2x), \qquad |z| < 1, \tag{4.10}$$

where ψ is any o.n. wavelet (cf. [3]).

We still have to verify that an s.o. wavelet ψ as in Definition 4.2 is a wavelet. To justify this claim, we simply verify that $\widetilde{\psi}$ defined by

$$\widehat{\widetilde{\psi}}(\omega) = \frac{\widehat{\psi}(\omega)}{\sum_{k\in\mathbf{Z}} |\widehat{\psi}(\omega + 2\pi k)|^2} \tag{4.11}$$

is the dual of ψ in the sense that $\widetilde{\psi}$ generates $\{\psi^{j,k}\}$ as in (4.8). Indeed, in the first place, it is clear from the definition (4.11) that $\widetilde{\psi}$ can be written as

$$\widetilde{\psi}(x) = \sum_{k\in\mathbf{Z}} a_k \psi(x + k),$$

where $\{a_k\} \in \ell^2$ is defined by

$$\sum_{k\in\mathbf{Z}} a_k e^{ik\omega} = \frac{1}{\sum_{k\in\mathbf{Z}} |\widehat{\psi}(\omega + 2\pi k)|^2}.$$

Hence, since ψ satisfies (4.5), it is sufficient to verify

$$\langle \widetilde{\psi}(\cdot - k), \psi(\cdot - \ell)\rangle = \delta_{k,\ell}, \qquad k, \ell \in \mathbf{Z}. \tag{4.12}$$

This is certainly valid, since

$$\begin{aligned}
\langle \widetilde{\psi}(\cdot - k), \psi\rangle &= \int_{-\infty}^{\infty} \widetilde{\psi}(x - k)\overline{\psi(k)}dx \\
&= \frac{1}{2\pi}\int_{-\infty}^{\infty} \widehat{\widetilde{\psi}}(\omega)\overline{\widehat{\psi}(\omega)}e^{-ik\omega}d\omega \\
&= \frac{1}{2\pi}\sum_{\ell\in\mathbf{Z}}\int_{2\pi\ell}^{2\pi(\ell+1)} \widehat{\widetilde{\psi}}(\omega)\overline{\widehat{\psi}(\omega)}e^{-ik\omega}d\omega \\
&= \frac{1}{2\pi}\sum_{\ell\in\mathbf{Z}}\int_{0}^{2\pi} \widehat{\widetilde{\psi}}(\omega + 2\pi\ell)\overline{\widehat{\psi}(\omega + 2\pi\ell)e^{-ik\omega}}d\omega \\
&= \frac{1}{2\pi}\int_0^{2\pi} \frac{\sum_{\ell\in\mathbf{Z}} |\widehat{\psi}(\omega + 2\pi\ell)|^2}{\sum_{m\in\mathbf{Z}} |\widehat{\psi}(\omega + 2\pi m)|^2} e^{-ik\omega}d\omega \\
&= \frac{1}{2\pi}\int_0^{2\pi} e^{-ik\omega}d\omega = \delta_{k,0}.
\end{aligned}$$

Finally, we must point out that there are both advantages and disadvantages in applying s.o. wavelets to signal analysis. With the extra degree of freedom gained by sacrificing orthogonality in the same scale (or frequency) levels, the compactly supported s.o. wavelets can be easily made symmetric or antisymmetric. In fact, the s.o. wavelets with minimum supports constructed by using finite B-spline series of arbitrary orders in [7] are automatically symmetric for splines of even orders and antisymmetric for the ones of odd orders. We will study the construction of these spline-wavelets in the next section. Unfortunately, when compactly supported s.o. wavelets are constructed, their duals cannot have compact support as shown in [8]. Hence, to require symmetry (or antisymmetry) as well as compact supports for both ψ and its dual $\widetilde{\psi}$, even orthogonality between levels must also be sacrificed. Such wavelets could be classified as *non-orthogonal wavelets*. But since ψ and $\widetilde{\psi}$ satisfy the duality condition:

$$\langle 2^{j/2}\psi(2^j \cdot -k), 2^{\ell/2}\widetilde{\psi}(2 \cdot -m)\rangle = \delta_{j,\ell} \cdot \delta_{k,m},$$

where $j, k, \ell, m \in \mathbb{Z}$, they are also called bi-orthogonal wavelets in the literature [12].

5. Spline-wavelets

The most popular method for constructing an o.n. wavelet is first to construct a "scaling function" ϕ that generates a so-called "multiresolution analysis" of L^2 with the property that $\{\phi(\cdot - k)\}$, $k \in \mathbb{Z}$, is an orthonormal family. Then by reversing the two-scaled sequence that describes the relation between $\phi(x)$ and $\phi(2x - n)$, $n \in \mathbb{Z}$, followed by an alternation of the signs of this sequence and a shift of $\phi(2x - n)$ by $\frac{1}{2}$, one gets an o.n. wavelet ψ. Since the notion of multiresolution analysis due to Mallat [17] and Meyer [18] is well documented (cf. [3,14,19]) and many papers in the wavelet literature are concerned with o.n. wavelets, we will not go into any details, except by mentioning that for each m, the mth order cardinal B-spline N_m generates a multiresolution analysis of L^2 and how it is orthonormalized to give the scaling function ϕ as mentioned above. To be more specific, let $\chi_{[0,1)}$ denote the characteristic function of the unit interval [0,1). Hence, $N_1 := \chi_{[0,1)}$ is the first order cardinal B-spline. For any positive integer m, the mth order cardinal B-spline N_m is then defined by the m-fold convolution of N_1 with itself. In other words, in terms of Fourier transforms, we have $\widehat{N}_m = (\widehat{N}_1)^m$. The scaling function ϕ that generates the same multiresolution analysis as N_m does, with the additional property that $\{\phi(\cdot - k)\}$ is an orthonormal family, is given in terms of its Fourier transform by

$$\hat{\phi}(\omega) = \frac{\widehat{N}_m(\omega)}{\left\{\sum_{k\in\mathbb{Z}} |\widehat{N}_m(\omega + 2\pi k)|^2\right\}^{1/2}}. \tag{5.1}$$

The route we chose in [6,7,8] was to work directly with N_m, without considering the orthogonalization procedure (5.1). We had two motives for doing so. First, to an approximation theorist, the B-spline N_m has all the nice properties we want in a basis, with the exception of orthogonality which is too restrictive. Properties such as "total positivity" [16] would be lost if the transformation (5.1) is applied. Secondly, in constructing wavelets out of splines in [6,7], we really wanted to obtain *explicit* formulas of the spline-wavelets; and by this, we mean formulas that allow easy implementation both in software and hardware, without any iterative computations, as required to yield the other wavelets, such as the Daubechies wavelets.

To describe the main idea behind our construction of the minimally supported spline wavelets in [7], we return to considering the nested sequence of closed subspaces $\{V_k^m : k \in \mathbb{Z}\}$ of L^2 that constitute the multiresolution analysis of L^2, namely:

$$V_j^m := \text{clos}_{L^2}\langle N_{m;j,k}\colon\ k \in \mathbb{Z}\rangle, \quad j \in \mathbb{Z}, \tag{5.2}$$

where

$$N_{m;j,k}(x) := N_m(2^j x - k). \tag{5.3}$$

Since $V_j^m \subset V_{j+1}^m$ for each $j \in \mathbb{Z}$, we may also consider the orthogonal complementary subspaces W_j^m of V_{j+1}^m, relative to V_j^m. In other words, we consider the subspaces W_j that satisfy:

$$\begin{cases} W_j^m \subset V_{j+1}^m; \\ W_j^m \perp V_j^m; \\ V_{j+1}^m = V_j^m + W_j^m. \end{cases} \tag{5.4}$$

The notation we will use to describe (5.4) is

$$V_{j+1}^m = V_j^m \oplus W_j^m.$$

It is easy to see that W_j^m, $j \in \mathbb{Z}$, are mutually orthogonal subspaces of L^2, and in fact, it can even be shown that

$$L^2 = \bigoplus_{j \in \mathbb{Z}} W_j^m = \cdots \oplus W_{-1}^m \oplus W_0^m \oplus \cdots. \tag{5.5}$$

Just as how the B-spline N_m generates all the nested spline spaces V_j^m, $j \in \mathbb{Z}$, as described in (5.2), we want to find a function $\psi_m \in W_0$ that generates all the mutually orthogonal spaces W_j^m, $j \in \mathbb{Z}$, namely:

$$W_j^m = \text{clos}_{L^2}\langle \psi_{m;j,k}\colon\ k \in \mathbb{Z}\rangle, \quad j \in \mathbb{Z}, \tag{5.6}$$

where

$$\psi_{m;j,k}(x) := \psi_m(2^j x - k).$$

Note that we have not required $\{\psi_m(\cdot - j)\}$ to be an o.n. basis of W_0; but instead, just as N_m has minimum support among all functions in V_0^m, we also want ψ_m to have minimum support in W_0^m. Observe that the intrinsic property in (5.5) already guarantees ψ_m to be an s.o. wavelet, as introduced in the previous section.

The two-scale equation that relates $N_m \in V_0^m$ and $N_{m;1,j} \in V_1^m$, $j \in \mathbb{Z}$, is easily shown to be

$$N_m(x) = \sum_{j \in \mathbb{Z}} p_j N_m(2x - j), \tag{5.7}$$

where

$$p_j := \begin{cases} 2^{-m+1}\binom{m}{j} & \text{for } 0 \le j \le m, \\ 0 & \text{otherwise.} \end{cases} \tag{5.8}$$

Now, to find our desirable $\psi_m \in W_0^m \subset V_1^m$, we must study the other two-scale equation:

$$\psi_m(x) = \sum_{j \in \mathbb{Z}} q_j N_m(2x - j), \tag{5.9}$$

where $\{q_j\}$ is necessarily an ℓ^2-sequence, and for ψ_m to have compact support, we actually want a finite sequence $\{q_j\}$. In fact, we want the shortest sequence to guarantee ψ_m to have minimum support among all functions in W_0. For normalization purposes, let us require $q_n = 0$ for $n < 0$ but $q_0 \neq 0$.

To continue our discussion, we need the notation of the symbol of a sequence, namely:

$$\begin{cases} P_m(z) = \sum_{j \in \mathbb{Z}} p_j z^j; \\ Q_m(z) = \sum_{j \in \mathbb{Z}} q_j z^j. \end{cases} \tag{5.10}$$

With this notation, and by setting

$$z := e^{-i\omega/2}, \tag{5.11}$$

the two governing equations (5.7) and (5.9) can be written as

$$\begin{cases} \widehat{N}_m(\omega) = \frac{1}{2} P_m(z) \widehat{N}_m\left(\frac{\omega}{2}\right); \\ \widehat{\psi}_m(\omega) = \frac{1}{2} Q_m(z) \widehat{N}_m\left(\frac{\omega}{2}\right). \end{cases} \tag{5.12}$$

Now, the requirement of $V_0^m \perp W_0^m$ is described by the following infinite system of equations:

$$\langle N_m(\cdot - k), \psi \rangle = 0, \qquad k \in \mathbb{Z}.$$

So, by applying the Plancherel identity and (5.12), we have:

$$\begin{aligned} 0 &= \langle N_m(\cdot - k), \psi \rangle \\ &= \frac{1}{2\pi} \int_{-\infty}^{\infty} \widehat{N}_m(\omega) \overline{\widehat{\psi}(\omega)} e^{-ik\omega} d\omega \\ &= \frac{1}{8\pi} \int_{-\infty}^{\infty} P_m(z) \overline{Q_m(z)} \left| \widehat{N}_m\left(\frac{\omega}{2}\right) \right|^2 e^{-ik\omega} d\omega \\ &= \frac{1}{8\pi} \sum_{j \in \mathbb{Z}} \int_{2\pi j}^{2\pi(j+1)} P_m(z) \overline{Q_m(z)} \left| \widehat{N}_m\left(\frac{\omega}{2}\right) \right|^2 e^{-ik\omega} d\omega \\ &= \frac{1}{8\pi} \sum_{j \in \mathbb{Z}} \int_0^{2\pi} P_m(e^{-i(\frac{\omega}{2} + j\pi)}) \overline{Q_m(e^{-i(\frac{\omega}{2} + j\pi)})} \\ &\quad \times \left| \widehat{N}_m\left(\frac{\omega}{2} + j\pi\right) \right|^2 e^{-ik\omega} d\omega. \end{aligned} \tag{5.13}$$

Next, by introducing the function

$$B_m(\omega) := \sum_{j\in\mathbf{Z}} \left|\widehat{N}_m\left(\frac{\omega}{2} + 2\pi j\right)\right|^2, \tag{5.14}$$

and breaking up the last summation in (5.13) into the sum of one over the even integers and the other over the odd ones, we arrive at

$$0 = \frac{1}{8\pi}\int_0^{2\pi} \{P_m(z)\overline{Q_m(z)}B_m(\omega) + P_m(-z)\overline{Q_m(-z)}B_m(\omega+\pi)\}e^{-ik\omega}d\omega.$$

Observe that since the function inside the braces is a 2π-periodic continuous function and all the Fourier coefficients of this function are equal to zero, we must have

$$P_m(z)\overline{Q_m(z)}B_m(\omega) + P_m(-z)\overline{Q_m(-z)}B_m(\omega+\pi) \equiv 0. \tag{5.15}$$

Let us first digress to study the function $B_m(\omega)$ introduced in (5.14). Observe that

$$\begin{aligned}&\int_{-\infty}^{\infty} N_m(x+j)\overline{N_m(x)}dx \qquad (5.16)\\ &= \int_{-\infty}^{\infty} N_m(x+j)N_m(x)dx \\ &= \int_{-\infty}^{\infty} N_{2m}(m+j-x)N_m(x)dx = N_{2m}(m+j).\end{aligned}$$

Hence, by applying the Poisson summation formula to (5.14) and referring to the relations (5.12) and (5.16), we have

$$B_m(\omega) = \sum_{j\in\mathbf{Z}} N_{2m}(m+j)z^j. \tag{5.17}$$

That is, $B_m(\omega)$ is the "Euler-Frobenius Laurent polynomial" of the $2m^{\text{th}}$ order B-spline N_{2m} (without normalizing its coefficients to be integers) (cf. [20]). In view of (5.17), we may write $B_m(\omega)$ as

$$B_m(\omega) = E_{2m-1}(z), \tag{5.18}$$

and conclude that $E_{2m-1}(z)$ satisfies

$$E_{2m-1}\left(\frac{1}{z}\right) \equiv E_{2m-1}(z)$$

and is z^{-m+1} multiple of a polynomial with real coefficients and of exact order $2m-1$ that does not vanish at $z=0$. Using this knowledge and the fact that $P_m(z) = 2^{-m+1}(1+z)^m$, we can solve for $Q_m(z)$ in the identity (5.15), which is equivalent to

$$P_m(z)\overline{Q_m(z)}E_{2m-1}(z) + P_m(-z)\overline{Q_m(-z)}E_{2m-1}(-z) = 0, \quad |z| = 1. \tag{5.19}$$

In fact, the polynomial solution with the lowest degree that does not vanish at $z = 0$ can be easily shown to be

$$Q_m(z) = z^{m-1}E_{2m-1}(-z)P_m(-z). \tag{5.20}$$

That is, we have

$$q_j = (-1)^j 2^{-m+1} \sum_{\ell=0}^{m} \binom{m}{\ell} N_{2m}(j-\ell+1). \tag{5.21}$$

This two-scale sequence, which has support given by $[0, 3m-2] \cap \mathbb{Z}$, determines the mth order B-spline-wavelet (or simply, B-wavelet) ψ_m which was introduced in our earlier work [7]. It is clear from (5.9) that

$$\operatorname{supp} \psi_m = [0, 2m-1]. \tag{5.22}$$

For more details, the reader is referred to [7].

6. Oscillating properties of spline-wavelets

Let m be an arbitrary positive integer and consider the mth order B-wavelet ψ_m. Then ψ_m is a function in W_0^m with minimum support such that $\{\psi_m(\cdot - k)\}$ is a Riesz basis of W_0^m. We will call any finite series

$$g(x) = \sum_{k=0}^{N} d_k \psi_m(x-k) \tag{6.1}$$

a *B-wavelet series*. This is the companion of a finite *B-spline series*:

$$f(x) = \sum_{k=0}^{N} c_k N_m(x-k). \tag{6.2}$$

In both (6.1) and (6.2), we assume, without loss of generality, that c_k, $d_k = 0$ for $k < 0$. Of course we could even assume that c_0, c_N, d_0, and d_N are nonzero. What is so special about the B-splines N_m is that they are *totally positive* (cf. [16]). As a consequence, the number of sign changes of the spline series $f(x)$ in (6.2) is controlled by the number of sign changes of its coefficient sequence $\{c_k\}$. This property, in turn, governs the shape characteristics of the graph of the function $f(x)$. This is why spline functions are very useful for curve designs. To be more specific, the standard notation S^- and S^+ will be used for counting strong and weak sign changes (cf. [16]), and when the sign changes of a function f are counted, we only consider the sign changes that occur in the interior of the support of f. For a B-spline series such as f in (6.2), it is well-known that

$$S^-\left(\sum_{k=0}^{N} c_k N_m(\cdot - k)\right) \le S^-(\{c_k\}). \tag{6.3}$$

Note that if all the coefficients c_k of the series are non-negative, then the B-spline series is also non-negative everywhere.

The behavior a B-wavelet series is just the opposite. No matter what coefficients are chosen, the series always oscillates quite a lot. In fact, we could even give a lower bound on the number of oscillations. For convenience, let us avoid zero components in the graph of the B-wavelet series $g(x)$ in (6.1); that is, we will always assume that

$$|d_\ell| + \cdots + |d_{\ell+2m-2}| > 0, \quad \ell = 0, \ldots, N - 2m + 2. \tag{6.4}$$

The following result was obtained in [9].

THEOREM 6.1. *Under the assumption (6.4), it follows that*

$$S^- \left(\sum_{k=0}^{N} d_k \psi_m(\cdot - k) \right) \geq N + 3m - 2. \tag{6.5}$$

In particular, the mth *order B-wavelet* ψ_m *has exactly* $3m - 2$ *simple zeros in* $(0, 2m - 1)$ *which is the interior of the support of* ψ_m.

Observe that for $d_k = \delta_{k,0}$, the series (6.1) becomes

$$g(x) = \psi_m(x) = \sum_{k=0}^{3m-2} q_k N_m(2x - k),$$

and hence, by (6.3), we have

$$S^-(\psi_m) \leq S^-(\{q_k\}) \leq 3m - 2.$$

So, it follows from (6.5) that

$$S^-(\psi_m) = S^-(\{q_k\}) = 3m - 2. \tag{6.6}$$

Consequently, not only ψ_m has exactly $3m - 2$ simple zeros in $(0, 2m - 1)$, the two-scale sequence must alternate in signs as already derived in (5.21).

The locations of the zeros of ψ_m are also estimated in [9]. Let $x_k^{(m)}$, $1 \leq k \leq 3m - 2$, be the (simple) zeros of ψ_m arranged in increasing order. Then we have

$$\begin{cases} x_k^{(m)} = 2m - 1 - x_{3m-1-k}^{(m)}; \\ \max\left(\dfrac{k}{2}, k - m\right) < x_k^{(m)} < \min\left(\dfrac{k + m - 1}{2}, k\right). \end{cases} \tag{6.7}$$

Of course a larger lower bound is plausible when the count S^+ of weak sign changes is considered. This is, however, a much more difficult problem, and so far we only have the following result for linear B-wavelets (cf. [9]).

THEOREM 6.2. *Under the assumption (6.4), it follows that*

$$2N + 4 \geq S^+ \left(\sum_{k=0}^{N} d_k \psi_2(\cdot - k) \right) \geq 2N + 4 - S^-(\{d_k\}). \tag{6.8}$$

In particular, if $S^-(\{d_k\}) = 0$*, then*

$$S^+\left(\sum_{k=0}^{N} d_k\psi_2(\cdot - k)\right) = 2N + 4. \tag{6.9}$$

It is tempting to generalize (6.8) to B-wavelet series of arbitrary orders. In this direction, we still do not have a very good conjecture. However, the property (6.9) indicates that the number of sign changes could become maximum, and the lower bounds in (6.5) and (6.8) tell us that any wavelet curve (i.e. the graph of a B-wavelet series) has to oscillate quite a lot. So, in contrast to the property of "Total Positivity" of the B-spline N_m, the corresponding B-wavelet ψ_m seems to possess a remarkable property which may be called "complete oscillation". With the total positivity property, splines are very useful for curve smoothing and shape design. On the other hand, the compete oscillating wavelets are very useful for detecting irregularities, choosing the essential information of the data curve for storage and transmittance, modeling transience and turbulence, etc.

Let us only consider a potential application to data reduction. It is well known that under certain conditions a bandpass band-limited signal can be completely recovered from its zero-crossings (or sign changes). The reason for this to hold is that an entire function of exponential type whose restriction to the real-axis is an L^2 function is characterized by its real zeros when some normalization condition is imposed. For any signal with finite energy, a spline-wavelet model is piecewise analytic, and hence, there seems to be some chance for a certain zero-crossing result to hold. In fact, in [9] we have derived such a result for linear spline-wavelets. It would be even better if the wavelet curves are characterized by their local maxima and minima, since reconstruction from the extrema is more stable. Problems of this type have very important applications to signal and image analyses. Recent results of S. Mallat and his group indicate that this is a very promising area of research.

7. Spline-wavelet algorithms

In Section 5, we have discussed the relationship between both $N_m \in V_0^m$ and $\psi_m \in W_0^m$ with the Riesz basis $\{N_m(2^{j+1} \cdot -k)\}$ of V_{j+1}^m, namely: by using the notation $N_{m;j,k}$ and $\psi_{m;j,k}$ for scaling by 2^j and translation by k, the two-scale equations (5.7) and (5.9) become

$$\begin{cases} N_{m;j,k} = \sum_\ell p_{\ell-2k} N_{m;j+1,\ell}; \\ \psi_{m;j,k} = \sum_\ell q_{\ell-2k} N_{m;j+1,\ell}, \end{cases} \tag{7.1}$$

where the finite sequences $\{p_k\}$ and $\{q_k\}$ are given in (5.8) and (5.21). This set of identities gives rise to a very efficient "reconstruction algorithm" as follows.

For any function $f_{j+1} \in V_{j+1}^m$, we know from (5.4) that it has a unique decomposition:

$$f_{j+1} = f_j + g_j, \tag{7.2}$$

where $f_j \in V_j^m$ and $g_j \in W_j^m$. We now derive an FIR (finite impulse response) algorithm for reconstructing f_{j+1} from f_j and g_j by using (7.1). In what follows, we will represent f_j and g_j by the coefficient sequences of their B-spline series and B-wavelet series representations, namely:

$$\begin{cases} f_j = \sum_k c_k^j N_{m;j,k}; \\ \mathbf{c}^j = \{c_k^j\}, k \in \mathbf{Z}, \end{cases} \tag{7.3}$$

and

$$\begin{cases} g_j = \sum_k d_k^j \psi_{m;j,k}; \\ \mathbf{d}^j = \{d_k^j\}, k \in \mathbf{Z}. \end{cases} \tag{7.4}$$

Hence, by (7.1), we have

$$\begin{aligned} f_j + g_j &= \sum_k \{c_k^j N_{m;j,k} + d_k^j \psi_{m;j,k}\} \\ &= \sum_k \left\{ c_k^j \sum_\ell p_{\ell-2k} N_{m;j+1,\ell} + d_k^j \sum_\ell q_{\ell-2k} N_{m;j+1,\ell} \right\} \\ &= \sum_\ell \left\{ \sum_k p_{\ell-2k} c_k^j + \sum_k q_{\ell-2k} d_k^j \right\} N_{m;j+1,\ell}. \end{aligned}$$

Since

$$f_{j+1} = \sum_\ell c_\ell^{j+1} N_{m;j+1,\ell}$$

and $\{N_{m;j+1,\ell}\}$, $\ell \in \mathbf{Z}$, is a Riesz basis, it follows from (7.2) that

$$c_\ell^{j+1} = \sum_k p_{\ell-2k} c_k^j + \sum_k q_{\ell-2k} d_k^j. \tag{7.5}$$

Before applying the FIR algorithm (7.5) to construct the B-spline coefficient sequence $\mathbf{c}^{j+1}$ from the B-spline and B-wavelet sequences $\mathbf{c}^j$ and $\mathbf{d}^j$, we note that the sequences $\mathbf{c}^j$ and $\mathbf{d}^j$ must first be "up-sampled" by filling in zeros in-between. In other words, we must first change $\mathbf{c}^j$ and $\mathbf{d}^j$ to $\tilde{\mathbf{c}}^j$ and $\tilde{\mathbf{d}}^j$, respectively, where

$$\tilde{c}_m^j = \begin{cases} c_{m/2}^j & \text{for even } m \\ 0 & \text{for odd } m \end{cases}$$

and

$$\tilde{d}_m^j = \begin{cases} d_{m/2}^j & \text{for even } m \\ 0 & \text{for odd } m. \end{cases}$$

Then the wavelet reconstruction algorithm (7.5) becomes the following FIR (or moving average) algorithm:

$$c_\ell^{j+1} = \sum_m p_{\ell-m} \tilde{c}_m^j + \sum_m p_{\ell-m} \tilde{d}_m^j. \tag{7.6}$$

(Cf. Figure 7.1 for the schematic diagram.)

$$\begin{array}{ccccccccc} \mathbf{d}^j & \longrightarrow & \tilde{\mathbf{d}}^j & & \mathbf{d}^{j+1} & \longrightarrow & \tilde{\mathbf{d}}^{j+1} & & \\ & & & \searrow & & & & \searrow & \cdots \\ \mathbf{c}^j & \longrightarrow & \tilde{\mathbf{c}}^j & \longrightarrow & \mathbf{c}^{j+1} & \longrightarrow & \tilde{\mathbf{c}}^{j+1} & \longrightarrow & \end{array}$$

Figure 7.1. (Reconstruction algorithm)

To obtain an algorithm for decomposing f_{j+1} into the sum of f_j and g_j in (7.2), we first need to study how the B-splines $N_m(2x-\ell)$, $\ell \in \mathbb{Z}$, are related to the basis $\{N_m(x-k), \psi_m(x-k): k \in \mathbb{Z}\}$ of the space V_1^m. Let us write

$$N_m(2x-\ell) = \sum_k \{a_{\ell-2k} N_m(x-k) + b_{\ell-2k}\psi_m(x-k)\} \tag{7.7}$$

and try to determine the sequences $\{a_n\}$ and $\{b_n\}$. In terms of Fourier transform, the formulas in (7.7) can be written as

$$\begin{cases} \frac{1}{2}\widehat{N}_m\left(\frac{\omega}{2}\right) = \frac{G_m(z)+G_m(-z)}{2}\widehat{N}_m(\omega) + \frac{H_m(z)+H_m(-z)}{2}\widehat{\psi}_m(\omega) \\ \frac{1}{2}\widehat{N}_m\left(\frac{\omega}{2}\right) = \frac{G_m(z)-G_m(-z)}{2}\widehat{N}_m(\omega) + \frac{H_m(z)-H_m(-z)}{2}\widehat{\psi}_m(\omega), \end{cases} \tag{7.8}$$

where even and odd values of ℓ in (7.7) are considered separately, $z := e^{-i\omega/2}$, and

$$\begin{cases} G_m(z) := \sum_n a_n z^{-n}; \\ H_m(z) := \sum_n b_n z^{-n}. \end{cases} \tag{7.9}$$

Hence, when (5.12) is substituted into (7.8) and $\widehat{N}_m\left(\frac{\omega}{2}\right)$ is canceled, we arrive at

$$\begin{cases} P_m(z)G_m(z) + Q_m(z)H_m(z) = 2; \\ P_m(z)G_m(-z) + Q_m(z)H_m(-z) = 0. \end{cases} \tag{7.10}$$

To solve (7.10) for $G_m(z)$ and $H_m(z)$, we need the following identity for the Euler-Frobenius polynomials $E_{2m-1}(z)$ defined in (5.18) and (5.17), namely:

$$\left(\frac{1+z}{2}\right)^{2m} E_{2m-1}(z) + (-1)^m \left(\frac{1-z}{2}\right)^{2m} E_{2m-1}(-z) \equiv z^m E_{2m-1}(z^2) \tag{7.11}$$

(cf. [6, Lemma 2]). First, by substituting (5.8) and (5.21) into (5.10), we have

$$\begin{cases} P_m(z) = 2^{-m+1}(1+z)^m; \\ Q_m(z) = 2^{-m+1}(-z)^{m-1}(1-z)^m E_{2m-1}(-z). \end{cases} \tag{7.12}$$

Then, by applying (7.11) and (7.12), the solution of (7.10) is easily seen to be

$$\begin{cases} G_m(z) = z^{-m}\left(\frac{1+z}{2}\right)^m E_{2m-1}(z)\frac{1}{E_{2m-1}(z^2)}; \\ H_m(z) = -z^{-2m+1}\left(\frac{1-z}{2}\right)^m \frac{1}{E_{2m-1}(z^2)}. \end{cases} \tag{7.13}$$

Hence, the sequences $\{a_n\}$ and $\{b_n\}$ in the decomposition formula (7.7) can be easily calculated by using (7.13) in the definition (7.9) of $G_m(z)$ and $H_m(z)$.

Let us now return to deriving an algorithm for decomposing f_{j+1} into $f_j + g_j$. Again the coefficient sequences of the B-spline and B-wavelet series are used to represent f_j and g_j. Analogous to (7.1), the decomposition formula (7.7) can be written as

$$N_{m;j+1,k} = \sum_{\ell}\{a_{k-2\ell}N_{m;j,\ell} + b_{k-2\ell}\psi_{m;j,\ell}\}. \tag{7.14}$$

Hence, by using the representation (7.3), we have

$$\begin{aligned} f_{j+1} &= \sum_k c_k^{j+1} N_{m;j+1,k} \\ &= \sum_k c_k^{j+1} \sum_\ell \{a_{k-2\ell}N_{m;j,\ell} + b_{k-2\ell}\psi_{m;j,\ell}\} \\ &= \sum_\ell \left\{\left(\sum_k a_{k-2\ell}c_k^{j+1}\right) N_{m;j,\ell} + \left(\sum_k b_{k-2\ell}c_k^{j+1}\right)\psi_{m;j,\ell}\right\}. \end{aligned}$$

Since $f_{j+1} = f_j + g_j$ and

$$f_j + g_j = \sum_\ell \{c_\ell^j N_{m;j,\ell} + d_\ell^j \psi_{m;j,\ell}\},$$

the linear independence of $\{N_{m;j,\ell}, \psi_{m;j,\ell} \colon \ell \in \mathbb{Z}\}$ now implies that

$$\begin{cases} c_\ell^j = \sum_k a_{k-2\ell}c_k^{j+1}; \\ d_\ell^j = \sum_k b_{k-2\ell}c_k^{j+1}. \end{cases} \tag{7.15}$$

Observe that both formulas in (7.15) are again moving-averages algorithms. However, the decomposed sequences $\mathbf{c}^j$ and $\mathbf{d}^j$ are obtained by "down-sampling" the moving-averaged outputs; that is, we first apply the algorithm

$$\begin{cases} \hat{c}_m^j = \sum_k a_{k-m}c_k^{j+1}; \\ \hat{d}_m^j = \sum_k b_{k-m}c_k^{j+1}, \end{cases} \tag{7.16}$$

and then obtain $\mathbf{c}^j$ and $\mathbf{d}^j$ by

$$\begin{cases} c_\ell^j = \hat{c}_{2\ell}^j; \\ d_\ell^j = \hat{d}_{2\ell}^j, \qquad \ell \in \mathbb{Z}. \end{cases} \tag{7.17}$$

(Cf. Figure 7.2 for the schematic diagram.)

$$\begin{array}{ccccccccccc} & & \hat{\mathbf{d}}^j & \longrightarrow & \mathbf{d}^j & & \hat{\mathbf{d}}^{j-1} & \longrightarrow & \mathbf{d}^{j-1} & & \\ & \nearrow & & & & \nearrow & & & & \nearrow & \cdots \\ \mathbf{c}^{j+1} & \longrightarrow & \hat{\mathbf{c}}^j & \longrightarrow & \mathbf{c}^j & \longrightarrow & \hat{\mathbf{c}}^{j-1} & \longrightarrow & \mathbf{c}^{j-1} & \longrightarrow & \end{array}$$

Figure 7.2. (Decomposition algorithm)

We must now point out that the moving averages in (7.16) are no longer FIR algorithms, since the decomposition sequences $\{a_n\}$ and $\{b_n\}$ are not finite sequences. Fortunately, these two sequences have very fast exponential decay, so that truncation of these sequences should not induce too much loss of information. Observe that in their z-transforms $G_m(z)$ and $H_m(z)$ as given in (7.13), the only term that gives us trouble is the factor

$$T_m(z^2) := \frac{1}{E_{2m-1}(z^2)}. \tag{7.18}$$

So, let us consider the Laurent expansion

$$T_m(z) = \sum_j \beta_j z^j \tag{7.19}$$

and its truncated Laurent polynomial:

$$T_m^N(z) = \sum_{j=-N}^{N} \beta_j z^j. \tag{7.20}$$

Then, instead of $G_m(z)$ and $H_m(z)$ in (7.13), we determine the coefficients $\{a_k^N\}$ and $\{b_k^N\}$ of the Laurent polynomials:

$$\begin{cases} G_m^N(z) = \sum_k a_k^N z^{-k} = z^{-m}\left(\frac{1+z}{2}\right)^m E_{2m-1}(z) T_m^N(z^2); \\ H_m^N(z) = \sum_k b_k^N z^{-k} = -z^{-2m+1}\left(\frac{1-z}{2}\right)^m T_m^N(z^2). \end{cases} \tag{7.21}$$

The FIR algorithms for wavelet decomposition we use are given by

$$\begin{cases} c_{N,\ell}^j = \sum_k a_{k-2\ell}^N c_k^{j+1}; \\ d_{N,\ell}^j = \sum_k b_{k-2\ell}^N c_k^{j+1}. \end{cases} \tag{7.22}$$

We must now study how much information is lost in applying (7.22) instead of the exact decomposition algorithm (7.15). In our recent work [10], we studied this approximation problem by comparing the exact reconstruction of $\mathbf{c}_N^j := \{c_{N,\ell}^j\}$ and $\mathbf{d}_N^j := \{d_{N,\ell}^j\}$ using the FIR algorithm (7.5) with the original sequence $\mathbf{c}^{j+1}$; that is, by considering

$$c_{N,\ell}^{*j+1} = \sum_k p_{\ell-2k} c_{N,k}^j + \sum_k q_{\ell-2k} d_{N,k}^j, \tag{7.23}$$

the approximation problem is to study

$$\mathcal{E}_N^{(m)}(\mathbf{c}^{j+1}) := \|\mathbf{c}^{j+1} - \mathbf{c}_N^{*j+1}\|_{\ell^2},$$

where $\mathbf{c}_N^{*j+1} = \{c_{N,\ell}^{*j+1}\}$ is the reconstructed sequence from the truncated decomposition of $\mathbf{c}^{j+1}$. To discuss the estimation result in [10], we need to recall the structure of the zeros of the Euler Frobenius polynomials

$$\Pi_{2m-1}(z) := z^{m-1} E_{2m-1}(z). \tag{7.24}$$

A good source of information is Schoenberg's monograph [20]. Observe that by multiplying z^{m-1} to $E_{2m-1}(z)$, $\Pi_{2m-1}(z)$ is then an algebraic polynomial of order $2m-1$ (or degree $2m-2$). All the $(2m-2)$ zeros of this polynomial are real and negative simple zeros. Let us arrange them in decreasing order, as

$$\lambda_{2m-2}^{(m)} < \lambda_{2m-3}^{(m)} < \cdots < \lambda_1^{(m)} < 0. \tag{7.25}$$

Then they come in reciprocal pairs, namely:

$$\lambda_1^{(m)} \lambda_{2m-2}^{(m)} = \lambda_2^{(m)} \lambda_{2m-3}^{(m)} = \cdots = \lambda_{m-1}^{(m)} \lambda_m^{(m)} = 1, \tag{7.26}$$

which implies that

$$|\lambda_{m-1}^{(m)}| < 1. \tag{7.27}$$

We also consider the companion sequence

$$\mu_k^{(m)} = \frac{(\lambda_k^{(m)})^{m-2}}{\Pi'_{2m-1}(\lambda_k^{(m)})}, \tag{7.28}$$

where $\Pi_{2m-1}(z)$ is defined in (7.24). The following result is proved in [10].

THEOREM 7.1. *Let m be any positive integer and $j \in \mathbb{Z}$. Then there exists an $N_0(m)$ such that for all $N \geq N_0(m)$ and all $\mathbf{c}^{j+1} \in \ell^2$,*

$$\mathcal{E}_N^{(m)}(\mathbf{c}^{j+1}) \leq \left(2 \sum_{j=1}^{m-1} \frac{\mu_j^{(m)}}{1 - \lambda_j^{(m)}} |\lambda_j^{(m)}|^{N+1} \right) \|\mathbf{c}^{j+1}\|_{\ell^2}. \tag{7.29}$$

Furthermore, $N_0(2), \ldots, N_0(4)$ can be chosen to be 0.

In particular, we have

$$\mathcal{E}_N^{(2)}(\mathbf{c}) \leq 2.732 \times (.268)^N \|\mathbf{c}\|_{\ell^2},$$
$$\mathcal{E}_N^{(3)}(\mathbf{c}) \leq 4.327 \times (.4306)^N \|\mathbf{c}\|_{\ell^2},$$

and

$$\mathcal{E}_N^{(4)}(c) \leq 7.838 \times (.5353)^N \|\mathbf{c}\|_{\ell^2},$$

where linear, quadratic, and cubic B-wavelets are used for FIR decomposition and reconstruction. For more details, the reader is referred to [10].

References

1. Beylin, G., Coifman, R., Daubechies, I., Mallat, S., Meyer, Y., Raphael, L. and Ruskia, B., eds., *Wavelets and Applications*, Jones and Bartlett, Publ., Boston, to appear.
2. Chui, C.K., 'Wavelets and spline interpolation', in *Wavelets, Subdivision Algorithms and Radial Functions*, ed. by W. Light, Oxford University Press, 1991, pp. 1-35.
3. Chui, C.K., *An Introduction to Wavelets*, Academic Press, Inc., Boston, 1992.
4. Chui, C.K., 'An overview of wavelets', in *Approximation Theory and Functional Analysis*, ed. by C.K. Chui, Academic Press, Inc., Boston, 1991 pp. 47-72.
5. Chui, C.K., ed., *Wavelets – A Tutorial*, Academic Press, Inc., Boston, 1992.
6. Chui, C.K. and Wang, J.Z., 'A cardinal spline approach to wavelets', Proc. Amer. Math. Soc., 1991, to appear.
7. Chui, C.K. and Wang, J.Z., 'On compactly supported spline wavelets and a duality principle', Trans. Amer. Math. Soc., 1991, to appear.
8. Chui, C.K. and Wang, J.Z., 'A general framework of compactly supported splines and wavelets', CAT Report #210, Texas A&M University, 1990.
9. Chui, C.K. and Wang, J.Z., 'An analysis of cardinal spline-wavelets', CAT Report #231, Texas A&M University, 1990.
10. Chui, C.K. and Wang, J.Z., 'Computational and algorithmic aspects of spline-wavelets', CAT Report #235, Texas A&M University, 1990.
11. Combes, J.M., Grossmann, A. and Tchamitchian, Ph., eds., *Wavelets: Time-Frequency Methods and Phase Space*, Springer-Verlag, N.Y., 1989; and 2nd edition, 1991.
12. Cohen, A., Daubechies, I. and Feauveau, J.C., 'Biorthogonal bases of compactly supported wavelets', Comm. Pure and Appl. Math., to appear.
13. Daubechies, I., 'Orthonormal bases of compactly supported wavelets', Comm. Pure and Appl. Math., **41** (1988), 909-996.
14. Daubechies, I., *Ten Lectures on Wavelets*, CBMS-NSF Series in Appl. Math., SIAM Publ., Philadelphia, to appear.
15. Grossmann, A. and Morlet, J., 'Decomposition of Hardy functions into square integrable wavelets of constant shape', SIAM J. Math. Anal., **15** (1984), 723-736.
16. Karlin, S., *Total Positivity* Vol. 1, Stanford University Press, 1968.
17. Mallat, S., 'Multiresolution approximations and wavelet orthonormal bases of $L^2(\mathbb{R})$', Trans. Amer. Math. Soc., **315** (1989), 69-87.

18. Meyer, Y., 'Ondelettes et functions splines, Seminaire Equations aux Derivees Partielles', Ecolé Polytechnique, Paris, France, Dec. 1986.
19. Meyer, Y., 'Ondelettes et Opérateurs, (two volumes), Hermann Publ., Paris, 1990.
20. Schoenberg, I.J., *Cardinal Spline Interpolation*, CBMS-NSF Series in Appl. Math. #12, SIAM Publ., Philadelphia, 1973.

PADE APPROXIMATION IN ONE AND MORE VARIABLES.

CUYT ANNIE, SENIOR RESEARCH ASSOCIATE NFWO
DEPARTMENT OF MATHEMATICS AND COMPUTER SCIENCE
UNIVERSITY OF ANTWERP (UIA)
UNIVERSITEITSPLEIN 1
B-2610 WILRIJK (BELGIUM)

Abstract.

We first recall results from univariate Padé approximation theory (UPA). The recursive ϵ-algorithm and the continued fraction representation obtained from the *qd*-algorithm are given for the normal case as well as for a non-normal table composed of square blocks. Convergence of UPA for meromorphic functions and continuity of the univariate Padé operator are discussed.

The same approximation problem is considered in the multivariate case. General order multivariate Padé approximants (MPA) are defined and a recursive computation scheme and a continued fraction representation are given, both for the normal case and for the case of a table of MPA with degenerate solutions. A de Montessus de Ballore convergence theorem is presented and the continuity of the multivariate Padé operator is considered.

1. Notations and definitions for UPA.

Consider a formal power series

$$f(x) = c_0 + c_1 x + c_2 x^2 + \dots \tag{1}$$

with $c_0 \neq 0$. In the sequel of the text we shall write ∂p for the exact degree of a polynomial p and ωp for the order of a power series p (i.e. the degree of the first nonzero term). The Padé approximation problem of order (n, m) for f consists in finding polynomials

$$p(x) = \sum_{i=0}^{n} a_i x^i$$

and

$$q(x) = \sum_{i=0}^{m} b_i x^i$$

such that in the power series $(fq - p)(x)$ the coefficients of x^i for $i = 0, \dots, n+m$ disappear, in other words

$$\begin{aligned} &\partial p \leq n \\ &\partial q \leq m \\ &\omega(fq - p) \geq n + m + 1 \end{aligned} \tag{2}$$

S. P. Singh (ed.), Approximation Theory, Spline Functions and Applications, 41–68.

Condition (2) is equivalent with the following two linear systems of equations

$$\begin{cases} c_0 b_0 = a_0 \\ c_1 b_0 + c_0 b_1 = a_1 \\ \vdots \\ c_n b_0 + c_{n-1} b_1 + \ldots + c_{n-m} b_m = a_n \end{cases} \tag{3a}$$

$$\begin{cases} c_{n+1} b_0 + c_n b_1 + \ldots + c_{n-m+1} b_m = 0 \\ \vdots \\ c_{n+m} b_0 + c_{n+m-1} b_1 + \ldots + c_n b_m = 0 \end{cases} \tag{3b}$$

with $c_i = 0$ for $i < 0$. In general a solution for the coefficients a_i is known after substitution of a solution for the b_i in the left hand side of (3a). So the crucial point is to solve the homogeneous system (3b) of m equations in the $m+1$ unknowns b_i. This system has at least one nontrivial solution because one of the unknowns can be chosen freely. Moreover all nontrivial solutions of (3) supply the same irreducible form. If $p(x)$ and $q(x)$ satisfy (3) we shall denote by $r_{n,m}(x) = (p_{n,m}/q_{n,m})(x)$ the irreducible form of p/q normalized such that $q_{n,m}(0) = 1$. This rational function $r_{n,m}(x)$ is called the Padé approximant of order (n, m) for f. By calculating the irreducible form, a polynomial may be cancelled in numerator and denominator of p/q. We shall therefore denote the exact degrees of $p_{n,m}$ and $q_{n,m}$ in $r_{n,m}$ respectively by n' and m'. Although $p_{n,m}$ and $q_{n,m}$ are computed from polynomials p and q that satisfy (2), it is not necessarily so that $p_{n,m}$ and $q_{n,m}$ satisfy (2) themselves. A simple example will illustrate this. Consider $f(x) = 1 + x^2$ and take $n = 1 = m$. A solution is given by $b_0 = 0 = a_0$ and $b_1 = 1 = a_1$. So $p(x) = x = q(x)$. Consequently $p_{n,m} = 1 = q_{n,m}$ with $\omega(fq_{n,m} - p_{n,m}) = 2 < n + m + 1$ and the corresponding equations (2) do not hold.

The Padé approximants $r_{n,m}$ for f can be ordered in a table for different values of n and m:

$$\begin{matrix} r_{0,0} & r_{0,1} & r_{0,2} & r_{0,3} & \cdots \\ r_{1,0} & r_{1,1} & r_{1,2} & r_{1,3} & \cdots \\ r_{2,0} & r_{2,1} & r_{2,2} & r_{2,3} & \cdots \\ \vdots & \vdots & \vdots & \ddots & \end{matrix}$$

This table is called the Padé table of f. The first column consists of the partial sums of f. The first row contains the reciprocals of the partial sums of $1/f$. We call a Padé approximant normal if it occurs only once in the Padé table. A criterion for the normality of an approximant is given in the next theorem.

THEOREM 1:
The Padé approximant $r_{n,m} = (p_{n,m}/q_{n,m})$ for f is normal if and only if the following three conditions are satisfied simultaneously:
(a) $n' = n$
(b) $m' = m$
(c) $\omega(fq_{n,m} - p_{n,m}) = n + m + 1$

Normality of a Padé approximant can also be guaranteed by the nonvanishing of certain determinants. We introduce the notation

$$D_{n,n+m} = \begin{vmatrix} c_n & c_{n-1} & \cdots & c_{n-m} \\ c_{n+1} & c_n & \cdots & c_{n-m+1} \\ \vdots & & \ddots & \vdots \\ c_{n+m} & c_{n+m-1} & \cdots & c_n \end{vmatrix}$$

The following result can be proved [PERR p. 243].

THEOREM 2:
The Padé approximant $r_{n,m} = (p_{n,m}/q_{n,m})$ for f is normal if and only if

$$D_{n,n+m-1} \neq 0$$
$$D_{n+1,n+m} \neq 0$$
$$D_{n,n+m} \neq 0$$
$$D_{n+1,n+m+1} \neq 0$$

2. Methods to compute normal Padé approximants.

In this section we suppose that every Padé approximant in the Padé table does at least itself satisfy condition (2). This is the case if for instance $\min(n-n', m-m') = 0$ for all n and m. A survey of algorithms for computing Padé approximants is given in [WUYT] and [BULT]. We discuss a limited number of them, mainly with the aim to generalize them to the multivariate case in the following sections.

2.1. Determinant formulas.

One can solve the system of equations (3b) explicitly and thus get a determinant representation for the Padé approximant. The following determinant formula for $q_{n,m}(x)$ can very easily be proved by solving (3b) using Cramer's rule after choosing $b_0 = D_{n,n+m-1}$. For $f(x) = \sum_{i=0}^{\infty} c_i x^i$ we write

$$F_k(x) = \sum_{i=0}^{k} c_i x^i$$

with $F_k(x) = 0$ for $k < 0$.

THEOREM 3:
If the Padé approximant of order (n, m) for f is given by $r_{n,m}(x) = (p_{n,m}/q_{n,m})(x)$ and if $D_{n,n+m-1} \neq 0$, then

$$p_{n,m}(x) = \frac{1}{D_{n,n+m-1}} \begin{vmatrix} F_n(x) & xF_{n-1}(x) & \dots & x^m F_{n-m}(x) \\ c_{n+1} & c_n & \dots & c_{n-m+1} \\ \vdots & & \ddots & \\ c_{n+m} & c_{n+m-1} & \dots & c_n \end{vmatrix}$$

and

$$q_{n,m}(x) = \frac{1}{D_{n,n+m-1}} \begin{vmatrix} 1 & x & \dots & x^m \\ c_{n+1} & c_n & \dots & c_{n-m+1} \\ \vdots & & \ddots & \\ c_{n+m} & c_{n+m-1} & \dots & c_n \end{vmatrix}$$

These determinant expressions are of course only useful for small values of n and m because the calculation of a determinant involves a lot of additions, multiplications and possible round-off. They merely exhibit closed form formulas for the solution.

2.2. The ϵ-algorithm.

Using these determinant representations for the Padé approximant, it can be proved that the elements in a normal Padé table satisfy the relationship [WYNN]

$$(r_{n,m+1} - r_{n,m})^{-1} + (r_{n,m} - r_{n-1,m})^{-1} = (r_{n+1,m} - r_{n,m})^{-1} + (r_{n,m} - r_{n,m-1})^{-1} \tag{4}$$

where we have defined

$$r_{n,-1} = \infty$$
$$r_{-1,m} = 0$$

The identity (4) is a star identity which relates

$$r_{n-1,m}(x) = N$$

$$r_{n,m-1}(x) = W \qquad r_{n,m}(x) = C \qquad r_{n,m+1}(x) = E$$

$$r_{n+1,m}(x) = S$$

and is often written as

$$(N - C)^{-1} + (S - C)^{-1} = (E - C)^{-1} + (W - C)^{-1}$$

If we introduce the following new notation for our Padé approximants

$$r_{n,m}(x) = \epsilon_{2m}^{(n-m)}$$

we obtain a table of ϵ-values where the subscript indicates a column and the superscript indicates a diagonal:

$$\begin{array}{cccc}
\epsilon_0^{(0)} & \epsilon_2^{(-1)} & \epsilon_4^{(-2)} & \dots \\
\epsilon_0^{(1)} & \epsilon_2^{(0)} & \epsilon_4^{(-1)} & \dots \\
\epsilon_0^{(2)} & \epsilon_2^{(1)} & \epsilon_4^{(0)} & \dots \\
\epsilon_0^{(3)} & \epsilon_2^{(2)} & \epsilon_4^{(1)} & \dots \\
\vdots & \vdots & \vdots &
\end{array}$$

The $\epsilon_0^{(n)}$ are the partial sums $F_n(x)$ of the Taylor series $f(x)$. Remark the fact that only even column-indices occur. The table can be completed with odd-numbered columns in the following way. We define elements

$$\epsilon_{2m+1}^{(n-m-1)} = \epsilon_{2m-1}^{(n-m)} + \frac{1}{\epsilon_{2m}^{(n-m)} - \epsilon_{2m}^{(n-m-1)}} \qquad n = 0,1,\dots \qquad m = 0,1,\dots \tag{5a}$$

From the star-identity (4) and with the aid of (5a) we can conclude by induction that also for the even-numbered columns [WYNN]

$$\epsilon_{2m}^{(n-m)} = \epsilon_{2m-2}^{(n-m+1)} + \frac{1}{\epsilon_{2m-1}^{(n-m+1)} - \epsilon_{2m-1}^{(n-m)}} \tag{5b}$$

The relations (5a) and (5b) are a means to calculate all the elements in the Padé table. This algorithm is very handy when one needs the value of a Padé approximant for a given x and one does not want to compute the coefficients of the Padé approximant explicitly. Computational difficulties can occur when the Padé table is not normal. Reformulations of the ϵ-algorithm in this case can be found in the next section.

2.3. The qd-algorithm.

Let us now consider the following sequence of elements on a descending staircase in the Padé table

$$T_k = \{r_{k,0}, r_{k+1,0}, r_{k+1,1}, r_{k+2,1}, \dots\} \qquad k \geq 0$$

and the following continued fraction

$$d_0 + d_1 x + \dots + d_k x^k + \frac{d_{k+1} x^{k+1} |}{| 1} + \frac{d_{k+2} x |}{| 1} + \frac{d_{k+3} x |}{| 1} + \dots \tag{6}$$

THEOREM 4:
If every three consecutive elements in T_k are different, then a continued fraction of the form (6) exists with $d_{k+i} \neq 0$ for $i \geq 1$ and such that the nth convergent equals the $(n+1)^{th}$ element of T_k.

In this way we are able to construct corresponding continued fractions for functions f analytic in the origin: if the nth convergent of (6) equals the $(n+1)^{th}$ element of T_0 then (6) is the corresponding continued fraction to the power series (2). By continued fractions of the form (6) one can only compute Padé approximants below the main diagonal in the Padé table. For the right upper half of the table one can use the reciprocal covariance property of Padé approximants. We now turn to the problem of the calculation of the coefficients d_{k+i} in (6) starting from the coefficients c_i of f. Consider the continued fraction $g_k(x)$, which is of the form (6), and which is given by

$$c_0+\ldots+c_k\,x^k+\frac{c_{k+1}x^{k+1}|}{|1}+\frac{-q_1^{(k+1)}x|}{|1}+\frac{-e_1^{(k+1)}x|}{|1}+\frac{-q_2^{(k+1)}x|}{|1}+\frac{-e_2^{(k+1)}x|}{|1}+\ldots \quad (7)$$

If the coefficients $q_\ell^{(k+1)}$ and $e_\ell^{(k+1)}$ are computed in order to satisfy theorem 4 then the convergents of g_k equal the successive elements of T_k. If we calculate the even part of $g_k(x)$ we get

$$c_0+\ldots+c_k\,x^k+\frac{c_{k+1}x^{k+1}|}{|1-q_1^{(k+1)}x}+\frac{-q_1^{(k+1)}e_1^{(k+1)}x^2|}{|1-(q_2^{(k+1)}+e_1^{(k+1)})x}+\frac{-q_2^{(k+1)}e_2^{(k+1)}x^2|}{|1-(q_3^{(k+1)}+e_2^{(k+1)})x}+\ldots$$

If we calculate the odd part of $g_{k-1}(x)$ we get

$$c_0+\ldots+c_k\,x^k+\frac{c_kq_1^{(k)}x^{k+1}|}{|1-(q_1^{(k)}+e_1^{(k)})x}+\frac{-e_1^{(k)}q_2^{(k)}x^2|}{|1-(q_2^{(k)}+e_2^{(k)})x}+\frac{-e_2^{(k)}q_3^{(k)}x^2|}{|1-(q_3^{(k)}+e_3^{(k)})x}-\ldots$$

The even part of $g_k(x)$ and the odd part of $g_{k-1}(x)$ are two continued fractions which have the same convergents $r_{k,0}, r_{k+1,1}, r_{k+2,2}, \ldots$ and which also have the same form. Hence the partial numerators and denominators must be equal, and we obtain for $k \geq 1$ and $\ell \geq 1$ [RUTI]

$$e_0^{(k)} = 0$$

$$q_1^{(k)} = \frac{c_{k+1}}{c_k}$$

$$e_\ell^{(k)} = e_{\ell-1}^{(k+1)} + q_\ell^{(k+1)} - q_\ell^{(k)} \quad (8a)$$

$$q_{\ell+1}^{(k)} = q_\ell^{(k+1)}\frac{e_\ell^{(k+1)}}{e_\ell^{(k)}} \quad (8b)$$

The numbers $q_\ell^{(k)}$ and $e_\ell^{(k)}$ are usually arranged in a table, where the superscript (k) indicates a diagonal and the subscript ℓ indicates a column. This table is called the *qd*-table:

$$
\begin{array}{cccccc}
e_0^{(1)} & & & & & \\
& q_1^{(1)} & & & & \\
e_0^{(2)} & & e_1^{(1)} & & & \\
& q_1^{(2)} & & q_2^{(1)} & & \\
e_0^{(3)} & & e_1^{(2)} & & e_2^{(1)} & \\
& q_1^{(3)} & & q_2^{(2)} & & \ddots \\
e_0^{(4)} & \vdots & e_1^{(3)} & \vdots & e_2^{(2)} & \\
\vdots & & \vdots & & \vdots &
\end{array}
$$

The formulas (8) can also be memorized as follows: $e_\ell^{(k)}$ is calculated such that in the following rhombus the sum of the two elements on the upper diagonal equals the sum of the two elements on the lower diagonal

$$
\begin{array}{ccccc}
& & q_\ell^{(k)} & & \\
& & & + & \\
e_{\ell-1}^{(k+1)} & & & & e_\ell^{(k)} \\
& + & & & \\
& & q_\ell^{(k+1)} & &
\end{array}
$$

and $q_{\ell+1}^{(k)}$ is computed such that in the next rhombus the product of the two elements on the upper diagonal equals the product of the two elements on the lower diagonal

$$
\begin{array}{ccccc}
& & e_\ell^{(k)} & & \\
& & & * & \\
q_\ell^{(k+1)} & & & & q_{\ell+1}^{(k)} \\
& * & & & \\
& & e_\ell^{(k+1)} & &
\end{array}
$$

Since the *qd*-algorithm computes the coefficients in (7), it can be used to compute the Padé approximants below the main diagonal in the Padé table. To calculate the Padé approximants in the right upper half of the table, the *qd*-algorithm itself can also be extended above the diagonal and the following results can be proved [HENR pp. 615-617]:

$$
\begin{array}{ccccccc}
 & q_1^{(0)} & & q_2^{(-1)} & & q_3^{(-2)} & \cdots \\
e_0^{(1)} & & e_1^{(0)} & & e_2^{(-1)} & & \\
 & q_1^{(1)} & & q_2^{(0)} & & q_3^{(-1)} & \cdots \\
e_0^{(2)} & & e_1^{(1)} & & e_2^{(0)} & & \\
 & q_1^{(2)} & & q_2^{(1)} & & q_3^{(0)} & \cdots \\
e_0^{(3)} & & e_1^{(2)} & & e_2^{(1)} & & \\
 & q_1^{(3)} & & q_2^{(2)} & & q_3^{(1)} & \cdots \\
e_0^{(4)} & \vdots & e_1^{(3)} & \vdots & e_2^{(2)} & \vdots & \\
\vdots & & \vdots & & \vdots & &
\end{array}
$$

Let $(1/f)(x) = w_0 + w_1 x + w_2 x^2 + \ldots$ and put

$$
\begin{aligned}
q_1^{(0)} &= \frac{-w_1}{w_0} \\
e_1^{(0)} &= \frac{w_2}{w_1}
\end{aligned}
$$

and for $k \geq 1$

$$
\begin{aligned}
q_{k+1}^{(-k)} &= 0 \\
e_{k+1}^{(-k)} &= \frac{w_{k+2}}{w_{k+1}}
\end{aligned}
$$

If the elements in the extended qd-table are all calculated by the use of (8) using the above starting values, then the continued fraction $h_k(x)$ given by

$$
\frac{1 \rfloor}{\lceil w_0 + \ldots + w_k x^k} + \frac{w_{k+1} x^{k+1} \rfloor}{\lceil 1} + \frac{-e_{k+1}^{(-k)} x \rfloor}{\lceil 1} + \frac{-q_{k+2}^{(-k)} x \rfloor}{\lceil 1} + \frac{-e_{k+2}^{(-k)} x \rfloor}{\lceil 1} + \frac{-q_{k+3}^{(-k)} x \rfloor}{\lceil 1} + \ldots
$$

supplies the Padé approximants on the staircase

$$
U_k = \{r_{0,k}, r_{0,k+1}, r_{1,k+1}, r_{1,k+2}, r_{2,k+2}, \ldots\}
$$

It is obvious that difficulties can arise if the division in (8b) cannot be performed by the fact that $e_\ell^{(k)} = 0$. This is the case if the Padé table is not normal since consecutive elements in T_k or U_k can then be equal. Reformulations of the qd-algorithm in this case are given in section 3. Other algorithms exist for the computation of Padé approximants in a row, column, diagonal, sawtooth or ascending staircase in the Padé table. We do not mention them here, but we refer to [BULTb], [LONG], [PIND] and [MCCA].

3. Block structure of the Padé table.

It is often the case that certain Padé approximants $r_{n,m}$ in the Padé table coincide. Then the table is not normal. In general the following result can be proved.

THEOREM 5:
Let the Padé approximant of order (n,m) for f be given by $r_{n,m} = p_{n,m}/q_{n,m}$. Let $n' = \partial p_{n,m}$ and $m' = \partial q_{n,m}$. Then
(a) $\omega(fq_{n,m} - p_{n,m}) = n' + m' + t + 1$ with $t \geq 0$
(b) for k and ℓ satisfying $n' \leq k \leq n' + t$ and $m' \leq \ell \leq m' + t : r_{k,\ell}(x) = r_{n,m}(x)$
(c) $n \leq n' + t$ and $m \leq m' + t$

The previous property is called the block structure of the Padé table: the table consists of square blocks of size $t+1$ containing equal Padé approximants. Theorem 5c in fact says that the same Padé approximant does not reappear outside the block. In a non-normal Padé table the algorithms presented in the previous section do not hold anymore. However, singular rules to deal with the square blocks, have been developed.

3.1. The ϵ-algorithm.

The star-identity (4) can be extended to the non-normal case as follows [CORD].

THEOREM 6:
Let the Padé table for f contain a block of size $t+1$ with corners $r_{n,m}$, $r_{n,m+t}$, $r_{n+t,m}$ and $r_{n+t,m+t}$. Then for $k = 1, 2, \ldots, t+1$:

$$(r_{n-1,m-1+k} - r_{n,m})^{-1} + (r_{n+t+1,m+t+1-k} - r_{n,m})^{-1} =$$
$$(r_{n-1+k,m-1} - r_{n,m})^{-1} + (r_{n+t+1-k,m+t+1} - r_{n,m})^{-1}$$

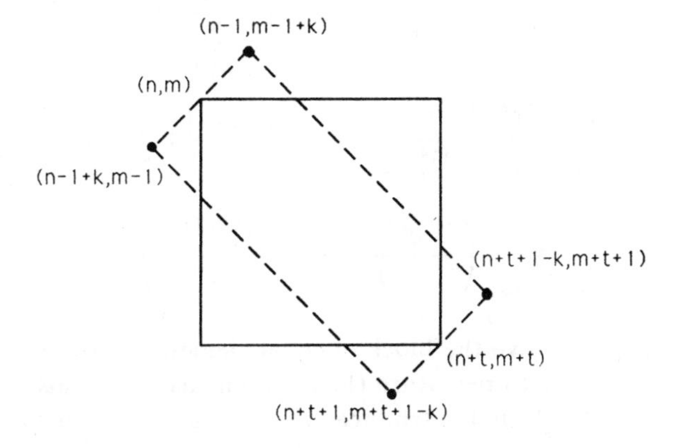

Starting from this generalized star-identity instead of from (4), it is possible to set up an algorithm generalizing the ϵ-algorithm, which makes it possible to compute the elements in a non-normal Padé table recursively.

3.2. The qd-algorithm.

Since several elements on a staircase in a non-normal Padé table might be equal, the usual representation (7) does not hold anymore. It is however possible to give other types of staircases, jumping over square blocks, whose elements can again be represented as the convergents of a continued fraction [CLAE]. Suppose there is a block of size $t+1$ having as corner elements $r_{n,m}$, $r_{n,m+t}$, $r_{n+t,m}$ and $r_{n+t,m+t}$. Consider for $k=1,\ldots,t$ the perturbed staircase

$$T^*_{n-m-t+k-1} = \{r_{n-m-t+k-1,0}, r_{n-m-t+k,0}, \ldots, r_{n,m+t-k}, \\ r_{n,m+t+1}, r_{n+1,m+t+1}, r_{n+2,m+t+1}, \ldots, r_{n+k+1,m+t-1}, r_{n+k+1,m+t}, \ldots\}$$

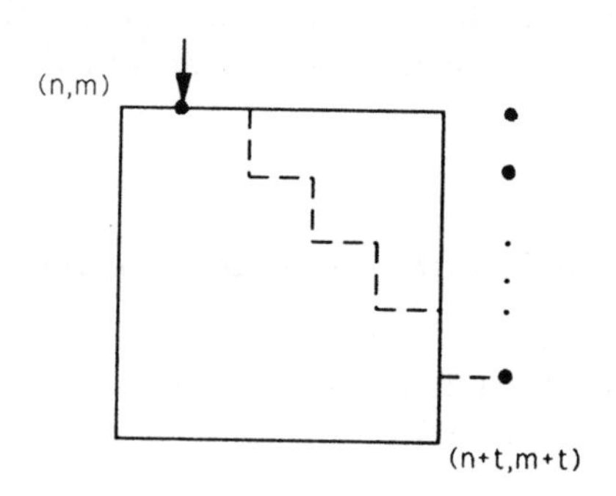

and its corresponding continued fraction

$$g_{n-m-t+k-1}(x) = c_0 + \ldots + c_s\, x^s + \frac{c_{s+1} x^{s+1}|}{|1} + \sum_{i=1}^{n-s-1} \left(\frac{-q_i^{(s+1)} x|}{|1} + \frac{-e_i^{(s+1)} x|}{|1} \right) + \\ \frac{-v_{k,1}^{(t+1)} x^{k+1}|}{|1 - v_{k,k+1}^{(t+1)} x - \ldots - v_{k,2}^{(t+1)} x^k} + \frac{-v_{k,k+2}^{(t+1)} x|}{|1} + \sum_{i=1}^{k} \frac{-v_{k,k+i+2}^{(t+1)} x|}{|1 + v_{k,k+i+2}^{(t+1)} x} + \\ \sum_{i=m+t+2}^{\infty} \left(\frac{-q_i^{(s+1)} x|}{|1} + \frac{-e_i^{(s+1)} x|}{|1} \right) \qquad s = n-m-t+k-1$$

This staircase passes above the block of equal elements in the Padé table and goes down column $m+t+1$ to recapture the old staircase. It is this vertical movement down column $m+t+1$ that introduces the v-values (v from "vertical"). Similar continued fractions can be constructed whose convergents are the elements on a special staircase passing below the block of equal elements and moving horizontally along row $n+t+1$ while introducing h-values. In general, when the Padé table consists of several blocks of different size each with equal elements, it remains possible using the same technique to construct continued fractions whose convergents are the elements of certain staircases in the Padé table. Let us now try to identify the new values $v_{k,i}^{(t+1)}$ for the continued fraction above.

THEOREM 7:
Let the Padé table for f contain a block of size $t+1$ with corners $r_{n,m}$, $r_{n,m+t}$, $r_{n+t,m}$ and $r_{n+t,m+t}$. Then the following relations hold:

(a)

$$
\begin{aligned}
v_{1,1}^{(t+1)} &= e_{m+t}^{(n-m-t)} q_{m+t+1}^{(n-m-t)} \\
v_{1,2}^{(t+1)} &= q_{m+t+1}^{(n-m-t)} \\
v_{1,3}^{(t+1)} &= e_{m+t+1}^{(n-m-t)} \\
v_{1,4}^{(t+1)} &= q_{m+t+2}^{(n-m-t)}
\end{aligned}
$$

(b) for $k > 1$ and $i = 2, 3, \ldots, 2k+1$:

$$
\begin{aligned}
v_{k,1}^{(t+1)} &= e_{m+t-k+1}^{(n-m-t+k-1)} v_{k-1,1}^{(t+1)} \\
v_{k,i}^{(t+1)} &= v_{k-1,i-1}^{(t+1)} \\
v_{k,2k+2}^{(t+1)} &= q_{m+t+2}^{(n-m-t+k-1)}
\end{aligned}
$$

(c)

$$
q_{m+t+1}^{(n-m-t)} \prod_{i=1}^{t+1} e_{m+i-1}^{(n-m-i+1)} = e_m^{(n-m+t+1)} \prod_{i=1}^{t+1} q_m^{(n-m+i)}
$$

(d)

$$
q_{m+t+1}^{(n-m-t)} + e_{m+t+1}^{(n-m-t)} = e_m^{(n-m+t+1)} + q_{m+1}^{(n-m+t+1)}
$$

(e) for $k = 1, 2, \ldots, t$:

$$
\begin{aligned}
q_{m+t+1}^{(n-m-t)} \prod_{i=1}^{k} e_{m+t-i+1}^{(n-m-t+i-1)} + e_{m+t+1}^{(n-m+t+2)} \prod_{i=1}^{k} q_{m+t+2}^{(n-m-t+i-1)} = \\
e_m^{(n-m+t+1)} \prod_{i=1}^{k} q_m^{(n-m+t-i+2)} + q_{m+1}^{(n-m+t+1)} \prod_{i=1}^{k} e_{m+i}^{(n-m+t-i+2)}
\end{aligned}
$$

(f)

$$
e_{m+t+1}^{(n-m+t+2)} \prod_{i=1}^{t+1} q_{m+t+2}^{(n-m-t+i-1)} = q_{m+1}^{(n-m+t+1)} \prod_{i=1}^{t+1} e_{m+i}^{(n-m+t-i+2)}
$$

Using these rules it is possible to compute $q_{m+t}^{(n-m-t)}$, $e_{m+t}^{(n-m-t)}$ and $q_{m+t+2}^{(n-m-t+i-1)}$ for $i = 1, 2, \ldots, t+1$.

4. Convergence and continuity.

It is clear that the poles of the Padé approximants in a sequence of elements from the Padé table will play an important role when studying the convergence of that sequence. A lot of information on the convergence of Padé approximants can be found in [BAKE]. We restrict ourselves to the convergence problem of Padé approximants for meromorphic functions f [MONT].

THEOREM 8:
Let f be analytic in $B(0,R) = \{z \in \mathbb{C} : |z| < R\}$ except in the poles $z_1, \ldots, z_k$ of f with

$$0 < |z_1| \leq \ldots \leq |z_k| < R$$

Let the pole at z_i have multiplicity μ_i and let the total multiplicity be $m = \sum_{i=1}^{k} \mu_i$. Then

$$lim_{n\to\infty} q_{n,m}(z) = \prod_{i=1}^{k} \left(1 - \frac{z}{z_i}\right)^{\mu_i}$$

and the column $(r_{n,m})_{n\in\mathbb{N}}$ of the Padé table converges uniformly to f on every closed and bounded subset of $B(0,r) \setminus \{z_1, \ldots, z_k\}$.

For the proof we refer to Saff's short and elegant proof which can for instance be found in [BAKE, pp. 252–254].

When we compute $r_{n,m}$ in finite precision arithmetic the computed result is not exactly the (n,m) Padé approximant, but it differs slightly from it by rounding errors and data perturbations. Since we can consider the computed result as the exact (n,m) Padé approximant of a slightly perturbed input power series, it is important to study the effect of such small perturbations on the operator $\mathcal{P}_{n,m}$ that associates with f its (n,m) Padé approximant. Since n and m are fixed here, we can adopt the notations

$$\mathcal{P} := \mathcal{P}_{n,m}$$
$$\mathcal{P}f := r_{n,m}$$

To measure the small perturbations we introduce a pseudo-norm for formal power series:

$$||c||_{n+m} = \max_{0\leq i\leq n+m} |c_i|$$

with $c = (c_0, \ldots, c_{n+m})$, and the supremum norm for continuous functions on an interval $[a,b]$:

$$||q|| = \sup_{a\leq x\leq b} |q(x)|$$

The Padé approximants $r_{n,m}(x) = (p_{n,m}/q_{n,m})(x)$ were normalized such that $q_{n,m}(0) = 1$. This implies the existence of an interval $[a, b]$ around the origin where $q_{n,m}(x)$ is strictly positive. For given $f(x) = \sum_{i=0}^{\infty} c_i x^i$ we call a neighbourhood U_δ of f, the set of power series $g(x) = \sum_{i=0}^{\infty} d_i x^i$ such that the pseudo-norm $||c - d||_{n+m} \leq \delta$. Under weaker conditions than normality it is possible to obtain a necessary and sufficient condition for the continuity of $\mathcal{P}$ [WERN].

THEOREM 9:
The Padé operator $\mathcal{P}$ is continuous in f, if and only if $\min(n - n', m - m') = 0$ *where n' and m' are respectively the exact degrees of numerator and denominator of $\mathcal{P}f$.*

5. Multivariate Padé approximants.

We have seen in the previous sections that univariate Padé approximants can be obtained in several equivalent ways: one can solve the system of defining equations explicitly and thus obtain a determinant expression, one can set up a recursive scheme such as the ϵ-algorithm or one can construct a continued fraction whose convergents lie on a descending staircase in the Padé table. In the past few years all these approaches have been generalized to the multivariate case [CHIS, HUGH, KARL, LEVI, LUTTb, MURP, SIEM] but mostly the equivalence between the different techniques was lost. However, for the following definition a lot of properties of the univariate Padé approximant remain valid, also the recursive computation and the continued fraction representation. We shall describe here a general framework that includes all types of definitions based on the use of a linear system of defining equations for the numerator and denominator coefficients of the multivariate Padé approximant. Definitions of this type can be found in [CHIS, CUYTd, HUGH, KARL, LEVI, LUTTa, LUTTb]. Other generalizations based on symbolic manipulation [CHAF] or using branched continued fractions instead of ordinary continued fractions [CUYTh] are not treated here. We restrict ourselves to the case of two variables because the generalization to functions of more variables is only notationally more difficult.

Given a Taylor series expansion

$$f(x, y) = \sum_{(i,j)\in\mathbb{N}^2} c_{ij} x^i y^j$$

with

$$c_{ij} = \frac{1}{i!}\frac{1}{j!}\frac{\partial^{i+j} f}{\partial x^i \partial y^j}\Big|_{(0,0)}$$

we shall compute an approximant $(p/q)(x, y)$ to $f(x, y)$ where $p(x, y)$ and $q(x, y)$ are determined by an accuracy–through–order principle. The polynomials $p(x, y)$

and $q(x,y)$ are of the form

$$p(x,y) = \sum_{(i,j)\in N} a_{ij}x^i y^j$$

$$q(x,y) = \sum_{(i,j)\in D} b_{ij}x^i y^j$$

where N and D are finite subsets of $\mathbb{N}^2$. The sets N and D indicate the degree of the polynomials $p(x,y)$ and $q(x,y)$. Let us denote

$$\partial p = N \qquad \#N = n+1$$

$$\partial q = D \qquad \#D = m+1$$

It is now possible to let $p(x,y)$ and $q(x,y)$ satisfy the following condition for the power series $(fq-p)(x,y)$, namely

$$(fq-p)(x,y) = \sum_{(i,j)\in \mathbb{N}^2\backslash E} d_{ij}x^i y^j \tag{9}$$

if, in analogy with the univariate case, the index set E is such that

$$N \subseteq E \tag{10a}$$

$$\#(E\backslash N) = m = \#D - 1 \tag{10b}$$

$$E \text{ satisfies the inclusion property} \tag{10c}$$

where (10c) means that when a point belongs to the index set E, then the rectangular subset of points emanating from the origin with the given point as its furthermost corner, also lies in E. Condition (10a) enables us to split the system of equations

$$d_{ij} = 0 \qquad (i,j) \in E$$

in an inhomogeneous part defining the numerator coefficients

$$\sum_{\mu=0}^{i}\sum_{\nu=0}^{j} c_{\mu\nu}b_{i-\mu,j-\nu} = a_{ij} \qquad (i,j)\in N$$

and a homogeneous part defining the denominator coefficients

$$\sum_{\mu=0}^{i}\sum_{\nu=0}^{j} c_{\mu\nu}b_{i-\mu,j-\nu} = 0 \qquad (i,j)\in E\backslash N \tag{11}$$

By convention $b_{k\ell} = 0$ if $(k,\ell) \notin D$. Condition (10b) guarantees the existence of a nontrivial denominator $q(x,y)$ because the homogeneous system has one equation less than the number of unknowns and so one unknown coefficient can be chosen freely. Condition (10c) finally takes care of the Padé approximation property, namely

$$q(0,0) \neq 0 \Longrightarrow (f - \frac{p}{q})(x,y) = \sum_{(i,j)\in \mathbb{N}^2 \setminus E} e_{ij} x^i y^j$$

For more information we refer to [CUYTc, CUYTd]. We denote this multivariate Padé approximant by $[N/D]_E$ and we can arrange successive Padé approximants in a table after fixing an enumeration of the points in the degree sets N and D and the equation set E. After numbering the points in $\mathbb{N}^2$, for instance as (0,0), (1,0), (0,1), (2,0), (1,1), (0,2), (3,0), ..., we can carry this enumeration over to the index sets N, D and E which are finite subsets of $\mathbb{N}^2$, to get:

$$N = \{(i_0, j_0), \ldots, (i_n, j_n)\} \tag{12a}$$

$$D = \{(d_0, e_0), \ldots, (d_m, e_m)\} \tag{12b}$$

$$E = N \cup \{(i_{n+1}, j_{n+1}), \ldots, (i_{n+m}, j_{n+m})\} \tag{12c}$$

With (12) we can set up descending chains of index sets, defining bivariate polynomials of lower degree and bivariate Padé approximation problems of lower order:

$$N = N_n \supset \ldots \supset N_k = \{(i_0, j_0), \ldots, (i_k, j_k)\} \supset \ldots \supset N_0 = \{(i_0, j_0)\} \qquad k = 0, \ldots, n \tag{13a}$$

$$D = D_m \supset \ldots \supset D_\ell = \{(d_0, e_0), \ldots, (d_\ell, e_\ell)\} \supset \ldots \supset D_0 = \{(d_0, e_0)\} \qquad \ell = 0, \ldots, m \tag{13b}$$

$$E = E_{n+m} \supset \ldots \supset E_{k+\ell} = \{(i_0, j_0), \ldots, (i_{k+\ell}, j_{k+\ell})\} \supset \ldots \supset E_0 = \{(i_0, j_0)\} \qquad k+\ell = 0, \ldots, n+m \tag{13c}$$

$$E_{k+1,k+\ell} = \{(i_{k+1}, j_{k+1}), \ldots, (i_{k+\ell}, j_{k+\ell})\} \qquad E \setminus N = E_{n+1,n+m}$$

We assume that the enumeration was such that each set $E_{k+\ell} \subset E_{n+m}$ satisfies the inclusion property in its turn. It was shown in [LEVIa] that a determinant representation for

$$p_k(x,y) = \sum_{(i,j)\in N_k} a_{ij} x^i y^j \qquad 0 \leq k \leq n$$

and

$$q_\ell(x,y) = \sum_{(i,j)\in D_\ell} b_{ij} x^i y^j \qquad 0 \le \ell \le m$$

satisfying

$$(fq_\ell - p_k)(x,y) = \sum_{(i,j)\in \mathbb{N}^2\setminus E_{k+\ell}} d_{ij} x^i y^j$$

is given by

$$p_k(x,y) = \begin{vmatrix} \sum_{(i,j)\in N_k} c_{i-d_0,j-e_0} x^i y^j & \dots & \sum_{(i,j)\in N_k} c_{i-d_\ell,j-e_\ell} x^i y^j \\ c_{i_{k+1}-d_0,j_{k+1}-e_0} & \dots & c_{i_{k+1}-d_\ell,j_{k+1}-e_\ell} \\ \vdots & & \vdots \\ c_{i_{k+\ell}-d_0,j_{k+\ell}-e_0} & \dots & c_{i_{k+\ell}-d_\ell,j_{k+\ell}-e_\ell} \end{vmatrix} \tag{14a}$$

$$q_\ell(x,y) = \begin{vmatrix} x^{d_0} y^{e_0} & \dots & x^{d_\ell} y^{e_\ell} \\ c_{i_{k+1}-d_0,j_{k+1}-e_0} & \dots & c_{i_{k+1}-d_\ell,j_{k+1}-e_\ell} \\ \vdots & & \vdots \\ c_{i_{k+\ell}-d_0,j_{k+\ell}-e_0} & \dots & c_{i_{k+\ell}-d_\ell,j_{k+\ell}-e_\ell} \end{vmatrix} \tag{14b}$$

where $c_{ij} = 0$ if $i < 0$ or $j < 0$. A solution of the original problem (10) is then given by $(p_n/q_m)(x,y)$ because $N_n = N$, $D_m = D$ and $E_{n+m} = E$. This formula is very analogous to the univariate formula given in section 3.

6. Computation of nondegenerate MPA.

From now on we assume that the homogeneous system of equations (11) has maximal rank. Then the Padé approximant $[N/D]_E$ is called nondegenerate.

6.1. The E-algorithm.

Let us rewrite the determinant formulas (14) as

$$p_k(x,y) = \begin{vmatrix} t_0(k) & \dots & t_\ell(k) \\ \Delta\, t_0(k) & \dots & \Delta\, t_\ell(k) \\ \vdots & & \vdots \\ \Delta\, t_0(k+\ell-1) & \dots & \Delta\, t_\ell(k+\ell-1) \end{vmatrix} \tag{15a}$$

$$q_\ell(x,y) = \begin{vmatrix} 1 & \dots & 1 \\ \Delta\, t_0(k) & \dots & \Delta\, t_\ell(k) \\ \vdots & & \vdots \\ \Delta\, t_0(k+\ell-1) & \dots & \Delta\, t_\ell(k+\ell-1) \end{vmatrix} \tag{15b}$$

with the series $t_0, \dots, t_\ell$ defined by:

$$t_0(0) = c_{i_0-d_0, j_0-e_0} x^{i_0-d_0} y^{j_0-e_0}$$

$$\Delta t_0(s-1) = t_0(s) - t_0(s-1) = c_{i_s-d_0, j_s-e_0} x^{i_s-d_0} y^{j_s-e_0} \qquad s = 1, \dots, k+\ell \quad (16a)$$

$$\Delta t_0(s-1) = 0 \qquad i_s < d_0 \text{ or } j_s < e_0$$

and for $r = 1, \dots, \ell$

$$t_r(0) = c_{i_0-d_r, j_0-e_r} x^{i_0-d_r} y^{j_0-e_r}$$

$$\Delta t_r(s-1) = t_r(s) - t_r(s-1) = c_{i_s-d_r, j_s-e_r} x^{i_s-d_r} y^{j_s-e_r} \qquad s = 1, \dots, k+\ell \quad (16b)$$

$$\Delta t_r(s-1) = 0 \qquad i_s < d_r \text{ or } j_s < e_r$$

This quotient of determinants can be computed using the E-algorithm given in [BREZb]:

$$E_0^{(k)} = t_0(k) \qquad k = 0, \dots, n+m$$

$$g_{0,\ell}^{(k)} = t_\ell(k) - t_{\ell-1}(k) \qquad \ell = 1, \dots, m \qquad k = 0, \dots, n+m$$

$$E_\ell^{(k)} = \frac{E_{\ell-1}^{(k)} g_{\ell-1,\ell}^{(k+1)} - E_{\ell-1}^{(k+1)} g_{\ell-1,\ell}^{(k)}}{g_{\ell-1,\ell}^{(k+1)} - g_{\ell-1,\ell}^{(k)}} \qquad k = 0, 1, \dots, n \qquad \ell = 1, 2, \dots, m$$

$$g_{\ell,r}^{(k)} = \frac{g_{\ell-1,r}^{(k)} g_{\ell-1,\ell}^{(k+1)} - g_{\ell-1,r}^{(k+1)} g_{\ell-1,\ell}^{(k)}}{g_{\ell-1,\ell}^{(k+1)} - g_{\ell-1,\ell}^{(k)}} \qquad r = \ell+1, \ell+2, \dots$$

The values $E_\ell^{(k)}$ and $g_{\ell,r}^{(k)}$ are stored as in the tables below.

$$\begin{array}{ccccc}
E_0^{(0)} & & & & \\
 & E_1^{(0)} & & & \\
E_0^{(1)} & & \ddots & & \\
 & E_1^{(1)} & & E_m^{(0)} & \\
E_0^{(2)} & \vdots & & \vdots & \ddots \\
\vdots & & & & E_{n+m}^{(0)} \\
 & & & & .\,^{.\,^{.}} \\
 & & & E_m^{(n)} & \\
 & E_1^{(n+m-1)} & & & \\
E_0^{(n+m)} & & & &
\end{array}$$

$$
\begin{array}{c|cc|c|cccc|c|cc}
g_{0,1}^{(0)} & g_{0,2}^{(0)} & & & g_{0,r}^{(0)} & & & & & g_{0,m}^{(0)} & \\
 & & g_{1,2}^{(0)} & & & g_{1,r}^{(0)} & & & & & \ddots \\
g_{0,1}^{(1)} & g_{0,2}^{(1)} & & & g_{0,r}^{(1)} & & \ddots & & & g_{0,m}^{(1)} & \\
 & & g_{1,2}^{(1)} & & & g_{1,r}^{(1)} & & g_{r-1,r}^{(0)} & & \vdots & g_{m-1,m}^{(0)} \\
g_{0,1}^{(2)} & g_{0,2}^{(2)} & \vdots & \cdots & g_{0,r}^{(2)} & \vdots & & \vdots & \cdots & & \vdots \\
\vdots & \vdots & & & \vdots & & & g_{r-1,r}^{(n+m-r+1)} & & & g_{m-1,m}^{(n+1)} \\
 & & & & & & \cdot^{\cdot^{\cdot}} & & & & \\
 & & g_{1,2}^{(n+m-1)} & & & g_{1,r}^{(n+m-1)} & & & & & \cdot^{\cdot^{\cdot}} \\
g_{0,1}^{(n+m)} & g_{0,2}^{(n+m)} & & & g_{0,r}^{(n+m)} & & & & & g_{0,m}^{(n+m)} &
\end{array}
$$

Finally with $k = n$ and $\ell = m$, this is with $N_k = N$, $D_\ell = D$ and $E_{k+\ell} = E$ we get $(p_n/q_m)(x,y) = E_m^{(n)}$ while intermediate values in the computation scheme are also multivariate Padé approximants since

$$E_\ell^{(k)} = \frac{p_k(x,y)}{q_\ell(x,y)}$$

and thus

$$q_\ell(0,0) \neq 0 \Longrightarrow f - E_\ell^{(k)} = \sum_{(i,j)\in\mathbb{N}^2\setminus E_{k+\ell}} e_{ij}x^iy^j$$

6.2. The qdg-algorithm.

In the same way as for UPA, these intermediate values can be used to build a table of multivariate Padé approximants:

$$\begin{array}{cccc}
[N_0/D_0]_{E_0} & [N_0/D_1]_{E_1} & [N_0/D_2]_{E_2} & \cdots \\
[N_1/D_0]_{E_1} & [N_1/D_1]_{E_2} & [N_1/D_2]_{E_3} & \cdots \\
[N_2/D_0]_{E_2} & [N_2/D_1]_{E_3} & [N_2/D_2]_{E_4} & \cdots \\
\vdots & \vdots & \vdots &
\end{array}$$

Let us consider descending staircases

$$\begin{array}{llll}
[N_s/D_0]_{E_s} & & & \\
[N_{s+1}/D_0]_{E_{s+1}} & [N_{s+1}/D_1]_{E_{s+2}} & & \\
 & [N_{s+2}/D_1]_{E_{s+3}} & [N_{s+2}/D_2]_{E_{s+4}} & \\
 & & \vdots & \cdots
\end{array}$$

in this table of MPA. It was proved in [CUYTn] that continued fractions of the form

$$[N_s/D_0]_{E_s} + \frac{[N_{s+1}/D_0]_{E_{s+1}} - [N_s/D_0]_{E_s} \rfloor}{\lfloor 1} + \frac{-q_1^{(s+1)} \rfloor}{\lfloor 1+q_1^{(s+1)}} + \frac{-e_1^{(s+1)} \rfloor}{\lfloor 1+e_1^{(s+1)}} + \frac{-q_2^{(s+1)} \rfloor}{\lfloor 1+q_2^{(s+1)}} + \frac{-e_2^{(s+1)} \rfloor}{\lfloor 1+e_2^{(s+1)}} + \ldots \tag{17}$$

can be constructed of which the successive convergents are the multivariate Padé approximants on this descending staircase. Here

$$[N_s/D_0]_{E_s} = \sum_{(i,j)\in N_s} c_{ij}x^iy^j$$

$$[N_{s+1}/D_0]_{E_{s+1}} = \sum_{(i,j)\in N_{s+1}} c_{ij}x^iy^j$$

and the partial numerators and denominators are obtained from an algorithm which is very qd-like: for $\ell \geq 2$

$$q_\ell^{(s+1)} = \frac{e_{\ell-1}^{(s+2)} q_{\ell-1}^{(s+2)}}{e_{\ell-1}^{(s+1)}} \quad \frac{g_{\ell-2,\ell-1}^{(s+\ell-1)} - g_{\ell-2,\ell-1}^{(s+\ell)}}{g_{\ell-2,\ell-1}^{(s+\ell-1)}} \quad \frac{g_{\ell-1,\ell}^{(s+\ell)}}{g_{\ell-1,\ell}^{(s+\ell)} - g_{\ell-1,\ell}^{(s+\ell+1)}} \tag{18a}$$

and for $\ell \geq 1$

$$e_\ell^{(s+1)} + 1 = \frac{g_{\ell-1,\ell}^{(s+\ell)} - g_{\ell-1,\ell}^{(s+\ell+1)}}{g_{\ell-1,\ell}^{(s+\ell)}} \left(q_\ell^{(s+2)} + 1\right) \tag{18b}$$

If we arrange the values $q_\ell^{(s+1)}$ and $e_\ell^{(s+1)}$ in a table as follows

$$\begin{array}{ccccccc}
q_1^{(1)} & & & & & & \\
& e_1^{(1)} & & & & & \\
q_1^{(2)} & & q_2^{(1)} & & & & \\
& e_1^{(2)} & & e_2^{(1)} & & & \\
q_1^{(3)} & & q_2^{(2)} & & \ddots & & \\
& e_1^{(3)} & & e_2^{(2)} & & & \\
q_1^{(4)} & & q_2^{(3)} & & \ddots & & \\
\vdots & e_1^{(4)} & \vdots & e_2^{(3)} & & & \\
& \vdots & & \vdots & & &
\end{array}$$

where subscripts indicate columns and superscripts indicate downward sloping diagonals, then (18a) links the elements in the rhombus

$$\begin{array}{ccc}
& e_{\ell-1}^{(s+1)} & \\
q_{\ell-1}^{(s+2)} & & q_\ell^{(s+1)} \\
& e_{\ell-1}^{(s+2)} &
\end{array}$$

and (18b) links two elements on an upward sloping diagonal

$$\begin{matrix} & e_\ell^{(s+1)} \\ q_\ell^{(s+2)} & \end{matrix}$$

Starting values for the algorithm are given by

$$q_1^{(s+1)} = \frac{\Delta t_0(s+1)}{\Delta t_0(s)} \quad \frac{g_{0,1}^{(s+1)}}{g_{0,1}^{(s+1)} - g_{0,1}^{(s+2)}}$$

In this way the univariate equivalence of the three main defining techniques for Padé approximants is also established for the multivariate case: algebraic relations, recurrence relations, continued fractions.

6.3. The multivariate ϵ*- and qd-algorithms.*

In [CUYTd, CUYTf] we introduced multivariate Padé approximants of order (ν, μ) using homogeneous polynomials. Numerator p and denominator q are of the form

$$p(x,y) = \sum_{i+j=\nu\mu}^{\nu\mu+\nu} a_{ij}x^iy^j$$

$$q(x,y) = \sum_{i+j=\nu\mu}^{\nu\mu+\mu} b_{ij}x^iy^j$$

and satisfy

$$(fq-p)(x,y) = \sum_{i+j=\nu\mu+\nu+\mu+1}^{\infty} d_{ij}x^iy^j$$

These multivariate Padé approximants satisfy a large number of the classical univariate properties but are as well a special case of the general order approximants defined above. The advantage of the "homogeneous" approximants is that they can be calculated recursively by means of the ϵ-algorithm [CUYTg] and that they can also be represented in continued fraction form using the classical qd-algorithm [CUYTi]. What concerns the ϵ-algorithm the starting values are again given by the partial sums of $f(x,y)$, namely

$$\epsilon_0^{(n)} = \sum_{i+j=0}^{n} c_{ij}x^iy^j$$

What concerns the multivariate qd-algorithm, the univariate qd-algorithm is first rewritten in a form such that it can immediately be generalized. If $r_{n,m}(x)$ is the $2m^{th}$ convergent of the continued fraction

$$\sum_{i=0}^{n-m} c_ix^i + \frac{c_{n-m+1}x^{n-m+1}|}{|1} + \frac{-q_1^{(n-m+1)}x|}{|1} + \frac{-e_1^{(n-m+1)}x|}{|1} + \ldots$$

then we can also say that $r_{n,m}(x)$ is the $2m^{th}$ convergent of the continued fraction

$$\sum_{i=0}^{n-m} c_i x^i + \frac{c_{n-m+1}x^{n-m+1} |}{| 1} + \frac{-Q_1^{(n-m+1)} |}{| 1} + \frac{-E_1^{(n-m+1)} |}{| 1} + \dots$$

with

$$Q_1^{(k)} = \frac{c_{k+1}x^{k+1}}{c_k x^k}$$
$$E_0^{(k)} = 0$$
$$E_\ell^{(k)} = E_{\ell-1}^{(k+1)} + Q_\ell^{(k+1)} - Q_\ell^{(k)}$$
$$Q_{\ell+1}^{(k)} = Q_\ell^{(k+1)} \frac{E_\ell^{(k+1)}}{E_\ell^{(k)}}$$

We have simply included the factor x in $Q_\ell^{(k)}$ and $E_\ell^{(k)}$. This last continued fraction can easily be generalized for a bivariate function: replace the expression $c_k x^k$ by an expression that contains all the terms of degree k in the bivariate series $\sum_{i+j=0}^{\infty} c_{ij} x^i y^j$. The starting values are then given by

$$Q_1^{(k)} = \frac{\sum_{i+j=k+1} c_{ij} x^i y^j}{\sum_{i+j=k} c_{ij} x^i y^j}$$

which are very analogous to the univariate starting formulas. Explicit determinant formulas for these homogeneous approximants, involving near-Toeplitz matrices, are given in [CUYTb].

7. Structure of the table of MPA.

For the multivariate Padé approximants discussed in the last paragraph of the previous section, the square block structure which is typical for univariate Padé approximants is preserved. This result is based on the fact that for the homogeneous Padé approximants different solutions to the same Padé approximation problem are equivalent and hence result in a unique irreducible form. If this irreducible form is given by $r_{n,m}(x,y) = (p_{n,m}/q_{n,m})(x,y)$ then we define

$$n' = \partial p_{n,m} - \omega q_{n,m}$$
$$m' = \partial q_{n,m} - \omega q_{n,m}$$

where ∂ and ω denote homogeneous degrees. We can prove that $n' \leq n$ and $m' \leq m$ because clearly n' and m' are an extension of the univariate definitions where $\omega q_{n,m} = 0$ because of the normalization.

THEOREM 10:
If the homogeneous Padé approximant of order (n,m) for $f(x,y)$ is given by $r_{n,m} = p_{n,m}/q_{n,m}$ with n' and m' as defined above, then:
(a) $\omega(fq_{n,m} - p_{n,m}) = \omega q_{n,m} + n' + m' + t + 1$ *with* $t \geq 0$
(b) *for* k *and* ℓ *satisfying* $n' \leq k \leq n' + t$ *and* $m' \leq \ell \leq m' + t$ *we have* $r_{k,\ell}(x,y) = r_{n,m}(x,y)$
(c) $n \leq n' + t$ *and* $m \leq m' + t$

As a conclusion we take a closer look at the meaning of the numbers $\omega q_{n,m}$, n' and m'. In the solution $p(x,y)$ and $q(x,y)$ the degrees have been shifted over nm. By taking the irreducible form of p/q part of that shift can disappear, but what remains in $p_{n,m}$, $q_{n,m}$ and $fq_{n,m} - p_{n,m}$ is a shift over $\omega q_{n,m}$ [CUYTf]. Now n' and m' play the same role as in the univariate case: they measure the exact degree of a polynomial by disregarding the shift over $\omega q_{n,m}$. The singular computation rules developed by Cordellier for the ϵ-algorithm and Claessens and Wuytack for the qd-algorithm for use in a non-normal table remain valid. The multivariate version is written down in the same way as was done in section 6 for the regular ϵ- and qd-algorithm. An extension of the E- and qdg-algorithms with singular rules can be found in [CUYTa].

8. Convergence and continuity.

The univariate theorem of de Montessus de Ballore deals with the case of simple poles as well as with the case of multiple poles. The former means that we have information on the denominator of the meromorphic function while the latter means that we also have information on the derivatives of that denominator.

By the set $N * D$ we denote the index set that results from the multiplication of a polynomial indexed by N with a polynomial indexed by D, namely $N * D = \{(i+k, j+\ell) | (i,j) \in N, (k,\ell) \in D\}$.

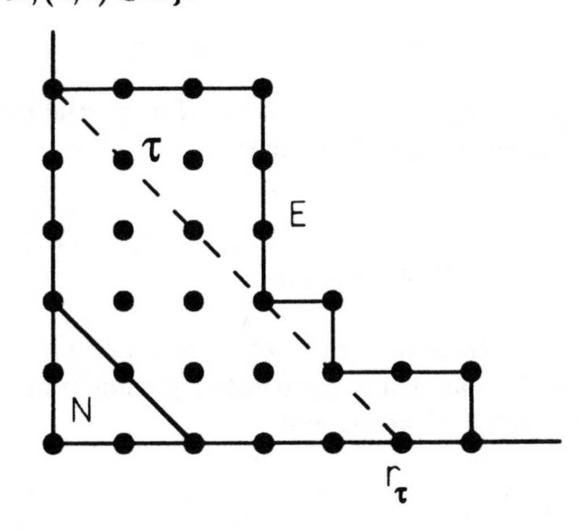

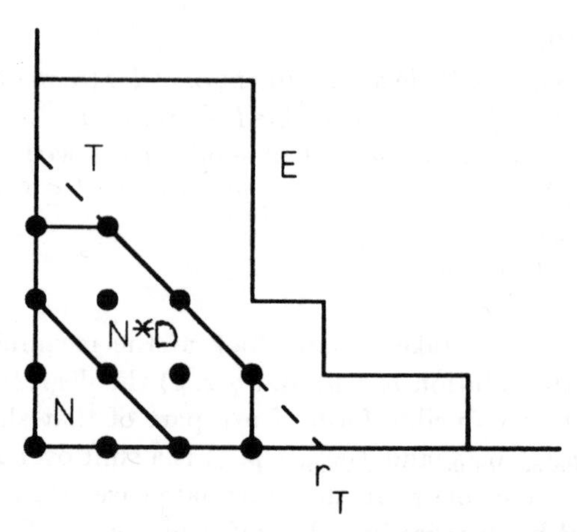

Since the set E satisfies the inclusion property we can inscribe isosceles triangles in E, with top in (0,0) and base along the antidiagonal. Let τ be the largest of these inscribed triangles. On the other hand, because $N * D$ is a finite subset of $\mathbb{N}^2$, we can circumscribe it with such triangles. Let T be the smallest of these circumscribing triangles. In both cases we call r_τ and r_T the "range" of the triangles τ and T respectively.

In what follows we discuss functions $f(x, y)$ which are meromorphic in a polydisc $B(0; R_1, R_2) = \{(x, y) : |x| < R_1, |y| < R_2\}$, meaning that there exists a polynomial

$$R_m(x, y) = \sum_{(d,e) \in D \subseteq \mathbb{N}^2} r_{de} x^d y^e = \sum_{i=0}^{m} r_{d_i e_i} x^{d_i} y^{e_i}$$

such that $(f R_m)(x, y)$ is analytic in the polydisc above. The denominator polynomial $R_m(x, y)$ can completely be determined by m zeros $(x_h, y_h) \in B(0; R_1, R_2)$ of $R_m(x, y)$

$$R_m(x_h, y_h) = 0 \qquad h = 1, \ldots, m \tag{19a}$$

or by a combination of zeros of R_m and some of its partial derivatives. For instance in the point (x_h, y_h) the partial derivatives

$$\left.\frac{\partial^{i_h + j_h} R_m}{\partial x^{i_h} \partial y^{j_h}}\right|_{(x_h, y_h)} = 0 \qquad (i_h, j_h) \in I_h \tag{19b}$$

can be given with I_h a finite subset of $\mathbb{N}^2$ of cardinality $\mu(h) + 1$ and satisfying the inclusion property. We can again enumerate the indices indicating the known and vanishing partial derivatives as follows:

$$I_h = \{(i_0^{(h)}, j_0^{(h)}), \ldots, (i_{\mu(h)}^{(h)}, j_{\mu(h)}^{(h)})\} \qquad (i_0^{(h)}, j_0^{(h)}) = (0, 0)$$

THEOREM 11:
Let $f(x,y)$ be a function which is meromorphic in the polydisc $B(0;R_1,R_2) = \{(x,y) : |x| < R_1, |y| < R_2\}$, meaning that there exists a polynomial

$$R_m(x,y) = \sum_{(d,e)\in D\subseteq \mathbb{N}^2} r_{de} x^d y^e = \sum_{i=0}^{m} r_{d_i e_i} x^{d_i} y^{e_i}$$

such that $(fR_m)(x,y)$ is analytic in the polydisc above. Further, we assume that $R_m(0,0) \neq 0$ so that necessarily $(0,0) \in D$. Let there also be given k zeros $(x_h, y_h) \in B(0;R_1,R_2)$ of $R_m(x,y)$ and k sets $I_h \subset \mathbb{N}^2$ with inclusion property, satisfying

$$(fR_m)(x_h, y_h) \neq 0 \qquad h = 1,\ldots,k \tag{20a}$$

$$\begin{cases} \left.\dfrac{\partial^{i_h+j_h} R_m}{\partial x^{i_h} \partial y^{j_h}}\right|_{(x_h,y_h)} = 0 & (i_h, j_h) \in I_h \qquad h = 1,\ldots,k \\ \sum_{h=1}^{k} (\mu(h)+1) = m & \#I_h = \mu(h)+1 \end{cases} \tag{20b}$$

and producing the nonzero determinant

$$\begin{vmatrix} x_1^{d_1} y_1^{e_1} & \cdots & x_1^{d_m} y_1^{e_m} \\ \vdots & & \vdots \\ \frac{d_1!}{(d_1-\mu(1))!} \frac{e_1!}{(e_1-\mu(1))!} x_1^{d_1-\mu(1)} y_1^{e_1-\mu(1)} & \cdots & \frac{d_m!}{(d_m-\mu(1))!} \frac{e_m!}{(e_m-\mu(1))!} x_1^{d_m-\mu(1)} y_1^{e_m-\mu(1)} \\ \vdots & & \vdots \\ x_k^{d_1} y_k^{e_1} & \cdots & x_k^{d_m} y_k^{e_m} \\ \vdots & & \vdots \\ \frac{d_1!}{(d_1-\mu(k))!} \frac{e_1!}{(e_1-\mu(k))!} x_k^{d_1-\mu(k)} y_k^{e_1-\mu(k)} & \cdots & \frac{d_m!}{(d_m-\mu(k))!} \frac{e_m!}{(e_m-\mu(k))!} x_k^{d_m-\mu(k)} y_k^{e_m-\mu(k)} \end{vmatrix} \tag{20c}$$

Then the $[N/D]_E = (p/q)(x,y)$ Padé approximant with D fixed by $R_m(x,y)$ and N and E growing, converges to $f(x,y)$ uniformly on compact subsets of

$$\{(x,y) : |x| < R_1, |y| < R_2, R_m(x,y) \neq 0\}$$

and its denominator

$$q(x,y) = \sum_{i=0}^{m} b_{d_i e_i} x^{d_i} y^{e_i}$$

converges to $R_m(x,y)$ under the following conditions for N and E: the range of the largest inscribed triangle in E and the range of the smallest triangle circumscribing

$N * D$ should both tend to infinity as the sets N and E grow along the column $[N/D]_E$ in the multivariate Padé table.

For homogeneous approximants also a continuity property was proved [CUYTj]. A more general result for general order approximants is under investigation. Let us define the operator $\mathcal{P}_{n,m}$ that associates with $f(x,y)$ its homogeneous (n,m) Padé approximant. Since n and m are fixed here, we again adopt the notations

$$\mathcal{P} := \mathcal{P}_{n,m}$$
$$\mathcal{P}f := r_{n,m}$$

If we write

$$C_k(x,y) = \sum_{i+j=k} c_{ij} x^i y^j$$

then C_k is a k-linear operator and we can introduce seminorms for the power series $f(x,y)$ as follows:

$$||f||_{n+m} = \max_{0 \leq k \leq n+m} ||C_k||$$

where $||C_k|| = \max_{||(x,y)||=1} |C_k(x,y)|$. Bivariate functions q continuous on a poly-interval $I = I_1 \times I_2$ are normed by the Chebyshev norm

$$||q|| = \sup_{(x,y) \in I} |q(x,y)|$$

THEOREM 12:
If $\min(n-n', m-m') = 0$ and $q_{n,m}(x,y) \neq 0$ for all (x,y) in a suitably chosen poly-interval I, then the Padé operator $\mathcal{P}$ is continuous in $f(x,y)$.

References.

[BAKE] Baker G. and Graves-Morris P., "Padé Approximants: Basic Theory", Encyclopedia of Mathematics and its Applications: vol 13, Addison-Wesley, Reading, 1981.

[BREZ] Brezinski C., *A general extrapolation algorithm*, Numer. Math. **35** (1980), 175–187.

[BULTa] Bultheel A., *Recursive algorithms for the Padé table: two approaches*, Wuytack L. ed., LNM **765** (1979), 211–230.

[BULTb] Bultheel A., *Division algorithms for continued fractions and the Padé table*, J. Comput. Appl. Math. **6** (1980), 259–266.

[CHAFF] Chaffy C., *$(Padé)_y$ of $(Padé)_x$ approximants of $F(x,y)$*, Nonlinear numerical methods and rational approximation, Cuyt A. ed., Reidel, Dordrecht.

[CHIS] Chisholm J. S. R., *N-variable rational approximants*, in [SAFF], 23–42.

[CLAE] Claessens G. and Wuytack L., *On the computation of non-normal Padé approximants*, J. Comput. Appl. Math. **5** (1979), 283–289.

[CORD] Cordellier F., *Démonstration algébrique de l'extension de l'identité de Wynn aux tables de Padé non normales*, Wuytack L. ed., LNM **765** (1979), 36–60.

[CUYTa] Cuyt A., *Rational Hermite interpolation in one and more variables*, In these proceedings.

[CUYTb] Cuyt A., *A comparison of some multivariate Padé approximants*, SIAM J. Math. Anal. **14** (1983), 194–202.

[CUYTc] Cuyt A., *A review of multivariate Padé approximation theory*, J. Comput. Appl. Math. **12 & 13** (1985), 221–232.

[CUYTd] Cuyt A., *Multivariate Padé approximants*, Journ. Math. Anal. Appl. **96** (1983), 238–243.

[CUYTe] Cuyt A., *A multivariate qd-like algorithm*, BIT **28** (1988), 98–112.

[CUYTf] Cuyt A., "Padé approximants for operators: theory and applications", LNM **1065**, Springer Verlag, Berlin, 1984.

[CUYTg] Cuyt A., *The epsilon-algorithm and multivariate Padé approximants*, Numer. Math. **40** (1982), 39–46.

[CUYTh] Cuyt A. and Verdonk B., *A review of branched continued fraction theory for the construction of multivariate rational approximants*, Appl. Numer. Math. **4** (1988), 263–271.

[CUYTi] Cuyt A. and Van der Cruyssen P., *Abstract Padé approximants for the solution of a system of nonlinear equations*, Comp. Math. Appl. **6** (1982), 445–466.

[CUYTj] Cuyt A., Werner H. and Wuytack L., *On the continuity of the multivariate Padé operator*, J. Comput. Appl. Math. **11** (1984), 95–102.

[HENR] Henrici P., "Applied and computational complex analysis: vol. 1 & 2", John Wiley, New York, 1976.

[HUGH] Hughes Jones R., *General rational approximants in n variables*, J. Approx. Theory **16** (1976), 201–233.

[KARL] Karlsson J. and Wallin H., *Rational approximation by an interpolation procedure in several variables*, in [SAFF], 83–100.

[LEVI] Levin D., *General order Padé type rational approximants defined from double power series*, J. Inst. Math. Appl. **18** (1976), 1–8.

[LONG] Longman I., *Computation of the Padé table*, Int. J. Comput. Math. **3** (1971), 53–64.

[LUTTa] Lutterodt C., *A two-dimensional analogue of Padé approximant theory*, Journ. Phys. A **7** (1974), 1027–1037.

[LUTTb] Lutterodt C., *Rational approximants to holomorphic functions in n dimensions*, J. Math. Anal. Appl. 53 (1976), 89–98.

[MCCA] Mc Cabe J., *The qd-algorithm and the Padé table: an alternative form and a general continued fraction*, Math. Comp. **41** (1983), 183–197.

[MONT] de Montessus de Ballore R., *Sur les fractions continues algébriques*, Rend. Circ. Mat. Palermo **19** (1905), 1–73.

[MURP] Murphy J. and O'Donohoe M., *A two-variable generalization of the Stieltjes-type continued fraction*, J. Comput. Appl. Math. **4** (1978), 181–190.

[PERR] Perron O., "Die Lehre von den Kettenbruchen II", Teubner, Stuttgart, 1977.

[PIND] Pindor M., *A simplified algorithm for calculating the Padé table derived from Baker and Longman schemes*, J. Comp. Appl. Math. **2** (1976), 25–258.

[RUTI] Rutishauser H., "Der Quotienten-Differenzen Algorithmus", Mitteilungen Institut für angewandte Mathematik (ETH) 7, Birkhäuser Verlag, Basel, 1957.

[SAFF] Saff E. and R. Varga, "Padé and rational approximation: theory and applications", Academic Press, New York, 1977.

[SIEM] Siemaszko W., *Branched continued fractions for double power series*, J. Comput. Appl. Math. **6** (1980), 121–125.

[WERN] Werner H. and Wuytack L., *On the continuity of the Padé operator*, SIAM J. Numer. Anal. **20** (1983), 1273–1280.

[WUYT] Wuytack L., *Commented bibliography on techniques for computing Padé approximants*, Wuytack L. ed., LNM **765** (1979), 375–392.

[WYNN] Wynn P., *On a device for computing the $e_m(S_n)$ transformation*, MTAC **10** (1956), 91–96.

RATIONAL HERMITE INTERPOLATION IN ONE AND MORE VARIABLES.

CUYT ANNIE, SENIOR RESEARCH ASSOCIATE NFWO
DEPARTMENT OF MATHEMATICS AND COMPUTER SCIENCE
UNIVERSITY OF ANTWERP (UIA)
UNIVERSITEITSPLEIN 1
B–2610 WILRIJK

Abstract.

In the first 4 sections we discuss topics from univariate rational Hermite interpolation (URI). These topics include the structure of the table of URI, a recursive computation scheme and a continued fraction representation both in the normal case and the non-normal case and a convergence theorem for rational Hermite interpolants of meromorphic functions.
In the next 4 sections these items are generalized to the multivariate case. We first introduce multivariate rational Hermite interpolants (MRI) for data sets satisfying the inclusion property or rectangle rule and give a recursive computation scheme and a non-branched continued fraction representation, both for the non-degenerate and the degenerate case. For general data sets only results for ordinary rational interpolation in the case of non-degeneracy were obtained in [CUYTd].

1. Notations and definitions for URI.

Consider a function f defined in a sequence of distinct points $(y_i)_{i\in\mathbb{N}}$ of the complex plane and let the derivatives $f^{(\ell)}(y_i)$ of the function f be given for $\ell = 0,\ldots,s_i-1$. We denote the exact degree of a polynomial p by ∂p and its order by ωp. The rational Hermite interpolation problem of order (n,m) for f consists in finding polynomials

$$p(x) = \sum_{i=0}^{n} a_i x^i$$

and

$$q(x) = \sum_{i=0}^{m} b_i x^i$$

such that for a particular j

$$n+m+1 = \sum_{i=0}^{j} s_i$$

and

$$f^{(\ell)}(y_i) = \left(\frac{p}{q}\right)^{(\ell)}(y_i) \qquad i=0,\ldots,j \qquad \ell = 0,\ldots,s_i-1 \tag{1}$$

S. P. Singh (ed.), Approximation Theory, Spline Functions and Applications, 69–103.

In this interpolation problem s_i interpolation points coincide with y_i, so s_i interpolation conditions must be fulfilled in y_i. Therefore this type of interpolation problem is also often referred to as the osculatory rational interpolation problem [WARN]. In case $s_i = 1$ for all $i \geq 0$ then the problem is identical to the ordinary rational interpolation problem. In case all the interpolation conditions must be satisfied in one single point y_0 then the osculatory rational interpolation problem is identical to the Padé approximation problem. Instead of solving problem (1) we consider the linear system of equations

$$(fq - p)^{(\ell)}(y_i) = 0 \qquad i = 0, \ldots, j \qquad \ell = 0, \ldots, s_i - 1 \tag{2}$$

Condition (2) is a homogeneous system of $n+m+1$ linear equations in the $n+m+2$ unknown coefficients a_i and b_i of p and q. Hence the system (2) always has at least one nontrivial solution. For different solutions of (2) the following equivalence can be proved. If the polynomials p_1, q_1 and p_2, q_2 both satisfy (2) then $p_1 q_2 = p_2 q_1$. Not all solutions of (2) also satisfy (1): it is very well possible that the polynomials p and q satisfying (2) are such that p/q is reducible. Nevertheless all solutions of (2) have the same irreducible form. For p and q satisfying (2) we shall denote by $r_{n,m}(x) = (p_{n,m}/q_{n,m})(x)$ the irreducible form of p/q where $q_{n,m}(x)$ is normalized such that $q_{n,m}(y_0) = 1$, and we shall call $r_{n,m}(x)$ the rational Hermite interpolant of order (n,m) for f. Although the terminology "interpolant" is used it may be that $r_{n,m}(x)$ does not satisfy the interpolation conditions (1) anymore [WUYT]. A simple example will illustrate this. Let $y_0 = 0$, $y_1 = 1$, $y_2 = 2$ and $f(y_0) = 0$, $f(y_1) = 3$, $f(y_2) = 3$. Take $n = m = 1$. A solution of this rational interpolation problem is $p(x) = 3x$ and $q(x) = x$. Thus $p_{1,1}(x) = 3$ and $q_{1,1}(x) = 1$. Clearly $(p_{1,1}/q_{1,1})(y_0) \neq f(y_0)$.
The problem of "unattainable" interpolation points is typical for the case of rational interpolation. Having computed the rational interpolant p/q from linear interpolation conditions, in other words conditions expressed for $fq - p$ instead of for $f - (p/q)$, it may occur that an interpolation point is also a common zero of p and q and hence that the rational function p/q is undefined in that interpolation point. Consequently the nonlinear interpolation condition cannot be satisfied in that interpolation point anymore, not even by the irreducible form of p/q. The interpolation point has become unattainable. As a conclusion we can say that the rational interpolation problem (1) has a solution if and only if $p_{n,m}(x)$ and $q_{n,m}(x)$ satisfy themselves the system of equations (2).
The rational Hermite interpolation problem can be reformulated as a Newton-Padé approximation problem. We introduce the following notations:

$$x_\ell = y_0 \qquad \ell = 0, \ldots, s_0 - 1$$

$$x_{d(i)+\ell} = y_i \qquad \ell = 0, \ldots, s_i - 1 \qquad d(i) = s_0 + s_1 + \ldots + s_{i-1} (i \geq 1)$$

$$c_{ij} = 0 \qquad i > j$$

$$c_{ij} = f[x_i, \ldots, x_j] \qquad i \leq j$$

with possible coalescence of points in the divided difference $f[x_i,\ldots,x_j]$. If we put

$$B_j(x) = \prod_{\ell=1}^{j}(x - x_{\ell-1})$$

with $B_0(x) = 1$ then formally

$$f(x) = \sum_{i=0}^{\infty} c_{0i}B_i(x)$$

This series is called the Newton series for f. Problem (2) is then equivalent [CLAEa] with the computation of polynomials

$$p(x) = \sum_{i=0}^{n} a_i B_i(x)$$

and

$$q(x) = \sum_{i=0}^{m} b_i B_i(x)$$

such that

$$(fq - p)(x) = \sum_{i \geq n+m+1} d_i B_i(x) \tag{3}$$

which is called the Newton-Padé approximation problem of order (n,m) for f. To determine solutions p and q of (3) the divided differences

$$d_i = (fq - p)[x_0,\ldots,x_i] \qquad i = 0,\ldots,n+m$$

must be calculated and put equal to zero. The following generalization of the Leibniz rule for differentiating a product of functions, is a useful tool to accomplish this [WARN]:

$$(fq)[x_0,\ldots,x_i] = \sum_{\ell=0}^{i} f[x_0,\ldots,x_\ell]q[x_\ell,\ldots,x_i]$$

Using this rule we can now write down the linear systems of equations that must be satisfied by the coefficients a_i and b_i in p and q:

$$\begin{cases} c_{00}b_0 = a_0 \\ c_{01}b_0 + c_{11}b_1 = a_1 \\ \vdots \\ c_{0n}b_0 + c_{1n}b_1 + \ldots + c_{mn}b_m = a_n \end{cases}$$

$$\begin{cases} c_{0,n+1}b_0 + \ldots + c_{m,n+1}b_m = 0 \\ \vdots \\ c_{0,n+m}b_0 + \ldots + c_{m,n+m}b_m = 0 \end{cases} \tag{4}$$

Since the problems (2) and (3) are equivalent, the rational function $r_{n,m}$ can as well be called the Newton-Padé approximant of order (n,m) to f. We shall see that many properties and algorithms valid for Padé approximants can be generalized for Newton-Padé approximants or rational Hermite interpolants [WARN]. The rational Hermite interpolants of order (n,m) for f can be ordered in a table:

$$\begin{array}{cccc} r_{0,0} & r_{0,1} & r_{0,2} & \cdots \\ r_{1,0} & r_{1,1} & r_{1,2} & \cdots \\ r_{2,0} & r_{2,1} & \cdots & \\ r_{3,0} & r_{3,1} & \cdots & \\ \vdots & \vdots & & \end{array}$$

In the first column one finds the polynomial interpolants for f and in the first row the inverses of the polynomial interpolants for $(1/f)$. If we define $n' = \partial p_{n,m}$ and $m' = \partial q_{n,m}$ then it can be shown that at least $n'+m'+t+1$ points $z_0, \ldots, z_{n'+m'+t}$ with $t \geq 0$ exist in $\{x_0, \ldots, x_{n+m}\}$ such that $r_{n,m}(z_i) = f(z_i)$. Again we call an entry of the table normal if it occurs only once in that table. A necessary condition for the normality of the rational interpolant $r_{n,m}(x)$ is formulated in the following theorem [WUYT].

THEOREM 1:
If the rational Hermite interpolant $r_{n,m} = p_{n,m}/q_{n,m}$ is normal and if $(fq_{n,m} - p_{n,m})(x_i) = 0$ for $i = 0, \ldots, n'+m'$, then
(a) $n' = n$ and $m' = m$
(b) $(fq_{n,m} - p_{n,m})(x_i) \neq 0$ for $i = n+m+1, n+m+2$.

Conclusion (b) in theorem 1 does not imply that $(fq_{n,m} - p_{n,m})(x_i) \neq 0$ for $i \geq n+m+1$. That the conditions (a) and (b) are not sufficient to guarantee the normality of $r_{n,m}(x)$ is illustrated in the following example. Let $x_i = i$ for $i = 0,1,2,\ldots$ and $f(x_0) = 0$, $f(x_1) = 1$, $f(x_2) = 3$, $f(x_3) = 4$, $f(x_i) = i$ for $i = 4,5,6,\ldots$. For $n = 0$ and $m = 1$ we find $r_{n,m}(x) = x$ with (a) and (b) of theorem 1 satisfied. But $r_{n,m}$ is not normal because $r_{n,m} = r_{k,\ell}$ for $k \geq 3$ and $\ell \geq 2$. However, it is possible to formulate a sufficient condition for the normality of $r_{n,m}$ [WUYT].

THEOREM 2:
If $r_{n,m} = p_{n,m}/q_{n,m}$ with $n = n'$, $m = m'$ and $(fq_{n,m} - p_{n,m})(x_i) = 0$ for at most $n + m + 1$ points from the sequence $(x_i)_{i\in \mathbb{N}}$, then $r_{n,m}$ is normal.

2. Methods to compute normal rational Hermite interpolants.

2.1. Determinant formulas.

First of all we give a determinant representation [CLAE] similar to that for the case of Padé approximation. Let us define

$$F_{i,j}(x) = \sum_{\ell=i}^{j} c_{i\ell} B_\ell(x) \qquad i \leq j$$

with $F_{i,j}(x) = 0$ if $i > j$.

THEOREM 3:
If the rank of the system of equations (4) is maximal, then (up to a normalization) $r_{n,m} = p_{n,m}/q_{n,m}$ is given by

$$p_{n,m}(x) = \begin{vmatrix} F_{0,n}(x) & F_{1,n}(x) & \dots & F_{m,n}(x) \\ c_{0,n+1} & c_{1,n+1} & \dots & c_{m,n+1} \\ c_{0,n+2} & c_{1,n+2} & \dots & c_{m,n+2} \\ \vdots & \vdots & \ddots & \vdots \\ c_{0,n+m} & c_{1,n+m} & \dots & c_{m,n+m} \end{vmatrix}$$

and

$$q_{n,m}(x) = \begin{vmatrix} B_0(x) & B_1(x) & \dots & B_m(x) \\ c_{0,n+1} & c_{1,n+1} & \dots & c_{m,n+1} \\ c_{0,n+2} & c_{1,n+2} & \dots & c_{m,n+2} \\ \vdots & \vdots & \ddots & \vdots \\ c_{0,n+m} & c_{1,n+m} & \dots & c_{m,n+m} \end{vmatrix}$$

One can see that in case all the interpolation points coincide with one single point, these determinant formulas reduce to the ones given in [CUYT] since the divided differences reduce to Taylor coefficients.

In the sequel of this section we suppose that every rational interpolant $r_{n,m}(x)$ itself satisfies the interpolation conditions (1). This is for instance satisfied if $\min(n - n', m - m') = 0$. In discussing algorithms for the calculation of rational Hermite interpolants we restrict ourselves to those computation schemes that reduce to well-known algorithms for the calculation of Padé approximants in case all the interpolation points coincide. Exactly those algorithms will be generalized to the multivariate case in the following sections. We do not discuss the construction of Thiele interpolating continued fractions using inverse or reciprocal differences, nor methods that construct rational Hermite interpolants on paths in the table different from descending staircases.

2.2. The generalized ϵ-algorithm.

It was proved in [CLAEd] that

$$(x - x_{n+m})^{-1}(r_{n-1,m} - r_{n,m})^{-1} + (x - x_{n+m+1})^{-1}(r_{n+1,m} - r_{n,m})^{-1} = (x - x_{n+m})^{-1}(r_{n,m-1} - r_{n,m})^{-1} + (x - x_{n+m+1})^{-1}(r_{n,m+1} - r_{n,m})^{-1}$$

Using this result it is possible to set up the following generalized ϵ- algorithm [CLAEd], in the same way as the ϵ-algorithm for Padé approximants was constructed from the star identity:

$$\epsilon_{-1}^{(n)} = 0 \qquad n = 0, 1, \dots$$

$$\epsilon_{2m}^{(-m-1)} = 0 \qquad m = 0, 1, \dots$$

$$\epsilon_0^{(n)} = r_{n,0}(x) \qquad n = 0, 1, \dots$$

$$\epsilon_{m+1}^{(n)} = \epsilon_{m-1}^{(n+1)} + \frac{1}{(x - x_{m+n+1})(\epsilon_m^{(n+1)} - \epsilon_m^{(n)})}$$

$$n = -\lfloor \tfrac{m}{2} \rfloor - 1, -\lfloor \tfrac{m}{2} \rfloor, \dots \qquad m = 0, 1, \dots$$

Finally

$$\epsilon_{2m}^{(n-m)} = r_{n,m}(x)$$

2.3. A generalization of the qd-algorithm.

Consider descending staircases in the table of rational Hermite interpolants

$$T_k = \{r_{k,0}, r_{k+1,0}, r_{k+1,1}, r_{k+2,1}, \dots\} \qquad k \geq 0$$

and continued fractions of the form

$$g_k(x) = c_0 + \sum_{i=1}^{k} c_i(x - x_0)(x - x_1)\dots(x - x_{i-1}) + \frac{c_{k+1}(x - x_0)\dots(x - x_k)|}{|1} + \frac{-q_1^{(k+1)}(x - x_{k+1})|}{|1 + q_1^{(k+1)}(x_0 - x_{k+1})} + \frac{-e_1^{(k+1)}(x - x_{k+2})|}{|1 + e_1^{(k+1)}(x_0 - x_{k+2})} + \frac{-q_2^{(k+1)}(x - x_{k+3})|}{|1 + q_2^{(k+1)}(x_0 - x_{k+3})} + \frac{-e_2^{(k+1)}(x - x_{k+4})|}{|1 + e_2^{(k+1)}(x_0 - x_{k+4})} + \dots \qquad (5)$$

THEOREM 4:
If every three consecutive elements in T_k are different, then a continued fraction of the form (5) exists with $c_{k+1} \neq 0$, $q_i^{(k+1)} \neq 0$, $e_i^{(k+1)} \neq 0$, $1+q_i^{(k+1)}(x_0 - x_{k+2i-1}) \neq 0$, $1+e_i^{(k+1)}(x_0 - x_{k+2i}) \neq 0$ for $i \geq 1$ and such that the nth convergent equals the $(n+1)^{th}$ element of T_k.

To calculate the coefficients $q_i^{(k+1)}$ and $e_i^{(k+1)}$ in (5) one can use the following recurrence relations. Compute the even part of the continued fraction $g_k(x)$ and the odd part of the continued fraction $g_{k-1}(x)$. These contractions have the same convergents $r_{k,0}, r_{k+1,1}, r_{k+2,2}, \ldots$ and they also have the same form. In this way one can check [CLAEc] that for $k \geq 1$

$$e_0^{(k)} = 0$$
$$q_1^{(k)} = \frac{f[x_0, \ldots, x_{k+1}]}{f[x_1, \ldots, x_{k+1}]}$$

and for $\ell \geq 1$ and $k \geq 1$

$$e_\ell^{(k)} = \frac{q_\ell^{(k+1)} - q_\ell^{(k)} + e_{\ell-1}^{(k+1)}\left[1 + q_\ell^{(k+1)}(x_0 - x_{k+2\ell-1})\right]}{1 + q_\ell^{(k)}(x_0 - x_{k+2\ell-1})}$$
$$q_{\ell+1}^{(k)} = \frac{e_\ell^{(k+1)} q_\ell^{(k+1)}\left[1 + e_\ell^{(k)}(x_0 - x_{k+2\ell})\right]}{e_\ell^{(k)}\left[1 + q_\ell^{(k+1)}(x_0 - x_{k+2\ell-1})\right] + e_\ell^{(k+1)}(e_\ell^{(k)} - q_\ell^{(k+1)})(x_0 - x_{k+2\ell+1})}$$

These coefficients are usually ordered as in the next table

$$\begin{array}{cccccc}
e_0^{(1)} & & & & & \\
 & q_1^{(1)} & & & & \\
e_0^{(2)} & & e_1^{(1)} & & & \\
 & q_1^{(2)} & & q_2^{(1)} & & \\
e_0^{(3)} & & e_1^{(2)} & & e_2^{(1)} & \cdots \\
 & q_1^{(3)} & & q_2^{(2)} & & \\
e_0^{(4)} & & e_1^{(3)} & & e_2^{(2)} & \cdots \\
 & \vdots & & \vdots & & \\
\vdots & & \vdots & & \vdots &
\end{array}$$

where the superscript denotes a diagonal in the table and the subscript a column.

3. Structure of the table of rational Hermite interpolants.

We already mentioned in section 1 that at least $n'+m'+t+1$ points $z_0, \ldots, z_{n'+m'+t}$ with $t \geq 0$ exist among $\{x_0, \ldots, x_{n+m}\}$ such that $r_{n,m}(z_i) = f(z_i)$. On the basis of this conclusion a property comparable with the block structure of the Padé table can be formulated. It is based on the following property.

THEOREM 5:
If the rank of the linear system (4) is $m - t$ then (up to a normalization) a unique solution $\overline{p}$ and $\overline{q}$ of (4) exists with

$$\partial \overline{p} \leq n - t$$
$$\partial \overline{q} \leq m - t$$

where at least one of the upper bounds is attained. Every other solution $p(x)$ and $q(x)$ of (4) can be written in the form

$$p(x) = \overline{p}(x)s(x)$$
$$q(x) = \overline{q}(x)s(x)$$

where $\partial s \leq t$.

Before describing the shape of the sets in the table of rational Hermite interpolants that contain equal elements, it is important to emphasize that the structure of the table can only be studied if the ordering of the interpolation points $\{x_i\}_{i \in \mathbb{N}}$ remains fixed once it is chosen. Since the polynomials $\overline{p}$ and $\overline{q}$ constructed in the previous theorem have the property that their degrees cannot be lowered simultaneously anymore unless some interpolation conditions are lost, we shall call them a minimal solution. This does not imply that $\overline{p}/\overline{q}$ is irreducible. However we still have $p_{n,m}\overline{q} = \overline{p}q_{n,m}$.

THEOREM 6:
Let $\overline{p}(x)$ and $\overline{q}(x)$ be the minimal solution of the Newton-Padé approximation problem of order (n, m) for f and let the rank of the linear system (4) be $m - t$.

(a) If $\partial \overline{p} = n - t - t_1$ then all the minimal solutions lying in the triangle with corner elements $(n - t - t_1, m - t)$, $(n - t - t_1, m + t + t_1)$ and $(n + t, m - t)$ are equal to $\overline{p}(x)$ and $\overline{q}(x)$.

(b) If $\partial \overline{q} = m - t - t_2$ then all the minimal solutions lying in the triangle with corner elements $(n - t, m - t - t_2)$, $(n - t, m + t)$ and $(n + t + t_2, m - t - t_2)$ are equal to $\overline{p}(x)$ and $\overline{q}(x)$.

(c) If

$$(f\overline{q} - \overline{p})(x) = \sum_{i \geq n+m-2t+t_3+1} d_i B_i(x)$$

with $d_{n+m-2t+t_3+1} \neq 0$ *then all the rational Hermite interpolants lying in the triangle with corner elements* $(n-t, m-t)$, $(n+t+t_3, m-t)$ *and* $(n-t, m+t+t_3)$ *are equal to* $\overline{p}(x)$ *and* $\overline{q}(x)$.

(d) *If* $\partial\overline{p} = n - s_1, \partial\overline{q} = m - s_2$ *and* $(f\overline{q} - \overline{p})(x) = \sum_{i \geq n+m+1+s_3} d_i B_i(x)$ *with* $d_{n+m+s_3+1} \neq 0$ *then all the rational interpolants lying in the square with corners* $(n - s_1, m - s_2)$ *and* $(n + s_2 + s_3, m + s_1 + s_3)$ *have the same irreducible form* $r_{n,m}(x)$.

(e) *If* $\partial\overline{p} = n-s_1, \partial\overline{q} = m-s_2, (f\overline{q}-\overline{p})(x) = \sum_{i \geq n+m+1} d_i B_i(x)$ *and* $r_{n-s_1,m-s_2}(x)$ *also satisfies the interpolation conditions in the points* $x_{n+m+1+\beta_j}$ *for* $j = 1, \ldots, s$ *and* $0 \leq \beta_1 < \ldots < \beta_s$, *then if* $\beta_j < 2j + s_1 + s_2$ *we have for* $\ell = \beta_j + 1, \ldots, 2j+s_1+s_2$: $r_{n+s_2+j,m-s_2+\ell-j}(x) = r_{n-s_1,m-s_2}(x) = r_{n-s_1+\ell-j,m+s_1+j}(x)$.

This theorem explains that the square block described in theorem 6c is only a starting point and that it can have a sort of tail concentrated along its main diagonal as illustrated in the next picture.

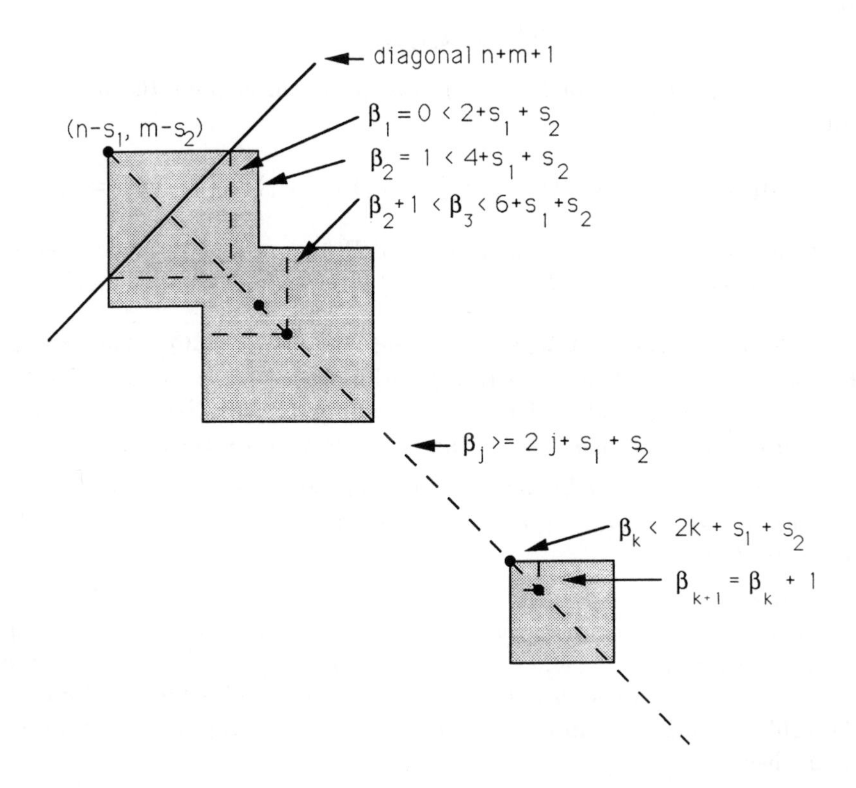

For the proof we refer to [CLAEf]. For a detailed study of the structure of the rational Hermite interpolation table we refer to [CLAEb]. Singular rules for the generalized ϵ- and qd-algorithm applicable in non-normal tables are under investigation.

4. Convergence of rational Hermite interpolants.

We shall now mention some results for the convergence of columns in the table of rational Hermite interpolants. Broadly speaking, the convergence of an arbitrary series of interpolation does not depend on the entire sequence of interpolation points x_i (as defined in the Newton-Padé approximation problem) but merely on its asymptotic character, as can be seen in the next theorem.

THEOREM 7:
Let the sequence of interpolation points $\{x_0, x_1, x_2, \ldots\}$ *be asymptotic to the sequence*

$$\{w_0, w_1, \ldots, w_j, w_0, w_1, \ldots, w_j, w_0, w_1, \ldots, w_j, \ldots\}$$

in the sense that

$$lim_{k\to\infty} x_{k(j+1)+i} = w_i$$

for $i = 0, \ldots, j$. *If the function* $f(z)$ *is analytic throughout the interior of the lemniscate*

$$B(w_0, \ldots, w_j, r) = \{z \in \mathbb{C} : |(z - w_0)(z - w_1) \ldots (z - w_j)| = r\}$$

then the $r_{n,0}$ *converge to* f *on the interior of* $B(w_0, \ldots, w_j, r)$. *The convergence is uniform on every closed and bounded subset interior to* $B(w_0, \ldots, w_j, r)$.

For the proof we refer to [WALS p. 61] and [DAVI pp. 90-91]. Let us now turn to the case of a meromorphic function f with poles $z_1, \ldots, z_m$ (counted with their multiplicity). Let the table of minimal solutions for the Newton-Padé approximation problem be normal. According to theorem 1 we then have $\partial \overline{q}_{n,m} = m$. Let $z_i^{(n)}$ for $i = 1, \ldots, m$ be the zeros of $q_{n,m}$ for $n = 0, 1, 2, \ldots$ and let $\rho_i = |(z_i - w_0)(z_i - w_1) \ldots (z_i - w_j)|$ with $0 < \rho_1 \leq \rho_2 \leq \ldots \leq \rho_m \leq \alpha r < r$ for a positive constant α.

THEOREM 8:
If the sequence of interpolation points $\{x_0, x_1, x_2, \ldots\}$ *is asymptotic to the sequence* $\{w_0, w_1, \ldots, w_j, w_0, w_1, \ldots, w_j, \ldots\}$, *if* f *is meromorphic in the interior of the lemniscate* $B(w_0, \ldots, w_j, r)$ *with poles* $z_1, \ldots, z_m$ *counted with their multiplicity and if the table of minimal solutions for the Newton-Padé approximation problem is normal, then*

$$lim_{n\to\infty} q_{n,m}(z) = \prod_{i=1}^{m} \left(\frac{z - w_i}{x_0 - w_i} \right) \quad \text{and} \quad lim_{n\to\infty} r_{n,m}(z) = f(z)$$

uniformly in every closed and bounded subset in the interior of $B(w_0, \ldots, w_j, r)$ *not containing the points* $z_1, \ldots, z_m$.

The proof is given in [CLAEb].

5. Multivariate rational Hermite interpolation problems.

For the sake of simplicity we restrict ourselves in the sequel of the text to the case of two variables. The generalization to the case of more than two variables will appear to be straightforward and only notationally more difficult. Let us first describe the conditions which have to be fulfilled by the multivariate data set before the interpolants can be constructed. Since we allow coalescence of interpolation points, we shall also point out how to deal with such a situation.
Consider for instance the following picture in $\mathbb{N}^2$ of the data set (x_i, y_j), where a circle indicates that in addition to $f_{ij} = f(x_i, y_j)$ also $\partial f / \partial x$ is given and a square indicates that also $\partial f / \partial x$, $\partial f / \partial y$ and $\partial^2 f / \partial y^2$ are provided.

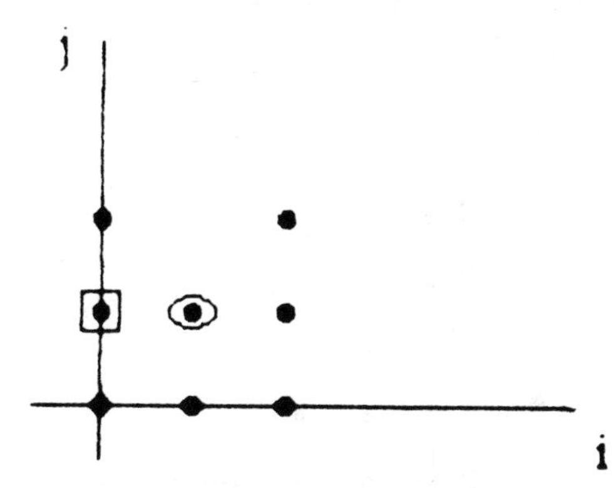

This situation can be considered as the limit situation of a data set with non-coalescent interpolation points where we let $x_3 \rightarrow x_0$, $x_4 \rightarrow x_1$, $y_3 \rightarrow y_1$ and $y_4 \rightarrow y_1$.

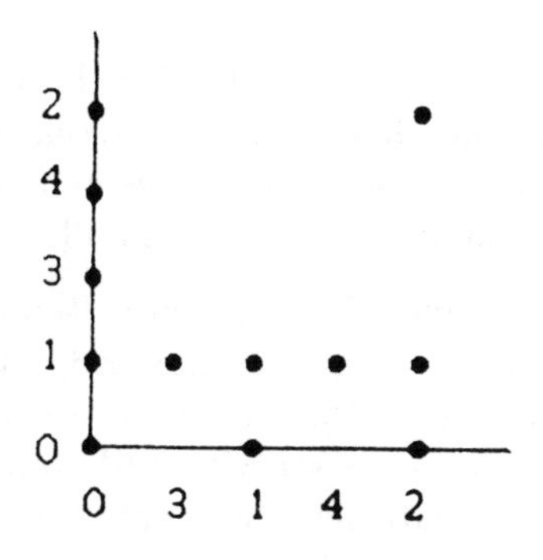

If we want to interpolate these (x_i, y_j, f_{ij}) by means of the techniques described below, then the data f_{ij} and the numbering of the x_i and y_j have to be such that

(a) x_0 is that x-coordinate for which the number of y-coordinates at which data are given is maximal, x_1 is the one of the leftover points for which the same is true, and so on

(b) y_0 is that y-coordinate for which the number of x-coordinates at which data are given is maximal, y_1 is the one of the leftover points for which the same is true, and so on

(c) the data set has the inclusion property, meaning that when a point belongs to the data set then the rectangular subset of points emanating from the origin with the given point as its furthermost corner also lies in the data set.

Note that (a) and (b) do not necessarily imply (c). We shall comment on the importance of condition (c) further on. For the picture above (c) is clearly not satisfied. So we try to renumber the interpolation points such that these three conditions are fulfilled. Let us introduce a new numbering (x'_i, y'_j) with $x'_0 = x_0, x'_1 = x_2, x'_2 = x_1, x'_3 = x_4, x'_4 = x_3$ and $y'_0 = y_1, y'_1 = y_0, y'_2 = y_2, y'_3 = y_4, y'_4 = y_3$. We then get the following picture in $\mathbb{N}^2$ of the data set.

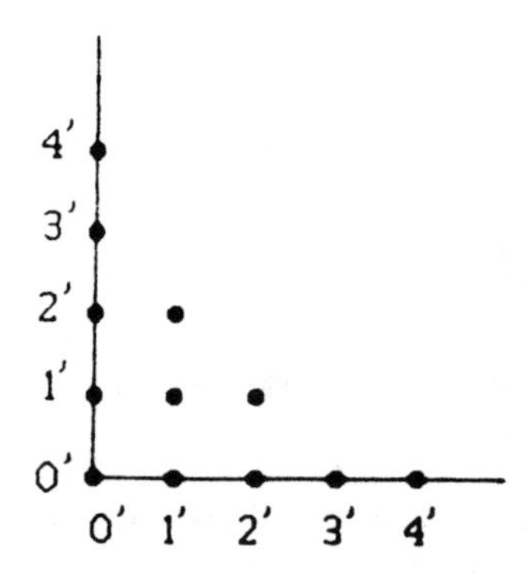

The interpolation problems that can be reduced to this situation are of course not the most general ones but they already represent quite a number of situations that can be dealt with. In the sections 5–7 we assume that the given data set is structured such that the conditions (a–c) are fulfilled.

Let the complex function values f_{ij} be given in the complex points (x_i, y_j) with $(i,j) \in I \subseteq \mathbb{N}^2$, where I satisfies the inclusion property or rectangle rule, meaning that when (i,j) belongs to I then (k,ℓ) belongs to I for $k \leq i$ and $\ell \leq j$. We know from the pictures above that a data set with coalescent interpolation points can be replaced by an intermediate data set where only function values are given. When certain interpolation points coincide, we must bear in mind that due to the renumbering these coalescent x- and y-coordinates are not necessarily consecutive. With the given interpolation points we define the following polynomial basis functions:

$$B_{ij}(x,y) = \prod_{k=1}^{i}(x - x_{k-1}) \prod_{\ell=1}^{j}(y - y_{\ell-1})$$

These basis functions are bivariate polynomials of degree $i+j$. With

$$c_{0i,0j} = f[x_0, \ldots, x_i][y_0, \ldots, y_j]$$

where coalescence of points in the divided difference is admitted [CUYT], we can now write in a purely formal manner [BERE]

$$f(x,y) = \sum_{(i,j)\in \mathbb{N}^2} c_{0i,0j} B_{ij}(x,y) \tag{6}$$

Hence we have constructed with the data a bivariate Newton interpolating series and we can start approximating it using bivariate rational functions. For the bivariate divided differences a Leibniz type product rule remains valid and will prove to be useful in the sequel:

$$(fq)[x_0, \ldots, x_i][y_0, \ldots, y_j] = \sum_{\mu=0}^{i}\sum_{\nu=0}^{j} f[x_0, \ldots, x_\mu][y_0, \ldots, y_\nu] q[x_\mu, \ldots, x_i][y_\nu, \ldots, y_j]$$

The definition of multivariate Newton-Padé approximant which we shall give is a very general one. It includes the univariate definition and a lot of the definitions for multivariate Padé approximants as a special case. With any finite subset D of $\mathbb{N}^2$ we associate a polynomial of which the coefficients and the basisfunctions are indexed by the indices in D. Given the double Newton series, we choose three subsets N, D and I of $\mathbb{N}^2$ and construct an $[N/D]_I$ Newton-Padé approximant to $f(x,y)$ as follows:

$$p(x,y) = \sum_{(i,j)\in N} a_{ij} B_{ij}(x,y) \tag{7a}$$

$$q(x,y) = \sum_{(i,j)\in D} b_{ij} B_{ij}(x,y) \tag{7b}$$

$$(fq-p)(x,y) = \sum_{(i,j)\in \mathbb{N}^2 \setminus I} d_{ij} B_{ij}(x,y) \tag{7c}$$

In analogy with the univariate case, we select N, D and I such that

D has $m+1$ elements, numbered $(d_0, e_0), \ldots, (d_m, e_m)$
$N \subset I$
I satisfies the rectangle rule
$\#(I \backslash N) = m$.

We will denote $\partial p = N$ and $\partial q = D$. Clearly condition (7c) is equivalent with

$$d_{ij} = (fq - p)[x_0, \ldots, x_i][y_0, \ldots, y_j] = 0 \qquad (i,j) \in I \tag{8}$$

Because $N \subset I$, the system of equations (8) can be divided into a non-homogeneous and a homogeneous part:

$$(fq)[x_0, \ldots, x_i][y_0, \ldots, y_j] = p[x_0, \ldots, x_i][y_0, \ldots, y_j] = a_{ij} \qquad (i,j) \in N \tag{9a}$$

$$(fq)[x_0, \ldots, x_i][y_0, \ldots, y_j] = 0 \qquad (i,j) \in I \backslash N \tag{9b}$$

Let's take a look at the conditions (9b). Suppose that I is such that the m homogeneous equations in (9b) are linearly independent and let us number the m elements in $I \backslash N$ indexing these equations by $(i_{n+1}, j_{n+1}), \ldots, (i_{n+m}, j_{n+m})$. By means of the Leibniz rule the homogeneous system (9b) of m equations in $m+1$ unknowns looks like

$$\begin{pmatrix} c_{d_0 i_{n+1}, e_0 j_{n+1}} & \cdots & c_{d_m i_{n+1}, e_m j_{n+1}} \\ \vdots & & \vdots \\ c_{d_0 i_{n+m}, e_0 j_{n+m}} & \cdots & c_{d_m i_{n+m}, e_m j_{n+m}} \end{pmatrix} \begin{pmatrix} b_{d_0, e_0} \\ \vdots \\ b_{d_m, e_m} \end{pmatrix} = \begin{pmatrix} 0 \\ \vdots \\ 0 \end{pmatrix} \tag{10}$$

As we suppose the rank of the coefficient matrix to be maximal, a solution $q(x,y)$ is given by

$$q(x,y) = \begin{vmatrix} B_{d_0 e_0}(x,y) & \cdots & B_{d_m e_m}(x,y) \\ c_{d_0 i_{n+1}, e_0 j_{n+1}} & \cdots & c_{d_m i_{n+1}, e_m j_{n+1}} \\ \vdots & & \vdots \\ c_{d_0 i_{n+m}, e_0 j_{n+m}} & \cdots & c_{d_m i_{n+m}, e_m j_{n+m}} \end{vmatrix} \tag{11a}$$

By the conditions (9a) we find as determinant representation for $p(x,y)$

$$p(x,y) = \begin{vmatrix} \sum\limits_{(i,j) \in N} c_{d_0 i, e_0 j} B_{ij}(x,y) & \cdots & \sum\limits_{(i,j) \in N} c_{d_m i, e_m j} B_{ij}(x,y) \\ c_{d_0 i_{n+1}, e_0 j_{n+1}} & \cdots & c_{d_m i_{n+1}, e_m j_{n+1}} \\ \vdots & & \vdots \\ c_{d_0 i_{n+m}, e_0 j_{n+m}} & \cdots & c_{d_m i_{n+m}, e_m j_{n+m}} \end{vmatrix} \tag{11b}$$

If for all $k, \ell \geq 0$ we have $q(x_k, y_\ell) \neq 0$ then with $e_{ij} = (1/q)[x_0, \ldots, x_i][y_0, \ldots, y_j]$ and I satisfying the inclusion property

$$(f - \frac{p}{q})(x,y) = \left[\frac{1}{q}(fq - p)\right](x,y) = \sum_{(i,j) \in \mathbb{N}^2 \backslash I} \tilde{d}_{ij} B_{ij}(x,y)$$

If I does not satisfy the inclusion property then

$$(fq-p)(x,y)=\sum_{(i,j)\in\mathbb{N}^2\setminus I} d_{ij}B_{ij}(x,y)$$

does not imply

$$(f-\frac{p}{q})(x,y)=\sum_{(i,j)\in\mathbb{N}^2\setminus I} \tilde{d}_{ij}B_{ij}(x,y)$$

since in that case $f-p/q$ also contains the terms that result from multiplying a "hole" in I by $(1/q)(x,y)$ [CUYTo]. From the determinant representations (11a) and (11b) we can easily obtain the determinant representation given in section 2 for univariate Newton-Padé approximants as a special case.

6. Methods for the computation of nondegenerate MRI.

In this section we continue to assume that the m equations in the homogeneous system (10) are linearly independent. Then the multivariate rational Hermite interpolation problem is called nondegenerate.

6.1. The E-algorithm.

Let us now introduce a numbering $r(i,j)$ of the points in $\mathbb{N}^2$, for instance the enumeration

$$(0,0),\ \underbrace{(1,0),(0,1)}_{\text{first diagonal}},\ \underbrace{(2,0),(1,1),(0,2)}_{\text{second diagonal}},\underbrace{(3,0),(2,1),(1,2),(0,3)}_{\text{third diagonal}},\ldots$$

and retain this order in N, D and I. If we denote $\#N=n+1$ then we can write

$$N=\bigcup_{k=0}^{n} N_k$$

with

$$\begin{gathered}\emptyset=N_{-1}\subset N_0\subset N_1\subset\ldots\subset N_{n-1}\subset N_n=N\\ \#N_k=k+1\\ N_k\setminus N_{k-1}=\{(i_k,j_k)\}\qquad k=0,1,\ldots,n\\ r(i_k,j_k)>r(i_s,j_s)\qquad k>s\end{gathered}$$

In other words, for each $k=0,\ldots,n$ we add to N_{k-1} the point (i_k,j_k) which is the next in line in $N\cap\mathbb{N}^2$ according to the enumeration given above. Denote $\#D=m+1$ and proceed in the same way. Hence

$$D=\bigcup_{\ell=0}^{m} D_\ell$$

with

$$D_{-1} = \emptyset \qquad D_\ell \setminus D_{\ell-1} = \{(d_\ell, e_\ell)\} \qquad \ell = 0, \ldots, m$$

We have assumed that the interpolation set I is such that the m homogeneous equations are linearly independent and hence we write for $I \setminus N$

$$I \setminus N = I_{n+1,n+m} = \bigcup_{\ell=1}^{m} I_{n+1,n+\ell}$$

with

$$I_{n+1,n} = \emptyset \qquad I_{n+1,n+\ell} \setminus I_{n+1,n+\ell-1} = \{(i_{n+\ell}, j_{n+\ell})\} \qquad \ell = 1, \ldots, m$$

To obtain a recursive algorithm, the determinant formulas (11) for the polynomials $p(x,y)$ and $q(x,y)$ are rewritten as follows. Multiply the $(\ell+1)^{th}$ row in $p(x,y)$ and $q(x,y)$ by $B_{i_{n+\ell}j_{n+\ell}}(x,y)$ ($\ell = 1, \ldots, m$), and then divide the $(\ell+1)^{th}$ column by $B_{d_\ell e_\ell}(x,y)$ ($\ell = 0, \ldots, m$). This respectively results for numerator and denominator in

$$\begin{vmatrix} \sum_{(i,j)\in N} c_{d_0 i, e_0 j} B_{d_0 i, e_0 j}(x,y) & \ldots & \sum_{(i,j)\in N} c_{d_m i, e_m j} B_{d_m i, e_m j}(x,y) \\ c_{d_0 i_{n+1}, e_0 j_{n+1}} B_{d_0 i_{n+1}, e_0 j_{n+1}}(x,y) & \ldots & c_{d_m i_{n+1}, e_m j_{n+1}} B_{d_m i_{n+1}, e_m j_{n+1}}(x,y) \\ \vdots & & \vdots \\ c_{d_0 i_{n+m}, e_0 j_{n+m}} B_{d_0 i_{n+m}, e_0 j_{n+m}}(x,y) & \ldots & c_{d_m i_{n+m}, e_m j_{n+m}} B_{d_m i_{n+m}, e_m j_{n+m}}(x,y) \end{vmatrix}$$

and

$$\begin{vmatrix} 1 & \ldots & 1 \\ c_{d_0 i_{n+1}, e_0 j_{n+1}} B_{d_0 i_{n+1}, e_0 j_{n+1}}(x,y) & \ldots & c_{d_m i_{n+1}, e_m j_{n+1}} B_{d_m i_{n+1}, e_m j_{n+1}}(x,y) \\ \vdots & & \vdots \\ c_{d_0 i_{n+m}, e_0 j_{n+m}} B_{d_0 i_{n+m}, e_0 j_{n+m}}(x,y) & \ldots & c_{d_m i_{n+m}, e_m j_{n+m}} B_{d_m i_{n+m}, e_m j_{n+m}}(x,y) \end{vmatrix}$$

where for $k \le i$ and $\ell \le j$

$$B_{ki,\ell j}(x,y) = \frac{B_{ij}(x,y)}{B_{k\ell}(x,y)} = (x - x_k)\ldots(x - x_{i-1})(y - y_\ell)\ldots(y - y_{j-1})$$

and for $k > i$ or $\ell > j$, $c_{ki,\ell j} = 0$. We can now easily construct $(m+1)$ series of which the successive partial sums can be found in the columns of $p(x,y)$. Take

$$t_0(n) = \sum_{(i,j)\in N} c_{d_0 i, e_0 j} B_{d_0 i, e_0 j}(x,y)$$

and

$$\Delta t_0(n+\ell-1) = t_0(n+\ell) - t_0(n+\ell-1)$$
$$= c_{d_0 i_{n+\ell}, e_0 j_{n+\ell}} B_{d_0 i_{n+\ell}, e_0 j_{n+\ell}}(x,y) \qquad \ell = 1,\ldots,m$$

for the first column of $p(x,y)$. Define for $r = 1,\ldots,m$

$$t_r(n) = \sum_{(i,j)\in N} c_{d_r i, e_r j} B_{d_r i, e_r j}(x,y)$$

and

$$\Delta t_r(n+\ell-1) = t_r(n+\ell) - t_r(n+\ell-1)$$
$$= c_{d_r i_{n+\ell}, e_r j_{n+\ell}} B_{d_r i_{n+\ell}, e_r j_{n+\ell}}(x,y) \qquad \ell = 1,\ldots,m$$

for the $(r+1)^{th}$ column of $p(x,y)$. Consequently

$$p(x,y) = \begin{vmatrix} t_0(n) & \ldots & t_m(n) \\ \Delta t_0(n) & \ldots & \Delta t_m(n) \\ \vdots & & \vdots \\ \Delta t_0(n+m-1) & \ldots & \Delta t_m(n+m-1) \end{vmatrix} \tag{12a}$$

$$q(x,y) = \begin{vmatrix} 1 & \ldots & 1 \\ \Delta t_0(n) & \ldots & \Delta t_m(n) \\ \vdots & & \vdots \\ \Delta t_0(n+m-1) & \ldots & \Delta t_m(n+m-1) \end{vmatrix} \tag{12b}$$

This quotient of determinants can easily be computed using the E-algorithm [BREZb]:

$$E_0^{(k)} = t_0(k) \qquad k = 0,\ldots,n+m$$
$$g_{0,\ell}^{(k)} = t_\ell(k) - t_{\ell-1}(k) \qquad \ell = 1,\ldots,m \qquad k = 0,\ldots,n+m$$
$$E_\ell^{(k)} = \frac{E_{\ell-1}^{(k)} g_{\ell-1,\ell}^{(k+1)} - E_{\ell-1}^{(k+1)} g_{\ell-1,\ell}^{(k)}}{g_{\ell-1,\ell}^{(k+1)} - g_{\ell-1,\ell}^{(k)}} \qquad k = 0,1,\ldots,n \qquad \ell = 1,2,\ldots,m \tag{13a}$$
$$g_{\ell,s}^{(k)} = \frac{g_{\ell-1,s}^{(k)} g_{\ell-1,\ell}^{(k+1)} - g_{\ell-1,s}^{(k+1)} g_{\ell-1,\ell}^{(k)}}{g_{\ell-1,\ell}^{(k+1)} - g_{\ell-1,\ell}^{(k)}} \qquad s = \ell+1, \ell+2, \ldots \tag{13b}$$

The values $E_\ell^{(k)}$ and $g_{\ell,s}^{(k)}$ are stored as in [BREZb]. We obtain $[N/D]_I = E_m^{(n)}$. Since the solution $q(x,y)$ of (7c) is unique, the value $E_m^{(n)}$ itself does not depend upon the numbering of the points within the sets N, D and H. But this numbering affects the interpolation conditions satisfied by the intermediate E-values [CUYTn].

THEOREM 10:
For $k = 0, \ldots, n$ and $\ell = 0, \ldots, m$

$$E_\ell^{(k)} = [N_k/D_\ell]_{N_k \cup I_{k+1,k+\ell}}$$

6.2. The qdg-algorithm.

If we suppose that the homogeneous system of equations (10) has maximal rank we can also write

$$I = \bigcup_{\ell=0}^{n+m} I_\ell$$

with

$$\begin{aligned} I_k = N_k \qquad & k = 0, \ldots, n \\ I_{n+\ell} \backslash I_{n+\ell-1} = \{(i_{n+\ell}, j_{n+\ell})\} \qquad & \ell = 1, \ldots, m \\ r(i_{n+\ell}, j_{n+\ell}) > r(i_s, j_s) \qquad & n + \ell > s \geq n + 1 \end{aligned}$$

With the subsets N_k, D_ℓ and $I_{k+\ell}$ rational interpolants $[N_k/D_\ell]_{I_{k+\ell}}$ can be constructed which satisfy only part of the interpolation conditions and which are of lower "degree". To this end we assume that the numbering $r(i_\ell, j_\ell)$ of the points in $\mathbb{N}^2$ is such that the inclusion property of the set I is carried over to the subsets I_ℓ. With these functions we can fill up a table of rational interpolants :

$$\begin{array}{cccc} [N_0/D_0]_{I_0} & [N_0/D_1]_{I_1} & [N_0/D_2]_{I_2} & \cdots \\ [N_1/D_0]_{I_1} & [N_1/D_1]_{I_2} & [N_1/D_2]_{I_3} & \cdots \\ [N_2/D_0]_{I_2} & [N_2/D_1]_{I_3} & [N_2/D_2]_{I_4} & \cdots \\ \vdots & \vdots & \vdots & \end{array}$$

where $[N/D]_I = [N_n/D_m]_{I_{n+m}}$. Our aim is to consider descending staircases of multivariate rational interpolants

$$\begin{array}{ccccc} [N_s/D_0]_{I_s} & & & & \\ [N_{s+1}/D_0]_{I_{s+1}} & [N_{s+1}/D_1]_{I_{s+2}} & & & \\ & [N_{s+2}/D_1]_{I_{s+3}} & [N_{s+2}/D_2]_{I_{s+4}} & & \\ & & \vdots & \cdots & \end{array} \tag{14}$$

and construct continued fractions of which the successive convergents equal the successive interpolants on the staircase. We restrict ourselves to the case where every three subsequent elements on the staircase are different. It was proved in [CUYTm] that given such a descending staircase, it is possible to construct a continued fraction of the form

$$C_s(x,y) = [N_s/D_0]_{I_s} + \frac{[N_{s+1}/D_0]_{I_{s+1}} - [N_s/D_0]_{I_s} \rfloor}{\lfloor 1} + \frac{-q_1^{(s+1)} \rfloor}{\lfloor 1+q_1^{(s+1)}} + \frac{-e_1^{(s+1)} \rfloor}{\lfloor 1+e_1^{(s+1)}} + \frac{-q_2^{(s+1)} \rfloor}{\lfloor 1+q_2^{(s+1)}} + \frac{-e_2^{(s+1)} \rfloor}{\lfloor 1+e_2^{(s+1)}} + \ldots \tag{15}$$

with this property. Here

$$[N_s/D_0]_{I_s} = \sum_{(i,j)\in N_s} c_{d_0 i, e_0 j} B_{d_0 i, e_0 j}(x,y)$$

$$[N_{s+1}/D_0]_{I_{s+1}} = \sum_{(i,j)\in N_{s+1}} c_{d_0 i, e_0 j} B_{d_0 i, e_0 j}(x,y)$$

and the coefficients $q_\ell^{(s+1)}$ and $e_\ell^{(s+1)}$ are computed using the following rules: for $\ell \geq 2$

$$q_\ell^{(s+1)} = \frac{e_{\ell-1}^{(s+2)} q_{\ell-1}^{(s+2)}}{e_{\ell-1}^{(s+1)}} \; \frac{g_{\ell-2,\ell-1}^{(s+\ell-1)} - g_{\ell-2,\ell-1}^{(s+\ell)}}{g_{\ell-2,\ell-1}^{(s+\ell-1)}} \; \frac{g_{\ell-1,\ell}^{(s+\ell)}}{g_{\ell-1,\ell}^{(s+\ell)} - g_{\ell-1,\ell}^{(s+\ell+1)}} \tag{16a}$$

and for $\ell \geq 1$

$$e_\ell^{(s+1)} + 1 = \frac{g_{\ell-1,\ell}^{(s+\ell)} - g_{\ell-1,\ell}^{(s+\ell+1)}}{g_{\ell-1,\ell}^{(s+\ell)}} \left(q_\ell^{(s+2)} + 1 \right) \tag{16b}$$

If we arrange the values $q_\ell^{(s+1)}$ and $e_\ell^{(s+1)}$ in a table as follows

$$\begin{array}{ccccc}
q_1^{(1)} & & & & \\
 & e_1^{(1)} & & & \\
q_1^{(2)} & & q_2^{(1)} & & \\
 & e_1^{(2)} & & e_2^{(1)} & \\
q_1^{(3)} & & q_2^{(2)} & & \ddots \\
 & e_1^{(3)} & & e_2^{(2)} & \\
q_1^{(4)} & & q_2^{(3)} & & \ddots \\
\vdots & e_1^{(4)} & \vdots & e_2^{(3)} & \\
 & \vdots & & \vdots &
\end{array}$$

where subscripts indicate columns and superscripts indicate downward sloping diagonals, then (16a) links the elements in the rhombus

$$
\begin{array}{ccc}
 & e_{\ell-1}^{(s+1)} & \\
q_{\ell-1}^{(s+2)} & & q_{\ell}^{(s+1)} \\
 & e_{\ell-1}^{(s+2)} &
\end{array}
$$

and (16b) links two elements on an upward sloping diagonal

$$
\begin{array}{cc}
 & e_{\ell}^{(s+1)} \\
q_{\ell}^{(s+2)} &
\end{array}
$$

If starting values for $q_{\ell}^{(s+1)}$ were known, all the values could be computed. These starting values are given by

$$q_1^{(s+1)} = \frac{E_1^{(s+1)} - E_0^{(s+1)}}{E_0^{(s+1)} - E_0^{(s)}} = \frac{\Delta t_0(s+1)}{\Delta t_0(s)} \quad \frac{g_{0,1}^{(s+1)}}{g_{0,1}^{(s+1)} - g_{0,1}^{(s+2)}} \tag{16c}$$

7. Structure of a degenerate table of MRI.

If the rank of the defining system of equations (10) is not maximal, we should look at $[N/D]_I$ as being a set of rational functions of which the numerator and denominator are given by (7a-b) and are satisfying (7c). A solution $[N/D]_I$ containing numerators and denominators of different "degrees" is called "degenerate". Let us denote the coefficient matrix of (10) by $C_{n+1,n+m}$. Note that the rows in $C_{n+1,n+m}$ are indexed by $I_{n+1,n+m}$.

In the univariate case and under certain conditions, the table of minimal solutions of the rational interpolation problem consists of triangles, once the numbering of the interpolation points is fixed [CLAEf]. The size of the triangles, as pointed out in section 3, is related to the rank deficiency of the interpolation problem. We shall now give a similar multivariate theorem and point out the differences with the univariate version. From this discussion it will also become clear why different solutions of the same rational interpolation problem are not necessarily equivalent anymore and hence not providing a unique irreducible form.

Theorem 11:
Let $p(x,y)$ and $q(x,y)$ be defined by (7). Let the rank of $C_{n+1,n+m}$ in (10) be given by $m-t$. Then for each pair (k,ℓ) with $0 \leq k \leq t$, $0 \leq \ell \leq t$, $k+\ell=t$ and the rank of $C_{n-k+1,n+m-t}$ equal to $m-\ell$ the following holds.

(a) *For $0 \leq i$, $0 \leq j$ and $i+j \leq t$, $[N_{n-k}/D_{m-\ell}]_{I_{n+m-t}}$ belongs to the solution set $[N_{n-k+i}/D_{m-\ell+j}]_{I_{n+m-t+i+j}}$, meaning that the (up to a multiplicative constant factor) unique rational function $[N_{n-k}/D_{m-\ell}]_{I_{n+m-t}}$ also solves the interpolation problems posed in $[N_{n-k+i}/D_{m-\ell+j}]_{I_{n+m-t+i+j}}$ where the solution set $[N_{n-k+i}/D_{m-\ell+j}]_{I_{n+m-t+i+j}}$ lies in the triangle of the table of rational interpolants with corner elements $[N_{n-k}/D_{m-\ell}]_{I_{n+m-t}}$, $[N_{n-k}/D_{m+k}]_{I_{n+m}}$ and $[N_{n+\ell}/D_{m-\ell}]_{I_{n+m}}$.*

(b) *If the solution $[N_{n-k}/D_{m-\ell}]_{I_{n+m-t}} = (p/q)(x,y)$ is such that $\partial p = N_{n-k-s_1}$ with $s_1 > 0$, then under the condition that the rank of $C_{n-k-s_1+1,n+m-t-s_1}$ is $m-\ell$, $[N_{n-k-s_1}/D_{m-\ell}]_{I_{n+m-t-s_1}}$ also solves $[N_{n-k-s_1+i}/D_{m-\ell+j}]_{I_{n+m-t-s_1+i+j}}$ for $0 \leq i$, $0 \leq j$ and $i+j \leq t+s_1$.*

(c) *If the solution $[N_{n-k}/D_{m-\ell}]_{I_{n+m-t}} = (p/q)(x,y)$ is such that $\partial q = D_{m-\ell-s_2}$ with $s_2 > 0$, then under the condition that the rank of $C_{n-k+1,n+m-t-s_2}$ is $m-\ell-s_2$, $[N_{n-k}/D_{m-\ell-s_2}]_{I_{n+m-t-s_2}}$ also solves $[N_{n-k+i}/D_{m-\ell-s_2+j}]_{I_{n+m-t-s_2+i+j}}$ for $0 \leq i$, $0 \leq j$ and $i+j \leq t+s_2$.*

(d) *If the solution $[N_{n-k}/D_{m-\ell}]_{I_{n+m-t}} = (p/q)(x,y)$ is such that*

$$(fq-p)(x,y) = \sum_{(i,j)\in \mathbb{N}^2 \setminus I_{n+m+t_3}} d_{ij} B_{ij}(x,y)$$

with $t_3 > 0$, then $[N_{n-k}/D_{m-\ell}]_{I_{n+m-t}}$ also solves $[N_{n-k+i}/D_{m-\ell+j}]_{I_{n+m-t+i+j}}$ where $0 \leq i$, $0 \leq j$ and $i+j \leq t+t_3$.

(e) *If the solution $[N_{n-k}/D_{m-\ell}]_{I_{n+m-t}} = (p/q)(x,y)$ is such that $\partial p = N_{n-k}$, $\partial q = D_{m-\ell}$ and*

$$(fq-p)(x,y) = \sum_{(i,j)\in \mathbb{N}^2 \setminus I_{n+m}} d_{ij} B_{ij}(x,y)$$

with $d_{i_{n+m+1} j_{n+m+1}} \neq 0$ then $[N_{n-k}/D_{m-\ell}]_{I_{n+m-t}} \in [N_i/D_j]_{I_{i+j}}$ if and only if (i,j) belongs to the triangle with corner elements $(n-k, m-\ell)$, $(n+\ell, m-\ell)$ and $(n-k, m+k)$.

(f) *For $i \geq 0$, $j \geq 0$ and $i+j \leq t$:*

$$\bigcap_{(i,j)} [N_{n+i}/D_{m+j}]_{I_{n+m+i+j}} \neq \emptyset$$

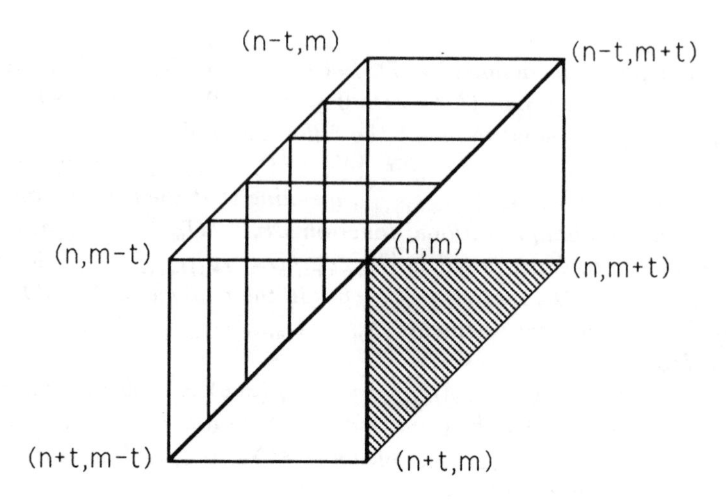

(g) Let $0 \leq k \leq t$ and the rank of $C_{n-k+1,n+m-t}$ be equal to $m-t+k$:

$$\bigcap_{j=0}^{t} [N_{n-k}/D_{m+j}]_{I_{n-k+m+j}} \neq \emptyset$$

$$\bigcap_{i=0}^{t} [N_{n+i}/D_{m-k}]_{I_{n+i+m-k}} \neq \emptyset$$

Let us now point out some differences between this theorem and its univariate counterpart in [CLAEf]. First of all, it is important to note that both the univariate and the multivariate theorem are proved under the same conditions. With the rank of $C_{n+1,n+m}$ equal to $m-t$, we are able in both cases to construct solutions p_1, q_1 of $[N_{n-t}/D_m]_{I_{n+m-t}}$ and p_2, q_2 of $[N_n/D_{m-t}]_{I_{n+m-t}}$ that are also contained in $[N_n/D_m]_{I_{n+m}}$. We have

$$\begin{aligned}(p_1q_2 - p_2q_1)(x,y) &= [q_1(fq_2 - p_2) - q_2(fq_1 - p_1)](x,y) \\ &= q_1(x,y) \sum_{(i,j)\in\mathbb{N}^2\setminus I_{n+m}} d_{ij}^{(2)} B_{ij}(x,y) - q_2(x,y) \sum_{(i,j)\in\mathbb{N}^2\setminus I_{n+m}} d_{ij}^{(1)} B_{ij}(x,y)\end{aligned}$$

from which we can conclude that $(p_1q_2 - q_1p_2)(x_i, y_j) = 0$ for all $(i,j) \in I_{n+m}$ with I_{n+m} satisfying the inclusion property. We also have

$$\partial(p_1q_2 - q_1p_2) = \{(i,j) = (r,s) + (t,u) \mid (r,s) \in N_n, (t,u) \in D_m\}$$

However, since we do not always have that

$$\partial(p_1q_2 - q_1p_2) \subset I_{n+m}$$

we cannot conclude that $(p_1q_2 - q_1p_2)(x,y)=0$ and hence we cannot prove as in [CLAEf] that it is also possible to construct a solution p_3, q_3 of $[N_n/D_m]_{I_{n+m}}$ with $\partial p_3 \subset N_{n-t}$ ánd $\partial q_3 \subset D_{m-t}$. In the univariate case however

$$N_n = \{(i,0) \mid 0 \leq i \leq n\}$$
$$D_m = \{(j,0) \mid 0 \leq j \leq m\}$$
$$\{(i+j,0) \mid i \in N_n, j \in D_m\} \subseteq I_{n+m} = \{(k,0) \mid 0 \leq k \leq n+m\}$$

and hence $p_1q_2 = p_2q_1$. Consequently in the univariate case the configuration described in theorem 11 can be enlarged with the triangle with corner elements $[N_{n-t}/D_{m-t}]_{I_{n+m-2t}}$, $[N_{n-1}/D_{m-t}]_{I_{n+m-t-1}}$ and $[N_{n-t}/D_{m-1}]_{I_{n+m-t-1}}$ resulting in the configuration described in section 3.

How is theorem 11 to be understood as a generalization of theorem 6? Clearly minimal solutions aren't uniquely determined anymore. In theorem 11 all solutions of the $(n-k, m-\ell)$ rational Hermite interpolation problem with $k+\ell=t$ are "minimal" in the sense that they use a minimal number of parameters and data to solve the (n,m) rational interpolation problem. Now each of the minimal solutions on the $(n+m-t)^{th}$ diagonal (with the restriction that the numerator and denominator "degree" must be less than or equal to n and m respectively) give rise to a triangular structure in the table. There's a whole triangle of rational interpolation problems that is solved by each minimal solution from the $(n+m-t)^{th}$ diagonal.

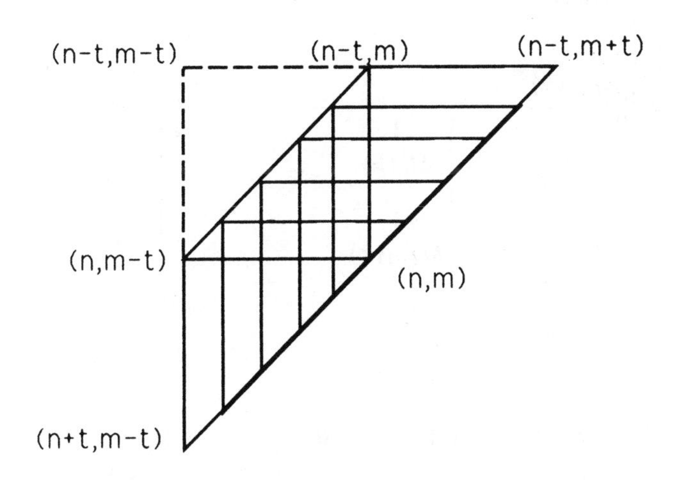

What's more, in the multivariate case a rational Hermite interpolation problem can have both a true irreducible minimal solution, a reducible minimal solution and a minimal solution with unattainable points. Note that in the multivariate case the solution must not be reducible in order to have unattainable interpolation points. This is a situation which is essentially different from the univariate one. In the univariate case theorem 6a and 6b never apply simultaneously [CLAEf] while this can be true in the multivariate case. The solution of $[N_n/D_m]$ common to all solution sets $[N_{n+i}/D_{m+j}]_{I_{n+m+i+j}}$ as described in theorem 11f could be called the "optimal solution" in the sense that it satisfies as many conditions as possible. If the rank of $C_{n+1,n+m+t}$ is still not maximal even more conditions can be added. So in the rational interpolation table a triangle emanating from $[N_n/D_m]_{I_{n+m}}$ can be filled with the optimal solution, while triangles emanating from $[N_{n-k}/D_{m-\ell}]_{I_{n+m-t}}$ with $k+\ell=t$ can be filled with minimal solutions. The rest of the hexagon is filled with the solutions constructed in the proof of theorem 11g.

7.1. Singular rules for the E-algorithm.

Let us introduce some new ratios of determinants. Let $E_{\ell,t}^{(k,u)}$ denote

$$E_{\ell,t}^{(k,u)} = \frac{\begin{vmatrix} t_0(k) & \dots & t_0(u-t) & t_0(u+1) & \dots & t_0(k+\ell+t) \\ \delta t_0(k) & \dots & & & & \\ \vdots & & & & & \\ \delta t_{\ell-1}(k) & \dots & & & & \end{vmatrix}}{\begin{vmatrix} 1 & \dots & 1 \\ \delta t_0(k) & \dots & \\ \vdots & & \\ \delta t_{\ell-1}(k) & \dots & \end{vmatrix}} \tag{17}$$

with

$$\delta t_j(i) = t_{j+1}(i) - t_j(i) \qquad j \geq 0 \qquad\qquad t_j(i) = 0 \qquad i < 0$$

These $E_{\ell,t}^{(k,u)}$ strongly resemble the E-values of the previous section (the classical values are obtained for $t=0$ and for $u \geq k+\ell+t$) and for fixed t and u they can be calculated recursively like the E-values [CUYTb] but now using help-entries

$$g_{h,\ell,t}^{(k,u)} = \frac{\begin{vmatrix} \delta t_\ell(k) & \dots & \delta t_\ell(u-t) & \delta t_\ell(u+1) & \dots & \delta t_\ell(k+h+t) \\ \delta t_0(k) & \dots & & & & \\ \vdots & & & & & \\ \delta t_{h-1}(k) & \dots & & & & \end{vmatrix}}{\begin{vmatrix} 1 & \dots & 1 \\ \delta t_0(k) & \dots & \\ \vdots & & \\ \delta t_{h-1}(k) & \dots & \end{vmatrix}} \tag{18}$$

THEOREM 12:
Let $p(x,y)$ and $q(x,y)$ be defined by (7) with I satisfying the inclusion property. Let the rank of the coefficient matrix $C_{n+1,n+m}$ in (10) be given by $m-t$. Let for each pair (k,ℓ) with $0 \le k \le t$, $0 \le \ell \le t$, $k+\ell = t$, the rank of $C_{n-k+1,n+m-t}$ equal its maximal rank $m-\ell$. Let the hexagonal block of degenerate solutions be isolated, which means that for $0 \le k \le t$ the coefficient matrices $C_{n-t+1,n+m-t+k}$ (top row), $C_{n-k+1,n+m-t}$ (leftmost antidiagonal), $C_{n+k+1,n+m-t+k}$ (leftmost column), $C_{n+t+1,n+m+k}$ (bottom row), $C_{n+t-k+1,n+m+t}$ (rightmost antidiagonal) and finally $C_{n-t+k+1,n+m+k}$ (rightmost column) all have maximal rank.
Then for $i = 1,\dots,t-1$ the following can be proved.

(a) *$E_{m,t}^{(n-t+i,n+m)}$ is well-defined and solves $[N_{n-t+i}/D_{m+t}]_{I_{n+m+i}}$.*
It also belongs to $[N_{n-t+i+k}/D_{m+t-k}]_{I_{n+m+i}}$ with $k = 0,\dots,t$, meaning that $E_{m,t}^{(n-t+i,n+m)}$ solving $[N_{n-t+i}/D_{m+t}]_{I_{n+m+i}}$ can be shifted downwards in the hexagonal block in the direction of the antidiagonal because it also solves the interpolation problems posed in $[N_{n-t+i+k}/D_{m+t-k}]_{I_{n+m+i}}$.

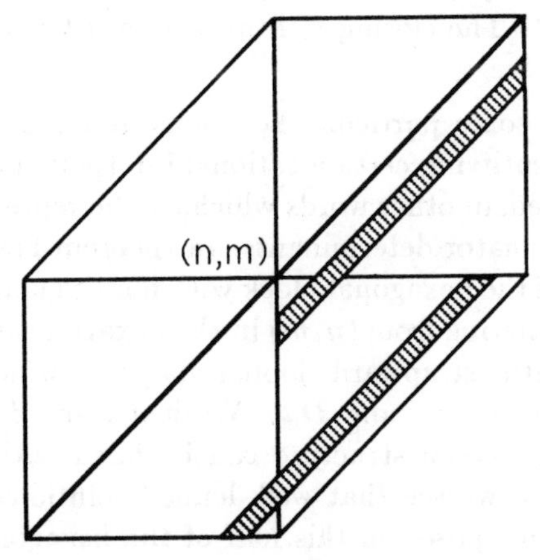

(b) *$E_{m-t+i,t}^{(n,n+m)}$ is well-defined and solves $[N_{n+t}/D_{m-t+i}]_{I_{n+m+i}}$.*

It also belongs to $[N_{n+t-k}/D_{m-t+i+k}]_{I_{n+m+i}}$ *with* $k = 0,\ldots,t$, *meaning that* $E^{(n,n+m)}_{m-t+i,t}$ *solving* $[N_{n+t}/D_{m-t+i}]_{I_{n+m+i}}$ *can be shifted upwards in the hexagonal block in the direction of the antidiagonal because it also solves the interpolation problems posed in* $[N_{n+t-k}/D_{m-t+i+k}]_{I_{n+m+i}}$.

(c) *On the rightmost upward sloping diagonal we have for* $i = 0,\ldots,t$:
$[N_{n+t-i}/D_{m+i}]_{I_{n+m+t}} = E^{(n,n+m)}_{m,t}$.

Note that the theorem provides us with a solution in the rightmost column of the isolated hexagonal block, column $m+t$, in the form of a ratio of determinants of size $m+1$, while the coefficient matrix $C_{n-t+i,n+m+i}$ is regular because the block is isolated, implying that its unique solution (up to a multiplicative constant) can also be represented as a ratio of determinants of size $m+t+1$. From this we can conclude that $E^{(n-t+i,n+m)}_{m,t}$ and $E^{(n-t+i)}_{m+t}$ differ only in a common multiplicative factor in numerator and denominator. When we run across such an isolated singular hexagonal block we want to know the values on the edges of the block, because from there on we can take up the nonsingular rules again and proceed with our recursive scheme. Let's walk around the block and try to identify the rational interpolants on all the edges. Remember that $[N_\ell/D_k]_{I_{\ell+k}}$ denotes the complete set of solutions while $E^{(\ell)}_k$ or $E^{(\ell,u)}_{k,t}$ denote a particular value from that set.

First there's the upward sloping diagonal with regular entries $[N_{n-t+i}/D_{m-i}]_{I_{n+m-t}}$ because $C_{n-t+i+1,n+m-t}$ has maximal rank for all $i = 0,\ldots,t$. Then we proved in [ALLO] that for $i = 0,\ldots,t$ the value $E^{(n-t)}_m$ also solves the rational interpolation problems posed in $[N_{n-t}/D_{m+i}]_{I_{n+m-t+i}}$ and analogously for $E^{(n)}_{m-t}$ and $[N_{n+i}/D_{m-t}]_{I_{n+m-t+i}}$. So this deals with the top row and leftmost column of our isolated block. The values in the rightmost column and on the bottom line of the hexagonal block were just respectively identified as $E^{(n-t+i,n+m)}_{m,t}$ and $E^{(n,n+m)}_{m-t+i,t}$ with $i = 1,\ldots,t-1$. The closing rightmost upward sloping diagonal is filled with $E^{(n,n+m)}_{m,t}$.

Let us now discuss some particular solutions at the interior of the hexagon. It is essential when identifying certain rational interpolants that we present solutions which are well-defined, in other words which can be represented as definite E-values with nonzero denominator determinants. In theorem 11a we mentioned how to fill the left upper half of the hexagonal block with nonsingular E-values. Using theorem 12c the triangle emanating from (n,m) in the hexagon can be filled with the regular values from the rightmost upward sloping diagonal, which are all equal and which have the correct degrees N_n and D_m. We just learned how the rest of the right lower half of the hexagonal structure can be filled with regular E-values. From theorem 12a and 12b we see that well-defined solutions for the rational Hermite interpolation problems posed in this half of the hexagon come from copies of the rightmost column or copies of the bottom line. Essentially this leaves us with the problem of computing these new E-values $E^{(n-t+i,n+m)}_{m,t}$ and $E^{(n,n+m)}_{m-t+i,t}$. When trying to provide a coherent computation scheme we must be careful not to involve intermediate singular values. Using the initialisation

$$\begin{aligned} E_{m-i,t}^{(n-t+i,n+m)} &= E_{m-i}^{(n-t+i)} && i=1,\ldots,t-1 \\ E_{m-i,t}^{(n,n+m)} &= E_{m-i}^{(n+t)} && i=1,\ldots,t-1 \\ g_{m-i,r,t}^{(n-t+i,n+m)} &= g_{m-i,r}^{(n-t+i)} && i=0,\ldots,t \\ g_{m-t+i,r,t}^{(n,n+m)} &= g_{m-t+i,r}^{(n+t)} && i=1,\ldots,t \end{aligned} \tag{19a}$$

and the rules

$$E_{\ell,t}^{(k,n+m)} = \frac{E_{\ell-1,t}^{(k,n+m)} g_{\ell-1,\ell,t}^{(k+1,n+m)} - E_{\ell-1,t}^{(k+1,n+m)} g_{\ell-1,\ell,t}^{(k,n+m)}}{g_{\ell-1,\ell,t}^{(k+1,n+m)} - g_{\ell-1,\ell,t}^{(k,n+m)}}$$

$$k=0,1,\ldots,n \qquad \ell=1,2,\ldots,m$$

$$g_{h,\ell,t}^{(k,n+m)} = \frac{g_{h-1,\ell,t}^{(k,n+m)} g_{h-1,h,t}^{(k+1,n+m)} - g_{h-1,\ell,t}^{(k+1,n+m)} g_{h-1,h,t}^{(k,n+m)}}{g_{h-1,h,t}^{(k+1,n+m)} - g_{h-1,h,t}^{(k,n+m)}}$$

$$\ell = h+1, h+2, \ldots$$

we can fill the following quasi-triangular table of values:

$$\begin{array}{ccccc} & & & E_{m-1,t}^{(n-t+1,n+m)} & E_{m,t}^{(n-t+1,n+m)} \\ & & \cdot^{\cdot^{\cdot}} & E_{m-1,t}^{(n-t+1,n+m)} & E_{m,t}^{(n-t+1,n+m)} \\ & E_{m-t+2,t}^{(n-2,n+m)} & & \vdots & \vdots \\ E_{m-t+1,t}^{(n-1,n+m)} & E_{m-t+2,t}^{(n-1,n+m)} & & & \\ E_{m-t+1,t}^{(n,n+m)} & E_{m-t+2,t}^{(n,n+m)} & \ldots & E_{m-1,t}^{(n,n+m)} & E_{m,t}^{(n,n+m)} \end{array} \tag{19b}$$

where the bottom row and rightmost column of (19b) respectively solve the interpolation problems in the bottom row and rightmost column of our hexagon. A proof for these rules can be constructed as in [CUYTn] for the case $t = 0$. The quasi-triangular table (19b) can on its turn only be filled completely if we do not encounter indefinite values at the interior of (19b). But even if we are unfortunate we should not despair. In [ALLO] we describe a more general iterative procedure to deal with singularities. This section in fact only describes the first iteration step, which may suffice if the hexagonal block is isolated and everything goes well. The initialisations in (19a) are easy to understand. The first is merely by notation: from the determinant representations for $E_{m-i}^{(n-t+i)}$ and $E_{m-i,t}^{(n-t+i,n+m)}$ one can see that these expressions are equal. The second initialisation follows from theorem 12b. The initialisations for the g-values are analogous. The first is by notation, the other by theorem 12b. How do we now get the starting E- and g-values on the bottom row of the hexagon? The bottom row of (19b) is computed from the nondegenerate rules with input values $E_{m-t}^{(n+t)}$ and $E_{m-t}^{(n+t+1)}$. The newly described extension together with its initialisations can then best be understood from the picture below:

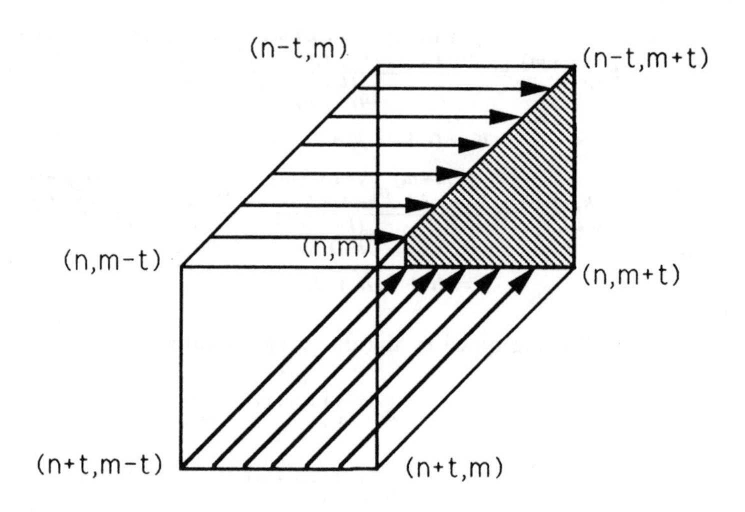

7.2. Singular rules for the qdg-algorithm.

It is clear that in the degenerate case, with the conditions of theorem 11 fullfilled, some of the continued fractions (15) are perturbed because the elements on the staircase (14) may not all be different. For a rank deficiency of t in $C_{n+1,n+m}$ it concerns the continued fractions $C_s(x,y)$ for $s = n-m-2t, \ldots, n-m+2t-1$. In this presentation the rank-deficiency t will be fixed because we focus our attention on a particular singular interpolation problem, namely $[N_n/D_m]_{I_{n+m}}$ with an isolated hexagonal block built around it. Hence we shall drop the index t in the notation of both $E_{k,t}^{(\ell,u)}$ and $g_{h,k,t}^{(\ell,u)}$. From now on when we refer to a particular element $[N_\ell/D_k]_{I_{\ell+k}}$ in the table, we only take the regular solutions in consideration. The hexagonal block of "size" t described above perturbs the partial numerators and denominators inside and on the border of the following octagonal structure from the qd-table:

$$\begin{array}{ccccccccc}
 & & & & q_{m+1}^{(n-m-t)} & \ldots & q_{m+t}^{(n-m-2t+1)} & & \\
 & & & e_m^{(n-m-t+1)} & & & & e_{m+t}^{(n-m-2t+1)} & \\
 & & & & & & & & q_{m+t+1}^{(n-m-2t+1)} \\
 & & & & & & & & \vdots \\
 & & \cdot^{\cdot^{\cdot}} & & & & & & \\
 & q_{m-t+1}^{(n-m+t)} & & & & & & & \\
e_{m-t}^{(n-m+t+1)} & & & & & & & & q_{m+t+1}^{(n-m-t)} \\
\vdots & & & & & & & e_{m+t}^{(n-m-t+1)} & \\
 & & & & & & \cdot^{\cdot^{\cdot}} & & \\
e_{m-t}^{(n-m+2t)} & & & & & & & & \\
 & q_{m-t+1}^{(n-m+2t)} & & & & q_{m+1}^{(n-m+t)} & & & \\
 & & e_{m-t+1}^{(n-m+2t)} & \ldots & e_m^{(n-m+t+1)} & & & &
\end{array} \tag{20}$$

In order to cope with the rank-deficient situation described in the theorem above, we introduce staircases T_s^* that coincide with T_s before entering the hexagonal block and after leaving it. The degenerate elements of T_s within the block are deleted and replaced by a number of nondegenerate elements, mostly from around the block. In general T_s^* is given by the following staircase. We distinguish between three cases. Let us use the shorter notation $[\frac{n}{m}]$ instead of $[N_n/D_m]_{I_{n+m}}$. The first case is where T_s enters the hexagonal block through the square in the right upper corner of the hexagon, in other words when s ranges from $n-m-2t$ to $n-m-t-1$:

$$T_s^* = \left\{ \begin{array}{lllll}
\left[\frac{s}{0}\right] & & & & \\
\left[\frac{s+1}{0}\right] & \ldots & & & \\
 & & \vdots & & \\
 & & \left[\frac{n-t}{n-s-t-1}\right] & \left[\frac{n-t}{m+t+1}\right] & \\
 & & & \left[\frac{n-t+1}{m+t+1}\right] & \\
 & & & \vdots & \\
 & & & \left[\frac{m+t+s+1}{m+t+1}\right] & \\
 & & & \left[\frac{m+t+s+2}{m+t+1}\right] & \left[\frac{m+t+s+2}{m+t+2}\right] \\
 & & & & \vdots
\end{array} \right\} \tag{21}$$

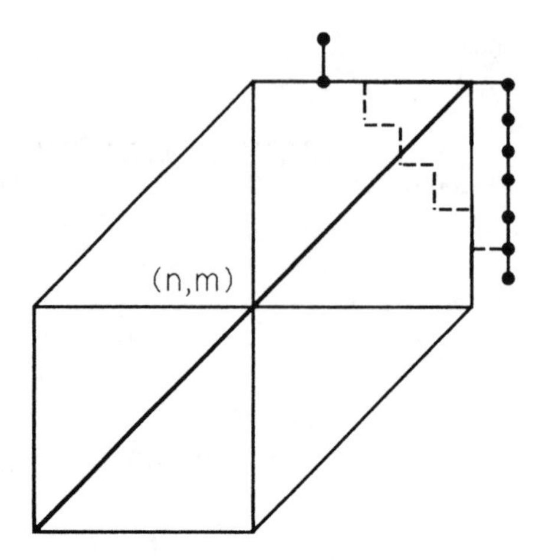

The second case is where T_s enters the hexagonal block by crossing the upward sloping diagonal, in other words when s ranges from $n-m-t$ to $n-m+t-1$. The third case is where T_s enters the hexagonal block through the square in the left bottom corner of the hexagon, in other words when s ranges from $n-m+t$ to $n-m+2t-1$. Because of the similarities between the different cases only the first case will be treated in detail here. If ℓ new elements are introduced by working with T_s^*, $\ell+2$ coefficients in $C_s(x,y)$ are perturbed and new rules must be given for the partial numerators and denominators in the new continued fraction $C_s^*(x,y)$ associated with T_s^*. Before proceeding to the continued fraction representation $C_s^*(x,y)$, we introduce some new quantities. We define

$$v_{m+1}^{(n-m)} = \frac{E_m^{(n+1)} - E_m^{(n)}}{E_m^{(n)} - E_m^{(n-1)}}$$

which links 3 consecutive elements in a column of the E-table, and

$$h_m^{(n-m+1)} = \frac{E_m^{(n)} - E_{m-1}^{(n)}}{E_{m-1}^{(n)} - E_{m-2}^{(n)}}$$

which links 3 consecutive elements in a row of the E-table. We also define for $n-t+1 \leq s+\ell \leq n, m-t+1 \leq \ell \leq m, n+m-t+1 \leq s+2\ell$

$$q_\ell^{*(s+1)} = \frac{E_\ell^{(s+\ell,n+m)} - E_{\ell-1}^{(s+\ell,n+m)}}{E_{\ell-1}^{(s+\ell,n+m)} - E_{\ell-1}^{(s+\ell-1,n+m)}}$$

for $n-t+1 \leq s+\ell+1 \leq n, m-t+1 \leq \ell \leq m, n+m-t+1 \leq s+2\ell+1$

$$e_\ell^{*(s+1)} = \frac{E_\ell^{(s+\ell+1,n+m)} - E_\ell^{(s+\ell,n+m)}}{E_\ell^{(s+\ell,n+m)} - E_{\ell-1}^{(s+\ell,n+m)}}$$

for $n-t+1 \leq s+\ell \leq n-1, m-t \leq \ell \leq m, s+2\ell-1 \geq n-m-t$

$$v_{\ell+1}^{*(s)} = \frac{E_{\ell}^{(s+\ell+1,n+m)} - E_{\ell}^{(s+\ell,n+m)}}{E_{\ell}^{(s+\ell,n+m)} - E_{\ell}^{(s+\ell-1,n+m)}}$$

and for $n-t \leq s+\ell \leq n, m-t+2 \leq \ell \leq m, s+2\ell-2 \geq n+m-t$

$$h_{\ell}^{*(s+1)} = \frac{E_{\ell}^{(s+\ell,n+m)} - E_{\ell-1}^{(s+\ell,n+m)}}{E_{\ell-1}^{(s+\ell,n+m)} - E_{\ell-2}^{(s+\ell,n+m)}}$$

It is clear that for definitions linking 3 elements from the E-table, certain transition rules apply when entering and leaving the hexagonal singular block. For $q_{\ell}^{*(s+1)}$, $e_{\ell}^{*(s+1)}$, $v_{\ell}^{*(s+1)}$ and $h_{\ell}^{*(s+1)}$ this transition is described in [ALLOb].

THEOREM 13:
The continued fraction representation $C_s^*(x,y)$ associated with T_s^* as in (21) is given by the following formula. The first and last line of the expression for C_s^* coincide with that for C_s while the middle part deals with the discrepancy between T_s^* and T_s:

$$\begin{aligned} C_s^*(x,y) = E_0^{(s)} &+ \frac{E_0^{(s+1)} - E_0^{(s)} \rvert}{\lvert 1} + \sum_{i=1}^{n-t-s-1} \left(\frac{-q_i^{(s+1)} \rvert}{\lvert 1 + q_i^{(s+1)}} + \frac{-e_i^{(s+1)} \rvert}{\lvert 1 + e_i^{(s+1)}} \right) \\ &+ \frac{-q_{n-t-s}^{*(s+1)} \rvert}{\lvert 1 + q_{n-t-s}^{*(s+1)}} + \frac{-e_{m+t+1}^{(n-m-2t)} \rvert}{\lvert 1 + e_{m+t+1}^{(n-m-2t)}} + \sum_{i=1}^{m-n+2t+s+1} \frac{-v_{m+t+2}^{(n-m-2t+i-1)} \rvert}{\lvert 1 + v_{m+t+2}^{(n-m-2t+i-1)}} \\ &+ \sum_{i=m+t+2}^{\infty} \left(\frac{-q_i^{(s+1)} \rvert}{\lvert 1 + q_i^{(s+1)}} + \frac{-e_i^{(s+1)} \rvert}{\lvert 1 + e_i^{(s+1)}} \right) \end{aligned}$$

To be able to use the continued fraction representation obtained in the previous theorem, we must find a coherent computation scheme for its partial numerators and denominators. Let us first introduce the following notations:

$$G_{\ell}^{(s)} = \frac{g_{\ell,\ell+1}^{(s)} - g_{\ell,\ell+1}^{(s+1)}}{g_{\ell,\ell+1}^{(s)}}$$

and

$$G_{\ell}^{(s,n+m)} = \frac{g_{\ell,\ell+1}^{(s,n+m)} - g_{\ell,\ell+1}^{(s+1,n+m)}}{g_{\ell,\ell+1}^{(s,n+m)}}$$

The next theorem lists the singular rules in the order they have to be implemented. This set of rules is complete and provides the partial numerators and denominators for all continued fractions disturbed by the degeneracy, not only for (21).

THEOREM 14:
We first concentrate on the values $v_{\ell+1}^{(s)}$, $v_{\ell+1}^{*(s)}$, $h_{\ell}^{(s+1)}$ *and* $h_{\ell}^{*(s+1)}$.

(a)

$$v_{\ell+1}^{(s)} = \frac{g_{\ell-1,\ell}^{(s+\ell-1)} - g_{\ell-1,\ell}^{(s+\ell)}}{g_{\ell-1,\ell}^{(s+\ell-1)}} \frac{e_{\ell}^{(s+1)}}{e_{\ell}^{(s)}} q_{\ell}^{(s+1)}$$

$$h_{\ell}^{(s+1)} = \frac{g_{\ell-1,\ell}^{(s+\ell)}}{g_{\ell-1,\ell}^{(s+\ell)} - g_{\ell-1,\ell}^{(s+\ell+1)}} e_{\ell-1}^{(s+2)}$$

(b)

$$v_{\ell+1}^{*(s)} = G_{\ell-1}^{(s+\ell,n+m)} \frac{e_{\ell}^{*(s+1)}}{e_{\ell}^{*(s)}} q_{\ell}^{*(s+1)}$$

$$h_{\ell}^{*(s+1)} = \frac{1}{G_{\ell-1}^{(s+\ell,n+m)}} e_{\ell-1}^{*(s+2)} \qquad s+\ell < n$$

$$h_{\ell}^{*(s+1)} = \frac{1}{G_{\ell-1}^{(n+s)}} e_{\ell-1}^{(n+s-\ell+2)} \qquad s+\ell = n$$

(c)

$$h_{m+1}^{*(n-m+t)} = \frac{1}{G_m^{(n+t)}} e_m^{(n-m+t+1)}$$

$$h_{m+2}^{*(n-m+t-2)} = \frac{G_{m-1}^{(n-1,n+m)}}{G_{m+1}^{(n+t-1)}} \frac{h_{m+1}^{*(n-m+t)} q_m^{*(n-m+1)}}{e_m^{*(n-m)}}$$

$$h_{m+t-u}^{*(n-m-t+2u+2)} = \frac{1}{G_{m+t-u-1}^{(n+u+1)}} h_{m+t-u-1}^{*(n-m-t+2u+4)} \qquad u = 0,\ldots,t-3$$

$$v_{m+2}^{*(n-m+t-2)} = \frac{1}{G_m^{(n+t)}} e_m^{(n-m+t+1)}$$

$$v_{m+t-u+1}^{*(n-m-t+2u)} = \frac{1}{G_{m+t-u-1}^{(n+u+1)}} v_{m+t-u}^{*(n-m-t+2u+2)} \qquad u = t-2,\ldots,1$$

$$v_{m+t+1}^{*(n-m-t)} = \frac{G_{m-1}^{(n-1,n+m)}}{G_{m+t-1}^{(n+1)}} \frac{v_{m+t}^{*(n-m-t+2)} q_m^{*(n-m+1)}}{e_m^{*(n-m)}}$$

We shall now concentrate on the octagonal gap in the qd-table due to the hexagonal block in the table of rational Hermite interpolants.

(d) *To fill the leftmost column of the octagonal singular block (20) in the qd-table in a bottom-up way, we compute for* $k = 1,\ldots,t$

$$e_{m-t}^{*(n-m+t+k)} = G_{m-t-1}^{(n+k-1)} q_{m-t}^{(n-m+t+k+1)} e_{m-t}^{*(n-m+t+k+1)}$$

with $e^{*(n-m+2t+1)}_{m-t} = e^{(n-m+2t+1)}_{m-t}$.

(e) *To fill the leftmost upward sloping diagonal of (20), we compute for* $k = t, \dots, 1$

$$q^{*(n-m-t+2k)}_{m-k+1} = \frac{G^{(n-t+k-1)}_{m-k-1}}{G^{(n-t+k,n+m)}_{m-k}} \frac{q^{(n-m-t+2k+1)}_{m-k} e^{*(n-m-t+2k+1)}_{m-k}}{e^{(n-m-t+2k)}_{m-k}}$$

and for $k = t-1, \dots, 0$

$$e^{*(n-m-t+2k+1)}_{m-k} + 1 = G^{(n-t+k)}_{m-k-1} \left(q^{*(n-m-t+2k+2)}_{m-k} + 1 \right)$$

with

$$q^{*(n-m-t)}_{m+1} = \frac{G^{(n-t-1)}_{m}}{G^{(n-t)}_{m+t}} \frac{q^{(n-m-t+1)}_{m} e^{*(n-m-t+1)}_{m}}{e^{(n-m-t)}_{m}}$$

(f) *To fill the top row of (20) together with the row immediately above the block since this uses "degenerate" values, we compute for* $k = 2, \dots, t+1$

$$q^{*(n-m-t-k+1)}_{m+k} = \frac{-g^{(n-t-1)}_{m+k-1,m+k}}{g^{(n-t)}_{m+k-1,m+k}} G^{(n-t-1)}_{m+k-2} q^{*(n-m-t-k+2)}_{m+k-1}$$

with $q^{*(n-m-2t)}_{m+t+1} = q^{(n-m-2t)}_{m+t+1}$, *and for* $k = 1, \dots, t$

$$e^{(n-m-t-k)}_{m+k} = \frac{-g^{(n-t)}_{m+k-1,m+k}}{g^{(n-t-1)}_{m+k-1,m+k}}$$

(g) *Column* $m+t+2$ *of q-values is the first to reappear in the continued fraction representations* $C^*_s(x,y)$. *It can be computed from the q-values and e-values with column index* $m+t+1$ *using the well-known non-singular rules. Column* $m+t+1$ *of e-values depends solely on column* $m+t+1$ *of q-values, so we focus on this last one. For* $k = 1, \dots, t-1$

$$q^{(n-m-2t+k)}_{m+t+1} = \frac{G^{(n-t+k-1,n+m)}_{m-1}}{G^{(n-t+k)}_{m+t}} \frac{e^{*(n-m-t+k+1)}_{m} q^{*(n-m-t+k+1)}_{m}}{e^{*(n-m-t+k)}_{m}}$$

where a band of q-values and e*-values is filled using rules constructed from the classical ones:*

$$q^{*(s+1)}_{\ell} = \frac{G^{(s+\ell-1,n+m)}_{\ell-2}}{G^{(s+\ell,n+m)}_{\ell-1}} \frac{e^{*(s+2)}_{\ell-1} q^{*(s+2)}_{\ell-1}}{e^{*(s+1)}_{\ell-1}}$$

$$\ell = m-t+1, \dots, m \quad s = n+m-t-2\ell+3, \dots, n+m-2\ell+1$$

and

$$e_{\ell}^{*(s+1)} + 1 = G_{\ell-1}^{(s+\ell,n+m)} \left(q_{\ell}^{*(s+2)} + 1 \right)$$

$$\ell = m - t + 1, \ldots, m \quad s = n + m - t - 2\ell + 2, \ldots, n + m - 2\ell$$

For $k = t$

$$q_{m+t+1}^{(n-m-t)} = \frac{G_{m-1}^{(n-1,n+m)}}{G_{m+t}^{(n)}} \prod_{i=1}^{t} \frac{1}{G_{m+t-i}^{(n+i)}} \frac{e_m^{(n-m+t+1)} q_m^{*(n-m+1)}}{e_m^{(n-m)}}$$

(h) To close the octagonal gap in the qd-table we now calculate the remaining elements. For $k = t, \ldots, 1$

$$q_{m+t-k+2}^{(n-m-t+2k-1)} = \frac{1}{G_{m+t-k+1}^{(n+k)}} e_{m+t-k+1}^{(n-m-t+2k)}$$

and inbetween

$$e_{m+t-k+2}^{(n-m-t+2k-2)} + 1 = G_{m+t-k+1}^{(n+k-1)} \left(q_{m+t-k+2}^{(n-m-t+2k-1)} + 1 \right)$$

On the bottom line we have for $k = 1, \ldots, t$

$$e_{m-t+k}^{(n-m+2t-k+1)} + 1 = G_{m-t+k-1}^{(n+t)} \left(q_{m-t+k}^{(n-m+2t-k+2)} + 1 \right)$$

Using theorem 14 the gap bordered in (20) involves the computation of the elements listed in the octagon below.

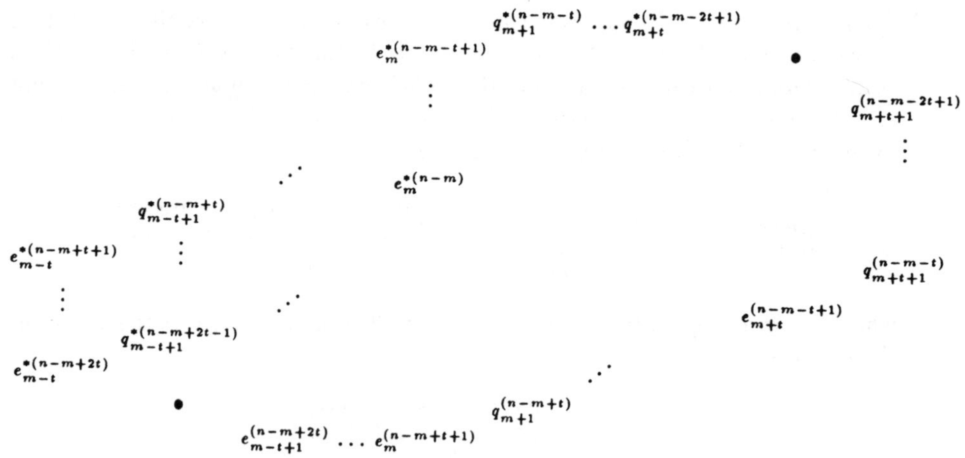

$$q_{m+1}^{*(n-m-t)} \quad \ldots \quad q_{m+t}^{*(n-m-2t+1)}$$

$$e_m^{*(n-m-t+1)} \qquad \bullet$$

$$\vdots \qquad q_{m+t+1}^{(n-m-2t+1)}$$

$$\cdot^{\cdot^{\cdot}} \qquad e_m^{*(n-m)} \qquad \vdots$$

$$q_{m-t+1}^{*(n-m+t)}$$

$$e_{m-t}^{*(n-m+t+1)} \qquad \vdots \qquad q_{m+t+1}^{(n-m-t)}$$

$$\vdots \qquad \cdot^{\cdot^{\cdot}} \qquad e_{m+t}^{(n-m-t+1)}$$

$$q_{m-t+1}^{*(n-m+2t-1)}$$

$$e_{m-t}^{*(n-m+2t)} \qquad \cdot^{\cdot^{\cdot}}$$

$$\bullet \qquad q_{m+1}^{(n-m+t)}$$

$$e_{m-t+1}^{(n-m+2t)} \quad \ldots \quad e_m^{(n-m+t+1)}$$

References.

[ALLOa] Allouche H. and Cuyt A., *A recursive computation scheme for singular tables of multivariate rational interpolants*, Numer. Algorithms (to appear).

[ALLOb] Allouche H. and Cuyt A., *Singular rules for a multivariate quotient--difference algorithm*, Numer. Algorithms (to appear).

[BERE] Berezin J. and Zhidkov N., "Computing methods I", Addison Wesley, New York, 1965.

[BREZ] Brezinski C., *A general extrapolation algorithm*, Numer. Math. **35** (1980), 175–187.

[CLAEa] Claessens G., *On the Newton-Padé approximation problem*, J. Approx. Theory **22** (1978), 150–160.

[CLAEb] Claessens G., "Some aspects of the rational Hermite interpolation table and its applications", Thesis, University of Antwerp, 1976.

[CLAEc] Claessens G., *A generalization of the qd-algorithm*, J. Comput. Appl. Math. **7** (1981), 237–247.

[CLAEd] Claessens G., *A useful identity for the rational Hermite interpolation table*, Numer. Math. **29** (1978), 227–231.

[CLAEe] Claessens G., *A new algorithm for osculatory rational interpolation*, Numer. Math. **27** (1976), 77–83.

[CLAEf] Claessens G., *On the structure of the Newton-Padé table*, J. Approx. Theory **22** (1978), 304–319.

[CUYTa] Cuyt A., *A multivariate qd-like algorithm*, BIT **28** (1988), 98–112.

[CUYTb Cuyt A., *A recursive computation scheme for multivariate rational interpolants*, SIAM J. Num. Anal. **24** (1987), 228–238.

[CUYTc] Cuyt A., "General order multivariate rational Hermite interpolants", Habilitation, University of Antwerp, 1986.

[CUYTd] Cuyt A. and Verdonk B., "Different techniques for the construction of multivariate Rational interpolants and Padé approximants", Monograph, University of Antwerp, 1988.

[CUYTe] Cuyt A. and Verdonk B., *Multivariate rational interpolation*, Computing **34** (1985), 41–61.

[WALS] Walsh J., "Interpolation and approximation by rational functions in the complex domain", Amer. Math. Soc., Providence Rhode Island, 1969.

[WARN] Warner D., "Hermite interpolation with rational functions", Ph. D., University of California, 1974.

[WUYT] Wuytack L., *On the osculatory rational interpolation problem*, Math. Comp. **29** (1975), 837–843.

THE METHOD OF ALTERNATING ORTHOGONAL PROJECTIONS

FRANK DEUTSCH†

ABSTRACT. The method of alternating orthogonal projections is discussed, and some of its many and diverse applications are described.

1. INTRODUCTION.

In this tutorial, we will give an outline of the method of alternating orthogonal projections, and mention some of its many and diverse applications. In its simplest abstract formulation, the method is due to von Neumann [1933]. We will give no proofs of any of the statements made; but we have provided enough references, we believe, to guide the interested reader who may want to trace a particular result for further details. We also hope that the tutorial will inspire further research in this area.

Let H be a Hilbert space and M a closed subspace of H. The orthogonal projection onto M will be denoted by P_M. In particular, P_M is linear, self-adjoint, idempotent:

$P_M^2 = P_M$, and $P_M(x)$ is the *best approximation* or nearest point to x from M:

$$\|x - P_M(x)\| = d(x, M),$$

where $d(x, M) = inf\{\|x - y\| \mid y \in M\}$. P_M is called the orthogonal projection onto M because of the characterizing property

$$< x - P_M(x), y >= 0 \qquad \forall\, y \in M.$$

That is, $x - P_M(x)$ is orthogonal to M. With the notation

$$M^\perp := \{y \in H \mid < y, x >= 0 \quad \forall\, x \in M\},$$

we see that $x - P_M(x) \in M^\perp$.

The key fact, easily proved, which motivated the first main result, may be stated as follows: *Let A and B be closed subspaces of H. Then $P_A P_B = P_B P_A$ if and only if $P_A P_B = P_{A\cap B}$.* In words, P_A and P_B commute if and only if their composition is also an orthogonal projection.

†Supported by NSF Grant DMS-9100228

S. P. Singh (ed.), Approximation Theory, Spline Functions and Applications, 105–121.

2. THE METHOD OF ALTERNATING PROJECTIONS.

Von Neumann was interested in what could be said in case P_A and P_B did not commute. He proved the following result.

2.1 THEOREM (von Neumann [1933]). *For each $x \in H$,*

(2.1.1)
$$\lim_{n\to\infty} (P_B P_A)^n x = P_{A\cap B} x.$$

The equation (2.1.1) suggests an algorithm, called the *method of alternating projections*, or the (MAP) for short. It is this: for any $x \in H$, set

$$x_0 = x \qquad \text{and}$$
$$x_n = P_B P_A x_{n-1} \qquad (= (P_B P_A)^n x)$$
$$\text{for } n = 1, 2, \ldots . \qquad \text{Then } x_n \to P_{A\cap B} x.$$

3. GEOMETRIC INTERPRETATION.

The geometric interpretation of the (MAP) is that to find the best approximation to x from $A \cap B$, one first projects x onto A; the resulting element is then projected onto B, and one continues to project the resultant alternately onto A and B. The sequence of elements thus generated converges to $P_{A\cap B}x$. The *practical* usefulness of the (MAP) stems from the fact that it is often much easier to compute the projections onto A or B individually than it is to compute the projection onto $A \cap B$.

4. AN EQUIVALENT FORMULATION.

If we replace A by $A^\perp$ and B by $B^\perp$ in Theorem 2.1, and use the readily verified facts that $P_{A^\perp} = I - P_A$, $P_{B^\perp} = I - P_B$, and $A^\perp \cap B^\perp = (\overline{A+B})^\perp$, we obtain the following (equivalent) reformulation of Theorem 2.1.

4.1 THEOREM. *For each $x \in H$,*

(4.1.1)
$$\lim_{n\to\infty} [(I - P_B)(I - P_A)]^n x = (I - P_{\overline{A+B}})x.$$

The extension of Theorem 2.1 (and hence of Theorem 4.1) to more than two subspaces is also valid. Let $A_1, A_2, \ldots, A_k$ be k closed subspaces in H.

4.2 THEOREM (Halperin [1962]). *For each $x \in H$,*

(4.2.1)
$$\lim_{n\to\infty} \left(P_{A_k} P_{A_{k-1}} \cdots P_{A_1}\right)^n x = P_{\bigcap_1^k A_i} x.$$

4.3 THEOREM. *For each $x \in H$,*

$$\lim_{n\to\infty} \left[(I - P_{A_k})(I - P_{A_{k-1}}) \cdots (I - P_{A_1})\right]^n x = (I - P_{\overline{A_1 + \cdots + A_k}})x. \tag{4.3.1}$$

It should be mentioned that the proof of Theorem 4.2 seems to require a different method of attack when $k \geq 3$ than that given by von Neumann for Theorem 2.1 when $k = 2$.

5. THE KACZMARZ METHOD.

It was noted in Deutsch [1983a] that Theorem 4.2 holds more generally when the A_i are closed *linear varieties* (i.e. translates of subspaces) with $\bigcap_1^k A_i \neq \emptyset$. From this observation, we can deduce the so-called *Kaczmarz method of iteration* for solving linear systems of equations.

Consider the linear system of equations

$$< a_i, x >= b_i \qquad (i = 1, 2, \ldots, k), \tag{5.1}$$

where

$$a_i = (a_{i1}, a_{i2}, \ldots, a_{im}) \in \mathbb{R}^m, \qquad x \in \mathbb{R}^m,$$

and $b_i \in \mathbb{R}$. Let

$$H_i := \{x \in \mathbb{R}^m \mid < a_i, x >= b_i\}$$

$(i = 1, 2, \ldots, k)$, and assume (5.1) is consistent; that is, $\bigcap_1^k H_i \neq \emptyset$.

To find a solution of (5.1) is equivalent to finding a point in $\bigcap_1^k H_i$. To find such a point, fix an arbitrary $x_0 \in \mathbb{R}^m$ and inductively define

$$x_n = \left(P_{H_k} P_{H_{k-1}} \cdots P_{H_1}\right) x_{n-1} \qquad (n = 1, 2, \ldots).$$

Then the (MAP) implies that

$$x_n \to P_{\bigcap_1^k H_i} x_0 =: y_0.$$

That is, y_0 satisfies equations (5.1) and is, in fact, the solution of (5.1) which is closest to x_0. In particular, taking $x_0 = 0$, we obtain that y_0 is the *minimum norm solution* to (5.1).

This iterative method for finding solutions to linear systems of equations is called *Kaczmarz's method* since it was first studied by Kaczmarz [1937]. Its practical

value stems from the well-known simple formula for finding best approximations from hyperplanes:

$$P_{H_i}y = y - [< a_i, y > - b_i]\frac{a_i}{< a_i, a_i >}.$$

There are two relevant comments concerning Kaczmarz's method that should be made:

1) **Only *one* row is operated on at a time, and**
2) **For "sparse" matrices $A = [a_{ij}]_{i=1,j=1}^{k,m}$, the computation of $P_{H_i}y$ is very fast.**

6. RATE OF CONVERGENCE.

The speed at which the (MAP) converges depends on the angles between the subspaces involved.

6.1 DEFINITION (Friedricks [1937]). *The angle $\alpha(A, B)$ between the subspaces A and B is the angle in $[0, \pi/2]$ whose cosine is given by*

$$\cos\alpha(A, B) := \sup\{< a, b > \mid a \in A\cap(A\cap B)^\perp, \|a\| \leq 1, \quad b \in B\cap(A\cap B)^\perp, \|b\| \leq 1\}.$$

It should be mentioned that some authors have used a different definition of angle. Namely, the factors $(A \cap B)^\perp$ were deleted in the above expression. Of course, when $A \cap B = \{0\}$, the two definitions agree.

6.2 LEMMA. *(1) (Lorch [1937]) $\alpha(A, B) > 0$ if and only if $A + B$ is closed.*
(2) $\alpha(A^\perp, B^\perp) = \alpha(A, B)$.

Part (2) of this lemma seems to be part of the folklore of the subject and we have not been able to determine when it was first proved.

6.3 THEOREM (Aronszajn [1950]). *For each $x \in H$, and for any integer $n \geq 1$,*

$$\|(P_B P_A)^n x - P_{A\cap B}x\| \leq c^{2n-1}\|x\|, \tag{6.3.1}$$

where $c = \cos\alpha(A, B)$.

Smith, Solmon, and Wagner [1977] obtained analogous bounds in the case of more than two subspaces.

6.4 THEOREM (Smith, Solmon, Wagner [1977]). *For each $x \in H$, and integer $n \geq 1$,*

$$\|(P_{A_k} P_{A_{k-1}} \cdots P_{A_1})^n x - P_{\bigcap_1^k A_i} x\| \leq c^n \|x\|,$$

where

$$c = \left[1 - \prod_1^{k-1} \sin^2 \Theta_i\right]^{1/2}$$

and Θ_i is the angle between the subspaces A_i and $\bigcap_{j=i+1}^{k} A_j$.

Kayalar and Weinert [1988] gave bounds even sharper than that of Theorem 6.4, but much more complicated to state. They also observed that the bound (6.3.1) was sharp.

Franchetti and Light [1986] showed that the convergence in Theorem 2.1 may be arbitrarily slow if $A + B$ is not closed (i.e. if the angle between A and B is zero). However, Gearhart and Koshy [1989] developed schemes for *accelerating* convergence of the (MAP). Dyer [1965] had earlier described a technique for accelerating the convergence of the Kaczmarz method.

Because of its usefulness and obvious geometric appeal, it is not surprising that the (MAP) has been continuously rediscovered. An incomplete list includes Aronszajn [1950], Nakano [1953], Wiener [1955], Powell [1970], Gordon, Bender, and Herman [1970], and Hounsfield [1973]—the Nobel Prize winning inventor of the EMI scanner.

7. AREAS OF APPLICATION.

In this section we will mention ten different areas of mathematics where the (MAP) (or slight variants thereof) has played an important role. We have already given some details of how the (MAP) can be used to solve linear equations in Section 5 (the Kaczmarz method).

We list below the different areas of application of the (MAP) along with some of the many authors who have contributed to these studies. It was not our intent to be complete. This would have been a nearly impossible task anyway owing to the huge number of people who have written about these and related matters. Rather, we have attempted to be *representative.*

7.1 Solving Linear Equations. We have already mentioned the Kaczmarz method (Kaczmarz [1937]) for solving linear systems in Section 5. Tanabe [1971] considered inconsistent systems also. Herman, Lent, and Lutz [1978] introduced relaxation parameters. The geometric interpretation is that instead of projecting onto a hyperplane at each step (as in the Kacsmarz method), one projects just short of, or just beyond, the hyperplane. Eggermont, Herman, and Lent [1981] extended Tanabe's result to include relaxation parameters. Many other writers have given extensions and generalizations of Kacsmarz's method.

7.2 Probability and Statistics. Wiener and Masani [1957] and Salehi [1967] have used the (MAP) in linear prediction theory. Burkholder and Chow [1961], Burkholder [1962], and Rota [1962] studied when norm convergence in $L_2(\mu)$ for the (MAP) could be replaced by almost sure convergence (viz., when μ is a probability measure). Dykstra [1983] essentially showed that the (MAP) holds in Euclidean n-space $\ell_2(n)$ when the subspaces are replaced by closed convex cones. He then used this to solve some restricted least-squares regression problems. Breiman and Friedman [1985] used the (MAP) to describe their "alternating conditional expectation" (or ACE) algorithm.

7.3 Dirichlet Problem. Schwarz [1870] described what he called the "Alternierende Verfahren" (= alternating method) to solve the Dirichlet problem on an irregular region in the plane which is the union of regular regions.

For example, suppose that D_1 and D_2 are two overlapping disks in the plane and $D = D_1 \cup D_2$. Further, let ∂S denote the boundary of any set S in the plane and suppose f is a real continuous function on ∂D.

The Dirichlet problem on D is to determine a function u on D such that

$$\begin{aligned} \Delta u = 0 \qquad & \text{in } D \\ u = f \qquad & \text{on } \partial D. \end{aligned}$$

Here $\Delta u := \frac{\partial^2 u}{\partial x_1^2} + \frac{\partial^2 u}{\partial x_2^2}$ is the Laplacian of u.

If $|f| \leq M$, we can extend f to a continuous function f_1 on ∂D_1 such that $|f_1| \leq M$. (This can be done, for example, by Tietze's theorem). We first solve the Dirichlet problem on D_1 with f_1 on ∂D_1. We next use this solution, u_1 say, to get a continuous extension f_2 of f on ∂D_2. We then solve the Dirichlet problem in D_2 with boundary value f_2 on ∂D_2 to get a harmonic solution v_1. We continue in this way, alternating back and forth between D_1 and D_2, to obtain a sequence $u_1, v_1, u_2, v_2, \ldots$ such that $\{u_n\}$ converges uniformly to a harmonic function u on D_1 , $\{v_n\}$ converges uniformly to a harmonic function v on D_2 and $u = v$ in $D_1 \cap D_2$. Then the function

$$w = \begin{cases} u & \text{on } D_1 \\ v & \text{on } D \backslash D_1 \end{cases}$$

solves the original Dirichlet problem on D.

7.4 Computing Bergman Kernels. In 1983 Skwarczynski (see Skwarczynski [1985a] and [1985b]) showed how to use the (MAP) to compute the Bergman kernel for the Hilbert space $L_2(D)$, where D is the region $\bigcup_1^m D_i$ and the sets D_i are regions in $\mathbb{C}^n$. It is assumed that the Bergman kernels K_{D_i} for the regions D_i are known. It is desired to compute the Bergman kernel K_D for the region D.

Let

$$F_i = \{f \in L_2(D) | \, f \text{ is holomorphic in } D_i\}$$

$(i = 1, 2, \ldots, m)$ so that F_i is a closed subspace in $L_2(D)$. Fix any $t \in D$, say $t \in D_1$, and define $f \in L_2(D)$ by

$$f(z) = \begin{cases} K_{D_1}(z,t) & \text{if } z \in D_1 \\ 0 & \text{if } z \in D \backslash D_1. \end{cases}$$

Then the sequence

$$f_1 = P_{F_1} f, \qquad f_2 = P_{F_2} f_1, \cdots, \qquad f_m = P_{F_m} f_{m-1},$$
$$f_{m+1} = P_{F_1} f_m, \qquad f_{m+2} = P_{F_2} f_{m+1}, \cdots$$

converges in $L_2(D)$ to $K_D(\cdot, t)$.

Ramadanov and Skwarczynski [1984a], [1984b] considered the rate of convergence of this method in the simplest case when $m = 2$, $n = 1$, and $D = \mathbb{C}$. They showed that if $int(D_1 \backslash D_2)$ and $int(D_2 \backslash D_1)$ are nonempty, then the angle γ between F_1 and F_2 is positive if the D_i are bounded by concentric circles, and $\gamma = 0$ when the D_i are bounded by parallel straight lines.

7.5 Approximating Multivariate Functions by Sums of Univariate Ones. Let $S = [s_1, s_2]$ and $T = [t_1, t_2]$ be intervals on the real line and let $H = L_2(S \times T)$ be the Hilbert space of all real measurable functions f on $S \times T$ which are square integrable:

$$||f||^2 = \int_S \int_T |f(s,t)|^2 ds\, dt < \infty.$$

Let $A = \{a \in H \mid a(s,t) = a(s)\}$ and $B = \{b \in H \mid b(s,t) = b(t)\}$. We want to find the best approximation to each $f \in H$ from the (closed) subspace $A + B$, i.e., we want $P_{A+B} f$. In words, we want the best approximation to any L_2-function of two variables by sums of functions of a single variable.

Deutsch [1983b] observed that in this case, $\cos \alpha(A, B) = 0$ and the (MAP) converges in one step to yield

$$P_{A+B}(f) = P_A(f) + P_B(f) - P_A P_B f.$$

Since it is easily verified that

$$P_A f(s,t) = \frac{1}{t_2 - t_1} \int_{t_1}^{t_2} f(s,t)dt$$

and similarly

$$P_B f(s,t) = \frac{1}{s_2 - s_1} \int_{s_1}^{s_2} f(s,t)ds,$$

we obtain

$$P_{A+B} f(s,t) = \frac{1}{t_2 - t_1} \int_{t_1}^{t_2} f(s,t)dt + \frac{1}{s_2 - s_1} \int_{s_1}^{s_2} f(s,t)ds - \frac{1}{(t_2 - t_1)(s_2 - s_1)} \int_{t_1}^{t_2} \int_{s_1}^{s_2} f(s,t$$

In words, the best approximation to f from $A + B$ is the sum of its T and S means minus its mean.

This result is also implicit in the work of Golomb [1959] who obtained even more general results using variational methods. Also, this result is contained in von Golitschek and Cheney [1979] who proved it using the characterization of best approximations from subspaces.

7.6 Least Change Secant Updates. A secant update is a certain iterative scheme for numerically solving $F(x) = 0$, where $F : \mathbb{R}^n \to \mathbb{R}^n$.

Powell [1970] used the (MAP) implicitly to derive a symmetric secant update. This was generalized by Dennis [1972] who showed how other symmetric updates could be derived from weighted least change secant updates. Further generalization was made by Dennis and Schnabel [1979]. Convergence theorems for least change secant methods in this setting were proved by Dennis and Walker [1981].

7.7 Multigrid Methods. The multigrid method is a numerical technique for the solution of partial differential equations. Gilbert and Light [1986] showed the explicit connection between the (MAP) and the multigrid method, and showed that a bound on the rate of convergence of the multigrid method could be determined from that of the (MAP). Braess [1981] used the (MAP) to solve the Poisson equation. Gatski, Grosch, and Rose [1982, 1988] studied numerical schemes to solve the Navier-Stokes equations. They used the Kaczmarz method, hence the (MAP), to solve the resulting difference equations.

7.8 Conformal Mapping. Under fairly general conditions on a region G in the complex plane, Wegmann [1989] used the (MAP) to numerically construct the conformal mapping of the unit disk $\{z|\ |z| < 1\}$ onto G. At each step of the iteration, the method requires only two complex Fourier transforms (and no conjugation). Moreover, the method handles more general regions G than had been considered before by other schemes.

7.9 Image Restoration. Youla [1978] described an iterative scheme for solving certain image restoration problems. Let A and B be closed subspaces of the Hilbert space H. The problem of image restoration considered may be formally stated as follows. Determine an element $x \in H$ given only that $x \in B$ and $y = P_A x$ is known.

Clearly, $x \in B$ implies that $P_B x = x$ and

$$y = P_A x = P_A P_B x = (I - P_{A^\perp}) P_B x = P_B x - P_{A^\perp} P_B x = x - P_{A^\perp} P_B x.$$

In particular, for a solution of the problem to exist, y must lie in the range of $I - P_{A^\perp} P_B$. Moreover, the above equation suggests the iterative scheme:

$$x_{n+1} = P_{A^\perp} P_B x_n + y \quad , \quad x_1 = y \ (n = 1, 2, \ldots). \tag{7.9.1}$$

Assuming that y is in the range of $I - P_{A^\perp} P_B$, say $y = x - P_{A^\perp} P_B x$, one can expand (7.9.1) and obtain

$$x_{n+1} = (P_{A^\perp} P_B)^n y + x - (P_{A^\perp} P_B)^{n-1} x. \tag{7.9.2}$$

By the (MAP) (namely, Theorem 2.1),

$$\lim x_{n+1} = P_{A^\perp \cap B} y + x - P_{A^\perp \cap B} x.$$

In particular, if $A^\perp \cap B = \{0\}$ (or equivalently, $\|P_{A^\perp} P_B\| < 1$), the iteration scheme (7.9.1) converges to x, a solution of the problem. In general, however, the iterative scheme may not converge.

This method was extended to more general convex sets, not necessarily subspaces, by Youla and Webb [1982].

7.10 Computed Tomography. There is a huge literature on computed tomography or ART (for algebraic reconstruction technique). Our description follows the approach of Smith, Solmon, and Wagner [1977].

For radiographic purposes, an object in $\mathbb{R}^m (m = 2 \text{ or } 3)$ is determined by its density function f, where $f(x)$ is the density of the object at point $x \in \mathbb{R}^m$. Let D denote the unit ball in $\mathbb{R}^m$, $f \in L_2(D)$, and $f = 0$ off D.

An x-ray (or radiograph) from the direction Θ provides a function (the Radon transform)

$$R_\Theta f(z) = \int_{-\infty}^{\infty} f(z + t\theta) dt \quad , \quad z \in \Theta^\perp$$

on the plane $\Theta^{\perp}$ whose value at $z \in \Theta^{\perp}$ is the total mass along the line through z parallel to Θ.

The reconstruction problem is this:

Recover the unknown function f from a finite number of its radiographs

$$R_{\Theta_1} f, \quad R_{\Theta_2} f, \ldots, R_{\Theta_k} f.$$

Let

$$\mathcal{N}(R_\Theta) := \{g \in L_2(D) \mid R_\Theta g = 0\}$$

and

$$A_\Theta := \mathcal{N}(R_\Theta) + f = \{g \in L_2(D) \mid R_\Theta g = R_\Theta f\}.$$

Observe that A_Θ is a closed linear variety,

$$\bigcap_{i=1}^{k} A_{\Theta_i} = \{g \in L_2(D) \mid R_{\Theta_i} g = R_{\Theta_i} f \qquad (i = 1, 2, \ldots, k)\},$$

and $f \in \bigcap_1^k A_{\Theta_i}$. Without any additional information about f other than its radiographs $R_{\Theta_i} f$, the best one can do is to choose some $g \in \bigcap_1^k A_{\Theta_i}$.

Let $P_i = P_{A_{\Theta_i}} \qquad (i = 1, 2, \ldots, k)$. Then it can be shown (Hamaker and Solmon [1978]) that

$$P_i h = h + \chi_D (R_{\Theta_i} f - R_{\Theta_i} h) / R_{\Theta_i} \chi_D$$

for any $h \in L_2(D)$, where χ_D denotes the characteristic function of D.

The (MAP) now implies that, starting with any $f_0 \in L_2(D)$ and setting

$$f_n = (P_k P_{k-1} \cdots P_1) f_{n-1} \qquad (n = 1, 2 \ldots),$$

we get that

$$f_n \to g_0 := P_{\bigcap_1^k A_{\Theta_i}} f_0.$$

In particular, taking $f_0 = 0$, we obtain the function $g_0 \in \bigcap_1^k A_{\Theta_i}$ which has minimal norm.

Finally, Hamaker and Solmon [1978] obtained a convergence rate for this iterative procedure in terms of the angles Θ_i.

The above method is based on knowledge of the complete Radon transform $R_\Theta f$. In practice, one often has available only a sampling of values of $R_\Theta f$ (i.e., a discretized version). For a review of iterative algorithms to handle this discretized version, see Censor and Herman [1987]. For a review of block-iterative methods to handle the discretized version, including block-Kaczmarz, see Censor [1988].

8. GENERALIZATIONS TO NON-HILBERT SPACES.

A natural question that arises is to what extent Theroems 2.1 and 4.1 can be extended to a non-Hilbert space setting.

If X is a strictly convex and reflexive Banach space and A is any closed subspace, then each $x \in X$ has a unique best approximation $P_A(x)$ in A (see e.g. Singer [1974], Theorem 3.17, p. 36).

8.1 THEOREM. *(Stiles [1965a]) Let X be strictly convex and reflexive, and have dimension at least 3. If*

$$\lim_{n\to\infty}(P_B P_A)^n x = P_{A\cap B}(x)$$

for each $x \in X$ and every pair of closed subspaces A, B in X, then X is a Hilbert space.

This shows that Theorem 2.1 has *no* global extension to more general spaces than Hilbert spaces. Interestingly enough, however, Theorem 4.1 *does* have a bona fide extension.

8.2 THEOREM. *(Stiles [1965b]) Let X be a finite-dimensional, smooth, and strictly convex space, and let A, B be any closed subspaces of X. Then*

$$\lim_{n\to\infty}[(I - P_B)(I - P_A)]^n x = (I - P_{\overline{A+B}})x \tag{8.2.1}$$

for all $x \in X$.

Stiles [1965b] also observed that smoothness cannot be dropped from Theorem 8.2. Several authors have given extensions of Theorem 8.2. They include Atlestam and Sullivan [1976], Franchetti [1973], Deutsch [1979], and Franchetti and Light [1984]. The most general result claimed to date is

8.3 THEOREM (Bosznay[1986]). *If X is a uniformly convex and uniformly smooth Banach space and A and B any closed subspaces, then*

$$\lim_{n\to\infty}[(I - P_B)(I - P_A)]^n x = (I - P_{\overline{A+B}})x$$

for all $x \in X$.

9. OTHER VARIANTS.

There are variants of the method of alternating projections which preserve some of its main features such as: i) they are iterative, ii) they alternate, and iii) they operate on only one set at a time.

Aharoni and Censor [1988] gave an algorithm for finding a point in the set $Q = \bigcap_{i \in I} Q_i$, where I is a finite index set and each Q_i is a finite collection of closed convex sets $\{C | C \in \Lambda_i\}$ in $\mathbb{R}^n$. Their algorithm requires knowledge of $P_C(x)$ for each $C \in Q_i$. (Thus, in practice, this would seem to limit the C's to be either half-spaces or hyperplanes.)

Censor [1984] reviews several such methods. In particular, he reviews the method of successive orthogonal projections to determine a point in $\bigcap_1^k Q_i$, where each $Q_i \subset \mathbb{R}^n$ is a closed convex set. The general scheme is of the form

$$x_{n+1} = x_n + \lambda_n [P_{Q_{i_n}}(x_n) - x_n],$$

where $\{\lambda_n\}$ is a sequence of relaxation parameters in $[0, 2]$.

Gubin, Polyak, and Raik [1967] had considered the same problem, but had allowed the collection of convex sets to be infinite.

Bregman [1965] gave a nice alternating scheme to find common points of convex sets. Under mild restrictions, the sequence $\{x_n\}$ generated converges *weakly* to some point in $C = \bigcap_{i \in I} C_i$. He applies his algorithm successfully to a number of problems including the fundamental problem of linear programming.

Cheney and Goldstein [1959] was one of the earliest papers to use an alternating iteration for a collection of convex sets.

Since finding a solution to a system of linear inequalities may be reformulated as finding a common point to a certain collection of convex sets (namely, half-spaces), the above methods apply in this situation also. Agmon [1954] and Motzkin and Schoenberg [1954] have attacked this problem directly. See also Spingarn [1985] and [1987].

10. THE DILIBERTO–STRAUS ALGORITHM.

Suppose that X is a normed linear space and A and B are "proximinal" subspaces. (Recall that a subspace M of X is *proximinal* if the set of best approximations to x,

$$P_M(x) := \{y \in M | \, \|x - y\| = \inf_{m \in M} \|x - m\|\},$$

is not empty for each $x \in X$.) In general, $P_A(x)$ and $P_B(x)$ contain more than one element. In this case, we seek "selections" for P_A and P_B for which the (MAP) holds. More precisely, let S_A and S_B be selections for P_A and P_B respectively. This means that $S_A, S_B : X \to X$ satisfy

$$S_A(x) \in P_A(x) \quad , \quad S_B(x) \in P_B(x)$$

for all $x \in X$. We ask "when is it true that

$$\lim_{n\to\infty} [(I - S_A)(I - S_B)]^n x = x - P(x)$$

for all $x \in X$, where P is a selection for $P_{\overline{A+B}}$?"

Diliberto and Straus [1951] considered the case when $X = C(S \times T)$ is the space of all real continuous functions on the product space $S \times T$, where $S = T = [0, 1]$, with the norm

$$\|x\| = \text{supp}_{(s,t)\in S\times T} \ |x(s,t)|,$$

and where

$$A = \{a \in C(S \times T) \mid a(s,t) = a(s)\}, \text{ and}$$
$$B = \{b \in C(S \times T) \mid b(s,t) = b(t)\}.$$

That is, they were seeking best approximations to continuous functions of two variables by sums of functions of a single variable. They observed that

$$S_A(x)(s) := \frac{1}{2}\left[\max_{t\in T} x(s,t) + \min_{t\in T} x(s,t)\right]$$

and

$$S_B(x)(t) := \frac{1}{2}\left[\max_{s\in S} x(s,t) + \min_{s\in S} x(s,t)\right]$$

were selections for P_A and P_B.

Starting with any $f \in C(S \times T)$, define

$$f_n = [(I - S_B)(I - S_A)]^n f \qquad (n = 1, 2, \ldots).$$

Diliberto and Straus [1951] showed that $\|f_n\|$ decreases to $d(f, A+B)$, the sequence $\{f_n\}$ has cluster points, and if f^* is any one of these, then $f - f^*$ is a best approximation to f from $A + B = \overline{A+B}$. Later, Aumann [1959] showed that the whole sequence converged. In other words,

$$\lim_{n\to\infty} [(I - S_B)(I - S_A)]^n f = (I - P)f$$

for each $f \in C(S \times T)$, for some selection P for P_{A+B}.

The rate of convergence of the Diliberto-Straus algorithm is studied in von Golitschek and Cheney [1979]. Von Gotitschek and Light [1987] showed that there is no *continuous* selection for P_{A+B}.

A generalization of the Diliberto-Straus algorithm was given by Golomb [1959]. Other aspects of the Diliberto-Straus algorithm have been considered by Rivlin and Sibner [1965], Flatto [1966], von Golitschek and Cheney [1979], Dyn [1980], Light and Cheney [1980], and von Golitschek and Cheney [1983]. The Diliberto-Straus algorithm in L_∞ was studied by Kelley [1981], while in L_1 it was studied by Light [1983] and Light and Holland [1984].

For a general survey of this and other aspects of the (MAP), see Papini [1988].

11. CONCLUSION.

We hope that the reader is by now convinced that the method of alternating projections has important applications in multiple branches of mathematics. In spite of this, we believe that its full potential has yet to be realized.

ACKNOWLEDGMENTS.

I have had the good fortune to have been able to communicate with several people about various aspects of the method of alternating projections. I am especially indebted to R. Lorentz for useful information about the multigrid method, and to W. Luxemburg for helpful discussions and lecture notes concerning Schwarz's "alternierende Verfahren."

ADDENDUM.

After this paper was written, a large number (more than 40) of additional papers, most of which appear to be applications or variants of the method of alternating projections, came to my attention through a miracle of modern technology. Namely, I had access to the MathSci Disc 1987–(June) 1991 which contains the contents of Mathematical Reviews on a computer disc. For example, there were several papers related to the Schwarz alternating method (P. L. Lions had three himself). Unfortunately, the deadline for having this manuscript ready for press did not allow me further time for reading these papers in detail and supplementing the present manuscript with all the ones which may have been relevant.

REFERENCES

S. Agmon, [1954], *The relaxation method for linear inequalitites*, Canadian J. Math. **6**, 382–392.

R. Aharoni and Y. Censor, [1988], *Block-iterative projection methods for parallel computation of solutions to convex feasibility problems.*, preprint.

N. Aronszajn, [1950], *Theory of reproducing kernels*, Trans. Amer. Math. Soc. **68**, 337–404.

B. Atlestam and F. Sullivan, [1976], *Iteration with best approximation operators*, Rev. Roumaine Math. Pures Appl. **21**, 125–131.

G. Aumann, [1959], Über approximative Nomographie, II, Bayer. Akad. Wiss., Math.–Natur. Kl. Sitzundsber, 103–109.

A. P. Bosznay, [1986], *A remark on the alternating algorithm*, Periodica Math. Hungarica **17**, 241–244.

D. Braess, [1981], *The contraction number of a multigrid method for solving the Poisson equation*, Numer. Math. **37**, 387–404.

L. M. Bregman, [1965], *The method of successive projection for finding a common point of convex sets*, Sov. Math. Dok. **6**, 688–692.

L. Breiman and J. H. Friedman, [1985], *Estimating optimal transformations for multiple regression and correlation*, J. Amer. Statist. **80**, 580–598.

D. L. Burkholder, [1962], *Successive conditional expectations of an integrable function*, Ann. Math. Statist. **33**, 887–893.

D. L. Burkholder and Y. S. Chow, [1961], *Iterates of conditional expectation operators*, Proc. Amer. Math. Soc. **12**, 490–495.

Y. Censor, [1984], *Iterative methods for the convex feasibility problem*, Annals of Discrete Math. **20**, 83–91.

———, [1988], *Parallel application of block-iterative methods in medical imaging and radiation therapy*, Math. Programming **42**, 307–325.

Y. Censor and G. T. Herman, [1987], *On some optimization techniques in image reconstruction from projections*, Appl. Numer. Math. **3**, 365–391.

W. Cheney and A. A. Goldstein, [1959], *Proximity maps for convex sets*, Proc. Amer. Math. Soc. **10**, 448–450.

J. E. Dennis, Jr., [1972], On some methods bases on Broyden's secant approximation to the Hessian, Numerical Methods for Non-linear Optimization, (F. A. Lootsma, ed.), Academic Press, London.

J. E. Dennis, Jr. and R. B. Schnabel, [1979], *Least change secant updates for quasi-Newton methods*, SIAM Review **21**, 443–459.

J. E. Dennis, Jr. and H. F. Walker, [1981], *Convergence theorems for least-change secant update methods*, SIAM J. Numer. Anal. **18**, 949–987.

F. Deutsch, [1979], *The alternating method of von Neumann,*, "Multivariate Approximation Theory", (W. Schempp and K. Zeller, eds.), Birkhauser-Verlag, Basel.

———, [1983a], *Applications of von Neumann's alternating projections algorithm*, "Mathematical Methods in Operations Research", (P. Kenderov, ed.), Sofia, Bulgaria, pp. 44–51.

———, [1983b], *Von Neumann's alternating method: The rate of convergence*, "Approximation Theory IV", (C. K. Chui, L. L. Schumaker, and J. D. Ward, eds.), Academic Press, New York, pp. 427–434.

D. P. Diliberto and E. G. Straus, [1951], *On the approximation of a function of several variables by the sum of functions of fewer variables.*, Pacific J. Math. **1**, 195–210.

J. Dyer, [1965], *Acceleration of the Convergence of the Kaczmarz Method and Iterated Homogeneous Transformations*, doctoral dissertation, Univ. Calif. at Los Angeles.

R. L. Dykstra, [1983], *An algorithm for restricted least squares regression*, J. Amer. Statist. Assoc. **78**, 837–842.

N. Dyn, [1980], *A straightforward generalization of Diliberto and Straus' algorithm does not work*, J. Approx. Theory **30**, 247–250.

P. P. B. Eggermont, G. T. Herman, and A. Lent, [1981], *Iterative algorithms for large partitioned linear systems, with applications to image reconstruction*, Linear Alg. and its Applic. **40**, 37–67.

L. Flatto, [1965], *The approximation of certain functions of several variables by sums of functions of fewer variables*, Amer. Math. Monthly **71**, 131–132.

C. Franchetti, [1973], *On the alternating approximation method*, Boll. Un. Mat. Ital. **7** no. (4), 169–175.

C. Franchetti and W. Light, [1984], *The alternating algorithm in uniformly convex spaces*, J. London Math. Soc. **29** no. (2), 545–555.

———, [1986], *On the von Neumann alternating algorithm in Hilbert space*, J. Math. Anal. Appl. **114**, 305–314.

K. Friedricks, [1937], *On certain inequalities and characteristic value problems for analytic functions and for functions of two variables*, Trans. Amer. Math. Soc. **41**, 321–364.

T. B. Gatski, C. E. Grosch, and M. E. Rose, [1982], *A numerical study of the two-dimensional Navier-Stokes equations in vorticity-velocity variables*, J. Comput. Physics **48**, 1–22.

———, [1988], *The numerical solution of the Navier-Stokes equations for 3-dimensional, unsteady, incompressible flows by compact schemes*, J. Comput. Physics **82**, 298–329.

W. B. Gearhart and M. Kosky, [1989], *Acceleration schemes for the method of alternating projections*, J. Comp. Appl. Math. **26**, 235–249.

J. Gilbert and W. A. Light, [1986], *Multigrid methods and the alternating algorithm.*

M. von Golitschek and E. W. Cheney, [1979], *On the algorithm of Diliberto and Straus for approximating bivariate functions by univariate ones*, Num. Funct. Anal. Approx. **1**, 341–363.

———, [1983], *Failure of the alternating algorithm for best approximation of multivariate functions*, J. Approx. Theory **38**, 139–143.

M. von Golitschek and W. A. Light, [1987], *Some properties of the Diliberto-Straus algorithms in $C(S \times T)$*, "Numerical Methods of Approximation Theory", (L. Collatz, G. Meinardus, and G. Nürnberger, eds.), vol. 8, Birkhäuser-Verlag, Basel.

M. Golomb, [1959], *Approximation by functions of fewer variables*, "On Numerical Approximation", (R. E. Langer, ed.), Univ. Wisconsin Press, Madison, pp. 275–327.

L. G. Gubin, B. T. Polyak, and E. V. Raik, [1967], *The method of projections for finding the common point of convex sets*, USSR Comput. Math. and Math. Phys. **7**, 1–24.

R. Gordon, R. Bender, and G. T. Herman, [1970], *Algebraic reconstruction techniques (ART) for three-dimensional electron microscopy and X-ray photography*, J. Theoretical Biol. **29**, 471–481.

I. Halperin, [1962], *The product of projection operators*, Acta Sci. Math. (Szeged) **23**, 96–99.

C. Hamaker and D. C. Solmon, [1978], *The angles between null spaces of X-rays*, J. Math. Anal. Appl. **62**, 1–23.

G. T. Herman, A. Lent, and P. H. Lutz, [1978], *Iterative relaxation methods for image reconstruction*, Communications of the ACM 21, 152–158.

G. N. Hounsfield, [1973], *Computerized transverse axial scanning (tomography): Part I Description of system*, British J. Radiol. **46**, 1016–1022.

S. Kaczmarz, [1937], *Angenäherte Auflösung von Systemen linearer Gleichungen*, Bull. Internat. Acad. Pol. Sci. Lett. **A 35**, 355–357.

S. Kayalar and H. L. Weinert, [1988], *Error bounds for the method of alternating projections*, Math. Control Signals Systems **1**, 43–59.

C. T. Kelley, [1981], *A note on the approximation of functions of several variables by sums of functions of one variable*, J. Approx. Theory **33**, 179–189.

W. A. Light, [1983], *The Diliberto-Straus algorithm in $L_1(X \times Y)$*, J. Approx. Theory **38**, 1–8.

W. A. Light and E. W. Cheney, [1980], *On the approximation of a bivariate function by the sum of univariate functions*, J. Approx. Theory **29**, 305–322.

W. A. Light and S. M. Holland, [1984], *The L_1-version of the Diliberto-Straus algorithm in $C(T \times S)$*, Proc. Edinburgh Math. Soc. **27**, 31–45.

T. S. Motzkin and I. J. Schoenberg, [1954], *The relaxation method for linear inequalities*, Canadian J. Math. **6**, 393–404.

H. Nakano, [1953], *Spectral theory in the Hilbert space*, Japan Soc. Promotion Sc., Tokyo.

J. von Neumann, [1933], *Functional Operators–Vol. II. The Geometry of Orthogonal Spaces (*, This is a reprint of mineographed lecture notes first distributed in 1933), Annals of Math. Studies #22, Princeton University Press, 1950.

P. L. Papini, [1988], *Alternating methods in approximation*, "Functional Analysis and Approximation", (P. L. Papini, ed.), Bagni di Lucca, Italy, pp. 219–229.

M. J. D. Powell, [1970], *A new algorithm for unconstrained optimization, Nonlinear Programming, (J. B. Rosen, O. L. Mangasarian, K. Ritter, eds.)*, Academic Press, New York.

I. P. Ramadanov and M. L. Skwarczynski, [1984a], Bull. Polish Acad. Sci. Math. **32**, 653–659.

———, [1984b], *Constructive Theory of Functions '84*, Sofia, 726–730.

T. J. Rivlin and R. J. Sibner, [1965], *The degree of approximation of certain functions of two variables by a sum of functions of one variable*, Amer. Math. Monthly **72**, 1101–1103.

G.-C. Rota, [1962], *An "alternierende Verfahren" for general positive operators*, Bull. Amer. Math. Soc. **68**, 95–102.

H. Salehi, [1967], *On the alternating projections theorem and bivariate stationary stochastic processes*, Trans. Amer. Math. Soc. **128**, 121–134.

H. A. Schwarz, [1870], *Ueber einen Grenzübergang durch alternirendes Verfahren*, Vierteljahrsschrift der Naturforschenden Gesellschaft in Zürich **15**, 272–286.

I. Singer, [1974], *The Theory of Best Approximation and Functional Analysis*, CBMS #13, SIAM, Philadelphia.

M. Skwarczynski, [1985a], *Alternating projections in complex analysis*, Complex Analysis and Applications '83, Sofia.

———, [1985b], *A general description of the Bergman projection*, Annales Polonici Math. **46**.

K. T. Smith, D. C. Solman and S. L. Wagner, [1977], *Practical and mathematical aspects of the problem of reconstructing objects from radiographs*, Bull. Amer. Math. Soc. **83**, 1227–1270.

J. E. Spingarn, [1985], *A primal-dual method for solving systems of linear inequalities*, Linear Alg. Applic. **65**, 45–62.

———, [1987], *A projection method for least-squares solutions to overdetermined systems of linear inequalities*, Linear Alg. Applic. **86**, 211–236.

W. J. Stiles, [1965a], *Closest-point maps and their products*, Nieuw Archief voor Wiskunde **13** no. (3), 19–29.

———, [1965b], *A solution to Hirschfeld's problem*, Nieuw Archief voor Wiskunde **13** no. (3), 116–119.

K. Tanabe, [1971], *Projection method for solving a singular system of linear equations and its applications*, Numer. Math. **17**, 203–214.

R. Wegmann, [1989], *Conformal mapping by the method of alternating projections*, Numer. Math. **56**, 291–307.

N. Wiener, [1955], *On the factorization of matrices*, Comment. Math. Helv. **29**, 97–111.

N. Wiener and P. Masani, [1957], *The prediction theory of multivariate stochastic processes, II: The linear predictor*, Acta Math. **93**, 95–137.

D. C. Youla, [1978], *Generalized image restoration by the method of alternating orthogonal projections*, IEEE Trans. Circuits Syst. **CAS–25**, 694–702.

D. C. Youla and H. Webb, [1982], *Image restoration by the method of convex projections: Part 1–Theory*, IEEE Trans. Medical Imaging **M1–1**, 81–94.

Department of Mathematics
The Pennsylvania State University
University Park, PA 16802
U.S.A.

SELECTIONS FOR METRIC PROJECTIONS

FRANK DEUTSCH†

Abstract. A review is given of conditions which characterize when the metric projection onto a proximinal subspace of a normed linear space has a selection which is continuous, (pointwise) Lipschitz continuous, or linear. Intrinsic characterizations of the subspaces in the particular spaces $C_0(T)$ or $L_p(\mu)$, $1 \leq p < \infty$, whose metric projections have one of these properties are also given.

1. INTRODUCTION.

It is our purpose to present an outline of what is known concerning various kinds of selections for the set valued metric projection. Since there have been a few surveys on metric selections given within the last eight years (e.g. Deutsch [1983] and Nürnberger and Sommer [1984]), it is our intention here to concentrate mainly on the more recent results, and to restrict our attention to those results which yield characterizing conditions, i.e., conditions which are both necessary and sufficient for the stated property.

Let X be a normed linear space and G a (linear) subspace. For any $x \in X$, the (possibly empty) set of all best approximations to x from G is defined by

$$P_G(x) := \{g \in G \mid \|x - g\| = d(x, G)\},$$

where $d(x, G) = \inf\{\|x - g\| \mid g \in G\}$. The mapping $P_G : X \to 2^G$ thus defined is called the *metric projection* onto G. G is called *proximinal* (resp. *Chebyshev*) provided $P_G(x)$ is not empty (resp. contains a single element) for each $x \in X$. When G is Chebyshev, we usually regard P_G as an ordinary (single-valued) function $P_G : X \to G$ and write $P_G(x) = g$ rather than $P_G(x) = \{g\}$.

†Supported by NSF Grant DMS-9100228

S. P. Singh (ed.), Approximation Theory, Spline Functions and Applications, 123–137.

It is well-known and easy to verify that G is proximinal if it is finite-dimensional or a closed subspace in a reflexive Banach space X. Moreover, a proximinal subspace of a strictly convex space X is always Chebyshev. Finally, if G is proximinal, then for any $x \in X$, $P_G(x)$ is a nonempty, closed, bounded, and convex subset of G. Also, $\|g\| \leq 2\|x\|$ for every $g \in P_G(x)$.

A mapping $p : X \to G$ is called a *selection* for P_G provided

$$p(x) \in P_G(x) \qquad \text{for all } x \in X.$$

(In this case, we sometimes say that G admits a metric selection.) Selections always exist by the axiom of choice. We will be interested in selections which have the additional property of continuity, (pointwise) Lipschitz continuity, or linearity. Note that linearity implies Lipschitz continuity since, from the last paragraph,

$$\|p(x) - p(y)\| = \|p(x - y)\| \leq 2\|x - y\|.$$

This problem is related, for example, to the stability of algorithms for computing best approximations. If it is known, say, that a continuous selection for P_G does not exist, then any algorithm for computing best approximations, being a selection, must be discontinuous or unstable. On the other hand, if it is known that a continuous selection for P_G exists, this provides us with a hunting license to find one.

We conclude the introduction with some necessary notation and terminology. The *kernel* of the metric projection P_G is the set

$$\ker P_G := P_G^{-1}(0) = \{x \in X \mid 0 \in P_G(x)\}.$$

A selection p for P_G is called *homogeneous* (resp. *additive modulo* G) provided

$$p(\alpha x) = \alpha p(x) \qquad \text{for all scalars } \alpha$$

$$(\text{ resp.} \qquad p(x + g) = p(x) + g \qquad \text{for all } x \in X,\ g \in G\ ).$$

The quotient map $Q = Q_G : X \to X/G$ is defined by

$$Q(x) = x + G \quad , \quad x \in X.$$

For any set $H \subset \ker P_G$, we write $X = G \oplus H$ to mean that each $x \in X$ has a unique representation in the form $x = g + h$, where $g \in G$ and $h \in H$. Further, H is called *homogeneous* if $\alpha H \subset H$ for any scalar α.

2. CONTINUOUS SELECTIONS.

2.1 THEOREM (Deutsch, Li, and Park [1989]). *Let G be a proximinal subspace of X and $Q : X \to X/G$ the quotient map. The following statements are equivalent.*

(1) *P_G has a continuous selection;*

(2) *P_G has a continuous selection which is homogeneous and additive modulo G;*

(3) *$\ker P_G$ contains a closed homogeneous subset H such that $X = G \oplus H$ and the mapping $p : G \oplus H \to G$, defined by $p(g + h) = g$, is continuous;*

(4) *$\ker P_G$ contains a closed homogeneous subset H such that $Q|_H$ is a homeomorphism between H and X/G.*

Moreover, the continuous selection is given by p if (3) holds, and by $x \mapsto x - (Q|_H)^{-1}(x + G)$ if (4) holds.

In case G is a Chebyshev subspace, the only subset H which can work in Theorem 2.1 is the full set $\ker P_G$. This is a consequence of

2.2 LEMMA (Deutsch, Li, and Park [1989]). *Let G be a Chebyshev subspace of $X, H \subset \ker P_G$, and $X = G \oplus H$. Then $H = \ker P_G$.*

From this lemma and Theorem 2.1 we immediately obtain the following corollary.

2.3 COROLLARY (Holmes [1972]). *Let G be a Chebyshev subspace of X. Then P_G is continuous if and only if $Q|_{\ker P_G}$ is a homeomorphism between $\ker P_G$ and X/G.*

2.4 THEOREM (Deutsch, Li, and Park [1989]). *Let G be a finite-dimensional subspace of X. Then P_G has a continuous selection if and only if $\ker P_G$ contains a closed homogeneous subset H such that $X = G \oplus H$.*

2.5 THEOREM (Deutsch, Li, and Park [1989]). *Let G be a proximinal subspace of finite codimension in X. Then P_G has a continuous selection if and only if $\ker P_G$ contains a boundedly compact homogeneous subset H with $X = G \oplus H$.*

Using Theorem 2.5, Lemma 2.2, and the fact (Cheney and Wulbert [1969]) that G is Chebyshev if and only if $X = G \oplus \ker P_G$, we obtain

2.6 THEOREM (Cheney and Wulbert [1969]). *Let G be a Chebyshev subspace of finite-codimension in a Banach space X. Then P_G is continuous if and only if $\ker P_G$ is boundedly compact.*

3. LIPSCHITZ CONTINUOUS SELECTIONS.

Our first result of this section includes characterizations of when the metric projection admits a Lipschitz continuous selection or a pointwise Lipschitz continuous selection.

3.1 THEOREM (Deutsch, Li, and Park [1989]). *Let G be a proximinal subspace which is complemented in X (e.g. $\dim G < \infty$ or codim $G < \infty$) and $Q : X \to X/G$ the quotient map. Then the following statements are equivalent:*

(1) *P_G has a (pointwise) Lipschitz continuous selection;*

(2) *P_G has a (pointwise) Lischitz continuous selection which is homogeneous and additive modulo G;*

(3) *$\ker P_G$ contains a closed homogeneous subset H such that $X = G \oplus H$ and the mapping $p : G \oplus H \to G$, defined by $p(g + h) = g$, is (pointwise) Lipschitz continuous;*

(4) *$\ker P_G$ contains a closed homogeneous subset H such that $Q|_H$ is a (pointwise) Lipschitz continuous homeomorphism between H and X/G.*

Moreover, the desired selection is given by p if (3) holds and by $x \mapsto x - (Q|_H)^{-1}(x + G)$ if (4) holds.

3.2 COROLLARY (Deutsch, Li, and Park [1989]). *Let G be a Chebyshev subspace which is complemented in the Banach space X. Then P_G is (pointwise) Lipschitz continuous if and only if $Q|_{\ker P_G}$ is a (pointwise) Lipschitz homeomorphism between $\ker P_G$ and X/G.*

The next result shows that for every closed subspace to admit a Lipschitz continuous metric selection is a rather strong condition.

3.3 THEOREM (Deutsch, Li, and Park [1989]). *Let X be a reflexive Banach space and suppose that each closed subspace admits a Lipschitz continuous metric selection. Then X is isomorphic to Hilbert space.*

4. LINEAR SELECTIONS.

The first result characterizes when a metric projection admits a linear selection. We already observed that a linear selection is a Lipschitz continuous one.

4.1 THEOREM (Deutsch [1982]). *Let G be a proximinal subspace of the normed linear space X and let $Q = Q_G$. Then the following statements are equivalent:*

(1) *P_G has a linear selection;*

(2) $\ker P_G$ *contains a closed subspace* H *such that* $X = G \oplus H (= G + H)$;

(3) $\ker P_G$ *contains a closed subspace* H *such that* $Q|_H$ *is an isometry of* H *and* X/G.

4.2 COROLLARY (Deutsch [1982]). *Let* G *be a proximinal subspace. Then* $\ker P_G$ *is a subspace if and only if* G *is Chebyshev and* P_G *is linear.*

4.3 COROLLARY (Holmes and Kripke [1968]). *Let* G *be a Chebyshev subspace. The following statements are equivalent:*

(1) P_G *is linear;*

(2) $\ker P_G$ *is a subspace;*

(3) $Q|_{\ker P_G}$ *is an isometry of* $\ker P_G$ *and* X/G.

4.4 EXAMPLE ($\ker P_G$ need *NOT* be a subspace for P_G to admit a linear selection). *Let* $X = \ell_\infty(2)$, *i.e.,* X *is* $\mathbb{R}^2$ *with the norm* $\|(x,y)\| = \max\{|x|, |y|\}$. *Let* $e_1 = (1,0), e_2 = (0,1)$, *and* $G =$ *span* $\{e_1\}$. *Note* $H :=$ *span* $\{e_2\} \subset \ker P_G, H \neq \ker P_G$, *and* $X = G \oplus H$ *so* P_G *has a linear selection (by Theorem 4.1) given by*

$$p(x) = x(1)e_1, \qquad x \in X.$$

4.5 COROLLARY (Aronszajn and Smith [1954]). *The metric projection onto a proximinal subspace of codimension one admits a linear selection.*

Using the complemented subspace theorem of Lindenstrauss and Tzafriri [1971], we can also deduce the following consequence of Theorem 4.1.

4.6 COROLLARY (Deutsch [1982]). *If each closed subspace of* X *is proximinal and its metric projection admits a linear selection, then* X *is isomorphic to Hilbert space.*

QUESTION. Assume $\dim X \geq 3$. Can the conclusion of Corollary 4.6 be strengthened to "then X is *isometric* to Hilbert space"?

The answer is affirmative if X is strictly convex (see James [1947], Hirschfeld [1958], and Rudin and Smith [1961]). The example in Deutsch [1982] (example 2.11) which purported to answer this question in the negative contains an error. (This was kindly pointed out to me by Dan Amir and Günther Nürnberger.)

5. APPLICATIONS IN $C_0(T)$.

In this section, we will state a number of *intrinsic* characterizations of finite-dimensional subspaces G of $C_0(T)$ whose metric projections admit selections of the type being considered. By "intrinsic" we mean that the characterization can be given entirely in terms of properties of G.

Throughout this section, T will denote a locally compact Hausdorff space and $C_0(T)$ will denote the Banach space of all real continuous functions f on T which "vanish at infinity" (i.e. $\{t \in T \mid |f(t)| \geq \varepsilon\}$ is compact for each $\varepsilon > 0$) and endowed with the supremum norm:

$$\|f\| = \max\{|f(t)| \; |t \in T\}.$$

If T is compact, then $C_0(T)$ reduces to all continuous functions on T and is usually denoted by $C(T)$.

For any set $F \subset C_0(T)$, let

$$Z(F) := \bigcap_{f \in F} \{t \in T \mid f(t) = 0\}.$$

For any subset A of T, let card (A), $bd\, A$, and $int\, A$ denote the cardinality, boundary, and interior of A respectively. Further, if G is a subspace of $C_0(T)$, let

$$G(A) := \{g \in G \mid g(t) = 0 \text{ for all } t \in A\},$$

and $G(\varnothing) = G$.

5.1 THEOREM. *(Li [1988a,b,c,d]) Let T be locally connected and G a finite-dimensional subspace of $C_0(T)$. Then P_G has a continuous selection if and only if, for each $g \in G$,*

(i) *card $(bd\, Z(g)) \leq \dim G(int\, Z(g)) =: r_g$*
(ii) *g has at most $r_g - 1$ zeros with sign changes.*

Various special cases of this theorem had been proved earlier by several authors. See Li [1991] for further details.

This theorem is false without the local connectedness condition. What is true is the following general result.

5.2 THEOREM (Li [1991]). *Let G be a finite-dimensional subspace of $C_0(T)$. Then P_G has a continuous selection if and only if G is a "regular weakly interpolating subspace"; that is, for any finite set $A = \{t_1, \ldots, t_r\} \subset T$ with $A \backslash int\, Z(G(A)) \neq \varnothing$ and any signs $\{\sigma_1, \ldots, \sigma_r\} \subset \{-1, 1\}$, there exists $g \in G$ and neighborhoods V_i of t_i, $1 \leq i \leq r$, such that*

$$\sigma_i\, g(t) \geq 0 \text{ for all } t \in V_i \qquad (i = 1, \ldots, r)$$

and $A \backslash int\, Z(g) \neq \varnothing$.

5.3 THEOREM. *Let G be a finite-dimensional subspace of $C[a,b]$. The following statements are equivalent:*

(1) *P_G has a continuous selection;*

(2) *P_G has a pointwise Lipschitz continuous selection;*

(3) *for each $g \in G$,*

(i) *card $(bd\, Z(g)) \leq r_g := \dim G(int\, Z(g))$*

(ii) *g has at most $r_g - 1$ zeros with sign changes.*

The equivalence (1) $\Leftrightarrow$ (3) is just Li's Theorem 5.1, while (1) $\Leftrightarrow$ (2) is from Blatt, Nürnberger, and Sommer [1981-82].

QUESTION. It would be interesting to know whether the equivalence (1) $\Leftrightarrow$ (2) holds in any space $C_0(T)$, not just in $C[a,b]$.

Recall that an n-dimensional subspace G of $C[a,b]$ is called "weak Chebyshev" if each $g \in G$ has at most $n-1$ sign changes. That is, there do not exist points $a \leq t_0 < t_1 < \cdots < t_n \leq b$ such that $g(t_i)g(t_{i-1}) < 0$ for $i = 1, 2, \ldots, n$.

5.4 THEOREM. *Let G be a finite-dimensional subspace of $C[a,b]$. Then P_G has a continuous selection if and only if*

(i) *G is weak Chebyshev, and*

(ii) *card $(bd\, Z(g)) \leq \dim G(int\, Z(g))$ for each $g \in G$.*

Nürnberger [1980] had verified that G having a continuous metric selection implies G is weak Chebyshev. The remainder of Theorem 5.4 is essentially due to Li [1991].

EXAMPLE. Let $g_1(t) = t$ and $G =$ span $\{g_1\}$ in $C[0,1]$. Then G is not Chebyshev, but P_G has a continuous selection by Theorem 5.4.

Using Theorem 5.4, the following beautiful result of Nürnberger and Sommer can be deduced.

5.5 THEOREM (Nürnberger and Sommer [1978]). *In $C[a,b]$, let $S_{n,k} = S_n\{t_1, \ldots, t_k\}$ denote the splines of degree n with k fixed knots $a < t_1 < \cdots < t_k < b$. Then $P_{S_{n,k}}$ has a continuous selection if and only if $k \leq n+1$.*

5.6 THEOREM (Li [1991]). *Let T be the unit circle in $\mathbb{R}^2$ and G an n-dimensional subspace in $C_{2\pi} := C_0(T)$. Then the following statements are equivalent:*

(1) *P_G has a continuous selection;*

(2) P_G *has a unique continuous selection;*

(3) *(i)* n *is odd*

(ii) Each $g \in G\backslash\{0\}$ *has no more than* n *zeros.*

Concerning Lipschitz continuous selections, we have the following results.

5.7 THEOREM (Li [1990]). *Let* G *be a finite-dimensional subspace of* $C_0(T)$. *Then the following statements are equivalent:*

(1) P_G *has a Lipschitz continuous selection;*

(2) $T\backslash Z(g)$ *is compact for each* $g \in G$.

5.8 COROLLARY. *Let* G *be a finite-dimensional subspace of* c_0. *Then* P_G *has a Lipschitz continuous selection if and only if there exists an integer* n *such that, for each* $g \in G$, $g(i) = 0$ *for all* $i \geq n$.

5.9 COROLLARY (Cline [1973], Berdyshev [1975]). *Let* T *be compact and infinite, and* G *be a finite-dimensional Chebyshev subspace. Then* P_G *is Lipschitz continuous if and only if* $\dim G = 1$.

5.10 COROLLARY. *Suppose* T *is a finite set. Then for each finite-dimensional subspace* G *of* $C(T)$, P_G *has a Lipschitz continuous selection. In particular, (Cline [1973]) every Chebyshev subspace in* $C(T)$ *has a Lipschitz continuous metric projection.*

EXAMPLE. (A Chebyshev subspace with a pointwise Lipschitz continuous metric projection which is not Lipschitz continuous.) Let $g_1 = (1, 1/2, 1/3, \ldots, 1/n, \ldots)$ and $G =$ span $\{g_1\}$. Then G is a Chebyshev subspace of $c_0 = C_0(\mathbb{N})$ (since g_1 has no zeros). Hence P_G is pointwise Lipschitz continuous (since strong uniqueness holds), but P_G is not Lipschitz continuous since $\mathbb{N}\backslash Z(g_1) = \mathbb{N}$ is not compact (Theorem 5.1).

The next example is a refinement of one given by Respess and Cheney [1982].

EXAMPLE. (A compact set T such that $C(T)$ contains a subspace of every finite dimension whose metric projections admit Lipschitz continuous selections.) Take $T = \{0\} \cup \left(\bigcup_{i=1}^{\infty} [1/2i, 1/(2i-1)]\right)$ and let $g_i = \mathcal{X}_{[1/2i, 1/(2i-1)]}$ $(i = 1, 2, \ldots)$. For any integer $n \geq 1$, let $G = G_n =$ span $\{g_1, g_2, \ldots, g_n\}$. That is, G is the space of functions which are constant as each subinterval $[1/2i, 1/(2i-1)]$ $(i = 1, 2, \ldots, n)$ and zero elsewhere. Clearly, for any $g \in G, T\backslash Z(g)$ is a finite union of compact intervals, hence compact. By Theorem 5.7, P_G has a Lipschitz continuous selection.

Note how this example contrasts with Corollary 5.9 which states that if G were a finite-dimensional *Chebyshev* subspace (of dimension at least 2), then P_G could never be Lipschitz continuous!

The final application in $C_0(T)$ is an intrinsic characterization of the finite-dimensional subspaces of $C_0(T)$ whose metric projection admits a linear selection.

5.11 THEOREM (Deutsch [1982] for $n = 1$, Lin [1985] for $n > 1$). *Let G be an n-dimensional subspace of $C_0(T)$. Then P_G has a linear selection if and only if G has a basis $\{g_1, g_2, \ldots, g_n\}$ such that card $(\text{supp}(g_i)) \leq 2$ for $i = 1, 2, \ldots, n$.*

Here $\text{supp}(g) = T \backslash Z(g)$.

6. APPLICATIONS IN $L_1(\mu)$.

In this section we give some applications in the space $L_1(\mu)$. Unlike the space $C_0(T)$, the results here are less complete.

Let (T, μ) be a measure space and let $L_1(\mu) = L_1(T, \mu)$ denote the Banach space of all measurable functions f on T with

$$\|f\| := \int_T |f| d\mu < \infty.$$

A measurable set A is called an *atom* if $\mu(A) > 0$ and if B is any measurable set with $B \subset A$, either $\mu(B) = 0$ or $\mu(B) = \mu(A)$. A union of a finite number of atoms is called a *unifat*. An element $g \in L_1(\mu)$ is said to satisfy the *Lazar condition* if whenever $B \subset \text{supp}(g) := T \backslash Z(g)$ satisfies $\int_B |g| d\mu = \int_{\text{supp}(g)\backslash B} |g| d\mu$, then either B or $\text{supp}(g) \backslash B$ must be a unifat.

6.1 THEOREM. *Let $G \in L_1(\mu) \backslash \{0\}$ and $G =$ span $\{g\}$. Then the following statements are equivalent:*

(1) *P_G has a continuous selection;*

(2) *P_G has a Lipschitz continuous selection;*

(3) *g satisfies the Lazar condition;*

(4) *there exists a unifat set $A \subset \text{supp}(g)$ such that either $B \subset A$ or $\text{supp}(g) \backslash B \subset A$ whenever $B \subset \text{supp}(g)$ and*

$$\int_B |g| d\mu = \int_{\text{supp}(g)\backslash B} |g| d\mu.$$

The equivalence of (1) and (3) is due to Deutsch, Indumathi, and Schnatz [1988]. The remaining equivalences are due to Deutsch and Li [1991]. We do not know whether there is a similar intrinsic characterization when $\dim G > 1$.

QUESTION. Does there exist an intrinsic characterization of the n-dimensional subspaces G of $L_1(\mu), n > 1$, such that P_G has a continuous selection? A *necessary* condition for P_G to have a continuous selection is that $\text{supp}(G) := \bigcup_{g \in G} \text{supp}(g)$ contain at least $\dim G$ atoms. Also, is it true that (1) $\Leftrightarrow$ (2) holds when $\dim G > 1$?

There is a *non*-intrinsic characterization of when P_G admits a continuous selection. Since it is the best result available at present, we include it here.

6.2 THEOREM. *(Li [1991a]) Let G be a finite-dimensional subspace of $L_1(T, \mu)$. Then P_G has a continuous selection if and only if, for each $f \in L_1(T, \mu)$, there exists $g^* \in P_G(f)$ such that the set*

$$\{t \in T \mid [f(t) - g(t)][g^*(t) - g(t)] > 0\}$$

is a unifat for every $g \in P_G(f)$.

For linear selections, we have the following characterization.

6.3 THEOREM (Deutsch [1982] for $n = 1$, Lin [1985] for $n > 1$). *Let G be an n-dimensional subspace of $L_1(\mu)$. Then P_G has a linear selection if and only if there exist n atoms $A_1, A_2, \ldots, A_n$ such that*

$$\int_T |g| d\mu \leq 2 \int_{\bigcup_1^n A_i} |g| d\mu$$

for every $g \in G$.

7. APPLICATIONS IN $L_p(\mu), 1 < p < \infty$.

In this section, we consider the Banach space $L_p(\mu) = L_p(T, \mu)$, for $1 < p < \infty$. Thus (T, μ) is a measure space and $L_p(\mu)$ consists of all measurable functions f on T with

$$\|f\| := \left[\int_T |f|^p d\mu\right]^{\frac{1}{p}} < \infty.$$

We first note that if $p = 2$, $L_2(\mu)$ is a Hilbert space so that every closed subspace G is Chebyshev and P_G is linear. (P_G is orthogonal projection onto G.) This is the strongest result possible so hereafter we assume $1 < p < \infty$ and $p \neq 2$.

Since $L_p(\mu)$ is uniformly convex, every closed subspace is Chebyshev. Moreover, if G is finite-dimensional, P_G is continuous. When $p > 2$, there is an even stronger result.

7.1 THEOREM (Holmes-Kripke [1968]). *Let G be a finite-dimensional subspace of $L_p(\mu)$, $p > 2$. Then for each $f \in L_p(\mu)\backslash G$, there exists a neighborhood U_f of f and a constant $\lambda = \lambda_f \geq 1$ such that*

$$\|P_G(h) - P_G(k)\| \leq \lambda \|h - k\|$$

for each $h \in U_f$ and each $k \in L_p(\mu)$.

Moreover, if $\dim L_p(\mu) < \infty$, *there is a single constant λ independent of f such that*

$$\|P_G(f) - P_G(h)\| \leq \lambda \|f - h\|$$

for all $f, h \in L_p(\mu)$.

In the same paper, Holmes and Kripke also gave examples to show that P_G need *not* be pointwise Lipschitz continuous in L_p, for $p > 2$.

An intrinsic characterization of when P_G is linear is as follows.

7.2 THEOREM. *Let G be an n-dimensional subspace of $L_p(\mu)$, $1 < p < \infty$, $p \neq 2$. Then the following statements are equivalent:*

(1) *P_G is linear;*

(2) *There is a basis $\{g_1, g_2, \ldots, g_n\}$ of G such that each* $\text{supp}(g_i)$ *is a unifat consisting of at most two atoms;*

(3) *There exist disjoint unifat sets $B_0, B_1, \ldots, B_k$ such that*

$$G = L_p(B_0) \oplus \left(\overset{k}{\underset{i=1}{\oplus}} H_i \right),$$

where H_i is a subspace of codimension one in $L_p(B_i)$.

Deutsch [1982] verified (1) $\Leftrightarrow$ (2) when $n = 1$; Lin [1985] proved (1) $\Leftrightarrow$ (3), and Park [1987] established (2) $\Leftrightarrow$ (3).

8. RELATED WORK.

In this section, we include a few related results which, for one reason or another, were not included in any of the previous sections.

8.1 THEOREM. *Let G be a finite-dimensional subspace of $X = C_0(T)$ or $L_1(\mu)$. Then P_G has a continuous selection if and only if P_G is "almost lower semicontinuous".*

Recall that P_G is called *almost lower semicontinuous* (alsc) if for each $x_0 \in X$ and each $\varepsilon > 0$ there exists a neighborhood U of x_0 such that

$$\bigcap_{x \in U} B_\varepsilon(P_G(x)) \neq \varnothing,$$

where $B_\varepsilon(P_G(x)) = \{g \in G \mid d(g, P_G(x)) < \varepsilon\}$.

It was verified by Deutsch and Kenderov [1983] (even for more general set-valued mappings) that: P_G is alsc if and only if, for each $\varepsilon > 0$, there exists a continuous function $p = p_\varepsilon : X \to G$ such that

$$d(p(x), P_G(x)) < \varepsilon$$

for each $x \in X$. In particular, if P_G has a continuous selection, it is alsc.

When $X = C_0(T)$, Theorem 8.1 was proved independently by Fischer [1987] and Li [1988]. When $X = L_1(\mu)$, the result is due to Li [1991a].

Furthermore, Theorem 8.1 holds if G is a 1-dimensional subspace in any normed linear space X. However, Brown [1989] and Zhivkov [1989] have independently constructed examples of finite-dimensional subspaces G with P_G alsc, yet P_G does not have a continuous selection.

Brown [1989] characterized when the metric projection onto a finite-dimensional subspace had a continuous selection in terms of derived mappings.

8.2 THEOREM (Brown [1989]; Theorem 4.3). *Let G be an n-dimensional subspace of the normed linear space X. Then P_G has a continuous selection if and only if $P_G^{(n)}(x) \neq \varnothing$ for all $x \in X$.*

Here the derived mapping $P_G^{(n)}$ is defined inductively as follows:

$$P_G'(x) := \bigcap_{\varepsilon>0} \bigcup_{U \in \mathcal{U}(x)} \bigcap_{z \in U} B_\varepsilon(P_G(z))$$

and

$$P_G^{(k+1)} = \left[P_G^{(k)}\right]' \quad , \qquad \text{where}$$

$\mathcal{U}(x)$ denotes the collection of all neighborhoods of x and

$$B_\varepsilon(P_G(z)) := \{g \in G \mid d(g,\, P_G(z)) < \varepsilon\}.$$

Related to this result is the following.

8.3 THEOREM (Deutsch, Indumathi, and Schnatz [1988]). *Let G be a finite-dimensional subspace of X. Then P_G has a continuous selection if and only if P_G has a lower semicontinuous submap.*

A mapping F is a *submap* of P_G provided $F(x) \subset P_G(x)$ for all $x \in X$.

Blatter [1990] gave an intrinsic characterization of those finite-dimensional subspaces of $C(T)$ whose metric projections admit *unique* continuous selections.

8.4 THEOREM (Blatter [1990]). *Let G be an n-dimensional subspace of $C(T), T$ compact. Then P_G has a unique continuous selection if and only if*

(i) *Each $g \in G \backslash \{0\}$ has at most n zeros,*
(ii) *G is weakly interpolating, and*
(iii) *For each set $\{t_1, t_2, \ldots, t_k\}$ of $k \leq n$ isolated points in T,*

$$\dim G(\{t_1, t_2, \ldots, t_k\}) \leq n - k.$$

Here $G(A) := \{g \in G \mid g = 0 \text{ on } A\}$. This result should be compared with those of section 5, especially Theorem 5.2.

Finally, we conclude with a result which describes a geometric condition on a normed linear space X which is equivalent to the metric projection onto each finite-dimensional subspace having a continuous selection.

8.5 DEFINITION. *A normed linear space X is said to have property (CS) if whenever x_0 and $z \neq 0$ are in X and $\|x_0\| = \|x_0 + z\| = \|x_0 - z\|$, then there exist $\varepsilon > 0$ and $\sigma \in \{-1, 1\}$ such that for each $x \in B_\varepsilon(x_0)$ either*

(i) *$\|x - \sigma z\| \leq \|x - c\sigma z\|$ for all $c \in \mathbb{R}$, or*
(ii) *$\|x - \sigma z\| < \|x - c\sigma z\|$ for all $c > 1$.*

Every strictly convex space has property (CS) as does every space with the property (P) of Brown [1964]. The main result concerning property (CS) is the next theorem.

8.6 THEOREM (Brown, Deutsch, Indumathi, and Kenderov [1988]). *A normed linear space X has property (CS) if and only if the metric projection onto each finite-dimensional subspace of X has a continuous selection.*

From this result can be deduced which spaces of type $C_0(T)$ or $L_1(\mu)$ have property (CS).

8.7 THEOREM (Brown, Deutsch, Indumathi, and Kenderov [1988]).

(1) *The space $C_0(T)$ has property (CS) if and only if T is discrete.*
In particular, if T is compact, then $C(T)$ has property (CS) if and only if it is finite-dimensional.
(2) *The space $L_1(\mu)$ has property (CS) if and only if it is finite-dimensional.*

References

N. Aronszajn and K. T. Smith, [1954], *Invariant subspaces of completely continuous operators*, Ann. of Math. **60**, 345–350.

V. I. Berdyshev, [1975], *Metric projection onto finite-dimensional subspaces of C and L*, Math. Zametki **18**, 473–488.

H.-P. Blatt, G. Nürnberger, and M. Sommer, [1981–1982], *A characterization of pointwise - Lipschitz - continuous selections for the metric projection*, Numer. Funct. Anal. and Optimiz. **4**, 101–121.

J. Blatter, [1990], *Unique continuous selections for metric projections of $C(X)$ onto finite-dimensional vector subspaces*, J. Approx. Theory **61**, 194–221.

A. L. Brown, [1964], *Best n-dimensional approximation to sets of functions*, Proc. London Math. Soc. **14**, 577–594.

———, [1989], *Set valued mappings, continuous selections, and metric projections*, J. Approx. Theory **57**, 48–68.

A. L. Brown, F. Deutsch, V. Indumathi, and P. Kenderov, [1988], *Lower semicontinuity, almost lower semicontinuity, and continuous selections for metric projections, (work in progress).*

E. W. Cheney and D. E. Wulbert, [1969], *The existence and unicity of best approximations*, Math. Scand. **24**, 113–140.

A. K. Cline, [1973], *Lipschitz conditions on uniform approximation operators*, J. Approx. Theory **8**, 160–172.

F. Deutsch, *[1982]*, *Linear selections for the metric projection*, J. Funct. Anal. **49**, 269–292.

———, [1983], *A survey of metric selections in "Fixed Points and Nonexpansive Mappings", (R. C. Sine, ed.)*, Contemporary Mathematics **18**, 49–71.

F. Deutsch, V. Indumathi, and K. Schnatz, [1988], *Lower semicontinuity, almost lower semicontinuity, and continuous selections for set-valued mappings*, J. Approx. Theory **53**, 266–294.

F. Deutsch and P. Kenderov, [1983], *Continuous selections and approximate selections for set-valued mappings and applications to metric projections*, SIAM J. Math. Anal. **14**, 185–194.

F. Deutsch and W. Li, [1991], *Strong uniqueness, Lipschitz continuity, and continuous selections for metric projections in L_1*, J. Approx. Theory **66**, 198–224.

F. Deutsch, W. Li, and S.-H. Park, [1989], *Characterizations of continuous and Lipschitz continuous metric selections in normed linear spaces*, J. Approx. Theory **58**, 297–314.

T. Fischer, [1987], *A continuity criterion for the existence of continuous selections for a set-valued mapping*, J. Approx. Theory **49**, 340–345.

R. A. Hirschfeld, [1958], *On best approximations in normed vector spaces, II*, Nieuw Arch. Wisk. **6**, 99–107.

R. B. Holmes, [1972], *On the continuity of best approximation operators, in "Proceedings Symp. finite Dimensional Topology"*, Annals of Mathematics Studies **69**, 137–157, Princeton Univ. Press, Princeton.

R. B. Holmes and B. R. Kripke, [1968], *Smoothness of approximation*, Michigan Math. J. **15**, 225–248.

R. C. James, [1947], *Inner products in normed linear spaces*, Bull. Amer. Math. Soc. **53**, 559–566.

W. Li, [1988], *The characterization of continuous selections for metric projections in $C(X)$*, Sci. Sinica **A:4**, 254–264.

———, [1988a], *Problems about continuous selections in $C(X)(I)$: Quasi-Haar subspaces*, Acta Math. Sinica **31**, 1–10.

———, [1988b], *Problems about continuous selections in $C(X)(II)$: Alternation signatures*, Acta Math. Sinica **31**, 11–20.

———, [1988c], *Problems about continuous selections in $C(X)(III)$: Local alternation elements*, Acta Math. Sinica **31**, 289–298.

———, [1988d], *Problems about continuous selections in $C(X)(IV)$: Characteristic description*, Acta Math. Sinica **31**, 299–308.

———, [1990], *Lipschitz continuous metric selections in $C_0(T)$*, SIAM J. Math. Anal. **21**, 205–220.

———, [1991], *Continuous Selections for Metric Projections and Interpolating Subspaces,*, "Approximation and Optimization" (B. Brosowski, F. Deutsch, and J. Guddat, eds.), Peter Lang, Frankfurt, Vol. 1, pp. 1–108.

———, [1991a], *Various continuities of metric projections in* $L_1(T, \mu)$, Progress in Approximation Theory, 583–607, Academic Press, New York.

P. K. Lin, [1985], *Remarks on linear selections for the metric projection*, J. Approx. Theory **43**, 64–74.

J. Lindenstrauss and L. Tzafriri, [1971], *On the complemented subspace problem*, Israel J. Math. **9**, 263–269.

G. Nürnberger, [1980], *Nonexistence of continuous selections for the metric projection*, SIAM J. Math. Anal. **11**, 460–467.

G. Nürnberger and M. Sommer, [1978], *Characterization of continuous selection for spline functions*, J. Approx. Theory **22**, 320–330.

———, [1984], *Continuous selections in Chebyshev approximation in "Parametric Optimization and Approximation" (B. Brosowski and F. Deutsch, eds.), ISNM,72*, Birkhäuser Verlag, Basel.

S.-H. Park, [1987], *Lipschitz continuous metric projections and selections*, doctoral dissertation, The Pennsylvania State University.

J. R. Respess and E. W. Cheney, [1982], *On Lipschitzian proximity maps in "Nonlinear Analysis and Applications" (ed. by J. H. Burry and S. P. Singh), Vol. 80*, Lecture Notes in Pure and Applied Math., Dekker, New York.

W. Rudin and K. T. Smith, [1961], *Linearity of best approximation : A characterization of ellipsoids*, Proc. Nederl. Akad. Wet. Ser. A **64**, 97–103.

N. V. Zhivkov, [1989], *A characterization of reflexive spaces by means of continuous approximate selections for metric projections*, J. Approx. Theory **56**, 59–71.

DEPARTMENT OF MATHEMATICS
THE PENNSYLVANIA STATE UNIVERSITY
UNIVERSITY PARK, PA 16802
U.S.A.

WEIGHTED POLYNOMIALS

M. v. Golitschek
Institut für Angewandte Mathematik und Statistik
der Universität Würzburg
8700 Würzburg
Germany

ABSTRACT. Weighted polynomials are functions of the form $w(x)^n P_n(x)$, $n \geq 1$, where $w(x)$ is a nonnegative continuous function on a closed set $A \subset \mathbb{R}$, and the P_n are algebraic polynomials of degree $\leq n$. Our main interest is to characterize those functions f which are uniformly approximable on A by the weighted polynomials.

1. INTRODUCTION

For the real algebraic polynomials $P_n \in \mathcal{P}_n$ of degree $\leq n$ and a continuous non-negative function w on a closed set $A \subset \mathbb{R}$, we consider the *weighted polynomials* on A given by

$$w(x)^n P_n(x), \quad n = 1, 2, \ldots. \tag{1.1}$$

We always assume that w is positive on some subinterval of A and that, if A is unbounded,

$$xw(x) \to 0, \quad x \to \pm\infty, \quad x \in A. \tag{1.2}$$

Example 1. The Incomplete Polynomials :

$$A := [0,1] \text{ and } w(x) := x^\sigma, \quad \sigma > 0.$$

(See Lorentz (1977), Saff and Varga (1978), v.Golitschek (1980).)

Example 2. Jacobi weights :

$$A := [-1,1] \text{ and } w(x) := (1+x)^{\sigma_1}(1-x)^{\sigma_2}, \quad \sigma_1, \sigma_2 > 0.$$

S. P. Singh (ed.), Approximation Theory, Spline Functions and Applications, 139–161.

(See Lanchance, Saff, Varga (1979), Saff, Ullman, Varga (1980).)

Example 3. Exponential Weights :

$$A := \mathbb{R} \text{ and } w(x) := W_\alpha(x) := \exp(-|x|^\alpha), \quad \alpha > 0.$$

By the substitution $x = n^{-1/\alpha}y$, the weighted polynomials $W_\alpha(x)^n P_n(x)$ become $W_\alpha(y)Q_n(y)$, $Q_n \in \mathcal{P}_n$. The results on the weighted polynomials $W_\alpha(x)^n P_n(x)$ are essential in recent developments in the theory of orthogonal polynomials on $\mathbb{R}$, that is, of the polynomials $p_n(W_\alpha^2)$, $n = 0, 1, \ldots$, which satisfy

$$\int_{-\infty}^{\infty} W_\alpha(x)^2 p_m(W_\alpha^2, x) p_n(W_\alpha^2, x) dx = \delta_{mn} \ , \quad m, n = 0, 1, \cdots,$$

(see Mhaskar and Saff (1984), Lubinsky, Mhaskar, Saff (1988)).

A detailed description of the theory of weighted polynomials has been given recently by v.Golitschek, Lorentz, Makovoz (1991). In the present paper, we reproduce parts of this theory introducing new and elementary proofs. Our main target are Weierstrass type theorems for weighted polynomials: we want to know which functions $f \in C(A)$ are approximable by the weighted polynomials $w^n P_n$, $n \geq 1$.

2. PROPERTIES OF THE WEIGHTED CHEBYSHEV POLYNOMIALS

If A is unbounded, then (1.2) implies that $w(x)^n P_n(x)$ is small for $x \to \pm\infty$: for each $0 < \epsilon < 1$ there exists a compact set $A_0 = A_0(\epsilon) \subset A$ so that for all $P_n \in \mathcal{P}_n$ and $n \geq 1$,

$$w(x)^n |P_n(x)| \leq \epsilon^n \|w^n P_n\|_{C(A)}, \quad x \in A \setminus A_0. \tag{2.1}$$

Indeed, let $I := [x_0 - c, x_0 + c] \subset A$ be a compact interval where w is positive, hence $\lambda := \min\{w(x) : x \in I\} > 0$. For $\phi := w^n P_n$ with $\|\phi\|_{C(A)} = 1$ it follows that $|P_n(x)| \leq \lambda^{-n}$, $x \in I$, and thus

$$|P_n(x)| \leq \lambda^{-n} |C_n((x - x_0)/c)|, \quad x \notin I.$$

Here, C_n denotes the ordinary Chebyshev polynomial of degree n. It is known that

$$|C_n(x)| \leq (|x| + \sqrt{|x|^2 - 1})^n \leq (2|x|)^n, \quad |x| \geq 1,$$

hence

$$|P_n(x)| \le (2\lambda^{-1}c^{-1}|x - x_0|)^n, \quad x \notin I. \tag{2.2}$$

Now take $A_0 := [x_0 - c_0, x_0 + c_0] \cap A$ with c_0 so large that
$2\lambda^{-1}c^{-1}|x - x_0|w(x) \le \epsilon$ for $x \in A \setminus A_0$. ∎

As a corollary of (2.1), if A is unbounded and if $A_0 := A_0(1/2)$ is the compact set in (2.1) for $\epsilon = 1/2$, then

$$\|w^n P_n\|_{C(A)} = \|w^n P_n\|_{C(A_0)}, \quad \text{for all } P_n \in \mathcal{P}_n, \quad n \ge 1. \tag{2.3}$$

A key role in the theory of weighted polynomials is played by the *weighted Chebyshev polynomials*: there exists a unique monic polynomial $Q_{w,n} \in \mathcal{P}_n$ which solves the minimum problem

$$\|w^n Q_{w,n}\|_{C(A)} = \min_{p \in \mathcal{P}_{n-1}} \|w(x)^n (x^n - p(x))\|_{C(A)} =: E_{w,n}. \tag{2.4}$$

The polynomial $C_{w,n} := C^A_{w,n} := Q_{w,n}/E_{w,n}$ is called the *weighted Chebyshev polynomial on A*. We have

Theorem 2.1. *(i) The weighted Chebyshev polynomial $w^n C_{w,n}$ has $n+1$ extrema $\xi_0^{(n)} < \xi_1^{(n)} < \cdots < \xi_n^{(n)}$ in A with alternating sign,*

$$w(\xi_i^{(n)})^n C_{w,n}(\xi_i^{(n)}) = (-1)^{n-i}, \quad i = 0, \ldots, n.$$

(ii) For each $P_n \in \mathcal{P}_n$ with $\|w^n P_n\|_{C(A)} \le 1$, $|P_n(x)| \le |C_{w,n}(x)|$ for $x \le \xi_0^{(n)}$ or $x \ge \xi_n^{(n)}$.

Proof. Suppose that $w^n Q_{w,n}$ has an alternation set $\xi_0^{(n)} < \xi_1^{(n)} < \cdots < \xi_r^{(n)}$ in A of length $r + 1 \le n$, but no alternation set of length $\ge r + 2$. If there are more than one alternation set of length $r + 1$, we take the following among them: As $\xi_0^{(n)}$ we take the smallest extreme point of $w^n Q_{w,n}$ in A. For $j = 1, 2, \cdots, r$, $\xi_j^{(n)}$ is the smallest extreme point of $w^n Q_{w,n}$ with the properties $\xi_j^{(n)} > \xi_{j-1}^{(n)}$ and $\text{sign}(Q_{w,n}(\xi_j^{(n)})) = -\text{sign}(Q_{w,n}(\xi_{j-1}^{(n)}))$. Let η_j, $j = 1, \cdots, r$, be the largest zero of $Q_{w,n}$ in the interval $(\xi_{j-1}^{(n)}, \xi_j^{(n)})$. Define $P \in \mathcal{P}_{n-1}$ by

$$P(x) := \text{sign}\left(Q_{w,n}(\xi_r^{(n)})\right) \prod_{j=1}^{r} (x - \eta_j).$$

One can easily prove that $\|w^n(Q_{w,n} - \epsilon P)\|_{C(A)} < \|w^n Q_{w,n}\|_{C(A)}$ for small $\epsilon > 0$, a contradiction to the minimality of $Q_{w,n}$. Hence, any minimal polynomial $w^n Q_{w,n}$

has an alternation set of length at least $n+1$. It cannot be larger than $n+1$ since $Q_{w,n}$ has at most n zeros on $\mathbb{R}$.

Since $Q_{w,n}$ has the leading coefficient 1 and n zeros in $(\xi_0^{(n)}, \xi_n^{(n)})$, it has no other zeros, and $Q_{w,n}(x) \to +\infty$ as $x \to +\infty$. Therefore $Q_{w,n}(\xi_n^{(n)}) > 0$.

(ii) is an elementary consequence of (i). ∎

A closed subset B of A is called an *essential set for* w if

$$\|w^n P_n\|_{C(B)} = \|w^n P_n\|_{C(A)} \tag{2.5}$$

for each weighted polynomial $w^n P_n$, $P_n \in \mathcal{P}_n$. An essential set B_0 for w is called a *minimal essential set for* w if it is contained in any other essential set. Clearly, the minimal essential set B_0, if it exists, is compact by (2.3) and unique by its definition.

For $P_n \in \mathcal{P}_n$ let $E_n(P_n)$ denote the set of all extrema of $w^n P_n$ on A. Obviously, B_0 is contained in the closure of the union of all sets $E_n(P_n)$. Conversely,

Theorem 2.2. (v.Golitschek, Lorentz, Makovoz (1991)) *The minimal essential set B_0 for w exists and is equal to*

$$B_0 = \overline{\bigcup_{P_n \in \mathcal{P}_n} E_n(P_n)} \ . \tag{2.6}$$

Proof. Let Ω be the set on the right hand side of (2.6). Clearly, $B_0 \subseteq \Omega$ if B_0 exists.

(i) For each P_n, $E_{2n}(P_n^2) = E_n(P_n)$.

(ii) For each $P_n \geq 0$ on A, $n \geq 2$, and each $x_0 \in E_n(P_n)$, there exists a $\tilde{P} \in \mathcal{P}_n$ with $E_n(\tilde{P}) = \{x_0\}$. For the proof, one takes $\tilde{P}(x) := P_n(x) - \lambda(x - x_0)^2$, with sufficiently small $\lambda > 0$.

From these properties it is obvious that Ω is a subset of any essential set B for w, hence $B_0 = \Omega$. ∎

Corollary 2.3. *For any $t_0 \in B_0$, $\epsilon > 0$ and $\delta > 0$ there exist an $x_0 \in U$, $U := (t_0 - \delta, t_0 + \delta) \cap A$, an integer $m \geq 1$ and a polynomial $P_m \in \mathcal{P}_m$ with the properties*

$$w(x_0)^m P_m(x_0) = \|w^m P_m\|_{C(A)} = 1, \tag{2.7}$$

$$w(x)^m |P_m(x)| \leq \epsilon, \quad x \in A \setminus U. \tag{2.8}$$

Proof. By the last theorem there exist an $x_0 \in U$ and a polynomial P_n with the property (2.7), and $m := n$. By (ii) we may assume that $E_n(P_n) = \{x_0\}$. For a sufficiently large integer r and $m := rn$, the polynomial $P_m := P_n^r$ satisfies (2.7) and (2.8). ■

Corollary 2.4. *(i) w is positive on B_0.*
(ii) The weighted Chebyshev polynomials for the weight w on A and on B_0 are the same: $C_{w,n}^A = C_{w,n}^{B_0}$.
(iii) All extreme points of $w^n C_{w,n}^A$ lie in B_0, all zeros of $C_{w,n}^A$ lie in the convex hull B_0^* of B_0.

Proof. (i) Let $t_0 \in B_0$. By the last theorem, there exist a sequence $t_k \to t_0$ in B_0 and polynomials $P_{n_k} \in \mathcal{P}_{n_k}$ so that $w(t_k)^{n_k} P_{n_k}(t_k) = \|w^{n_k} P_{n_k}\|_{C(A)} = 1$.

Let $I = [x_0 - c, x_0 + c]$ be an interval where w is positive, and let $\lambda > 0$ be the minimum of w on I. We may take I and the sequence (t_k) so that they are disjoint. Then it follows from (2.2) that

$$w(t_k)^{-1} = |P_{n_k}(t_k)|^{1/n_k} \le 2\lambda^{-1} c^{-1} |t_k - x_0|.$$

Hence, for $k \to \infty$, $w(t_0)^{-1} \le 2\lambda^{-1} c^{-1} |t_0 - x_0|$ which implies that $w(t_0)$ is positive. (ii) and (iii) are obvious. ■

For $w = 1$ on $A = [-1, 1]$, the extreme points of the ordinary Chebyshev polynomials C_n, $n = 1, 2, \ldots$, on A are dense in A. One might conjecture that, stronger than in Theorem 2.2, B_0 is the closure of the extreme points of the weighted Chebyshev polynomials $w^n C_{w,n}$, $n \ge 1$. This has been shown by Mhaskar and Saff (1985) under restrictive assumptions on the weight function w. Their proofs use difficult methods from potential theory. In the next three theorems, we shall prove some of their results under more general assumptions, and with elementary tools.

Theorem 2.5. *Under the assumptions of §1, let $B_0^* = [a_0, b_0]$ be the convex hull of the minimal essential set B_0 for w. Then the least and the largest alternation points of the weighted Chebyshev polynomials $w^n C_{w,n}$ converge,*

$$\lim_{n\to\infty} \xi_0^{(n)} = a_0 \;, \quad \lim_{n\to\infty} \xi_n^{(n)} = b_0. \tag{2.9}$$

Proof. By Theorem 2.2, $\xi_0^{(n)} \ge a_0$, $\xi_n^{(n)} \le b_0$ for all $n \ge 1$. It suffices to prove the first of the relations (2.9).

Suppose to the contrary that $\xi_0^{n_k} \ge a_0 + \delta_0$ is valid for some $\delta_0 > 0$ and infinitely many n_k. Since $a_0 \in B_0$, we have $w(a_0) > 0$ by Corollary 2.4 (i) and thus, for some $0 < \delta \le \delta_0$ and $U := [a_0, a_0 + \delta) \cap A$, the number

$\gamma_0 := \inf\{1; w(x) : \ x \in U\}$ is positive. In addition, with $\epsilon := 1/8$, there exists an $x_0 \in U$, an integer $m \geq 1$ and a polynomial $P_m \in \mathcal{P}_m$ so that (2.7) and (2.8) are valid.

We define the positive number $\gamma_1 := \max\{1; \|w\|_{C(A)}\}$. Observe that m, γ_0 and γ_1 are independent of n.

Let $n = n_k$ be large. Let $r \in \mathbb{N}$ and $0 \leq q \leq m-1$ be defined by $n = rm+q$. Then we have

$$w(x_0)^n P_m(x_0)^r \geq \gamma_0^{m-1} w(x_0)^{rm} P_m(x_0)^r = \gamma_0^{m-1}.$$

For any $x \in A \setminus U$,

$$|w(x)^n P_m(x)^r| \leq \gamma_1^{m-1} |w(x)^{rm} P_m(x)^r| \leq \gamma_1^{m-1} 8^{-r} \leq 8\gamma_1^{m-1} \rho^{-n},$$

where $\rho := 8^{1/m} > 1$ is independent of n. We consider the polynomials

$$R_n := C_{w,n} - 2\gamma_0^{1-m} P_m^r \operatorname{sign}(C_{w,n}(\xi_0^{(n)})).$$

For large $n = n_k$ they satisfy

$$\operatorname{sign}(R_n(\xi_i^{(n)})) = \operatorname{sign}(C_{w,n}(\xi_i^{(n)})), \quad i = 0, 1, \ldots, n,$$

and $\operatorname{sign}(R_n(x_0)) = -\operatorname{sign}(C_{w,n}(\xi_0^{(n)}))$. Since $x_0 < \xi_0^{(n)}$, R_n has $n+1$ real zeros, a contradiction. ∎

The minimal essential set B_0 can have isolated points, for example if A contains such points and w is there positive. An important subset of B_0 is

$$B_{00} := \{x \in B_0 : x \text{ is not an isolated point of } B_0\}.$$

Theorem 2.6. *Under the assumptions of §1,*

$$\lim_{n\to\infty} \max_{x\in B_{00}} \min_{0\leq i\leq n} |x - \xi_i^{(n)}| = 0. \tag{2.10}$$

Proof. Suppose to the contrary that there exist some $\delta_0 > 0$, some increasing subsequence n_k, $k = 1, 2, \ldots$, and points $t_k \in B_{00}$ for which

$$\min_{0\leq i\leq n_k} |t_k - \xi_i^{(n_k)}| \geq \delta_0.$$

The set B_{00} is compact. Hence, taking a subsequence of the n_k, if necessary, we may suppose that the sequence t_k converges to some point $\tau_0 \in B_{00}$. In addition, since τ_0 is not isolated in B_0, there exists $\tau_1 \in B_0$ with $0 < |\tau_1 - \tau_0| < \delta_0/2$. We may assume that $\tau_0 < \tau_1$. We set $\delta := (\tau_1 - \tau_0)/4$ and define the two open disjoint intervals $U_\nu := (\tau_\nu - \delta, \tau_\nu + \delta)$, $\nu = 0, 1$. It follows that for sufficiently large $n = n_k$ all alternation points $\xi_i^{(n)}$, $i = 0, 1, \ldots, n$, lie outside of $J := [\tau_0 - \delta, \tau_1 + \delta]$.

By Corollary 2.3, there exist points $x_\nu \in U_\nu$, $\nu = 0, 1$, and polynomials $P_{m_\nu} \in \mathcal{P}_{m_\nu}$ with the properties (2.7) and (2.8), with $\epsilon := 1/8$, $U := U_\nu$, $m := m_\nu$. We may assume that $m := m_0 = m_1$. (Otherwise we replace P_{m_0} by $P_{m_0}^{m_1}$ and P_{m_1} by $P_{m_1}^{m_0}$ and have $m := m_0 m_1$.) For large $n = n_k$ and $r := [n/m]$ we consider the polynomials

$$R_n^+ := C_{w,n} + 2(P_{m_0}^r - P_{m_1}^r), \quad R_n^- := C_{w,n} - 2(P_{m_0}^r - P_{m_1}^r).$$

Using the arguments of the proof of Theorem 2.5 it follows that at least one of the two polynomials R_n^+ and R_n^- has $n+2$ real zeros, a contradiction. ■

If $t_0 \in B_0$ is some interior point of A, and if w is differentiable at t_0, then t_0 is not isolated in B_0. Indeed, let $x_0 := t_0$ and P_m be as in Corollary 2.3, with some $\epsilon > 0$, $\delta > 0$, and let $B_0 \cap (x_0 - \delta, x_0 + \delta) = \{x_0\}$. Then, for all sufficiently small $\gamma > 0$,

$$\phi_1(x) := w(x)^m (P_m(x) \pm \gamma), \quad \phi_2(x) := w(x)^m (P_m(x) \pm \gamma x)$$

must have their unique extrema at x_0, hence

$$\phi_1'(x_0) = m w(x_0)^{m-1} w'(x_0) \{P_m(x_0) \pm \gamma\} + w(x_0)^m P_m'(x_0) = 0.$$

This implies that $w'(x_0) = 0$, $P_m'(x_0) = 0$ and $\phi_2'(x_0) = \pm\gamma w(x_0)^m$. From $\phi_2'(x_0) = 0$ we get $w(x_0) = 0$, a contradiction to Corollary 2.4(i). ■

In many important cases the function

$$q(x) := -\log w(x), \quad x \in A$$

is convex on A. For example, if
(i) $w(x) = x^\sigma$, $\sigma > 0$ on $A := [0, 1]$,
(ii) $w(x) = (1+x)^{\sigma_1}(1-x)^{\sigma_2}$, $\sigma_1, \sigma_2 > 0$ on $A := [-1, 1]$,
(iii) $w(x) = e^{-|x|^\alpha}$, $\alpha \geq 1$, on $A = \mathbb{R}$ or on $A = [0, \infty)$.

Then one knows (Mhaskar and Saff (1985)) that the minimal essential set B_0 is an interval. We shall prove this important result in a simple way:

Theorem 2.7. *If A is an interval and if $q := -\log w$ is convex on A, then the minimal essential set B_0 for w is a compact interval.*

Proof. Suppose that B_0 is not an interval. There exist then two point $\eta_0 < \eta_1$ in B_0 so that the open interval $J := (\eta_0, \eta_1)$ is disjoint with B_0. It follows from Theorems 2.2 and 2.5 that for all large n and some $j = j(n)$, $0 \le j \le n-1$,

$$\xi_j^{(n)} \le \eta_0 < \eta_1 \le \xi_{j+1}^{(n)}. \tag{2.11}$$

Let $x_j^{(n)}$ be the unique zero of $C_{w,n}$ in the interval $J_n := (\xi_j^{(n)}, \xi_{j+1}^{(n)})$ and let $Q_{w,n} = E_{w,n}C_{w,n}$ be the extreme monic polynomial in (2.4). We define the functions

$$p_n(x) := \frac{1}{n} \sum_{i=1, i\neq j}^{n} \log |x - x_i^{(n)}| \quad , \quad v_n(x) := -q(x) + p_n(x).$$

Obviously,

$$w(x)^n |Q_{w,n}(x)| = |x - x_j^{(n)}| e^{n v_n(x)}. \tag{2.12}$$

By differentiation, we get for $x \in J_n$

$$p_n''(x) = -\frac{1}{n} \sum_{i=1, i\neq j}^{n} \frac{1}{(x - x_i^{(n)})^2} \le -\frac{1}{2(b_0 - a_0)^2} =: -\kappa, \tag{2.13}$$

where $[a_0, b_0] := B_0^*$ is the convex hull of B_0. In addition, for $r := j$ and $r := j+1$, it follows from (2.12) and from $w(\xi_r^{(n)})^n Q_{w,n}(\xi_r^{(n)}) = \pm E_{w,n}$ that

$$\begin{aligned} v_n(\xi_r^{(n)}) &= \frac{1}{n} \log E_{w,n} - \frac{1}{n} \log |\xi_r^{(n)} - x_j^{(n)}| \\ &\ge \frac{1}{n} \log E_{w,n} - \frac{1}{n} \log (b_0 - a_0) =: \lambda_n. \end{aligned} \tag{2.14}$$

Let $u_n \in \mathcal{P}_2$ be the quadratic polynomial which satisfies $u_n'' = -\kappa$ and $u_n(\xi_j^{(n)}) = u_n(\xi_{j+1}^{(n)}) = 0$. There exists a positive number ρ (independent of n) and some subinterval $I_n \subset J_n$ of length $\ge (\eta_1 - \eta_0)/2$ so that $u_n(x) \ge \rho$, $x \in I_n$. Since q is convex, and by (2.13) and (2.14), $v_n(x) \ge \lambda_n + u_n(x)$ for $x \in J_n$, hence

$$v_n(x) \ge \lambda_n + \rho, \quad x \in I_n.$$

This implies that

$$w(x)^n |Q_{w,n}(x)| \ge |x - x_j^{(n)}| e^{n\lambda_n + n\rho}, \quad x \in I_n$$

and thus

$$\|w^n Q_{w,n}\|_{C(I_n)} \geq \frac{\eta_1 - \eta_0}{4(b_0 - a_0)} E_{w,n} e^{n\rho}.$$

This is impossible for large n, a contradiction. ∎

Theorem 2.8. *Let $A = [a, b]$ be a compact interval and let $q := -\log w$ be convex on A. Then, uniformly and geometrically on each compact subinterval of $A \setminus B_0$,*

$$\lim_{n\to\infty} w(x)^n C_{w,n}(x) = 0. \tag{2.15}$$

Proof. By the last theorem, the minimal essential set B_0 for w is an interval, $B_0 = [a_0, b_0]$. We assume that $b_0 < b$. Let $n \geq 1$ be fixed and let $\xi_n^{(n)} \leq x \leq b$. The function

$$r_n(x) := \frac{1}{n} \log |Q_{w,n}(x)| = \frac{1}{n} \sum_{i=1}^{n} \log |x - x_i^{(n)}|$$

satisfies

$$r_n''(x) = -\frac{1}{n} \sum_{i=1}^{n} \frac{1}{(x - x_i^{(n)})^2} \leq -\frac{1}{(b - a_0)^2} =: -2\gamma < 0, \quad \xi_n^{(n)} \leq x \leq b.$$

Since

$$\phi_n(x) := \frac{1}{n} \log |w(x)^n Q_{w,n}(x)| \leq \frac{1}{n} \log |w(\xi_n^{(n)})^n Q_{w,n}(\xi_n^{(n)})| = \frac{1}{n} \log E_{w,n}$$

and since q is convex, it follows that

$$\phi_n(x) \leq \frac{1}{n} \log E_{w,n} - \gamma (x - \xi_n^{(n)})^2, \quad \xi_n^{(n)} \leq x \leq b,$$

and thus that

$$w(x)^n |Q_{w,n}(x)| \leq E_{w,n} e^{-\gamma(x - b_0)^2}, \quad b_0 \leq x \leq b. \tag{2.16}$$

The proof of (2.16) for $a \leq x \leq a_0$ is similar. ∎

Because of (2.3), Theorem 2.8 is also valid if $A = \mathbb{R}$ or $A = [0, \infty)$ and if w satisfies (1.2).

3. A WEIERSTRASS THEOREM FOR ANALYTIC WEIGHTS

In this section we assume that
(i) $A = [a, b]$ is a compact interval,
(ii) $q := -\log w$ is convex on A and
(iii) q is analytic on the minimal essential set B_0 for w.

We want to know which functions $f \in C(A)$ are approximable on A by the weighted polynomials $w^n P_n$, $n \geq 1$.

By Theorem 2.7, $B_0 = [a_0, b_0]$ is an interval, and by Theorems 2.8 and 2.1, each sequence of weighted polynomials $(w^n P_n)_1^\infty$ which is uniformly bounded on A, converges to zero outside of B_0. Hence, a necessary condition for the approximability is

$$f(x) = 0, \quad x \in A \setminus B_0. \tag{3.1}$$

Theorem 3.1. *Under the assumptions (i)-(iii), each function $f \in C(A)$ which vanishes outside of (a_0, b_0), satisfies*

$$\lim_{n\to\infty} \min_{P_n \in \mathcal{P}_n} \|f - w^n P_n\|_{C(A)} = 0. \tag{3.2}$$

Proof. The assumption (iii) implies that, for some $\delta > 0$, q is even analytic on the rectangle

$$\mathcal{R} := \{z = x + iy : a_0 \leq x \leq b_0,\ -\delta \leq y \leq \delta\}. \tag{3.3}$$

Hence, for $x \in B_0$ and $-\delta \leq y \leq \delta$, all partial derivatives of $u(x,y) := \Re e\ q(x+iy)$ exist and are continuous. In addition, since

$$\frac{\partial u}{\partial y}(x,0) = \frac{\partial^3 u}{\partial y^3}(x,0) = 0, \quad \frac{\partial^2 u}{\partial y^2}(x,0) = -q''(x), \quad x \in B_0$$

it follows by the convexity of q on B_0 that

$$\Re e\ q(x+iy) - q(x) \leq K\ y^4\ , \quad x + iy \in \mathcal{R}, \tag{3.4}$$

where $K := \frac{1}{4!}\|\frac{\partial^4 u}{\partial y^4}\|_{C(\mathcal{R})}$.

We may assume that $\delta > 0$ is so small that $16K\delta^2(b_0 - a_0)^2 \leq 1$ and

$$\log\left(1 + \frac{\delta^2}{2(b_0 - a_0)^2}\right) \geq \frac{\delta^2}{4(b_0 - a_0)^2}.$$

For $n \geq 1$, the $n+1$ extrema $\{\xi_i^{(n)}\}_{i=0}^n$ of $w^n C_{w,n}$ lie in B_0 and the n zeros $\{x_i^{(n)}\}_{i=1}^n$ of $C_{w,n}$ and the extreme points interlace, that is,

$$a_0 \leq \xi_0^{(n)} < x_1^{(n)} < \xi_1^{(n)} < \cdots < x_n^{(n)} < \xi_n^{(n)} \leq b_0. \tag{3.5}$$

In addition, we have (2.9) and $B_{00} = B_0$, hence the application of Theorem 2.6 yields that the maximal distance $h_n := \max\{\xi_i^{(n)} - \xi_{i-1}^{(n)} : i = 1, \ldots, n\}$ satisfies

$$\lim_{n\to\infty} h_n = 0 \ . \tag{3.6}$$

Let $\epsilon > 0$ be fixed. Since $f(a_0) = f(b_0) = 0$, there exists an entire function $g = g(f, \epsilon)$ so that

$$\|f(x) - (x - a_0)(x - b_0)g(x)\|_{C(B_0)} < \epsilon/2.$$

By (2.9) it follows that $\|(x - a_0)(x - b_0) - (x - \xi_0^{(n)})(x - \xi_n^{(n)})\|_{C(B_0)} \to 0$. Hence the entire functions $f_n(x) := (x - \xi_0^{(n)})(x - \xi_n^{(n)})g(x)$ satisfy

$$\|f - f_n\|_{C(B_0)} < \epsilon, \quad n \geq n_0 = n_0(f, \epsilon). \tag{3.7}$$

The rectangles $\mathcal{R}_n := \{z = x + iy : \xi_0^{(n)} \leq x \leq \xi_n^{(n)}, \ -\delta \leq y \leq \delta\}$, are contained in $\mathcal{R}$. Their boundary curves are denoted by Γ_n. They intersect the x-axis in the extreme points $\xi_0^{(n)}$ and $\xi_n^{(n)}$ of $w^n C_{w,n}$. It follows from the definitions of f_n and Γ_n that

$$\frac{|f_n(\eta)|}{|\eta - x|} \leq C \ \text{ for all } \ \eta \in \Gamma_n, \ \ x \in A, \tag{3.8}$$

with a constant C independent of n.

Next we introduce the functions

$$H_n := f_n/w^n \ , \ \ G_n(x, \eta) := \frac{H_n(\eta)\{C_{w,n}(\eta) - C_{w,n}(x)\}}{(\eta - x)C_{w,n}(\eta)} \ .$$

H_n is analytic on $\mathcal{R}$. For fixed $x \in \mathbb{R}$, $G_n(x, \eta)$ is an analytic function of η on $\mathcal{R} \setminus \{x_1^{(n)}; \ldots; x_n^{(n)}\}$. At the $x_j^{(n)}$, $G_n(x, \ )$ has poles with the residues

$$\operatorname{Res}(G_n(x, \ ), x_j^{(n)}) = H_n(x_j^{(n)}) \prod_{k=1, k\neq j}^{n} \frac{x - x_k^{(n)}}{x_j^{(n)} - x_k^{(n)}}.$$

By the Residue Theorem

$$\frac{1}{2\pi i}\int_{\Gamma_n} G_n(x,\eta)d\eta = \sum_{j=1}^{n} \mathrm{Res}(G_n(x,\ \ ), x_j^{(n)})$$
$$= \sum_{j=1}^{n} H_n(x_j^{(n)}) \prod_{k=1,k\neq j}^{n} \frac{x - x_k^{(n)}}{x_j^{(n)} - x_k^{(n)}} =: L_n(x).$$

We realize that L_n is the polynomial of degree $\leq n-1$ which interpolates H_n at the n zeros $x_j^{(n)}$ of $C_{w,n}$.

For $\xi_0^{(n)} < x < \xi_n^{(n)}$ Cauchy's integral formula yields

$$H_n(x) = \frac{1}{2\pi i}\int_{\Gamma_n} \frac{H_n(\eta)d\eta}{\eta - x} = \frac{1}{2\pi i}\int_{\Gamma_n} \frac{H_n(\eta)C_{w,n}(\eta)d\eta}{(\eta - x)C_{w,n}(\eta)}$$

and thus

$$L_n(x) - H_n(x) = \frac{1}{2\pi i}\int_{\Gamma_n} G_n(x,\eta)d\eta - \frac{1}{2\pi i}\int_{\Gamma_n} \frac{H_n(\eta)C_{w,n}(\eta)d\eta}{(\eta - x)C_{w,n}(\eta)}$$
$$= \frac{1}{2\pi i}\int_{\Gamma_n} \frac{H_n(\eta)C_{w,n}(x)d\eta}{(x-\eta)C_{w,n}(\eta)} \ . \tag{3.9}$$

If $x > \xi_n^{(n)}$ or $x < \xi_0^{(n)}$ then

$$\frac{1}{2\pi i}\int_{\Gamma_n} \frac{H_n(\eta)C_{w,n}(\eta)d\eta}{(\eta - x)C_{w,n}(\eta)} = 0,$$

hence

$$L_n(x) = \frac{1}{2\pi i}\int_{\Gamma_n} G_n(x,\eta)d\eta = \frac{1}{2\pi i}\int_{\Gamma_n} \frac{H_n(\eta)C_{w,n}(x)d\eta}{(x-\eta)C_{w,n}(\eta)} \ . \tag{3.10}$$

We set

$$M_n := \frac{1}{2\pi}\int_{\Gamma_n} \frac{|d\eta|}{|w(\eta)^n C_{w,n}(\eta)|} \ .$$

Since $\|w^n C_{w,n}\|_{C(A)} = 1$ it follows from (3.8), (3.9) and (3.10) that

$$|w(x)^n L_n(x) - f_n(x)| \leq \frac{1}{2\pi}\int_{\Gamma_n} \frac{|H_n(\eta)||d\eta|}{|x-\eta||C_{w,n}(\eta)|}$$
$$\leq CM_n \ , \ \xi_0^{(n)} \leq x \leq \xi_n^{(n)}, \tag{3.11}$$

$$|w(x)^n L_n(x)| \leq CM_n \ , \ x \leq \xi_0^{(n)}, \ x \geq \xi_n^{(n)}, \ x \in A. \tag{3.12}$$

Hence the relation (3.2) is true if

$$\lim_{n\to\infty} M_n = 0. \tag{3.13}$$

It remains to prove (3.13): We take $n \in \mathbb{N}$ so large that $h_n \le \delta^2/(4(b_0 - a_0))$.

Let $\xi_{j-1}^{(n)} \le x \le \xi_j^{(n)}$, $j = 1, 2, \ldots, n$. For each zero $x_k^{(n)}$ of $C_{w,n}$,

$$\frac{|x - x_k^{(n)} \pm i\delta|^2}{|\xi_j^{(n)} - x_k^{(n)}|^2} = \frac{(x - x_k^{(n)})^2 + \delta^2}{(\xi_j^{(n)} - x_k^{(n)})^2} \ge \frac{(\xi_j^{(n)} - x_k^{(n)})^2 - 2h_n(b_0 - a_0) + \delta^2}{(\xi_j^{(n)} - x_k^{(n)})^2}$$

$$\ge \frac{(\xi_j^{(n)} - x_k^{(n)})^2 + \delta^2/2}{(\xi_j^{(n)} - x_k^{(n)})^2} \ge 1 + \frac{\delta^2}{2(b_0 - a_0)^2}$$

and thus

$$\frac{|C_{w,n}(\xi_j^{(n)})|}{|C_{w,n}(x \pm i\delta)|} = \prod_{k=1}^{n} \frac{|\xi_j^{(n)} - x_k^{(n)}|}{|x - x_k^{(n)} \pm i\delta|} \le \left(1 + \frac{\delta^2}{2(b_0 - a_0)^2}\right)^{-n/2}.$$

Similarly, for $j = 0$, $j = n$ and $-\delta \le y \le \delta$,

$$\frac{|C_{w,n}(\xi_j^{(n)})|}{|C_{w,n}(\xi_j^{(n)} + iy)|} = \prod_{k=1}^{n} \frac{|\xi_j^{(n)} - x_k^{(n)}|}{|\xi_j^{(n)} - x_k^{(n)} + iy|} \le \left(1 + \frac{y^2}{2(b_0 - a_0)^2}\right)^{-n/2}.$$

Next we apply (3.4): for $a_0 \le x \le b_0$, $0 \le y \le \delta$, one has

$$\frac{w(x)}{|w(x \pm iy)|} = |\exp\left(q(x \pm iy) - q(x)\right)| \le \exp\left(Ky^4\right). \tag{3.14}$$

In addition, for $\xi_{j-1}^{(n)} \le x \le \xi_j^{(n)}$, $j = 1, 2, \cdots, n$,

$$\frac{w(\xi_j^{(n)})}{w(x)} = \exp\left(q(x) - q(\xi_j^{(n)})\right)| \le \exp\left(\omega(q; h_n)\right) \tag{3.15}$$

where $\omega(q; \;)$ is the modulus of continuity of q on $B_0 = [a_0, b_0]$. Hence, for $\xi_{j-1}^{(n)} \le x \le \xi_j^{(n)}$, $j = 1, 2, \cdots$,

$$\frac{1}{|w(x \pm i\delta)^n C_{w,n}(x \pm i\delta)|} = \frac{|w(\xi_j^{(n)})^n C_{w,n}(\xi_j^{(n)})|}{|w(x \pm i\delta)^n C_{w,n}(x \pm i\delta)|}$$

$$\le \left(1 + \frac{\delta^2}{2(b_0 - a_0)^2}\right)^{-n/2} \exp\{nK\delta^4 + n\omega(q; h_n)\}$$

$$\le \exp\{nK\delta^4 + n\omega(q; h_n) - \frac{n\delta^2}{8(b_0 - a_0)^2}\}$$

$$\le \exp\{n\omega(q; h_n) - \frac{n\delta^2}{16(b_0 - a_0)^2}\}.$$

Similarly, for $j=0$, $j=n$ and $0\le y\le\delta$,

$$\frac{1}{|w(\xi_j^{(n)}\pm iy)^n C_{w,n}(\xi_j^{(n)}\pm iy)|}=\frac{|w(\xi_j^{(n)})^n C_{w,n}(\xi_j^{(n)})|}{|w(\xi_j^{(n)}\pm iy)^n C_{w,n}(\xi_j^{(n)}\pm iy)|}$$

$$\le \exp(nKy^4)\left(1+\frac{y^2}{2(b_0-a_0)^2}\right)^{-n/2}\le \exp\{-\frac{ny^2}{16(b_0-a_0)^2}\}.$$

The last two relations imply (3.13) since $h_n\to 0$ and thus $\omega(q;h_n)\to 0$. ■

In our proof we have used the Hermite remainder formula (3.9) for Lagrange interpolation. Earlier, Lubinsky and Saff (1988) applied this formula to prove Weierstrass theorems for the weighted polynomials with exponential weights $w(x)=e^{-|x|^\alpha}$, $\alpha>0$, on $A=\mathbb{R}$ and He and Li (1989) for the Jacobi weights.

4. WEIERSTRASS THEOREMS FOR THE EXPONENTIAL WEIGHTS

In this section we prove *Weierstrass Theorems* for the exponential weights
$W_\alpha(x):=\exp(-|x|^\alpha)$, $\alpha>0$, on $A:=[0,\infty)$ and $A:=\mathbb{R}$.
In the proof of the next theorem we use again Hermite's remainder formula (3.9) for Lagrange interpolation; the paths of integration Γ_n will be the boundaries of sectors of disks of radius $\xi_n^{(n)}$ and center 0. These sectors are contained in the Riemann surface $\mathbb{C}_\alpha$ of the analytic function $q(z)=z^\alpha$, where $q(0)=0$ and $q(x)=-\log W_\alpha(x)$ for $x>0$.

Theorem 4.1. *Let $A:=[0,\infty)$ and $\alpha>0$. (i) The minimal essential set B_0 for $w:=W_\alpha$ is an interval of the form $B_0=[0,b_\alpha^*]$, for some b_α^* which is given later. (ii) For each continuous function f on A, with $f(0)=0$ and $f(x)=0$ for $x\ge b_\alpha^*$, one has*

$$\lim_{n\to\infty}\min_{P_n\in\mathcal{P}_n}\|f-W_\alpha^n P_n\|_{C[0,\infty)}=0. \tag{4.1}$$

Proof. (i) Since $w=W_\alpha>0$ has its maximum at 0, the weighted polynomial w^nP_n, $P_n=1$, has its unique extremum at zero, hence $0\in B_0$ and the convex hull of B_0 is of the form $B_0^*=[0,b_\alpha^*]$, for some $b_\alpha^*>0$. If $\alpha\ge 1$, q is convex on A, hence $B_0=B_0^*$ by Theorem 2.7. If $0<\alpha<1$, we will first prove (ii). From (ii) we then conclude easily that $B_0=B_0^*$: take any point $x_0\in(0,b_\alpha^*)$ and a continuous function $f\in C(A)$ which has a peak at x_0 and is zero outside of a small neighborhood U_δ of x_0. For large n, the minimal weighted polynomials in (4.1) have their extrema in this neighborhood. Since $\delta>0$ is arbitrary, $x_0\in B_0$ by Theorem 2.2.

(ii) Let $\alpha > 0$ and $w := W_\alpha$. Let $0 \le \xi_0^{(n)} < x_1^{(n)} < \ldots < \xi_n^{(n)} \le b_\alpha^*$ be the alternation points and zeros of the weighted Chebyshev polynomials $w^n C_{w,n}$, $n \ge 1$. We claim that

$$\xi_0^{(n)} = 0 \ , \ \lim_{n\to\infty} \xi_n^{(n)} = b_\alpha^*. \tag{4.2}$$

The second relation follows from Theorem 2.5. Since $|C_{w,n}|$ and w are strictly monotone decreasing on $[0, \xi_0^{(n)}]$ if $\xi_0^{(n)} > 0$,

$$w(0)^n |C_{w,n}(0)| > w(\xi_0^{(n)})^n |C_{w,n}(\xi_0^{(n)})| = 1,$$

a contradiction. This yields the first relation in (4.2).

Let $\epsilon > 0$ be fixed. Let $g = g(f, \epsilon)$ be an entire function which satisfies $\|f - x(x - b_\alpha^*)g(x)\|_{C(B_0^*)} < \epsilon/2$ and let $f_n(z) := z(z - \xi_n^{(n)})g(z)$. Then, for large n,

$$\|f - f_n\|_{C(B_0^*)} < \epsilon. \tag{4.3}$$

In addition, there exists some $C > 0$ independent of n, x, r, t, so that for all $x \ge 0$

$$\begin{aligned} &|f_n(\xi_n^{(n)} e^{it})| \le C|\xi_n^{(n)} e^{it} - x| \ , \ t \in \mathbb{R}, \\ &|f_n(re^{it})| \le C|re^{it} - x| \ , \ 0 < r \le b_\alpha^*, \ \pi/2 \le t \le 3\pi/2. \end{aligned} \tag{4.4}$$

1. Let $\alpha \ge 1/2$. We define the positive number ϕ by $\phi := \pi/(2\alpha)$ for $1/2 \le \alpha \le 1$, $\phi := \pi/2$ for $1 < \alpha \le 3$, $\phi := (2k+1)\pi/(2\alpha)$ for $2k - 1 \le \alpha \le 2k+1$, $k = 2, 3, \ldots$. This implies that

$$\pi/2 \le \phi \le \pi \ , \ \cos(\alpha\phi) \le 0. \tag{4.5}$$

The path of integration is $\Gamma_n := \Gamma_n^1 \cup \Gamma_n^2 \cup \Gamma_n^3 \subset \mathbb{C}_\alpha$ where

$$\begin{aligned} \Gamma_n^1 &:= \{\eta(t) = \xi_n^{(n)} e^{it} \ , \ -\phi \le t \le \phi\}, \\ \Gamma_n^2 &:= \{\eta(t) = re^{i\phi} \ , \ \xi_n^{(n)} \ge r \ge 0\}, \\ \Gamma_n^3 &:= \{\eta(t) = re^{-i\phi} \ , \ 0 \le r \le \xi_n^{(n)}\}. \end{aligned}$$

We proceed as in the proof of Theorem 3.1: we take $H_n := f_n/w^n$, and we take the algebraic polynomial L_n of degree $\le n - 1$ which interpolates H_n at the n zeros $x_j^{(n)}$ of $C_{w,n}$. Then we have (as in (3.9))

$$L_n(x) - H_n(x) = \frac{1}{2\pi i} \int_{\Gamma_n} \frac{H_n(\eta) C_{w,n}(x) d\eta}{(x - \eta) C_{w,n}(\eta)} \tag{4.6}$$

for all $0 < x < \xi_n^{(n)}$, that is, since $\|w^n C_{w,n}\|_{C[0,\infty)} = 1$,

$$|w(x)^n L_n(x) - f_n(x)| \leq \frac{1}{2\pi} \left| \int_{\Gamma_n} \frac{H_n(\eta) d\eta}{(x-\eta) C_{w,n}(\eta)} \right|. \tag{4.7}$$

Similarly, we obtain (as in (3.10))

$$L_n(x) = \frac{1}{2\pi i} \int_{\Gamma_n} \frac{H_n(\eta) C_{w,n}(x) d\eta}{(x-\eta) C_{w,n}(\eta)} \tag{4.8}$$

for all $x > \xi_n^{(n)}$, hence

$$|w(x)^n L_n(x)| \leq \frac{1}{2\pi} \left| \int_{\Gamma_n} \frac{H_n(\eta) d\eta}{(x-\eta) C_{w,n}(\eta)} \right|. \tag{4.9}$$

We have $w(0) = |C_{w,n}(0)| = 1$, $\pi/2 \leq \phi \leq \pi$ and thus, for $r \geq 0$,

$$\begin{aligned} \frac{1}{|C_{w,n}(re^{\pm i\phi})|} &= \frac{|C_{w,n}(0)|}{|C_{w,n}(re^{\pm i\phi})|} \leq \prod_{k=1}^{n} \frac{x_k^{(n)}}{|ir + x_k^{(n)}|} \\ &\leq \left(1 + (r/b_\alpha^*)^2\right)^{-n/2}. \end{aligned} \tag{4.10}$$

From (4.5) it follows that

$$|w(re^{\pm i\phi})|^{-1} = |\exp(r^\alpha e^{\pm i\alpha\phi})| = \exp(r^\alpha \cos(\alpha\phi)) \leq 1. \tag{4.11}$$

The inequalities (4.10), (4.4) and (4.11) imply that

$$\begin{aligned} &\sup_{x \geq 0} \left| \int_{\Gamma_n^2 \cup \Gamma_n^3} \frac{H_n(\eta) d\eta}{(x-\eta) C_{w,n}(\eta)} \right| \\ &\leq 2 \sup_{x \geq 0} \int_{\Gamma_n^2} \frac{|f_n(\eta)||d\eta|}{|x-\eta||w(\eta)^n C_{w,n}(\eta)|} \\ &\leq 2C \int_0^{\xi_n^{(n)}} \left(1 + (r/b_\alpha^*)^2\right)^{-n/2} dr \to 0, \end{aligned} \tag{4.12}$$

as $n \to \infty$. For $\eta = \xi_n^{(n)} e^{it} \in \Gamma_n^1$ we have

$$\frac{|C_{w,n}(\xi_n^{(n)})|}{|C_{w,n}(\eta)|} = \prod_{k=1}^{n} \frac{|\xi_n^{(n)} - x_k^{(n)}|}{|\xi_n^{(n)} e^{it} - x_k^{(n)}|} \leq 1. \tag{4.13}$$

Hence we get for all $x \geq 0$

$$\begin{aligned}\left|\int_{\Gamma_n^1} \frac{H_n(\eta)d\eta}{(x-\eta)C_{w,n}(\eta)}\right| &\leq \int_{\Gamma_n^1} \frac{|H_n(\eta)|w(\xi_n^{(n)})^n C_{w,n}(\xi_n^{(n)})|d\eta|}{|(x-\eta)C_{w,n}(\eta)|} \\ &\leq \int_{\Gamma_n^1} \frac{|f_n(\eta)|}{|x-\eta|}\frac{w(\xi_n^{(n)})^n|d\eta|}{|w(\eta)|^n} \\ &\leq C\xi_n^{(n)}\int_{-\phi}^{\phi} |\exp\{n(\xi_n^{(n)})^\alpha(e^{i\alpha t}-1)\}|dt \\ &= C\xi_n^{(n)}\int_{-\phi}^{\phi} \exp\{-2n(\xi_n^{(n)})^\alpha \sin^2(\alpha t/2)\}dt.\end{aligned}$$

This yields

$$\sup_{x\geq 0}\left|\int_{\Gamma_n^1} \frac{H_n(\eta)d\eta}{(x-\eta)C_{w,n}(\eta)}\right| \to 0, \tag{4.14}$$

as $n \to \infty$. Since $\epsilon > 0$ is arbitrary, we get (4.1) by combining the above inequalities (4.3), (4.7), (4.9), (4.12) and (4.14).

2. Let $0 < \alpha < 1/2$. This case is only a little more difficult. The path of integration Γ_n in $\mathbf{C}_\alpha$ is the same as for $\alpha \geq 1/2$, with $\phi := \pi$. Instead of $H_n = f_n/w^n$ we now take

$$H_n(z) := f_n(z)\sum_{j=-m}^{m} \frac{1}{w(ze^{2\pi ij})^n} = f_n(z)\sum_{j=-m}^{m} \exp(nz^\alpha e^{2\pi i\alpha j}), \tag{4.15}$$

where $m \geq 1$ is the least positive integer which satisfies

$$\cos(\alpha\pi(2m+1)) \leq 0. \tag{4.16}$$

For example, we have $m = 1$ if $1/6 \leq \alpha < 1/2$. In general, m is uniquely determined by the inequalities $4m-2 < \alpha^{-1} \leq 4m+2$.

The numbers $j\alpha$, $j = 1, 2, \cdots, m$, are not integers, hence

$$\lambda_j := \sin^2(j\alpha\pi) > 0 \ , \ j = 1, 2, \cdots, m.$$

H_n is analytic on $\mathbf{C}_\alpha \setminus \{0\}$, continuous on $\mathbf{C}_\alpha$ with $H_n(0) = 0$ since $f_n(0) = 0$.

For $x \geq 0$,

$$
\begin{aligned}
|f_n(x) - w(x)^n H_n(x)| &= |f_n(x)| \exp(-nx^\alpha) \left| \sum_{j=-m, j\neq 0}^{m} \exp\left(nx^\alpha e^{2\pi i j\alpha}\right) \right| \\
&\leq 2|f_n(x)| \exp(-nx^\alpha) \sum_{j=1}^{m} \exp\left(nx^\alpha \cos(2\pi j\alpha)\right) \\
&= 2|f_n(x)| \sum_{j=1}^{m} \exp\left(-2nx^\alpha \sin^2(\pi j\alpha)\right) \\
&= 2|f_n(x)| \sum_{j=1}^{m} \exp\left(-2nx^\alpha \lambda_j\right).
\end{aligned}
$$

Since all λ_j are positive and since $f_n(x) = x(x - \xi_n^{(n)})g(x)$ with $\xi_n^{(n)} \to b_\alpha^*$, it follows that

$$
\lim_{n\to\infty} \|f_n - w^n H_n\|_{C(B_0^*)} = 0. \tag{4.17}
$$

For $r \geq 0$,

$$
\begin{aligned}
&|H_n(re^{i\pi}) - H_n(re^{-i\pi})| \\
&= |f_n(-r)| \left| \sum_{j=-m}^{m} \left\{ \exp\left(nr^\alpha e^{\pi i\alpha(2j+1)}\right) - \exp\left(nr^\alpha e^{\pi i\alpha(2j-1)}\right) \right\} \right| \\
&= |f_n(-r)| \left| \exp\left(nr^\alpha e^{\pi i\alpha(2m+1)}\right) - \exp\left(nr^\alpha e^{\pi i\alpha(-2m-1)}\right) \right| \\
&= 2|f_n(-r)| \exp\left(nr^\alpha \cos\{\pi\alpha(2m+1)\}\right).
\end{aligned}
$$

By (4.16) we therefore obtain

$$
|H_n(re^{i\pi}) - H_n(re^{-i\pi})| \leq 2|f_n(-r)| \ , \ r \geq 0. \tag{4.18}
$$

We proceed now as in the case $\alpha \geq 1/2$: let L_n be the algebraic polynomial of degree $\leq n-1$ which interpolates H_n at the n zeros $x_j^{(n)}$ of the Chebyshev polynomial $C_{w,n}$. Then we have (4.6) and (4.7) for $0 < x < \xi_n^{(n)}$, (4.8) and (4.9) for $x > \xi_n^{(n)}$. It follows from the orientation of Γ_n^2 and Γ_n^3 and from (4.18) that

$$\left|\int_{\Gamma_n^2\cup\Gamma_n^3}\frac{H_n(\eta)d\eta}{(x-\eta)C_{w,n}(\eta)}\right|$$
$$=\left|e^{i\pi}\int_{\xi_n^{(n)}}^{0}\frac{H_n(re^{i\pi})dr}{(x-re^{i\pi})C_{w,n}(re^{i\pi})}+e^{-i\pi}\int_0^{\xi_n^{(n)}}\frac{H_n(re^{-i\pi})dr}{(x-re^{-i\pi})C_{w,n}(re^{-i\pi})}\right|$$
$$=\left|\int_0^{\xi_n^{(n)}}\frac{(H_n(re^{i\pi})-H_n(re^{-i\pi}))dr}{(x+r)C_{w,n}(-r)}\right|$$
$$\leq 2\int_0^{\xi_n^{(n)}}\frac{|f_n(-r)|dr}{|(x+r)C_{w,n}(-r)|}\,.$$

We have instead of (4.10), for all $r > 0$,

$$\frac{1}{|C_{w,n}(-r)|}=\frac{|C_{w,n}(0)|}{|C_{w,n}(-r)|}\leq\prod_{k=1}^{n}\frac{x_k^{(n)}}{-r+x_k^{(n)}}\leq(1+(r/b_\alpha^*))^{-n}\,. \tag{4.19}$$

This, (4.18) and (4.4) imply

$$\sup_{x\geq 0}\left|\int_{\Gamma_n^2\cup\Gamma_n^3}\frac{H_n(\eta)d\eta}{(x-\eta)C_{w,n}(\eta)}\right|\leq 2C\int_0^{\xi_n^{(n)}}(1+(r/b_\alpha^*))^{-n}\,dr\to 0, \tag{4.20}$$

as $n\to\infty$.

Using (4.13) and (4.4), we get for all $x \geq 0$

$$\left|\int_{\Gamma_n^1}\frac{H_n(\eta)d\eta}{(x-\eta)C_{w,n}(\eta)}\right|\leq\int_{\Gamma_n^1}\frac{|H_n(\eta)|w(\xi_n^{(n)})^nC_{w,n}(\xi_n^{(n)})|d\eta|}{|(x-\eta)C_{w,n}(\eta)|}$$
$$\leq\sum_{j=-m}^{m}\int_{\Gamma_n^1}\frac{|f_n(\eta)|}{|x-\eta|}w(\xi_n^{(n)})^n|\exp(n\eta^\alpha e^{2\pi i\alpha j})||d\eta|$$
$$\leq C\xi_n^{(n)}\sum_{j=-m}^{m}\int_{-\pi}^{\pi}|\exp\{n(\xi_n^{(n)})^\alpha(e^{2\pi i\alpha j+i\alpha t}-1)\}|dt$$
$$=C\xi_n^{(n)}\sum_{j=-m}^{m}\int_{-\pi}^{\pi}\exp\{-2n(\xi_n^{(n)})^\alpha\sin^2(\pi\alpha j+\alpha t/2)\}dt.$$

This yields (4.14) as $n\to\infty$. Since $\epsilon > 0$ is arbitrary, we get (4.1) by combining the above inequalities. ■

Corollary 4.2. *Let $A = \mathbb{R}$ and $\alpha > 0$. (i) The minimal essential set for W_α is the interval $[-b_\alpha, b_\alpha]$ where $b_\alpha := 2^{-1/\alpha}\sqrt{b^*_{\alpha/2}}$.*
(ii) Each function $f \in C(\mathbb{R})$ which vanishes at the origin and outside of $(-b_\alpha, b_\alpha)$ satisfies

$$\lim_{n\to\infty} \min_{P_n \in \mathcal{P}_n} \|f - W_\alpha^n P_n\|_{C(-\infty,\infty)} = 0. \tag{4.21}$$

Proof. We set $\beta := \alpha/2$ and use the substitution $y = 2^{1/\beta}x^2$, $y \geq 0$, which maps $\mathbb{R}$ onto $[0,\infty)$ and $[-b_\alpha, b_\alpha]$ onto $[0, b^*_\beta]$, where $b^*_\beta = 2^{1/\beta}b_\alpha^2$. It follows that $[-b_\alpha, b_\alpha]$ is the minimal essential set for W_α on $\mathbb{R}$.

Let $f \in C(\mathbb{R})$ be an even function which satisfies the conditions of the corollary. Since W_α and f are even, the weighted polynomials of best approximation for f are even. Hence f is uniformly approximable on $\mathbb{R}$ by the $W_\alpha^n P_n$, $n \in \mathbb{N}$, if and only if $F(y) := f(x)$ is uniformly approximable on $[0,\infty)$ by the $W_\beta^n(y)P_n(y)$. The application of the last theorem to F and W_β yields (4.21).

If $f \in C(\mathbb{R})$ is odd and has the properties of the corollary, we may assume that $g(x) := W_\alpha(x)^{-2}f(x)/x$ is also continuous. Since g is even, we have (4.21) for g, hence

$$\lim_{n\to\infty} \min_{P_{2n} \in \mathcal{P}_{2n}} \|f(x) - W_\alpha(x)^{2n+2}xP_{2n}(x)\|_{\mathbb{R}} = 0. \tag{4.22}$$

In the proof of (4.22) we use that $\lim W_\alpha(x)^{2n+2}xP_{2n}(x) = 0$, $n \to \infty$, uniformly for large $|x|$.

For an arbitrary $f \in C(\mathbb{R})$ which vanishes at $x = 0$ and outside of $(-b_\alpha, b_\alpha)$ we apply (4.21) and (4.22) for the even and odd parts of f, respectively. ∎

Theorem 4.1 and Corollary 4.2 are due to Lubinsky and Saff (1988). They use also the Hermite remainder formula for Lagrange interpolation, but their proofs are more difficult. In addition, they show that the condition $f(0) = 0$ can be omitted in Corollary 4.2 if $\alpha \geq 1$.

Mhaskar and Saff (1984) (see also Rahmanov (1984)) have proved that

$$b_\alpha = \left(\frac{2^{\alpha-2}\Gamma(\alpha/2)^2}{\Gamma(\alpha)}\right)^{1/\alpha}$$

and that

$$b^*_\alpha = \left(\frac{2^{2\alpha-1}\Gamma(\alpha)^2}{\Gamma(2\alpha)}\right)^{1/\alpha}.$$

It has been described by v.Golitschek, Lorentz, Makovoz (1991, §5) how the minimal essential sets B_0 can be computed for general weights w if the minimal essential set B_0 for w is an interval.

5. A WEIERSTRASS THEOREM VIA THE ALTERNATION THEOREM

The classical Weierstrass Theorem for algebraic polynomials $\mathcal{P}_n$, $n \geq 1$, on compact intervals $A = [a, b]$ has many proofs. One of these proofs uses the Alternation Theorem of Chebyshev: If $f \in C[a, b]$, the polynomial P_n^* of best uniform approximation to f from $\mathcal{P}_n$ is characterized by the existence of an alternation set

$$a \leq t_1^{(n)} < t_2^{(n)} < \cdots < t_{n+2}^{(n)} \leq b$$

of length $n + 2$, that is, for $\eta_n := +1$ or $\eta_n := -1$, one has for $j = 1, 2, \ldots, n + 2$

$$\eta_n(-1)^j \{f(t_j^{(n)}) - P_n^*(t_j^{(n)})\} = \|f - P_n^*\|_{C[a,b]} =: d_n(f). \tag{5.1}$$

Proof of the Weierstrass Theorem for $\mathcal{P}_n$:

Assume to the contrary that some $f \in C[a, b]$ is not uniformly approximable, that is, $d_n(f) \geq \rho > 0$ for all $n \geq 1$ and some $\rho > 0$. We define $\delta > 0$ so that the modulus of continuity of f satisfies $\omega(f, \delta) = \rho$.

Let n be fixed. If for some j

$$t_{j+1}^{(n)} - t_j^{(n)} \leq \delta, \tag{5.2}$$

then

$$\begin{aligned} &\eta_n(-1)^j \{P_n^*(t_{j+1}^{(n)}) - P_n^*(t_j^{(n)})\} \\ &= 2d_n(f) + \eta_n(-1)^j \{f(t_{j+1}^{(n)}) - f(t_j^{(n)})\} \geq 2d_n(f) - \rho \geq \rho. \end{aligned} \tag{5.3}$$

If (5.2) holds for all $j = 1, 2, \ldots, n+1$, the difference $P_n^*(t_{j+1}^{(n)}) - P_n^*(t_j^{(n)})$ alternates $n+1$ times in sign, hence P_n^* has n local extrema in (a, b) and thus P_n' has n zeros in (a, b), a contradiction. Hence there must exist intervals $I_j := [t_j^{(n)}, t_{j+1}^{(n)}]$, of length $> \delta$. Their number $N = N(n)$ is $< (b - a)/\delta$. We choose one of them, say I_{j_1}.

Let n be sufficiently large. For $m := 2N + 2$, there exist polynomials R_n of degree $\leq n$ with the following properties:

(i) $|R_n(x)| \leq \rho/3$, for all $x \in [a, b] \setminus I_{j_1}$.

(ii) R_n has in I_{j_1} at least m alternating local maxima and minima of values $> 2\|f\|_{C[a,b]}$ and $< -2\|f\|_{C[a,b]}$, respectively.

We leave it to the reader to confirm that the polynomial $S_n := P_n^* + R_n$ has at least n local extrema in (a, b), a contradiction. ■

Using these ideas, another Weierstrass Theorem for weighted polynomials has been proved in v.Golitschek, Lorentz, Makovoz (1991). For its formulation we need some notations:

Let $A = [a, b]$ be a compact interval, let $w \in C(A)$ be positive on (a, b). A point $x_0 \in A$ is called an *approximate zero* of the weight w if for each $\epsilon > 0$, there is a neighborhood U of x_0 with the property

$$|w(x)^n P_n(x)| \leq \epsilon \|w^n P_n\|_{C(A)}, \quad x \in U, \quad P_n \in \mathcal{P}_n, \quad n \geq n_0(\epsilon).$$

Plainly, $x_0 \notin B_0$. For example, each zero of w is an approximate zero. By z we denote *the number of approximate zeros* among the endpoints a, b of A.

Theorem 5.1. *If no weighted polynomial $w^n P_n \neq 0$, $P_n \in \mathcal{P}_n$, $n = 1, 2, \ldots$, can have $n + z$ local extrema in (a, b), then, for each function $f \in C(A)$ which vanishes on $A \setminus B_0$,*

$$\liminf_{n \to \infty} \min_{P_n \in \mathcal{P}_n} \|f - w^n P_n\|_{C(A)} = 0.$$

For example, the incomplete polynomials on $A = [0, 1]$, the Jacobi weights on $A = [-1, 1]$ and the exponential weights W_α on $A = \mathbb{R}$, $\alpha = 2, 4, 6, \ldots$, have the properties of the last theorem, but not W_α for the other values of $\alpha > 0$. Results similar to the last theorem are in Borwein and Saff (1990).

REFERENCES

1. Borwein, P. and Saff, E.B. (1990) 'On the denseness of weighted incomplete approximation', Preprint.

2. v.Golitschek, M. (1980) 'Approximation with incomplete polynomials', J. Approximation Theory, 28, 155-160.

3. v.Golitschek, M., Lorentz, G.G. and Makovoz, Y. (1991) 'Asymptotics of weighted approximation'. In: The first US-USSR Conference on Approximation Theory, Tampa, Florida, March 1990.

4. He, X. and Li, X. (1989) 'Uniform convergence of polynomials associated with varying Jacobi weights'. ICM 89-009, Tampa.

5. Lachance, M.A., Saff, E.B. and Varga,R.S. (1979) 'Bounds for incomplete polynomials vanishing at both endpoints of an interval'. In: Constructive Approaches to Mathematical Models (C. V. Coffman and G. J. Fix, eds.), Academic Press, New York, pp.421-437.

6. Lorentz, G.G. (1977) 'Approximation by incomplete polynomials (problems and results)'. In: *Padé and Rational Approximation: Theory and Applications,* (E.B. Saff and R.S. Varga, eds.), Academic Press, New York, pp.289-302.

7. Lubinsky, D.S., Mhaskar, H.N. and Saff, E.B. (1988) 'A proof of Freud's conjecture for exponential weights.' Constructive Approximation, 4 , 65 - 84.

8. Lubinsky, D.S. and Saff, E.B. (1988) 'Uniform and mean approximation by certain weighted polynomials, with applications', Constructive Approximation, 4 , 21-64.

9. Mhaskar,H.N. and Saff, E.B. (1984) 'Extremal problems for polynomials with exponential weights', Trans. Amer. Math. Soc., 285 , 203-234.

10. Mhaskar,H.N. and Saff, E.B. (1985) 'Where does the sup norm of a weighted polynomials live? (A generalization of incomplete polynomials)', Constructive Approximation, 1, 71-91.

11. Mhaskar, H.N. and Saff, E.B. (1985) 'A Weierstrass type theorem for certain weighted polynomials'. In: Approximation Theory and Applications (S.P. Singh, ed.), pp. 115-123, Pitman Publishing Ltd..

12. Rahmanov, E.A. (1984) 'On asymptotic properties of polynomials orthogonal on the real axis', Math. USSR.-Sb., 47 , 155-193.

13. Saff, E.B., Ullman, J.L. and Varga, R.S. (1980) 'Incomplete polynomials: an electrostatic approach'. In: Approximation Theory III, (E.W. Cheney, ed.), pp. 769-782, Academic Press, New York.

14. Saff, E.B. and Varga, R.S. (1978) 'Uniform approximation by incomplete polynomials', Internat. J. Math. Sci. 1, 407-420.

SOME ASPECTS OF RADIAL BASIS FUNCTION APPROXIMATION

W. A. Light
Mathematics Department
University of Leicester
University Road
Leicester LE1 7RH
England

ABSTRACT. This paper deals with three basic aspects of radial basis approximation. A typical example of such an approximation is the following. A function f in $C(\mathbb{R}^n)$ is to be approximated by a linear combination of 'easily computable' functions $g_1, \ldots, g_m$. For these functions the simplest choice in the radial basis context is to define g_i by $x \mapsto \|x - x_i\|_2$ for $x \in \mathbb{R}^n$ and $i = 1, 2, \ldots, m$. Here $\|\cdot\|_2$ is the usual Euclidean norm on $\mathbb{R}^n$. These functions are certainly easily computable, but do they form a flexible approximating set? There are various ways of posing the question of flexibility, and we consider here three possible criteria by which the effectiveness of such approximations may be judged. These criteria are labelled *density, interpolation* and *order of convergence* in the exposition.

1 Introduction

In this paper we will consider largely the space $C(\mathbb{R}^n)$, consisting of continuous, real-valued functions on $\mathbb{R}^n$. Occasionally, the domain of definition of the functions under consideration will be restricted to the compact subset D contained in $\mathbb{R}^n$. Suppose $x_1, \ldots, x_m$ are points in $\mathbb{R}^n$, and ϕ is a continuous function from $\mathbb{R}^+$ to $\mathbb{R}$. Our approximating subspace will be the span of the functions

$$x \mapsto \phi(\|x - x_i\|_2), \quad i = 1, 2, \ldots, m.$$

Here $\|\cdot\|_2$ is the usual Euclidean norm. We will address three basic problems.

Problem 1 (Density) Let D be a compact subset of $\mathbb{R}^n$ and let G be the subspace of $C(D)$ defined by

$$G = \text{span}\{x \mapsto \phi(\|x - y\|_2) : y \in D\}.$$

Is G dense in $C(D)$?

S. P. Singh (ed.), Approximation Theory, Spline Functions and Applications, 163–190.

Problem 2 (Interpolation) Suppose $x_1, x_2, \ldots, x_m \in \mathbb{R}^n$ and data $d_1, d_2, \ldots, d_m$ are prescribed. Does there exist a unique set of scalars $a_1, a_2, \ldots, a_m$ such that

$$\sum_{j=1}^{m} a_j \phi(\|x_i - x_j\|_2) = d_i, \quad i = 1, 2, \ldots, m?$$

Of course, in examining this question, we want to be able to identify functions ϕ for which this problem has a unique solution *independent* of the data $d_1, \ldots, d_m$ *and* independent of the location of the points $x_1, \ldots, x_m$.

Problem 3 (Convergence Rates) When the answer to Problem 1 is affirmative, we would like some estimation of the convergence rate. This is not an easy task, and so we consider the following model problem. Let W_h be the subspace of $C(\mathbb{R}^n)$ consisting of all elements of the form

$$w_h(x) = \sum_{z \in \mathbb{Z}^n} \phi\left(\left\|\frac{x}{h} - z\right\|_2\right) c^h(z), \quad x \in \mathbb{R}^n.$$

Here $c^h \in \ell_\infty(\mathbb{Z}^n)$, so that the real numbers $c^h(z)$, $z \in \mathbb{Z}^n$ are uniformly bounded; h is a positive real number; and the above series is assumed to be absolutely convergent for each value of x. We wish to try to identify functions ϕ for which $\operatorname{dist}(f, W_h) = \mathcal{O}(h^k)$ for all functions f having sufficient smoothness. Here, k is a natural number, and the distance is measured using the usual supremum norm.

The emphasis in each of these problems is on finding a reasonably large class of functions ϕ which admit a solution. However, we are not really at liberty to pick and choose for ϕ in an arbitrary fashion, because examples of such functions have been finding favour in multivariate interpolation for at least 20 years now. We would naturally like our theory to cover these examples. Common choices for ϕ are

$$\left.\begin{array}{ll} \phi(r) = r & \text{linear} \\ \phi(r) = r^3 & \text{cubic} \\ \phi(r) = r^2 \log r & \text{thin plate spline} \\ \phi(r) = \exp(-r^2) & \text{Gaussian} \\ \phi(r) = (r^2 + c^2)^{1/2} & \text{multiquadric} \\ \phi(r) = (r^2 + c^2)^{-1/2} & \text{inverse multiquadric} \end{array}\right\}, \; r \geq 0.$$

The multiquadrics were introduced by Hardy [12] in 1971, and the work of Franke [10] in 1982 demonstrated that the technique of approximating data by interpolation using radial basis functions had a number of very attractive features. All the choices for ϕ listed above were examined in [10], and the multiquadrics received particular commendation.

It should be pointed out at the outset of this discussion that we are dipping into the subject. More comprehensive accounts of the theory can be found in Dyn [7, 8] and Powell [20]. The choice of material for this paper has more to do with the predilections and current research of the author than with any desire to provide a comprehensive account of the field.

It will turn out that Problems 1 and 2 can be approached by a similar analysis. The version we use, which captures the density results as well as those of interpolation is due to Brown [2]. However, theorems concerning interpolation have a long history. The first results are due to Schoenberg [22], who showed that the interpolation problem can always be solved in the linear case, $\phi(r) = r$. Schoenberg was really interested in the following type of question. Let $X = \{x_1, \ldots, x_m\}$ be a finite set and let (X, d) be a metric space. When can (X, d) be embedded in ℓ_2? The papers of Micchelli [19] and Madych and Nelson [17] address directly the subject of interpolation, and consider rather more general situations than those given in Problem 2. We refer the reader to Micchelli [19] for the full story. Our approach uses a basic argument from functional analysis plus some very beautiful classical theory about completely monotone functions. Problem 3 has been cast in such a way that it is amenable to Fourier analysis, and our approach will be essentially a generalisation of well-known results due to Fix and Strang [9]. But this part of our exposition is more properly viewed as a synthesis of the ideas of de Boor and Jia [1], Cheney and Light [4], Jia and Lei [16] and Halton and Light [11]. In 1986 Jia [15] pointed out an error in the original work of Fix and Strang. In the case that the function ϕ which generates the approximation w_h has compact support, this error was corrected by de Boor and Jia [1]. Cheney and Light [4] then showed how the assumption of compact support may be weakened to one where ϕ has 'rapid decay' at infinity. In the process, they gave a modified version of the de Boor–Jia theory. The papers of Jia and Lei [16] and Halton and Light [11] attempt to recover in different ways the spirit of the original de Boor–Jia theory.

The remainder of our account divides into two sections, the first treating Problems 1 and 2, while the second treats Problem 3.

2 Density and Interpolation

Throughout this section D is assumed to be a compact subset of $\mathbb{R}^n$, and M denotes the L_∞–closure of the span of the set $\{\phi(\| \cdot - y\|_2) : y \in D\}$. The question of density is whether $M = C(D)$. An elementary application of the Hahn–Banach theorem shows that $M = C(D)$ if and only if the only functional $\mu \in [C(D)]^*$ which satisfies

$$\mu(\phi(\| \cdot - y\|_2)) = 0 \quad \text{for all } y \in D$$

is the zero functional. The Riesz representation theorem shows that each $\mu \in [C(D)]^*$ corresponds to a signed, regular, Borel measure on D, which we will denote again by μ. Thus $M = C(D)$ if and only if

$$\int_D \phi(\|x - y\|_2)\, d\mu(x) = 0 \quad \text{for all } y \in D \tag{1}$$

implies that μ is the trivial measure.

So far we have referred only to the property of density. Let us now briefly consider the interpolation problem, and get a feeling for how these two questions are linked. Suppose $x_1, x_2, \ldots, x_m$ are fixed, distinct points in D, and $d_1, d_2, \ldots, d_m$ are the prescribed data at these points. The linear system

$$\sum_{j=1}^{m} a_j \phi(\|x_i - x_j\|_2) = d_i, \qquad i = 1, 2, \ldots, m$$

has a unique solution for the unknowns $a_1, a_2, \ldots, a_m$ if and only if the matrix A with elements

$$A_{ij} = \phi(\|x_i - x_j\|_2), \qquad i,j = 1,2,\ldots,m,$$

is invertible. We call this matrix the interpolation matrix. Now recall that a signed, regular, Borel measure μ on D can have mass 1, for example, concentrated at each of the points $x_1, x_2, \ldots, x_m$ in D. In this case (1) would read

$$\int_D \phi(\|x - y\|_2)\, d\mu(x) = \sum_{j=1}^{m} \phi(\|x_j - y\|_2) \quad \text{for all } y \in D.$$

By setting y equal in turn to $x_1, x_2, \ldots, x_m$ we can obtain the quantity

$$\sum_{j=1}^{m} \phi(\|x_j - x_i\|_2)$$

for $i = 1,2,\ldots,m$. This expression now provides information about the interpolation matrix (and in fact could be used to calculate the operator norm of A, when the ℓ_∞ or ℓ_1 norm is employed in $\mathbb{R}^n$).

Lemma 2.1 *Let $x \in \mathbb{R}^n$. Then*

$$\exp(-\|x\|_2^2) = \pi^{-n/2} \int_{\mathbb{R}^n} e^{2iux} e^{-\|u\|_2^2}\, du.$$

Proof. Let $f_n : \mathbb{R}^n \to \mathbb{R}$ be defined by $f_n(x) = \exp(-\|x\|_2^2/2)$, $x \in \mathbb{R}^n$. Note that f_1 satisfies the differential equation

$$y' + xy = 0.$$

Taking Fourier transforms of each side of this equation gives

$$t\widehat{y}(t) + (\widehat{y})'(t) = 0.$$

Now

$$\left(\frac{\widehat{f_1}}{f_1}\right)'(x) = \frac{(\widehat{f_1})'(x) f_1(x) - f_1'(x)\widehat{f_1}(x)}{f_1^2(x)} = \frac{x\widehat{f_1}(x) f_1(x) - x f_1(x)\widehat{f_1}(x)}{f_1^2(x)} = 0.$$

Therefore $\widehat{f_1}/f_1$ is a constant. Since $f_1(0) = 1$ and

$$\widehat{f_1}(0) = \int_{\mathbb{R}} f_1(x)\, dx = \int_{\mathbb{R}} e^{-x^2/2}\, dx = \sqrt{2\pi},$$

it follows that $\widehat{f_1} = \sqrt{2\pi} f_1$. Since

$$f_n(x) = \prod_{i=1}^{n} f_1(x_i), \quad \text{where } x = (x_1, x_2, \ldots, x_n) \in \mathbb{R}^n,$$

it follows that

$$\widehat{f}_n(t) = \prod_{i=1}^{n} \widehat{f}_1(t_i) = (2\pi)^{n/2} \prod_{i=1}^{n} f_1(t_i) = (2\pi)^{n/2} f_n(t), \quad t \in \mathbb{R}^n.$$

Finally,

$$\begin{aligned} \pi^{-n/2} \int_{\mathbb{R}}^{n} e^{2iux} e^{-\|u\|_2^2}\, du &= \pi^{-n/2} \int_{\mathbb{R}}^{n} e^{\sqrt{2}iux} e^{-\|u\|_2^2/2} \left(\frac{1}{\sqrt{2}}\right)^n du \\ &= (2\pi)^{-n/2} \widehat{f}_n(-\sqrt{2}x) \\ &= (2\pi)^{-n/2} (2\pi)^{n/2} f_n(-\sqrt{2}x) \\ &= \exp(-\|x\|_2^2). \quad \blacksquare \end{aligned}$$

In the next few results we are going to refer to the Fourier transform of a regular, signed, Borel measure on D. If μ is such a measure, then by $\widehat{\mu}$ we mean the (continuous) function from $\mathbb{R}^n$ to $\mathbb{R}$ given by

$$\widehat{\mu}(t) = \int_{\mathbb{R}^n} e^{-ixt}\, d\mu(x), \quad t \in \mathbb{R}^n.$$

Note that there is no problem with the existence of this integral, since we may extend μ to a measure $\overline{\mu}$ on all of $\mathbb{R}^n$ by defining $\overline{\mu}(E) = \mu(E \cap D)$ for all Borel measurable sets E in $\mathbb{R}^n$. We will not distinguish between μ and $\overline{\mu}$ throughout the rest of this section.

Lemma 2.2 *Let μ be a non-trivial, regular, signed, Borel measure on D and let $g : \mathbb{R}^n \to \mathbb{R}$ be defined by*

$$g(y) = \int_D e^{2iyx}\, d\mu(x), \quad y \in \mathbb{R}^n.$$

Then $g \in C(D)$ and there is a point $z \in \mathbb{R}^n$ such that $g(z) \neq 0$.

Proof. Note that in the terminology established prior to the lemma, $g(y) = \widehat{\mu}(-2y)$, $y \in \mathbb{R}^n$. Since $\mu \not\equiv 0$, there is an $f \in C(D)$ such that

$$\int_D f(-y)\, d\mu(y) \neq 0. \tag{2}$$

The function f can be extended using Tietze's theorem to the whole of $\mathbb{R}^n$ in such a way that the extension $\tilde{f}$ is a continuous function with compact support. The measure μ is extended in the manner outlined prior to the Lemma. Defining the convolution in the usual way, viz:

$$(\tilde{f} * \mu)(x) = \int_{\mathbb{R}^n} \tilde{f}(x-y)\, d\mu(y), \quad x \in \mathbb{R}^n,$$

it is easy to see that the normal rule for the Fourier transform of a convolution holds: $(\tilde{f} * \mu)\widehat{\ }(x) = (\widehat{f}.\widehat{\mu})(x)$, $x \in \mathbb{R}^n$. Now if $\widehat{\mu}(x) = 0$ for all $x \in \mathbb{R}^n$, then $(\tilde{f} * \mu)\widehat{\ }(x) = 0$, for all $x \in \mathbb{R}^n$. Note that $\tilde{f} * \mu$ is a continuous, compactly supported function and so the Plancherel theorem states that

$$\|\tilde{f} * \mu\|_2 = (2\pi)^{-n/2} \|(f * \mu)\widehat{\ }\|_2,$$

and thus it would follow that $\tilde{f} * \mu \equiv 0$. However,

$$(\tilde{f} * \mu)(0) = \int_D f(-y)\, d\mu(y) \neq 0$$

by (2), and since $\tilde{f} * \mu$ is continuous, it follows that $\tilde{f} * \mu \neq 0$. Hence we cannot have $\widehat{\mu}(x) = 0$ for all $x \in \mathbb{R}^n$, which establishes the claim of the Lemma. ■

Theorem 2.3 *Let μ be a non-trivial, regular, signed, Borel measure on the compact set D contained in $\mathbb{R}^n$, and let α be a positive real number. Then*

$$\iint_{D\times D} e^{-\alpha\|x-y\|_2^2}\, d\mu(x)\, d\mu(y) > 0.$$

Proof. Using **2.1** we have

$$\begin{aligned}\iint_{D\times D} e^{-\alpha\|x-y\|_2^2}\, d\mu(x)\, d\mu(y) &= \iint_{D\times D} e^{-\|\alpha(x-y)\|_2^2}\, d\mu(x)\, d\mu(y)\\ &= \iint_{D\times D} \pi^{-n/2}\int_{\mathbb{R}^n} e^{2iu\alpha(x-y)}e^{-\|u\|_2^2}\, du\, d\mu(x)\, d\mu(y).\end{aligned}$$

Note that $|e^{2iuz}| = 1$ for all $z \in D$ and $u \in \mathbb{R}^n$. Also the function $u \to e^{-\|u\|_2^2}$ is absolutely integrable over $\mathbb{R}^n$. Hence, by Fubini's theorem,

$$\begin{aligned}\iint_{D\times D} e^{-\alpha\|x-y\|_2^2}\, d\mu(x)\, d\mu(y) &= \pi^{-n/2}\int_{\mathbb{R}^n} e^{-\|u\|_2^2}\iint_{D\times D} e^{2iu\alpha(x-y)}\, d\mu(x)\, d\mu(y)\, du\\ &= \pi^{-n/2}\int_{\mathbb{R}^n} e^{-\|u\|_2^2}\int_D e^{2iu\alpha x}\, d\mu(x)\int_D e^{-2iu\alpha y}\, d\mu(y)\, du\\ &= \pi^{-n/2}\int_{\mathbb{R}^n} e^{-\|u\|_2^2}\left|\int_D e^{2iu\alpha x}\, d\mu(x)\right|^2 du.\end{aligned}$$

Now **2.2** shows that the integrand

$$e^{-\|u\|_2^2}\left|\int_D e^{2iu\alpha x}\, d\mu(x)\right|^2, \quad u \in \mathbb{R}^n,$$

is a non-trivial, continuous function of u whenever μ is non-trivial. The result now follows. ■

The next two corollaries provide us with our first answers to Problems 1 and 2.

Corollary 2.4 *Let $\phi : \mathbb{R}^n \to \mathbb{R}$ be defined by $\phi(t) = e^{-t^2}$, $t \in \mathbb{R}$, and let D be a compact subset of $\mathbb{R}^n$. Then the span of the set of functions $M = \{\phi(\|\cdot - y\|_2) : y \in D\}$ is dense in $C(D)$.*

Proof. Suppose μ is a regular, signed, Borel measure on D satisfying

$$\int_D \phi(\|x-y\|_2)\, d\mu(x) = 0 \quad \text{for all } y \in D.$$

As was pointed out at the start of this section, we need to show that μ is the trivial measure. From the above equation it follows that

$$\iint_{D\times D} \phi(\|x-y\|_2)\, d\mu(x)\, d\mu(y) = 0,$$

and this is enough to conclude from **2.3** that μ is the trivial measure. ■

A closer examination of the proof of **2.3** yields the following result about interpolation.

Corollary 2.5 *Let $\phi : \mathbb{R}^n \to \mathbb{R}$ be defined by $\phi(t) = e^{-t^2}$, $t \in \mathbb{R}$. Let $x_1, x_2, \ldots, x_m$ be distinct points in $\mathbb{R}^n$. Then the interpolation matrix A, whose ij^{th} element is $\phi(\|x_i - x_j\|_2^2)$, $i, j = 1, 2, \ldots, m$, is positive definite.*

Proof. The proof of **2.3** shows that for any non-trivial, regular, signed, Borel measure μ and any compact subset D in $\mathbb{R}^n$,

$$\iint_{D\times D} e^{-\|x-y\|_2^2}\, d\mu(x)\, d\mu(y) > 0. \tag{3}$$

Suppose now the set D is chosen so that it contains the points $x_1, x_2, \ldots, x_m$. Choose the non-trivial measure μ to have masses c_i concentrated at x_i, $i = 1, 2, \ldots, m$ and no mass elsewhere. Then **2.2** takes the form

$$\sum_{i,j=1}^{m} c_i c_j \phi(\|x_i - x_j\|_2) > 0.$$

This shows that the interpolation matrix is positive definite as required. ∎

The nice results of **2.4** and **2.5** do not appear to be very general in character, but the surprising fact is that, through the theory of completely monotone functions, these results provide rather easily some results with wide applicability. Everything relies on a characterisation theorem due to Bernstein.

Definition 2.6 *A function $g \in C^\infty(0,\infty)$ is said to be completely monotonic if it satisfies $(-1)^k f^{(k)}(x) \geq 0$ for all $x \in (0,\infty)$ and $k = 0, 1, 2, \ldots$.*

Theorem 2.7 *A necessary and sufficient condition that $g \in C[0,\infty) \cap C^\infty(0,\infty)$ be completely monotonic is that there exists a finite, non-negative, Radon measure γ such that*

$$g(t) = \int_0^\infty e^{-tu}\, d\gamma(u) \quad \textit{for } 0 < t < \infty.$$

Note that a *Radon measure* γ on $(0,\infty)$ is a Borel measure which satisfies

(a) $\gamma(K) < \infty$ for every compact set K contained in $(0,\infty)$

(b) each Borel measurable set $B \subset (0,\infty)$ can be measured in the following way:

$$\gamma(B) = \sup\{\gamma(K) : \ K \subset B \text{ and } K \text{ is compact}\}.$$

We are now in a position to get at our first elegant extension of **2.4** and **2.5**. We will consider a function ψ which satisfies

(i) $\psi \in C[0,\infty) \cap C^\infty(0,\infty)$

(ii) ψ is completely monotonic, but not constant on $(0,\infty)$.

Condition (ii) denies the possibility that the measure associated with ψ has all its mass concentrated at the origin.

Lemma 2.8 *Suppose ψ satisfies the conditions (i) and (ii) above and μ is a signed, regular, non-trivial, Borel measure on the compact set D in $\mathbb{R}^n$. Then*

$$\iint_{D\times D} \psi(\|x-y\|_2^2)\,d\mu(x)\,d\mu(y) > 0.$$

Proof. Using **2.7**, we can write

$$\iint_{D\times D} \psi(\|x-y\|_2^2)\,d\mu(x)\,d\mu(y) = \iint_{D\times D}\int_0^\infty e^{-u\|x-y\|_2^2}\,d\gamma(u)\,d\mu(x)\,d\mu(y).$$

The integrand in the integral on the right-hand side of this inequality is absolutely integrable, and so

$$\iint_{D\times D} \psi(\|x-y\|_2^2)\,d\mu(x)\,d\mu(y) = \int_0^\infty \left\{\iint_{D\times D} e^{-u\|x-y\|_2^2} d\mu(x)\,d\mu(y)\right\} d\gamma(u).$$

Now by **2.3**, since μ is non-trivial, we have, for $u > 0$,

$$\iint_{D\times D} e^{-u\|x-y\|_2^2}\,d\mu(x)\,d\mu(y) > 0,$$

and so

$$\iint_{D\times D} \psi(\|x-y\|_2^2)\,d\mu(x)\,d\mu(y) > 0,$$

as required. ■

The following theorem can be established in the same way as **2.4** and **2.5**.

Theorem 2.9 *Let $\psi \in C[0,\infty)\cap C^\infty(0,\infty)$ be completely monotonic but not constant on $(0,\infty)$.*

(i) Let D be a compact subset of $\mathbb{R}^n$. Then the span of the set of functions $M = \{\psi(\|\cdot - y\|_2^2) : y \in D\}$ is dense in $C(D)$.

(ii) Let $x_1, x_2, \ldots, x_m$ be distinct points in $\mathbb{R}^n$. Then the interpolation matrix A whose ij^{th} element is $\psi(\|x_i - x_j\|_2^2)$, $i,j = 1,2,\ldots,m$, is positive definite.

Examples. The theory developed so far covers the cases

(i) $\phi(r) = e^{-r^2}$ where $\psi(r) = e^{-r}$

(i) $\phi(r) = (r^2+c^2)^{-1/2}$ where $\psi(r) = (r+c^2)^{-1/2}$.

Our final class of functions is those whose derivative is completely monotone. We consider functions ψ which satisfy

(i) $\psi \in C[0,\infty)\cap C^\infty(0,\infty)$

(ii) ψ' is completely monotonic, but not constant on $(0,\infty)$.

(iii) $\psi(0) \geq 0$.

To get a representation theorem for ψ, we suppose that $\theta \in C[0,\infty)$ is a completely monotone function on $(0,\infty)$. Then there is a finite Radon measure γ such that

$$\theta(t) = \int_0^\infty e^{-tu}\, d\gamma(u) \quad \text{for } 0 < t < \infty.$$

Integrating this expression over the interval $[\epsilon, s]$, where $\epsilon > 0$ gives

$$\int_\epsilon^s \theta(t)\, dt = \int_\epsilon^s \int_0^\infty e^{-tu}\, d\gamma(u) = \int_0^\infty (e^{-\epsilon u} - e^{-su})u^{-1}\, d\gamma(u)$$

providing that the order of integration can be reversed. In order to ensure this, we must assume that $\int_1^\infty u^{-1}\, d\gamma(u) < \infty$. Then, since

$$(e^{-\epsilon u} - e^{-su})u^{-1} \leq (1 - e^{-su})u^{-1}, \quad u \in [0,1),$$

we see that integrability over $[0,1)$ is guaranteed, while our condition $\int_1^\infty u^{-1}\, d\gamma(u) < \infty$ ensures integrability over the remaining range $[1,\infty)$. Letting $\epsilon \to 0$ gives

$$\int_0^s \theta(t)\, dt = \int_0^\infty (1 - e^{-su})u^{-1}\, d\gamma(u).$$

The function ψ defined by $\psi(s) = \int_0^s \theta(t)\, dt$, $s \in [0,\infty)$ is now in $C[0,\infty) \cap C^\infty(0,\infty)$ and its derivative is the completely monotone function θ. This argument can be reversed to conclude that the above representation is a characterisation of functions whose derivative is completely monotone up to the possibility of an additive constant. If we deny the possibility that ψ' is constant, then this again simply prevents the measure γ from having mass exclusively at the origin.

Lemma 2.10 *Suppose ψ satisfies conditions (i)–(iii) above and μ is a signed, regular, non-trivial, Borel measure on a compact subset D contained in $\mathbb{R}^n$. If $\mu(D) = 0$ then*

$$\iint_{D\times D} \psi(\|x-y\|_2^2)\, d\mu(x)\, d\mu(y) < 0.$$

Proof. Using the representation for ψ, there exists a Radon measure γ, whose mass is not exclusively concentrated at the origin, such that $\int_1^\infty u^{-1}\, d\gamma(u) < \infty$ and

$$\psi(t) = \psi(0) + \int_0^\infty (1 - e^{-st})s^{-1}\, d\gamma(s), \quad t \in (0,\infty).$$

Now using Fubini's theorem, and recalling that $\mu(D) = 0$,

$$\begin{aligned}&\iint_{D\times D} \psi(\|x-y\|_2^2)\, d\mu(x)\, d\mu(y)\\ &\quad = \iint_{D\times D} \left\{\psi(0) + \int_0^\infty (1 - e^{-s\|x-y\|_2^2})s^{-1}\, d\gamma(s)\right\} d\mu(x)\, d\mu(y)\end{aligned}$$

$$= \int_0^\infty \Big\{ \iint_{D\times D} \psi(0)\, d\mu(x)\, d\mu(y)$$
$$+ \iint_{D\times D} (1 - e^{-s\|x-y\|_2^2}) s^{-1}\, d\mu(x)\, d\mu(y) \Big\}\, d\gamma(s)$$
$$= - \int_0^\infty \Big\{ \iint_{D\times D} e^{-s\|x-y\|_2^2}\, d\mu(x)\, d\mu(y) \Big\}\, s^{-1}\, d\gamma(s).$$

Now **2.3** shows that

$$\iint_{D\times D} e^{-s\|x-y\|_2^2}\, d\mu(x)\, d\mu(y) > 0,$$

and from this the result follows. ■

The application of this result is somewhat delicate, and it is easier to begin with the result on interpolation.

Theorem 2.11 *Let $\psi \in C[0,\infty) \cap C^\infty(0,\infty)$ be a function whose derivative is completely monotonic but not constant on $(0,\infty)$, and suppose $\psi(0) \geq 0$. Let $x_1, x_2, \ldots, x_m$ be distinct points in $\mathbb{R}^n$. Then the interpolation matrix whose elements are $\psi(\|x_i - x_j\|_2^2)$, $i,j = 1,2,\ldots,m$ is non-singular.*

Proof. Choose the compact set D to contain $x_1, x_2, \ldots, x_m$ and then apply **2.10** to the non-trivial measure μ with mass c_i at x_i, $i = 1,2,\ldots,m$. This gives

$$0 > \iint_{D\times D} \psi(\|x-y\|_2^2)\, d\mu(x)\, d\mu(y) = \sum_{i,j=1}^{m} c_i c_j \psi(\|x_i - x_j\|_2^2) = c^T A c,$$

where $A = (A_{ij}) = (\psi(\|x_i - x_j\|_2^2))$ and $c = (c_1, c_2, \ldots, c_m)$. However, there is the additional restriction on μ that

$$0 = \mu(D) = \int_D d\mu(x) = \sum_{i=1}^{m} c_i.$$

Now suppose A has eigenvalues $\lambda_1 \geq \lambda_2 \geq \cdots \geq \lambda_m$. Let V be the subspace of $\mathbb{R}^m$ defined by $\{c \in \mathbb{R}^m : c = (c_1, c_2, \ldots, c_m) \text{ and } \sum_1^m c_i = 0\}$. By the Courant–Fisher theorem,

$$\lambda_2 = \min_{\dim U = m-1} \max_{\substack{u \in U \\ \|u\|_2 = 1}} u^T A u \leq \max_{\substack{c \in V \\ \|c\|_2 = 1}} c^T A c < 0.$$

Now $\psi(0) \geq 0$ forces trace$(A) \geq 0$ and so we conclude that $\lambda_1 > 0$. Hence A is non-singular. ■

Theorem 2.12 *Suppose $\psi \in C[0,\infty) \cap C^\infty(0,\infty)$ is a function whose derivative is completely monotone but not constant on $(0,\infty)$ and $\psi(t) > 0$ for $t > 0$. Let D be a compact subset of $\mathbb{R}^n$. Then the span of the set of functions $M = \{\psi(\|\cdot - y\|_2^2) : y \in D\}$ is dense in $C(D)$.*

Proof. Set $V = \text{span}\{\psi(\|\cdot - y\|_2^2) : y \in D\}$. We shall show that if $\mu \in [C(D)]^*$ and $\mu(v) = 0$ for all $v \in V$, then μ is the zero functional. We do not distinguish between the functional μ and its corresponding regular, signed, Borel measure throughout the remainder of the proof. Our assumption is that

$$\int_D \psi(\|x - y\|_2^2)\, d\mu(x) = 0 \quad \text{for all } y \in D.$$

Fix z in D and define a new measure ν by the equation $\mu = \mu(D)\delta_z + \nu$. Here δ_z is the functional (regular, signed, Borel measure) such that $\delta_z(f) = f(z)$ for all $f \in C(D)$. Notice that

$$\nu(D) = \mu(D) - \mu(D)\delta_z(D) = \mu(D) - \mu(D).1 = 0.$$

Hence,

$$\begin{aligned}
0 &= \iint_{D\times D} \psi(\|x - y\|_2^2)\, d\mu(x)\, d\nu(y) \\
&= \iint_{D\times D} \psi(\|x - y\|_2^2)[\mu(D)d\delta_z(x) + d\nu(x)]\, d\nu(y) \\
&= \mu(D)\iint_{D\times D} \psi(\|x - y\|_2^2)\, d\delta_z(x)\, d\nu(y) + \iint_{D\times D} \psi(\|x - y\|_2^2)\, d\nu(x)\, d\nu(y) \\
&= \mu(D)\int_D \psi(\|z - y\|_2^2)\, d\nu(y) + \iint_{D\times D} \psi(\|x - y\|_2^2)\, d\nu(x)\, d\nu(y) \\
&= \mu(D)\int_D \psi(\|z - y\|_2^2)[d\mu(y) - \mu(D)d\delta_z(y)] + \iint_{D\times D} \psi(\|x - y\|_2^2)\, d\nu(x)\, d\nu(y) \\
&= -[\mu(D)]^2\psi(0) + \iint_{D\times D} \psi(\|x - y\|_2^2)\, d\nu(x)\, d\nu(y).
\end{aligned}$$

Now $\psi(0) \geq 0$ and **2.10** shows

$$\iint_{D\times D} \psi(\|x - y\|_2^2)\, d\nu(x)\, d\nu(y) < 0,$$

unless ν is the trivial measure. This forces the conclusion that ν is indeed the trivial measure, and so $\mu = \mu(D)\delta_z$. Then for all $y \in D$, we have

$$0 = \int_D \psi(\|x - y\|_2^2)\, d\mu(x) = \int_D \psi(\|x - y\|_2^2)\mu(D)\, d\delta_z(x) = \mu(D)\psi(\|z - y\|_2^2).$$

Since $\psi(\|z - y\|_2^2) \neq 0$ for $y \neq z$, we have $\mu(D) = 0$, and so μ is trivial as required. ∎

Examples Our new theory adds the cases

(i) $\phi(r) = r$ when $\psi(r) = \sqrt{r}$

(ii) $\phi(r) = (r^2 + c^2)^{1/2}$ when $\psi(r) = \sqrt{r + c^2}$.

The two examples not covered in our list are the cubic and thin plate spline cases. In each of these situations, the interpolating or approximating functions need to be augmented by low degree polynomials (in fact by linear polynomials). The theory is very similar — no major new ideas are introduced — and we leave the interested reader to explore this aspect for herself. She is advised to consult [20].

3 Convergence Rates

The discussions of this section differ widely in character from those of the previous section. Firstly, as can be seen in Section 1, the problem studied has no direct computational relevance. Secondly, the techniques of this section are those of Fourier transforms and Sobolev spaces. The Sobolev space $W_p^k(\mathbb{R}^n)$ consists of all functions $f : \mathbb{R}^n \to \mathbb{R}$ for which $\|f\|_{k,p} < \infty$ where

$$\|f\|_{k,p} = \left(\sum_{j \le k} |f|_{j,p}^p \right)^{1/p} \quad \text{and} \quad |f|_{j,p} = \sum_{|\alpha|=j} \|D^\alpha f\|_p,$$

with

$$\|f\|_{k,\infty} = \sum_{j \le k} |f|_{j,\infty} \quad \text{and} \quad |f|_{j,\infty} = \sum_{|\alpha|=j} \|D^\alpha f\|_\infty.$$

Here, $\alpha = (\alpha_1, \alpha_2, \ldots, \alpha_n) \in \mathbb{Z}_+^n$ is a multi-integer and $|\alpha| = \sum_{j=1}^n \alpha_j$, $\alpha! = \prod_{j=1}^n \alpha_j!$. If $x = (x_1, x_2, \ldots, x_n) \in \mathbb{R}^n$ then $x^\alpha = \prod_{j=1}^n x_j^{\alpha_j}$ and $\|x\| = \max_{1 \le j \le n} |x_j|$. It is convenient to have a notation for the normalised monomials

$$V_\alpha(x) = x^\alpha / \alpha!, \quad x \in \mathbb{R}^n.$$

Then π_k denotes the space of polynomials of total degree at most k, so that

$$\{V_\alpha : 0 \le |\alpha| \le k\}$$

is a basis for π_k. It helps if we regard π_{-1} as the trivial subspace. We usually use Rudin [21] as our guide in matters relating to the Fourier transform, except that we define

$$\hat{f}(x) = \int_{\mathbb{R}^n} f(y) e^{ixy}\, dy, \qquad x \in \mathbb{R}^n.$$

We also set

$$(Bf)(x) = f(-x); \quad (T_x f)(y) = f(y - x), \qquad (x, y \in \mathbb{R}^n);$$

and

$$(S_h f)(x) = f(hx), \qquad (h > 0, x \in \mathbb{R}^n).$$

Whenever a convolution appears it is now the following discrete convolution:

$$(f * g)(x) = \sum_{\nu \in \mathbb{Z}^n} f(x - \nu) g(\nu).$$

We denote by e_y the function $e_y(x) = e^{ixy}$, $x, y \in \mathbb{R}^n$, where the product xy is the usual scalar product in $\mathbb{R}^n$.

In the following developments, there are a considerable number of technical details which must be checked. These are left to the reader in the form of exercises. Now fix $n, k \in \mathbb{N}$ and $\lambda \in (0,1)$. The space E will consist of all functions $f \in C(\mathbb{R}^n)$ such that

$$\sup_{x \in \mathbb{R}^n} \{|f(x)|(1 + \|x\|)^{n+k+\lambda}\} < \infty.$$

Our first main result now follows.

Theorem 3.1 *Let $\{\psi_\alpha\}_{|\alpha|<k}$ be a set of functions in E such that $\widehat{\psi}_0(0) = 1$ and*

$$\sum_{\beta\leq\alpha} V_\beta(-iD)\widehat{\psi}_{\alpha-\beta}(2\pi\nu) = 0, \quad \nu \in \mathbb{Z}^n \setminus \{0\},\ |\alpha| < k.$$

Then there exist compactly supported functions b_α, $|\alpha| < k$ such that the function $\phi = \sum_{|\alpha|<k} \psi_\alpha * b_\alpha$ belongs to E and satisfies

$$\|f - S_{1/h}(\phi * S_h f)\|_\infty \leq Ch^k |f|_{k,\infty},$$

for all $f \in W^k_\infty(\mathbb{R}^n)$.

This Theorem forms one half of the results of Cheney and Light [4], except for the additional information that the functions b_α are compactly supported. The weaker result that $b_\alpha \in E$ for $|\alpha| < k$ was all that was established in [4]. The additional information about the compact support of the b_α comes from Jia and Lei [16] and we present some of their arguments in the following exposition. The substance of the other half of [4] is that the conditions given in **3.1** on the Fourier transforms of the ψ_α (often referred to as the Strang–Fix conditions) are not only sufficient for the stated order of convergence but also necessary. The approximation

$$w_h = S_{1/h}(\phi * S_h f)$$

is often called a quasi–interpolant. With this nomenclature, the result of [4] can be paraphrased by saying that the Strang–Fix conditions hold if and only if there is a quasi–interpolant w_h such that $\|f - w_h\|_\infty = \mathcal{O}(h^k)$ as $h \to \infty$.

Theorem **3.1** will be established by a series of exercises and lemmata.

Exercises 3.1

3.1.1 Show that the space E is invariant under the shift operator and under the operator B.

3.1.2 Show that $|V_\alpha(x)| \leq \|x\|^{|\alpha|}$ for all $x \in \mathbb{R}^n$.

3.1.3 If $f \in E$ and $|\alpha| \leq k$ show that $V_\alpha f \in L_1(\mathbb{R}^n)$.

The next lemma is a formal manipulation with Fourier transforms.

Lemma 3.2 *If $|\alpha| \leq k$ and $f \in E$, then*

$$(V_\alpha T_x B f)\widehat{\ } = \sum_{\beta\leq\alpha} V_\beta(x) e_{-x} B V_{\alpha-\beta}(-iD)\widehat{f}.$$

Proof. By **3.1.1** E is invariant under T_x and B. Thus $T_x B f \in E$. By **3.1.3**, $V_\alpha T_x B f \in L_1(\mathbb{R}^n)$, and thus has a Fourier transform. Now

$$\begin{aligned}
(V_\alpha T_x B f)\widehat{\ } &= V_\alpha(iD)(T_x B f)\widehat{\ } \\
&= V_\alpha(iD)[e_{-x}(Bf)\widehat{\ }] \\
&= i^{|\alpha|} V_\alpha(D)[e_{-x} B\widehat{f}] \\
&= i^{|\alpha|} \sum_{\beta\leq\alpha} [V_\beta(D) e_{-x}][V_{\alpha-\beta}(D) B\widehat{f}].
\end{aligned}$$

Note that the use of the Liebnitz rule is justified since the second bracketed term is

$$[V_{\alpha-\beta}B\widehat{f}] = i^{-|\alpha-\beta|}[V_{\alpha-\beta}Bf]\widehat{}$$

and this latter term is certainly well-defined. Now

$$\begin{aligned}(V_\alpha T_x Bf)\widehat{} &= \sum_{\beta\leq\alpha} i^{|\alpha|}V_\beta(-ix)e_{-x}(-1)^{|\alpha|-|\beta|}BV_{\alpha-\beta}(D)\widehat{f} \\ &= \sum_{\beta\leq\alpha} V_\beta(x)e_{-x}BV_{\alpha-\beta}(-iD)\widehat{f}. \qquad \blacksquare\end{aligned}$$

Exercises 3.2

3.2.1 The number of elements $\nu \in \mathbb{Z}^n$ satisfying $\|\nu\| = j$ is not greater than $c(1+j)^{n-1}$. If $j \neq 0$ then this bound can be replaced by cj^{n-1}.

3.2.2 If $f \in E$ and $|\alpha| \leq k$ then the convolutions $|f| * |V_\alpha|$ and $|V_\alpha| * |f|$ exist.

We need the following version of the Poisson summation formula, whose proof may be found in Jackson [14].

Lemma 3.3 *Let $f \in C(\mathbb{R}^n)$ be such that $|f(x)|$ is $\mathcal{O}(\|x\|^{-n-\theta})$ for some $\theta > 0$ as $\|x\| \to \infty$, and such that $\sum_{\nu\in\mathbb{Z}^n} |\widehat{f}(2\pi\nu)| < \infty$. Then $\sum_{\nu\in\mathbb{Z}^n} f(\nu) = \sum_{\nu\in\mathbb{Z}^n} \widehat{f}(2\pi\nu)$.*

Theorem 3.4 *Suppose functions $\{\psi_\alpha\}_{|\alpha|<k}$ are such that $\widehat{\psi}_0(0) = 1$ and*

$$\sum_{\beta\leq\alpha} V_\beta(-iD)\widehat{\psi}_{\alpha-\beta}(2\pi\nu) = 0, \qquad (\nu \in \mathbb{Z}^n \setminus \{0\},\ |\alpha| < k).$$

Then

$$V_\alpha - \sum_{\beta\leq\alpha} \psi_{\alpha-\beta} * V_\beta \in \pi_{|\alpha|-1}, \qquad (|\alpha| < k).$$

Proof. We want to apply **3.3**. Verification that the hypotheses of **3.3** are satisfied is left to the reader. We have, using **3.3** and **3.2**,

$$\begin{aligned}\sum_{0\leq\beta\leq\alpha} (\psi_{\alpha-\beta} * V_\beta)(x) &= \sum_{0\leq\beta\leq\alpha}\sum_{\nu\in\mathbb{Z}^n} \psi_{\alpha-\beta}(x-\nu)V_\beta(\nu) \\ &= \sum_{\nu\in\mathbb{Z}^n}\sum_{0\leq\beta\leq\alpha} (V_\beta T_x B\psi_{\alpha-\beta})(\nu) \\ &= \sum_{\nu\in\mathbb{Z}^n}\sum_{0\leq\beta\leq\alpha} (V_\beta T_x B\psi_{\alpha-\beta})\widehat{}(2\pi\nu) \\ &= \sum_{\nu\in\mathbb{Z}^n}\sum_{0\leq\beta\leq\alpha}\sum_{0\leq\gamma\leq\beta} V_\gamma(x)e_{-x}(2\pi\nu)[BV_{\beta-\alpha}(-iD)\widehat{\psi}_{\alpha-\beta}](2\pi\nu) \\ &= \sum_{\nu\in\mathbb{Z}^n}\sum_{0\leq\gamma\leq\alpha} V_\gamma(x)e^{-2\pi ix\nu}\sum_{\gamma\leq\beta\leq\alpha} [V_{\beta-\gamma}(-iD)\widehat{\psi}_{\alpha-\beta}](-2\pi\nu) \\ &= \sum_{\nu\in\mathbb{Z}^n}\sum_{0\leq\gamma\leq\alpha} V_\gamma(x)e^{-2\pi ix\nu}\sum_{0\leq\delta\leq\alpha-\gamma} [V_\delta(-iD)\widehat{\psi}_{\alpha-\gamma-\delta}](-2\pi\nu)\end{aligned}$$

$$= \sum_{0\le\gamma\le\alpha} V_\gamma(x) \sum_{0\le\delta\le\alpha-\gamma} [V_\delta(-iD)\widehat{\psi}_{\alpha-\gamma-\delta}](0)$$

$$= V_\alpha(x)\widehat{\psi}_0(0) + \sum_{\substack{\gamma\le\alpha \\ \gamma\ne\alpha}} V_\gamma(x) \sum_{0\le\delta\le\alpha-\gamma} [V_\delta(-iD)\widehat{\psi}_{\alpha-\gamma-\delta}](0)$$

$$= V_\alpha(x) + p(x)$$

where $p \in \pi_{|\alpha|-1}$. ■

Exercises 3.3

3.3.1 In the above analysis, check that

$$\sum_{\nu\in\mathbb{Z}^n} \Big| \sum_{0\le\beta\le\alpha} (V_\beta T_x B\psi_{\alpha-\beta})\widehat{\ }(2\pi\nu)\Big| < \infty.$$

(Hint: many of the manipulations in the proof above should be helpful!)

3.3.2 Show that any $r+1$ translates of the univariate function $t \mapsto t^r$ span the univariate polynomial space π_r.

3.3.3 Fix $\gamma = (k, k, \ldots, k) \in \mathbb{Z}^n$. For each $0 \le \alpha \le \gamma$ there is a compactly supported $c_\alpha : \mathbb{Z}^n \to \mathbb{R}$ such that $V_\gamma * c_\alpha = V_\alpha$.

Lemma 3.5 *Let $\{\psi_\alpha\}_{|\alpha|<k}$ be a set of functions in E such that*

$$\sum_{0\le\beta\le\alpha} \psi_{\alpha-\beta} * V_\beta - V_\alpha \in \pi_{|\alpha|-1}, \quad |\alpha| < k.$$

Then there exist finitely supported functions $c_\beta : \mathbb{Z}^n \to \mathbb{R}$ such that

$$\sum_{|\beta|<k} (\psi_\beta * c_\beta) * V_\alpha - V_\alpha \in \pi_{|\alpha|-1}, \quad |\alpha| < k.$$

Proof. Let $\gamma = (k, k, \ldots, k)$. By **3.3.3**, there exist compactly supported functions $c_\beta : \mathbb{Z}^n \to \mathbb{R}$ such that $V_\gamma * c_\beta = V_{\gamma-\beta}$. If $0 \le \alpha, \beta \le \gamma$ then

$$V_{\alpha-\beta} = D^{\gamma-\alpha} V_{\gamma-\beta} = D^{\gamma-\alpha}(V_\gamma * c_\beta) = (D^{\gamma-\alpha} V_\gamma) * c_\beta.$$

Since c_β is compactly supported, the convolutions $\psi_\beta * c_\beta$ are finite sums. Select $S \subset \mathbb{Z}^n$ so that $c_\beta(\nu) = 0$ if $\nu \notin S$ and $0 \le \beta \le \alpha$. Then

$$|(\psi_\beta * c_\beta)(x)| \le \sum_{\nu\in S} |\psi(x-\nu)c_\beta(\nu)| \le M \max_{\nu\in S} |\psi_\beta(x-\nu)|,$$

where $M = \|c_\beta\|_1$. This shows that $\psi_\beta * c_\beta \in E$. It follows from Exercise **3.2.2** that $(\psi_\beta * c_\beta) * V_\alpha$ exists, and is defined by an absolutely convergent series. Now

$$\sum_{|\beta|<k} \psi_\beta * c_\beta * V_\alpha = \sum_{|\beta|<k} \psi_\beta * V_{\alpha-\beta} = \sum_{0\le\beta\le\alpha} \psi_\beta * V_{\alpha-\beta} = \sum_{0\le\delta\le\alpha} \psi_{\alpha-\delta} * V_\delta = V_\alpha + p$$

for some $p \in \pi_{|\alpha|-1}$. ■

The next two lemmas owe their inspiration to [16].

Lemma 3.6 *Let $\psi \in E$ and suppose the formula $Lp = \psi * p$ defines an operator $L : \pi_k \to \pi_k$. Then $T_x L = L T_x$ for all $x \in \mathbb{R}^n$.*

Proof. Fix $y \in \mathbb{R}^n$ and take $\alpha \geq 0$, $|\alpha| < k$. Then, for any $x \in \mathbb{R}^n$, $(T_x L V_\alpha)(y)$ is a polynomial in x of degree at most k. Furthermore, by using the binomial theorem,

$$\begin{aligned}(LT_x V_\alpha)(y) &= \sum_{\nu \in \mathbb{Z}^n} \psi(y-\nu) V_\alpha(\nu - x) \\ &= \sum_{\nu \in \mathbb{Z}^n} \psi(y-\nu) \sum_{0 \leq \beta \leq \alpha} V_\beta(\nu) V_{\alpha-\beta}(-x) \\ &= \sum_{0 \leq \beta \leq \alpha} V_{\alpha-\beta}(-x) \sum_{\nu \in \mathbb{Z}^n} \psi(y-\nu) V_\beta(\nu).\end{aligned}$$

Hence, $(LT_x V_\alpha)(y)$ is also a polynomial in x of degree at most k. It will now suffice to establish that $(LT_\mu V_\alpha)(y) = (T_\mu L V_\alpha)(y)$ for all $\mu \in \mathbb{Z}^n$. We have

$$\begin{aligned}(LT_\mu V_\alpha)(y) &= \sum_{\nu \in \mathbb{Z}^n} \psi(y-\nu) V_\alpha(\nu - \mu) \\ &= \sum_{\nu \in \mathbb{Z}^n} \psi(y-\nu-\mu) V_\alpha(\nu) \\ &= \sum_{\nu \in \mathbb{Z}^n} T_\mu \psi(y-\nu) V_\alpha(\nu) \\ &= (T_\mu L V_\alpha)(y).\end{aligned}$$

Thus for each $y \in \mathbb{R}^n$ and each $\alpha \geq 0$, $|\alpha| \leq k$, $(LT_x V_\alpha)(y) = (T_x L V_\alpha)(y)$. Since the monomials $\{V_\alpha : |\alpha| \leq k\}$ form a basis for π_k, it follows that $LT_x p = T_x L p$ for all $p \in \pi_k$. ■

Lemma 3.7 *Let $\psi \in E$ be such that $\psi * V_\alpha - V_\alpha \in \pi_{|\alpha|-1}$ for $1 \leq |\alpha| \leq k$. Then there exists a function $b : \mathbb{Z}^n \to \mathbb{R}$ supported on $\{\nu \in \mathbb{Z}^n : |\nu| \leq k-1\}$ such that $\psi * b * p = p$ for all $p \in \pi_k$.*

Proof. Define a linear operator L on π_k by $Lp := \psi * p$, $p \in \pi_k$. Then $(I - L)$ is a degree reducing operator on π_k, so that $(I-L)^{k+1} p = 0$ for all $p \in \pi_k$. Suppose now $p \neq 0$. Then $Lp \neq 0$, otherwise the following contradiction would arise:

$$\text{degree}(p) = \text{degree}((I-L)p) < \text{degree}(p).$$

Hence, L is an injection, and so is thus a bijection. Consequently, L is invertible. From **3.6** it follows that $T_x L = L T_x$ for all $x \in \mathbb{R}^n$, and so $T_x L^{-1} = L^{-1} T_x$ for all $x \in \mathbb{R}^n$. Now from [5] the functionals

$$\tilde{\mu}(p) := p(\mu), \qquad \mu \in \mathbb{Z}^n, \mu \geq 0, |\mu| \leq k$$

form a basis for π_k^*. Define $\phi \in \pi_k^*$ by

$$\phi(p) = (L^{-1}p)(0), \qquad p \in \pi_k.$$

Then there exist real numbers a_α, $\alpha \geq 0$, $|\alpha| \leq k$, such that

$$\phi(p) = \sum_{|\alpha|\leq k} a_\alpha \widetilde{\alpha}(p).$$

Now fix $x \in \mathbb{R}^n$. Then

$$p(x) = (L^{-1}Lp)(x) = (T_{-x}L^{-1}Lp)(0) = (L^{-1}T_{-x}Lp)(0) = \sum_{|\alpha|\leq k} a_\alpha (T_{-x}Lp)(\alpha).$$

Set $b(-\alpha) = a_\alpha$, $\alpha \geq 0$, $|\alpha| \leq k$ and $b(-\alpha) = 0$ otherwise. Then

$$\begin{aligned} p(x) &= \sum_{|\alpha|\leq k} a_\alpha \sum_{\nu\in\mathbb{Z}^n} \psi(x+\alpha-\nu)p(\nu) \\ &= \sum_{\nu\in\mathbb{Z}^n}\sum_{\alpha\in\mathbb{Z}^n} \psi(x+\alpha-\nu)b(-\alpha)p(\nu) \\ &= \sum_{\nu\in\mathbb{Z}^n}\sum_{\alpha\in\mathbb{Z}^n} \psi(x-\alpha-\nu)b(\alpha)p(\nu) \\ &= \sum_{\nu\in\mathbb{Z}^n} (\psi * b)(x-\nu)p(\nu) \\ &= (\psi * b * p)(x). \qquad \blacksquare \end{aligned}$$

This result allows us to establish the existence of a "quasi–interpolant" with the desired rate of convergence. We prefer initially *not* to have the dilation factor present in the argument.

Theorem 3.8 *Let ϕ be an element of E such that $\phi * p = p$ for all $p \in \pi_{k-1}$. Then there is a constant C such that for all $f \in W^k_\infty(\mathbb{R}^n)$, $\|f - \phi * f\|_\infty \leq C|f|_{k,\infty}$.*

Proof. Fix $x \in \mathbb{R}^n$ and let p be the Taylor polynomial of degree $k-1$ for f at x. Let $r = f - p$. Then we can write

$$r(y) = \sum_{|\alpha|=k} V_\alpha(y-x)(D^\alpha f)(\xi_{\alpha y}).$$

Hence we have

$$\begin{aligned} |f(x) - (\phi * f)(x)| &= |p(x) - (\phi * f)(x)| \\ &= |(\phi * p)(x) - (\phi * f)(x)| \\ &= |(\phi * r)(x)| \\ &\leq \sum_{\nu\in\mathbb{Z}^n} |\phi(x-\nu)||r(\nu)| \\ &\leq \sum_{\nu\in\mathbb{Z}^n} |\phi(x-\nu)| \sum_{|\alpha|=k} |V_\alpha(\nu-x)||(D^\alpha f)(\xi_{\alpha\nu})| \\ &\leq \max_{|\alpha|=k} \|D^\alpha f\|_\infty \sum_{|\alpha|=k}\sum_{\nu\in\mathbb{Z}^n} |\phi(x-\nu)||V_\alpha(x-\nu)|. \end{aligned}$$

Now consider the expression $A(x) := \sum_{|\alpha|=k} \sum_{\nu\in\mathbb{Z}^n} |\phi(x-\nu)||V_\alpha(x-\nu)|$. It is an elementary observation that for $\mu \in \mathbb{Z}^n$,

$$A(x+\mu) = \sum_{|\alpha|=k} \sum_{\nu\in\mathbb{Z}^n} |\phi(x+\mu-\nu)||V_\alpha(x+\mu-\nu)| = \sum_{|\alpha|=k} \sum_{\nu\in\mathbb{Z}^n} |\phi(x-\nu)||V_\alpha(x-\nu)| = A(x).$$

Hence,

$$\begin{aligned}
\|A\| &= \sup_{\|x\|\le 1} |A(x)| \\
&\le \sum_{|\alpha|=k} \sum_{\nu\in\mathbb{Z}^n} (1+\|x-\nu\|)^{-n-k-\lambda}(1+\|x-\nu\|)^k \\
&\le B \sup_{\|x\|\le 1} \sum_{|\alpha|=k} \sum_{\nu\in\mathbb{Z}^n} (1+\|x-\nu\|)^{-n-\lambda} \\
&< \infty.
\end{aligned}$$

Thus, for $x \in \mathbb{R}^n$,

$$|f(x) - (\phi * f)(x)| \le C|f|_{k,\infty}. \qquad \blacksquare$$

We are now in a position to prove Theorem **3.1**. The simple proof is contained in the following exercises.

Exercises 3.4

3.4.1 Let $L : W^k_\infty(\mathbb{R}^n) \to L_\infty(\mathbb{R}^n)$ be a linear operator such that

$$\|f - Lf\|_\infty \le A|f|_{k,\infty} \quad \text{for all } f \in W^k_\infty(\mathbb{R}^n).$$

Set $L_h = S_{1/h}LS_h$ where $(S_hf)(x) = f(hx)$, $h > 0$. Show that

$$\|f - L_hf\| \le Ah^k|f|_{k,\infty} \quad \text{for } h > 0 \text{ and } f \in W^k_\infty(\mathbb{R}^n).$$

3.4.2 Prove Theorem **3.1** by stringing together the appropriate results!

Theorem **3.1** admits the following important generalisation. We can write (in the terminology of **3.1**),

$$\begin{aligned}
\phi * S_hf &= \left(\sum_{|\alpha|<k} \psi_\alpha * b_\alpha\right) * S_hf \\
&= \sum_{|\alpha|<k} \left(\sum_{\psi\in\Psi} a^\alpha_\psi \psi\right) * (b_\alpha * S_hf) \\
&= \sum_{\psi\in\Psi} \psi \left(\sum_{|\alpha|<k} a^\alpha_\psi b_\alpha * S_hf\right) \\
&= \sum_{\psi\in\Psi} \psi * c^h_\psi,
\end{aligned}$$

where $c^h_\psi = \sum_{|\alpha|<k} a^\alpha_\psi b_\alpha * S_hf$, $h > 0$, $\psi \in \Psi$. These coefficient functions satisfy an interesting condition.

Lemma 3.9 *With the above setup, and for $0 < h < 1$, there exist constants A, r independent of h such that*

(i) $\|c_\psi^h\|_\infty \le A\|f\|_\infty$ *for all* $\psi \in \Psi$, $f \in W_\infty^k(\mathbb{R}^n)$

(ii) $c_\psi^h(\nu) = 0$ *whenever* $dist(\nu h, suppf) > r$.

Proof. By definition, for $f \in W_\infty^k(\mathbb{R}^n)$ we have

$$\begin{aligned} |c_\psi^h(\nu)| &\le \sum_{|\alpha|<k} |a_\psi^\alpha| \sum_{\mu\in\mathbb{Z}^n} |b_\alpha(\nu-\mu)f(h\mu)| \\ &\le \|f\|_\infty \sum_{|\alpha|<k} |a_\psi^\alpha| \sum_{\mu\in\mathbb{Z}^n} |b_\alpha(\nu-\mu)| \\ &\le \|f\|_\infty \sum_{|\alpha|<k} \|b_\alpha\|_1 |a_\psi^\alpha|. \end{aligned}$$

Setting $A = \max_{\psi\in\Psi} \sum_{|\alpha|<k} \|b_\alpha\|_1 |a_\psi^\alpha|$ gives $\|c_\psi^h\|_\infty \le A\|f\|_\infty$. Now in addition, suppose the support of each b_α is contained within a ball of radius ρ. Then

$$|c_\psi^h(\nu)| = \sum_{|\alpha|<k} |a_\psi^\alpha| \sum_{\mu\in\mathbb{Z}^n} |b_\alpha(\mu)f(h\nu-\mu)| = \sum_{|\alpha|<k} |a_\psi^\alpha| \sum_{\|\mu\|\le\rho} |b_\alpha(\mu)f(h\nu-\mu)|.$$

If $\text{dist}(\nu h, \text{supp} f) > \rho$, then for $0 < h < 1$ it follows that $h\nu - h\mu \notin \text{supp} f$ for $\|\mu\| \le \rho$. Hence $c_\psi^h(\nu) = 0$ whenever $\text{dist}(\nu h, \text{supp} f) > \rho$. ■

This brings us to the following important definition.

Definition 3.10 *Let Ψ be a finite set of functions in E. Then Ψ provides local, controlled approximation of order k if there exist constants A, B such that for $0 < h < 1$ and $f \in W_\infty^k(\mathbb{R}^n)$ one can find $c_\psi^h : \mathbb{Z}^n \to \mathbb{R}$, $\psi \in \Psi$, such that*

(i) $\|f - S_{1/h} \sum_{\psi\in\Psi} \psi * c_\psi^h\|_\infty \le Ah^k|f|_{k,\infty}$

(ii) $\|c_\psi^h\|_\infty \le B\|f\|_\infty$

(iii) there exists a constant r independent of h such that if $dist(\nu h, suppf) > r$ then $c_\psi^h(\nu) = 0$ for all $\psi \in \Psi$.

Note that some sort of restriction on the range of h permissible here is necessary. If we remove the requirement that $0 < h < 1$ and set

$$f_h(x) = S_{1/h} \sum_{\psi\in\Psi} \psi * c_\psi^h,$$

then conditions (i) and (iii) above claim that for all $h > 0$

$$\|f - f_h\|_\infty \le Ah^k|f|_{k,\infty}$$

and

$$c_\psi^h(\nu) = 0 \qquad \text{whenever } \text{dist}(\nu h, \text{supp} f) > r.$$

If $\nu \neq 0$, then $\|\nu\| \geq 1$ and so as $h \to \infty$, $\|\nu h\| \to \infty$. Thus if f has compact support, it is possible to locate a value h_0 such that $c_\psi^h(\nu) = 0$ for all $\nu \in \mathbb{Z}^n \setminus \{0\}$ and $h > h_0$. Then we require

$$\left| f(x) - \sum_{\psi \in \Psi} \psi\left(\frac{x}{h}\right) c_\psi^h(0) \right| \leq A h^k |f|_{k,\infty}, \qquad x \in \mathbb{R}^n,\ h > h_0.$$

For simplicity, let us assume Ψ consists of a single function. Then we require

$$\left| f(x) - \psi\left(\frac{x}{h}\right) c_\psi^h(0) \right| \leq A h^k |f|_{k,\infty}, \qquad x \in \mathbb{R}^n,\ h > h_0.$$

We now construct a counterexample to this inequality. Take $n = 1$ and $k = 2$. Consider the function f which is piecewise linear with $f(x) = 0$ for all $x \leq -1$ and $x \geq 1$ and $f(0) = 1$. Then $f \in W_\infty^2(\mathbb{R})$, and $|f|_{2,\infty} = \|f''\|_\infty = 0$. This in turn forces

$$f(x) = \psi\left(\frac{x}{h}\right) c_\psi^h(0) \qquad \text{for all } h > h_0 \text{ and } x \in \mathbb{R}.$$

Setting $x = 0$ in the above inequality gives

$$1 = \psi(0) c_\psi^h(0) \qquad \text{for all } h > h_0,$$

and so $\psi(0) \neq 0$ and $c_\psi^h(0) = [\psi(0)]^{-1}$ for all $h > h_0$. Now take $x = 2$. Then

$$0 = \psi\left(\frac{2}{h}\right) c_\psi^h(0) = \psi\left(\frac{2}{h}\right) [\psi(0)]^{-1} \qquad \text{for all } h > h_0.$$

Letting $h \to \infty$ gives the contradiction

$$0 = \lim_{h \to \infty} \psi\left(\frac{2}{h}\right) [\psi(0)]^{-1} = \psi(0)[\psi(0)]^{-1} = 1.$$

Theorem **3.1** may now be rephrased as follows.

Theorem 3.11 *Let Ψ be a finite set of functions in E and suppose there exist functions $\{\psi_\alpha\}_{|\alpha|<k}$ in Ψ such that $\widehat{\psi}_0(0) = 1$ and*

$$\sum_{\beta \leq \alpha} V_\beta(-iD) \widehat{\psi}_{\alpha-\beta}(2\pi\nu) = 0, \qquad \nu \in \mathbb{Z}^n \setminus \{0\},\ |\alpha| < k.$$

Then Ψ provides local, controlled approximation of order k.

Theorem **3.11** is almost the same as that given by de Boor and Jia [1], with two important exceptions. Firstly, the setting in **3.11** is more general, in that Ψ consists of functions in E, rather than functions having compact support. Secondly, condition (ii) in **3.10** does not appear in de Boor and Jia's theory. In that paper there is only the concept of local approximation, where the meaning of the adjective local is that **3.10**(iii) holds. The significance of control is that condition **3.10**(ii) holds – there is a control on the size of the coefficients in the approximation by the size of the norm of the function to be approximated. This mismatch is perturbing at first sight, since the compactly supported case should be contained neatly in **3.11**. In fact, this turns out to be a rather delicate matter. It is possible to give a version of **3.11** for functions in E, which truly contains the theory of de Boor and Jia. Thus when the functions in Ψ are in E and have compact support, the control condition becomes vacuous. This is the substance of the paper by Halton and Light [11]. In that paper, one is rapidly brought to the conclusion that *unless* Ψ consists solely of compactly supported functions, then the condition on control *cannot* be dispensed with. In its present form, Definition **3.10** and Theorem **3.11** follow Jia and Lei [16], although the main techniques of proof come from [4] and [11]. The object of our treatment is to get as clean an exposition as possible. The interested reader may refer to the original papers for an overview of the field. It should also be mentioned that Jia and Lei cover the additional case of L_p-norms.

The following Lemma (or something similar) is crucial to all the correct versions of the Strang–Fix theory.

Lemma 3.12 *Let Ψ be finite subset of E and let $f \in W_\infty^k(\mathbb{R}^n)$ have compact support. If $f_h = S_{1/h}(\sum_{\psi\in\Psi} \psi * c_\psi^h)$ provides local, controlled approximation of order k to f, then there exist constants c and ρ independent of h such that*

$$|f_h(x)| \leq ch^{k+\lambda}\|x\|^{-n-k-\lambda} \qquad \textit{whenever } \|x\| > \rho.$$

Proof. Since f_h provides local, controlled approximation to f and f has compact support, we can assume ρ is chosen so that $f(x) = 0$ for all $\|x\| > \rho/2$ and $c_\psi^h(\nu) = 0$ for all $\|\nu h\| > \rho/2$. Then for any $x \in \mathbb{R}^n$,

$$\begin{aligned}
|f_h(x)| &= \Big|S_{1/h}\Big(\sum_{\psi\in\Psi} \psi * c_\psi^h\Big)(x)\Big| \\
&\leq \sum_{\psi\in\Psi}\sum_{\nu\in\mathbb{Z}^n} \Big|\psi\Big(\frac{x}{h}-\nu\Big)c_\psi^h(\nu)\Big| \\
&\leq \sum_{\psi\in\Psi}\sum_{\|\nu h\|\leq\rho} \Big(1+\Big\|\frac{x}{h}-\nu\Big\|\Big)^{-n-k-\lambda} |c_\psi^h(\nu)| \\
&\leq B\|f\|_\infty h^{n+k+\lambda}\sum_{\psi\in\Psi}\sum_{\|\nu h\|\leq\rho} (h+\|x-\nu h\|)^{-n-k-\lambda}.
\end{aligned}$$

For $\|x\| > \rho$ we have $h + \|x-\nu h\| \geq \|x\|/2$ and so

$$\begin{aligned}
|f_h(x)| &\leq B\|f\|_\infty h^{n+k+\lambda}\sum_{\psi\in\Psi}\sum_{\|\nu h\|\leq\rho} (\|x\|/2)^{-n-k-\lambda} \\
&\leq B'\|f\|_\infty h^{n+k+\lambda}\|x\|^{-n-k-\lambda}\sum_{\|\nu h\|\leq\rho} 1 \\
&\leq ch^{k+\lambda}\|x\|^{-n-k-\lambda}. \qquad \blacksquare
\end{aligned}$$

What is needed at this stage is a lot of simple manipulations with Fourier transforms. These are consigned to a set of exercises.

Exercises 3.5

3.5.1 If, in addition to the hypotheses of **3.12**, $\|f - f_h\| = \mathcal{O}(h^k)$ as $h \to 0$, then for $|\alpha| < k$, $\|V_\alpha(D)(\widehat{f} - \widehat{f}_h)\|_\infty$ is also $\mathcal{O}(h^k)$ as $h \to 0$. Hint: write

$$|V_\alpha(iD)(\widehat{f} - \widehat{f}_h)| = \left(\int_{\|y\|>2\rho} + \int_{\|y\|\le 2\rho}\right) |V_\alpha(y)||f(y) - f_h(y)|\, dy$$

where $\mathrm{supp}(f) \subset \{y \in \mathbb{R}^n : \|y\| \le \rho\}$.

3.5.2 Assume the hypotheses of **3.12**. Then

$$\widehat{f}_h = h^n \sum_{\psi\in\Psi} S_h\widehat{\psi} \sum_{\nu\in\mathbb{Z}^n} e_{-\nu h} c_\psi^h(\nu), \quad h > 0.$$

3.5.3 Assume the hypotheses of **3.12**. Show that

$$V_\alpha(D)\widehat{f}_h = h^{n+|\alpha|}(-i)^{|\alpha|} \sum_{\psi\in\Psi} \sum_{0\le\beta\le\alpha} [S_h V_\beta(iD)\widehat{\psi}] \sum_{\nu\in\mathbb{Z}^n} V_{\alpha-\beta}(\nu) e_{-h\nu} c_\psi^h(\nu).$$

Now we introduce the compactly supported function

$$u(x) := \prod_{i=1}^n M_{k+1}(x_i), \qquad x = (x_1, x_2, \ldots, x_n) \in \mathbb{R}^n,$$

where M_{k+1} is a univariate B-spline of order k having the property that

$$\widehat{u}(x) = \prod_{i=1}^n \left[\frac{\sin(x_i/2)}{x_i/2}\right]^{k+1}, \qquad x \in \mathbb{R}^n.$$

In [1] one also finds the property

$$[V_\alpha(D)\widehat{u}](x/h) = o(h^k) \qquad (x \ne 0,\ h \to 0,\ |\alpha| < k).$$

Exercises 3.6

3.6.1 Let u be the function defined above and suppose $u_h = S_{1/h} \sum_{\psi\in\Psi} \psi * c_\psi^h$ provides local, controlled approximation of order k to u. Show, using 3.5.1 and 3.5.3, that for $|\alpha| < k$ and $\mu \in \mathbb{Z}^n \setminus \{0\}$,

$$\lim_{h\to 0} h^n \sum_{\psi\in\Psi} \sum_{0\le\beta\le\alpha} [V_\beta(iD)\widehat{\psi}](2\pi\mu) \sum_{\nu\in\mathbb{Z}^n} V_{\alpha-\beta}(\nu) c_\psi^h(\nu) = 0.$$

Our final theorem, which ties up the exposition of the Strang–Fix theory, is the converse of **3.11**.

Theorem 3.13 *Let Ψ be a finite subset in E which provides local, controlled approximation of order k. Then there exists a sequence $\{\psi_\alpha\}_{|\alpha|<k}$ in Ψ such that $\widehat{\psi}_0(0) = 1$ and*

$$\sum_{0\le\beta\le\alpha} V_\beta(iD)\widehat{\psi}_{\alpha-\beta}(2\pi\nu) = 0, \quad (\nu \in \mathbb{Z}^n \setminus \{0\},\ |\alpha| < k).$$

Proof. We work entirely with the function u defined previously. Firstly, there must exist $\psi \in \Psi$ such that $\widehat{\psi}(0) \neq 0$, because if not, then Exercise 3.5.2 gives

$$\widehat{u}_h(0) = h^n \sum_{\psi \in \Psi} (S_h \widehat{\psi})(0) \sum_{\nu \in \mathbb{Z}^n} e_{-\nu h}(0) c_\psi^h(\nu) = h^n \sum_{\psi \in \Psi} \widehat{\psi}(0) c_\psi^h(\nu) = 0.$$

Then Exercise 3.5.1 shows that $\|\widehat{u} - \widehat{u}_h\|_\infty = \mathcal{O}(h^k)$ and since $\widehat{u}(0) = 1$, this implies the contradictory conclusion $\lim_{h\to 0} \widehat{u}_h(0) = 1$.

Since u_h is unchanged if each element in Ψ is replaced by elements in $\operatorname{span}\Psi$, as long as the overall span is identical, we may assume that there is a χ in Ψ with $\widehat{\chi}(0) = 1$ and $\widehat{\psi}(0) = 0$ for all $\psi \in \Psi$ with $\psi \neq \chi$. Then, from Exercise 3.5.2,

$$1 = \lim_{h\to 0} \widehat{u}_h(0) = \lim_{h\to 0} h^n \sum_{\psi \in \Psi} \widehat{\psi}(0) \sum_{\nu \in \mathbb{Z}^n} c_\psi^h(\nu) = \lim_{h\to 0} h^n \sum_{\nu \in \mathbb{Z}^n} c_\chi^h(\nu).$$

Let S be the set of all vectors $w = (w_{\psi,\gamma})$ where $\psi \in \Psi$ and $|\gamma| < k$, satisfying

$$\lim_{h\to 0} \sum_{\psi \in \Psi} \sum_{|\gamma|<k} w_{\psi,\gamma} h^n \sum_{\nu \in \mathbb{Z}^n} c_\psi^h(\nu) V_\gamma(\nu) = 0.$$

We claim $S^\perp$ contains the vector w' with $w'_{\chi,0} = 1$. If not, then $v_{\chi,0} = 0$ for all $v \in S^\perp$. It then follows that $(S^\perp)^\perp$ contains the vector $\delta_{\chi,\psi}\delta_{0,\sigma}$, $\psi \in \Psi$, $|\sigma| < k$, which in turn lies in S. This gives rise to the contradiction

$$\begin{aligned} 0 &= \lim_{h\to 0} \sum_{\psi \in \Psi} \sum_{|\sigma|<k} \delta_{\chi,\psi}\delta_{0,\sigma} h^n \sum_{\nu \in \mathbb{Z}^n} c_\psi^h(\nu) V_\sigma(\nu) \\ &= \lim_{h\to 0} h^n \sum_{\nu \in \mathbb{Z}^n} c_\chi^h(\nu) \\ &= 1. \end{aligned}$$

Define

$$\psi_\gamma = (-1)^{|\gamma|} \sum_{\psi \in \Psi} w'_{\psi,\gamma} \psi, \qquad |\gamma| < k.$$

Then

$$\widehat{\psi}_0(0) = \sum_{\psi \in \Psi} w'_{\psi,0} \widehat{\psi}(0) = w'_{\chi,0} \widehat{\chi}(0) = 1.$$

Furthermore, by Exercise 3.6.1,

$$\lim_{h\to 0} h^n \sum_{\psi \in \Psi} \sum_{0 \le \beta \le \alpha} [V_\beta(iD)\widehat{\psi}](2\pi\mu) \sum_{\nu \in \mathbb{Z}^n} V_{\alpha-\beta}(\nu) c_\psi^h(\nu) = 0,$$

for $|\alpha| < k$ and $\mu \neq 0$. Making the change of variable $\alpha - \beta = \gamma$ gives

$$\lim_{h\to 0} h^n \sum_{\psi \in \Psi} \sum_{0 \le \gamma \le \alpha} [V_{\alpha-\gamma}(iD)\widehat{\psi}](2\pi\mu) \sum_{\nu \in \mathbb{Z}^n} V_\gamma(\nu) c_\psi^h(\nu) = 0,$$

for $|\alpha| < k$ and $\mu \neq 0$. This shows that

$$\{[V_{\alpha-\gamma}(iD)\widehat{\psi}](2\pi\mu)\}_{\psi,\gamma}$$

is in S for $|\alpha| < k$ and $\mu \neq 0$. Hence

$$\sum_{\psi\in\Psi}\sum_{|\gamma|<k} w'_{\psi,\gamma}[V_{\alpha-\gamma}(iD)\widehat{\psi}](2\pi\mu) = 0, \quad \mu \neq 0,\ |\alpha| < k.$$

Finally, for $\mu \neq 0$,

$$\begin{aligned}
\sum_{0\le\beta\le\alpha} V_\beta(-iD)\widehat{\psi}_{\alpha-\beta}(2\pi\mu) &= \sum_{0\le\beta\le\alpha}(-1)^{|\beta|}V_\beta(iD)\left[(-1)^{|\alpha-\beta|}\sum_{\psi\in\Psi} w'_{\psi,\alpha-\beta}\widehat{\psi}\right](2\pi\mu) \\
&= (-1)^{|\alpha|}\sum_{\psi\in\Psi}\sum_{0\le\beta\le\alpha} w'_{\psi,\alpha-\beta}[V_\beta(iD)\widehat{\psi}](2\pi\mu) \\
&= (-1)^{|\alpha|}\sum_{\psi\in\Psi}\sum_{|\gamma|<k} w'_{\psi,\gamma}[V_{\alpha-\gamma}(iD)\widehat{\psi}](2\pi\mu) \\
&= 0,
\end{aligned}$$

for $\mu \neq 0$ and $|\alpha| < k$. ■

We conclude this account with a brief and sketchy outline of how the Strang–Fix conditions are applied to radial basis approximation. Our treatment comes from Jackson [14]. We will confine our attention to the case where Ψ consists of a single function ψ. The only radial function in our list which has sufficiently rapid decay to permit a direct application of the theory is the Gaussian. However, there is in general a possibility that the convolution approximation $f_h = S_{1/h}(\psi * c_\psi^h)$ can be formed from a radial function $\phi(\|\cdot\|_2)$ by setting

$$\psi(x) = \sum_{\mu\in I} a_\mu \phi(\|x-\mu\|_2) \tag{4}$$

where we seek the index set I, and the coefficients a_μ such that

(i) $\psi \in E$

(ii) $\widehat{\psi}(0) = 1$ and $[V_\beta(-D)\widehat{\psi}](2\pi\nu) = 0$, $\nu \in \mathbb{Z}^n \setminus \{0\}$.

Thus there is a possibility (at least) that the Strang–Fix theory will be applicable to radial basis functions. Suppose initially that $\phi \in L^1(\mathbb{R}^n)$. Then,

$$\widehat{\psi}(t) = \sum_{\mu\in I} a_\mu [T_\mu\phi]\widehat{\ }(t) = \widehat{\phi}(t)\sum_{\mu\in I} a_\mu e^{-i\mu t}. \tag{5}$$

Now, using **2.1** we have that if $\phi(r) = e^{-r^2}$ then

$$\widehat{\phi}(t) = \pi^{n/2}e^{-\pi^2\|t\|_2^2}.$$

Thus our equation $\widehat{\psi}(0) = 1$ demands that

$$1 = \widehat{\psi}(0) = \widehat{\phi}(0)\sum_{\mu\in I} a_\mu e^{-i\mu.0} = \pi^{n/2}\sum_{\mu\in I} a_\mu. \tag{6}$$

In addition, we must ensure at least that for $\nu \in \mathbb{Z}^n \setminus \{0\}$,

$$0 = \widehat{\psi}(2\pi\nu) = \widehat{\phi}(2\pi\nu)\sum_{\mu\in I} a_\mu e^{-i\mu.2\pi\nu} = \pi^{n/2}e^{-\pi^2\|2\pi\nu\|_2^2}\sum_{\mu\in I} a_\mu. \tag{7}$$

Thus (6) and (7) provide conflicting requirements on $\sum_{\mu\in I} a_\mu$, and we therefore draw the rather surprising conclusion that the Gaussian choice for ϕ in this context is extremely poor.

Further progress on these questions demands the use of distributions, since none of the radial basis functions apart from the Gaussian has a Fourier transform in the classical sense. We want to conclude this part of our discussion with an *informal* treatment of one example of the distributional case. We consider the case $\phi(r) = r$. Then the function $x \mapsto \phi(\|x\|_2)$, $x \in \mathbb{R}^n$, is a (tempered) distribution (Rudin [21]), and hence has a Fourier transform in the distributional sense. This transform can be identified with the function $g : \mathbb{R}^n \to \mathbb{R}$ given by

$$g(t) = \begin{cases} B\|t\|^{-n-1}, & t \neq 0 \\ 0, & t = 0 \end{cases},$$

where B is a known constant. We now seek to manufacture a function ψ of the form given in equation (4) satisfying the conditions (i) and (ii) given immediately after that equation. If we are successful in obtaining ψ in E, it will follow (see [21]) that $\widehat{\psi}$ is in $C(\mathbb{R}^n)$. This in turn forces some behaviour on the choice of I and $\{a_\mu : \mu \in I\}$ via (5) as follows. For $t \neq 0$, expanding by Taylor series,

$$\begin{aligned} \widehat{\psi}(t) &= B\|t\|^{-n-1}\Big(\sum_{\mu\in I} a_\mu e^{-i\mu t}\Big) \\ &= B\|t\|^{-n-1}\sum_{\mu\in I} a_\mu \sum_{m=0}^{\infty} \frac{(-i\mu t)^m}{m!} \\ &= B\|t\|^{-n-1}\sum_{m=0}^{\infty} \frac{(-i)^m}{m!}\sum_{\mu\in I} a_\mu(\mu t)^m. \end{aligned} \tag{8}$$

If we examine (5) in the special case we are considering, and recall that we expect $\widehat{\psi}$ to be in $C(\mathbb{R}^n)$, then it is plain from the form of $\widehat{\psi}$ that the only problem with continuity is at the point $t = 0$. For $t \neq 0$ we have,

$$\widehat{\psi}(t) = B\|t\|_2^{-n-1}\Big(\sum_{\mu\in I} a_\mu e^{-i\mu t}\Big),$$

and expanding by Taylor series when t is small, we get

$$\begin{aligned} \widehat{\psi}(t) &= B\|t\|_2^{-n-1}\sum_{\mu\in I} a_\mu\left[\sum_{m=0}^{n+1} \frac{(-i\mu t)^m}{m!} + o(\|t\|_2^{n+1})\right] \\ &= B\|t\|^{-n-1}\Big(\sum_{m=0}^{n+1} \frac{(-i)^m}{m!}\sum_{\mu\in I} a_\mu(\mu t)^m + o(\|t\|_2^{n+1})\Big). \end{aligned}$$

In this last expression, we see that for small t, $\widehat{\psi}(t)$ is essentially $\|t\|_2^{-n-1}$ multiplied by a polynomial in t of degree at most $n+1$. The continuity of $\widehat{\psi}$ forces n to be odd so that

$\|t\|_2^{-n-1}$ is the reciprocal of a polymonial of degree $n+1$. This restriction on the dimension is a recurring problem for all choices of ϕ as long as the index set I is restricted to be finite. For ψ to be continuous at zero with $\widehat{\psi}(0) = 1$, we see that we must have

$$\sum_{m=0}^{\infty} \frac{(-i)^m}{m!} \sum_{\mu\in I} a_\mu(\mu t)^m = B^{-1}\{\|t\|_2^{n+1} + o(\|t\|_2^{n+1})\} \quad \text{as } t \to 0. \tag{9}$$

To investigate the condition $(D^\beta\widehat{\psi})(2\pi\nu) = 0$, let us set $a(t) = B\|t\|^{-n-1}$ and $b(t) = \sum_{\mu\in I} a_\mu e^{-i\mu t}$ for $t \in \mathbb{R}^n \setminus \{0\}$. Then, by the Liebnitz rule and the periodicity of b,

$$\begin{aligned} (V_\beta(D)\widehat{\psi})(2\pi\nu) &= \sum_{0\le\alpha\le\beta} (V_{\beta-\alpha}(D)a)(2\pi\nu)(V_\alpha(D)b)(2\pi\nu) \\ &= \sum_{0\le\alpha\le\beta} (V_{\beta-\alpha}(D)a)(2\pi\nu)(V_\alpha(D)b)(0), \quad \nu \in \mathbb{Z}^n \setminus \{0\}. \end{aligned}$$

Now (9) shows that b has a zero of order n at 0, and so we can easily obtain $(D^\beta\widehat{\psi})(2\pi\nu) = 0$ for $0 \le |\beta| \le n$ and $\nu \in \mathbb{Z}^n \setminus \{0\}$. The only question remaining is the rate of decrease of $|\psi(x)|$ as $\|x\| \to \infty$. If we now examine the Strang–Fix conditions we see that there is a possibility that we can obtain an order of convergence of h^{n+1} as $h \to 0$. The only question remaining is the rate of decrease of $|\psi(x)|$ as $\|x\| \to \infty$. To obtain such an order of convergence, we need $|\psi(x)| \sim \|x\|^{-2n-1}$ as $\|x\| \to \infty$. Suppose now that (9) holds. Then it is plain from the form of $\widehat{\psi}$ that $\widehat{\psi} \in L^1(\mathbb{R}^n)$. Furthermore, it follows from (8) that the function $\widehat{\psi}$ may be differentiated as often as we please for $t \ne 0$, and that such differentiations cause the rate of decay of the derivatives as $\|t\| \to \infty$ to be at least as rapid as that of $\widehat{\psi}$ itself. The problem is the behaviour at $t = 0$. If we want to secure $|\psi(x)| \sim \|x\|^{2n+1}$ as $\|x\| \to \infty$, we have to strengthen (9). Thus we assume that the a_μ for $\mu \in I$ have been chosen so that

$$\sum_{m=0}^{2n+1} \frac{(-i)^m}{m!} \sum_{\mu\in I} a_\mu(\mu t)^m = B^{-1}\|t\|_2^{n+1} p_n(t) \tag{10}$$

where p_n is a polynomial of degree at most n, satisfying $p_n(0) = 1$. We can then write, for small t,

$$\begin{aligned} \widehat{\psi}(t) &= B\|t\|_2^{-n-1} \sum_{m=0}^{\infty} \frac{(-i)^m}{m!} \sum_{\mu\in I} a_\mu(\mu t)^m \\ &= B\|t\|_2^{-n-1}\{B^{-1}\|t\|^{n+1}p_n(t) + \frac{(-i)^{2n+2}}{(2n+2)!} \sum_{\mu\in I} a_\mu(\mu t)^{2n+2} + o(\|t\|_2^{2n+2})\} \\ &= p_n(t) + B\frac{(-i)^{2n+2}}{(2n+2)!} \frac{1}{\|t\|^{n+1}} \sum_{\mu\in I} a_\mu(\mu t)^{2n+2} + o(\|t\|_2^{n+1}). \end{aligned}$$

Exercises 3.7

3.7.1 With the above assumptions on $\widehat{\psi}$, show that $D^\alpha\widehat{\psi}$ is integrable in a neighbourhood of the origin for $|\alpha| \le 2n+1$. (Recall that n is odd, so $\|t\|_2^{n+1}$ is in fact a polynomial, and also that $\|t\|^j$ is integrable in a neighbourhood of the origin for $j \ge -n+1$.)

We can now write $(V_\alpha(-D)\widehat{\psi})\widehat{\ } = V_\alpha(\widehat{\psi})\widehat{\ }$. Since $V_\alpha(-D)\widehat{\psi} \in L^1(\mathbb{R}^n)$ it follows by the Riemann–Lebesgue Lemma that $|V_\alpha(\widehat{\psi})\widehat{\ }(x)| \to 0$ as $\|x\| \to \infty$. By a slightly illegal argument, this allows us to conclude $|(V_\alpha\psi)(x)| \to 0$ as $\|x\| \to \infty$. There remains only the question of choosing the coefficients a_μ and the index set I in accordance with (10). The interested reader can refer to Jackson [14] or Powell [20] for more details on this aspect of the theory. We prefer to close with a theorem summarising the above results.

Theorem 3.14 *Let $f \in W^n_\infty(\mathbb{R}^n)$ and define $\psi : \mathbb{R}^n \to \mathbb{R}$ by $\psi(x) = \|x\|$. Then there exist $c^h : \mathbb{Z}^n \to \mathbb{R}$ such that*

$$\|f - S_{1/h}(\psi * c^h)\|_\infty \leq Ah^{n+1}|f|_{n,\infty}.$$

Acknowledgements

It is a pleasure to acknowledge research collaboration with Ward Cheney and Julie Halton. In addition, Martin Buhmann helped greatly in clarifying my thinking on a number of matters, and Rick Beatson pointed out a number of errors in a first draft of this manuscript.

References

[1] C. de Boor and R.Q. Jia, *Controlled approximation and a characterisation of the local approximation order*, Proc. Amer. Math. Soc. **95** (1985), 547-553.

[2] A.L. Brown, *Uniform Approximation by Radial Basis Functions* Appendix B in *Radial Basis Functions in 1990* – see [20].

[3] M.D. Buhmann, *Multivariable Interpolation using Radial Basis Functions*, Ph.D. Dissertation, University of Cambridge, 1989.

[4] E.W. Cheney and W.A. Light, *Quasi-interpolation with base functions having non-compact support*, Constr. Approx. (to appear).

[5] K.C. Chung and T.H. Yao, *On lattices admitting unique Lagrange interpolation* SIAM J. Num. Anal. **14** (1977), 735-741.

[6] J. Duchon, *Splines minimizing rotation-invariant seminorms in Sobolev spaces*, in *Constructive Theory of Functions of Several Variables, Lecture Notes in Mathematics 571*, eds. W. Schempp and K. Zeller, Springer–Verlag (Berlin), 1977, 85-100.

[7] N. Dyn, *Interpolation of scattered data by radial functions*, in *Topics in multivariate approximation*, eds. C.K. Chui, L.L. Schumaker and F. Utreras, Academic Press (New York), 1987, 47–61.

[8] N. Dyn, *Interpolation and approximation by radial and related functions*, in *Approximation Theory VI: Volume 1*, eds. C.K. Chui, L.L. Schumaker and J.D. Ward, Academic Press (New York), 1989, 211–234.

[9] G. Fix and G. Strang, *Fourier analysis of the finite element method in Ritz-Galerkin theory*, *Stud. Appl. Math.*, Vol. 48, 1969, 265-273.

[10] R. Franke, *Scattered data interpolation: tests of some methods*, *Math. Comp.*, Vol. 38, 1982, 181-200.

[11] E.J. Halton and W.A. Light, *On Local and Controlled Approximation Order*, J. Approx. Th. (to appear).

[12] R.L. Hardy, *Multiquadric equations of topography and other irregular surfaces*, *J. Geophys. Res.*, Vol. 76, 1971, 1905-1915.

[13] R.L. Hardy, *Theory and applications of the multiquadric-biharmonic method*, *Comput. Math. Applic.*, Vol. 19, 1990, 163-208.

[14] I.R.H. Jackson, *Radial Basis Function Methods for Multivariable Approximation*, Ph.D. Dissertation, University of Cambridge, 1988.

[15] R.-Q. Jia, *A counterexample to a result concerning controlled approximation*, Proc. Amer. Math. Soc. **97** (1986), 647-654.

[16] R.-Q. Jia and J. Lei, *Approximation by multiinteger translates of functions having non-compact support*, Preprint, 1990.

[17] W.R. Madych and S.A. Nelson, *Multivariate interpolation and conditionally positive definite functions*, *Approx. Theory Appl.*, Vol. 4, 1988, 77-89.

[18] W.R. Madych and S.A. Nelson, *Multivariate interpolation and conditionally positive definite functions II*, *Math. Comp.*, Vol. 54, 1990, 211-230.

[19] C.A. Micchelli, *Interpolation of scattered data: distance matrices and conditionally positive definite functions*, *Constr. Approx.*, Vol. 2, 1986, 11-22.

[20] M.J.D. Powell, *Radial Basis Functions in 1990* in *Advances in Numerical Analysis Volume II – Wavelets, Subdivision Algorithms and Radial Basis Functions* Oxford University Press, 1991, 105-210.

[21] W. Rudin, *Functional Analysis* 2nd ed., McGraw-Hill, 1973.

[22] I.J. Schoenberg, *Contributions to the problem of approximation of equi-distant data by analytic functions*, A. B. Quart. Appl. Math. **4** (1946), 45-99 and 112-141.

[23] G. Strang and G. Fix, *A Fourier analysis of the finite–element variational method* in *Constructive aspects of functional analysis*, (G. Geymonat, ed.), C.I.M.E., (1973), 793-840.

A TUTORIAL ON MULTIVARIATE WAVELET DECOMPOSITION

CHARLES A. MICCHELLI[1]
IBM Research Center
P. O. Box 218
Yorktown Heights, New York 10598
e-mail address: CAM@YKTVMZ. bitnet

1 Introduction

We shall document here four one-hour lectures delivered at the NATO ASI on Approximation Theory, Spline Functions and Applications held at Acquafredda di Maratea, Italy, Spring 1991. In the style of the Nato ASI, we have made these lectures nearly self-contained. The published results we draw upon appear in the three papers [1, 2, 4]. Nearly all of the material described here comes from joint work with R. Q. Jia.

Our subject is wavelet decomposition. Generally speaking, we view this as a study of orthogonal decomposition of L^2 on some euclidean space generated by two basic operations: shift and scale. This leads us to consider questions concerning matrices over the ring of trigonometric series. We begin in the first section by reviewing basic concepts and establish some notational conventions to be used later.

2 Preliminary Facts and Multiresolution

The basic operations which we consider are

$$(sh^y f)(x) = f(x-y),\ x, y \in \mathbb{R}^s \tag{2.1}$$

and

$$(sc^k f)(x) = f(2^k x),\ x \in \mathbb{R}^s, k \in \mathbb{Z}. \tag{2.2}$$

Certainly, these are bounded linear operators acting on $L^2(\mathbb{R}^s)$ equipped with the scalar product

$$(f, g) = \int_{\mathbb{R}^s} f(t)\overline{g}(t)dt,\ f, g \in L^2(\mathbb{R}^s). \tag{2.3}$$

[1] **Partially supported by a DARPA grant and an SERC visiting fellowship to the University of Cambridge.**

S. P. Singh (ed.), Approximation Theory, Spline Functions and Applications, 191–212.

Within $L^2(\mathbb{R}^s)$ we will consider subspaces generated by integer shifts of one fixed function ϕ. To this end, we set

$$[c,\phi] := \sum_{\alpha\in\mathbb{Z}^s} c_\alpha sh^\alpha\phi \tag{2.4}$$

where $c = (c_\alpha : \alpha \in \mathbb{Z}^s)$ (whenever this sum is convergent, *a.e.* $\mathbb{R}^s$). Associated with the function ϕ is the space

$$V(\phi) := \{[c,\phi] : c \in \ell^2(\mathbb{Z}^s)\} \tag{2.5}$$

and its scaled counterparts

$$V_k(\phi) := sc^k V(\phi). \tag{2.6}$$

To study these scale of spaces we introduce a Banach subspace of $L^2(\mathbb{R}^s)$ as follows. For every $\phi \in L^2(\mathbb{R})$ we set

$$\phi^0(x) = \sum_{\alpha\in\mathbb{Z}} |\phi(x-\alpha)|. \tag{2.7}$$

Whenever ϕ^0 is defined on $\mathbb{R}^s$ it represents there is a one-periodic nonnegative function. We define a new norm

$$|\phi|_2 := \|\phi^0\|_{L^2([0,1]^s)} \tag{2.8}$$

and let $\mathcal{L}^2$ be the set of all Lebesgue measurable functions with $|\phi|_2 < \infty$. Clearly $\mathcal{L}^2 \subset L^2$ and in fact since

$$\begin{aligned}\|\phi\|_2^2 &= \int_{\mathbb{R}^s} |\phi(x)|^2 dx \\ &= \int_{[0,1]^s} \sum_{\alpha\in\mathbb{Z}^s} |\phi(x+\alpha)|^2 dx \le \int_{[0,1]^s} (\sum_{\alpha\in\mathbb{Z}^s} |\phi(x+\alpha)|)^2 dx\end{aligned}$$

we get

$$\|\phi\|_2 \le |\phi|_2. \tag{2.9}$$

Conversely, it is easy to see that every $\phi \in L^2(\mathbb{R}^s)$ of compact support is in $\mathcal{L}^2$. Specifically, if n is a positive integer such that $\phi(x) = 0$, whenever $x = (x_1, \ldots, x_s)$ satisfies

$$|x|_\infty = \max\{|x_i| : 1 \le i \le s\} > n, \tag{2.10}$$

then

$$\begin{aligned}|\phi|_2^2 &= \int_{[0,1]^s} (\sum_{\alpha\in\mathbb{Z}^s} |\phi(x-\alpha)|)^2 dx \\ &= \int_{[0,1]^s} (\sum_{\alpha\in J} |\phi(x-\alpha)|)^2 dx\end{aligned}$$

where

$$J := \{\alpha : \exists x \in [0,1]^s, \text{ with } |x - \alpha|_\infty \leq n\}.$$

Since $\#J \leq m_0 := (2n+2)^s$ we get

$$|\phi|_2 \leq m_0^{\frac{1}{2}} \|\phi\|_2. \tag{2.11}$$

Next, we need to review some operations on biinfinite vectors. First, we recall the process of convolving two biinfinite vectors a and b,

$$(a * b)_\alpha := \sum_{\beta \in \mathbb{Z}^s} a_\beta b_{\alpha - \beta}, \ \alpha \in \mathbb{Z}^s. \tag{2.12}$$

Hand in hand with convolution goes the association of a biinfinite vector with a trigonometric series,

$$(\text{trig } a)(w) := \sum_{\alpha \in \mathbb{Z}^s} a_\alpha e^{-i\alpha \cdot w}. \tag{2.13}$$

The mapping trig has several useful properties. It is a nonexpansive mapping from $\ell^1(\mathbb{Z}^s)$ into $C[-\pi,\pi]^s$ (with the maximum norm). Moreover, as a map on $\ell^2(\mathbb{Z}^s)$ it acts as an isometry from ℓ^2 onto $L^2[-\pi,\pi]^s$, with normalized Lebesgue measure. In other words,

$$\begin{aligned} \|a\|_2^2 : &= \sum_{\alpha \in \mathbb{Z}^s} |a_\alpha|^2 = (2\pi)^{-s} \int_{[-\pi,\pi]^s} |(\text{ trig } a)(w)|^2 dw \\ : &= \|\text{ trig } a\|_2^2. \end{aligned} \tag{2.14}$$

Convolution and trig combine to form an important, but easily verified, identity

$$\text{trig } a * b = \text{ trig } a \cdot \text{ trig } b. \tag{2.15}$$

A useful and well-known inequality follows from these facts. We have in mind that

$$\begin{aligned} \|a * b\|_2 &= \|\text{trig } a * b\|_2 \\ &= \|\text{trig } a \cdot \text{trig } b\|_2 \\ &\leq \|\text{trig } b\|_\infty \cdot \|\text{trig } a\|_2. \end{aligned}$$

Therefore, we obtain

$$\|a * b\|_2 \leq \|b\|_1 \|a\|_2, \tag{2.16}$$

an inequality which we make use of in our first proposition.

Proposition 2.1

$$\|[c,\phi]\|_2 \leq |\phi|_2 \cdot \|c\|_1, \tag{2.17}$$

$$\|[c,\phi]\|_2 \leq |\phi|_2 \cdot \|c\|_2. \tag{2.18}$$

Proof.

$$
\begin{aligned}
([c,\phi])^0(x) &= \sum_{\alpha\in\mathbb{Z}^s} |\sum_{\beta\in\mathbb{Z}^s} c_\beta\phi(x-\alpha-\beta)| \\
&\le \|c\|_1\phi^0(x),\ x\in\mathbb{R}^s.
\end{aligned}
$$

This gives the first inequality above. For the second, we observe that

$$
\begin{aligned}
\|[c,\phi]\|_2^2 &= \sum_{\beta\in\mathbb{Z}^s}\int_{[0,1]^s+\beta} |[c,\phi](x)|^2 dx \\
&= \int_{[0,1]^s}\sum_{\beta\in\mathbb{Z}^s} |[c,\phi](x+\beta)|^2 dx.
\end{aligned}
$$

Using (2.16) with $a=c$ and $b=\phi(x+\cdot)$ gives

$$
\begin{aligned}
\sum_{\beta\in\mathbb{Z}^s} |[c,\phi](x+\beta)|^2 &\le \|b\|_1^2\|c\|_2^2 \\
&= \phi^0(x)\|c\|_2^2,\ x\in\mathbb{R}^s.
\end{aligned}
$$

Using this inequality in the above equation proves the remaining claim.

Remark 2.1. From (2.18) we see that if $\phi\in\mathcal{L}^2$ and

$$\|[c,\phi]\|_2 \ge m\|c\|_2$$

for some $m>0$ then $V(\phi)$ is a closed subspace of $L^2(\mathbb{R}^s)$.

We now review the multiresolution point of view as it is fundamental to our wavelet decomposition of $L^2(\mathbb{R}^s)$. We start with a $\phi\in L^2(\mathbb{R}^s)$. We say ϕ admits multiresolution whenever

(i) There exists positive constants m, K such that

$$m\|c\|_2 \le \|[c,\phi]\|_2 \le K\|c\|_2.$$

(We say ϕ has ℓ^2 – stable integer translates)

(ii)

$$V_k(\phi)\subset V_{k+1}(\phi),\ k\in\mathbb{Z}.$$

(iii)

$$\cap_{k\in\mathbb{Z}}V_k(\phi)=0,\ \overline{\cup_{k\in\mathbb{Z}}V_k(\phi)}=L^2(\mathbb{R}^s).$$

Our first theorem gives sufficient conditions for ϕ to admit multiresolution.

Theorem 2.1 *Let ϕ be in $\mathcal{L}^2$. Suppose there is a positive constant $m>0$ such that*

$$m\|c\|_2 \le \|[c,\phi]\|_2 \tag{2.19}$$

and a sequence $a=(a_\alpha:\alpha\in\mathbb{Z}^s)\in\ell^1(\mathbb{Z}^s)$ such that

$$\phi = sc[a,\phi]. \tag{2.20}$$

Then ϕ admits multiresolutions.

Proof.
(i) follows from (2.19) and (2.18) by choosing $K = |\phi|_2$. For (ii) we first note the following formula

$$[c, [a, \phi]] = [a * c, \phi]. \tag{2.21}$$

valid for $c \in \ell^2(\mathbf{Z}^s)$, $a \in \ell^1(\mathbf{Z}^s)$ and $\phi \in \mathcal{L}^2$. Therefore, according to (2.16) both sides of (2.21) represent a function in $V(\phi)$ when $a \in \ell^1(\mathbf{Z}^s)$ and $c \in \ell^2(\mathbf{Z}^s)$. Now given $k \in \mathbf{Z}^s$ and $c \in \ell^2(\mathbf{Z}^s)$ we have by (2.20)

$$\begin{aligned} sc^k[c,\phi] &= sc^k[c, sc[a,\phi]] \\ &= sc^{k+1}[c,[a,\phi]] \\ &= sc^{k+1}[c * a, \phi] \\ &\in V_{k+1} \end{aligned}$$

which proves (ii).

The proof of (iii) is longer. For the first part of this claim, we choose $f \in \cap_{k \in \mathbf{Z}} V_k$. Then for each $j \in \mathbf{Z}$ there is a $d \in \ell^2(\mathbf{Z}^s)$ such that

$$sc^j f = [d, \phi].$$

By our hypothesis (2.19) we have

$$\|d\|_2 \leq m^{-1}\|sc^j f\|_2 = m^{-1}2^{-js/2}\|f\|_2. \tag{2.22}$$

Also, for each $x \in \mathbf{R}^s$

$$|[d,\phi](x)| \leq \|d\|_\infty \phi^0(x) \leq \|d\|_2 \phi^0(x)$$

and so

$$|(sc^j f)(x)| \leq \|d\|_2 \phi^0(x).$$

Pick any ball $B \subset \mathbf{R}^s$ centered at the origin and integrate the above pointwise inequality over the set $2^{-j}B$ to obtain

$$\begin{aligned} \|f\|_{L^2(B)} &= 2^{sj/2}\|sc^j f\|_{L^2(2^{-j}B)} \\ &\leq 2^{sj/2}\|d\|_2\|\phi^0\|_{L^2(2^{-j}B)}. \end{aligned}$$

Combining this with (2.22) gives

$$\|f\|_{L^2(B)} \leq m^{-1}\|f\|_2\|\phi^0\|_{L^2(2^{-j}B)}.$$

Since this is valid for all j and $\phi^0 \in L^2([0,1]^s)$ we conclude, by sending $j \to \infty$, that $f = 0$.

For (iii), part two, we consider the linear operator

$$T_k f := sc^k[ssc^{-k}f, \phi]$$

where

$$(ssc^{-k}f)(\alpha) := (sc^{-k}f)(\alpha) = f(2^{-k}\alpha),\ \alpha \in \mathbf{Z}^s.$$

Our first observation about this operator is that it is bounded in the $L^2(\mathbb{R}^s)$ norm independent of $k \in \mathbb{Z}$ whenever $f \in C_0(\mathbb{R}^s)$. To see this we choose an integer $r > \frac{1}{2}$ such that $f(x) = 0$ whenever $|x|_\infty > r$. Then by (2.18),

$$\begin{aligned}
\|T_k f\|_2^2 &= 2^{-ks}\|[ssc^{-k}f,\phi]\|_2^2 \\
&\leq 2^{-ks}|\phi|_2^2\|ssc^{-k}f\|_2^2 \\
&\leq 2^{-ks}\#\{\alpha : |\alpha|_\infty \leq 2^k r\}|\phi|_2^2\|f\|_\infty^2 \\
&= (2r + 2^{-k})^s|\phi|_2^2\|f\|_\infty^2 \\
&\leq 4^s r^s|\phi|_2^2\|f\|_\infty^2,
\end{aligned}$$

that is,

$$\|T_k f\|_2 \leq (4r)^{s/2}|\phi|_2\|f\|_\infty, \; k \in \mathbb{Z}. \tag{2.23}$$

Using this bound we will next show that

$$\lim_{k\to\infty} T_k f = \hat{\phi}(0)f \tag{2.24}$$

weakly in $L^2(\mathbb{R}^s)$ for every $f \in C_0(\mathbb{R}^s)$. In view of (2.23), it suffices to prove that

$$\lim_{k\to\infty} (T_k f, g) = \hat{\phi}(0)(f, g) \tag{2.25}$$

for a dense family of functions g in $L^2(\mathbb{R}^s)$. For this purpose, we introduce the Fourier transform

$$\hat{f}(w) := \int_{\mathbb{R}^s} f(x)e^{-iw\cdot x}dx$$

and restrict g so that $\hat{g} \in C_0(\mathbb{R}^s)$. Then by the Plancherel formula the left hand side of (2.25) becomes

$$\lim_{k\to\infty} (2\pi)^{-s}(\widehat{T_k f}, \hat{g}). \tag{2.26}$$

However

$$\widehat{T_k f} = 2^{-ks}sc^{-k}(\text{trig } ssc^{-k}f)sc^{-k}\hat{\phi}$$

and therefore

$$\begin{aligned}
\|\widehat{T_k f}\|_\infty &\leq 2^{-ks}\|\text{trig } ssc^{-k}f\|_\infty\|\hat{\phi}\|_\infty \\
&\leq 2^{-ks}\#\{\alpha : |\alpha|_\infty \leq r2^k\}\|\phi\|_1\|f\|_\infty \\
&\leq (4r)^s|\phi|_2\|f\|_\infty, \; k \in \mathbb{Z}.
\end{aligned}$$

Also, since $f \in C_0(\mathbb{R}^s)$ it is straightforward to see that

$$\lim_{k\to\infty} \hat{T}_k f = \hat{\phi}(0)\hat{f}$$

pointwise on $\mathbb{R}^s$. Thus by our choice of g it follows by the bounded convergence theorem that (2.26) equals

$$\hat{\phi}(0)(2\pi)^{-s}(\hat{f}, \hat{g}).$$

By another application of Plancherel's formula this equals the right hand side of (2.25) thereby establishing this identity.

It should be clear now the importance of the nonvanishing of $\hat{\phi}(0)$. Indeed, if that were the case then we have

$$\begin{aligned} C_0(\mathbb{R}^s) &\subseteq \text{weak closure } \cup_{k\in\mathbb{Z}} V_k \\ &= \text{strong closure } \cup_{k\in\mathbb{Z}} V_k \end{aligned}$$

which finishes the proof of Theorem 2.1.

To establish that $\hat{\phi}(0) \neq 0$ we take a somewhat roundabout approach. Nevertheless through our detour, we accumulate interesting facts about ϕ and a when they satisfy the hypothesis of Theorem 2.1. We begin with the claim that

$$\hat{\phi}(2\pi\alpha) = 0,\ \alpha \in \mathbb{Z}^s\backslash\{0\}. \tag{2.27}$$

This is a consequence of the refinement equation (2.20). To see this we rewrite (2.20) in an equivalent form in the transform domain, viz.,

$$\begin{aligned} \hat{\phi} &= (sc[a,\phi])^\wedge = 2^{-s}sc^{-1}([a,\phi]^\wedge) \\ &= 2^{-s}sc^{-1}(\text{trig } a\cdot\hat{\phi}) \\ &= 2^{-s}sc^{-1}(\text{trig } a)\cdot sc^{-1}\hat{\phi}. \end{aligned}$$

Using this functional equation for $\hat{\phi}$ successively gives for each positive integer m

$$\hat{\phi} = \pi_{j=1}^{m}(2^{-s}sc^{-j}(\text{trig } a))sc^{-m}\hat{\phi}. \tag{2.28}$$

We consider two possibilities. The first is that

$$|2^{-s}(\text{trig } a)(0)| < 1. \tag{2.29}$$

Since $a \in \ell^1(\mathbb{Z}^s)$ we have, as noted earlier, that trig $a \in C(\mathbb{R}^s)$. Hence for every $\xi \in \mathbb{R}^s$ there is a $j_0 \in \mathbb{Z}$ such that

$$|2^{-s}(\text{trig } a)(2^{-j}\xi)| < 1,\ \text{ if } j \geq j_0.$$

Looking back to (2.28) and letting $m \to \infty$ and keeping in mind that $\phi \in L^1(\mathbb{R}^s)$ proves that $\hat{\phi} = 0$. Thus, certainly in this case (2.27), is true (of course, our condition (2.19) rules out the possibility that $\phi = 0$).

Alternately, when

$$|2^{-s}(\text{trig } a)(0)| \geq 1 \tag{2.30}$$

we return to (2.28) and evaluate both sides at $2^{m+1}\pi\alpha$. Simplifying the right-hand side of this formula by using the 2π-periodicity of trig a gives

$$\begin{aligned} |\hat{\phi}(2^{m+1}\pi\alpha)| &= |2^{-s}(\text{trig } a)(0)|^m|\hat{\phi}(2\pi\alpha)| \\ &\geq |\hat{\phi}(2\pi\alpha)|. \end{aligned}$$

Therefore, sending $m \to \infty$ and using the Riemann Lebesgue lemma gives (2.27) even in this case.

To complete our argument that $\hat{\phi}(0)$ is nonzero, we find it useful to reexpress the hypothesis (2.19) in the Fourier transform domain. At this point, it seems appropriate to introduce another basic operation. For every $f, g, \in \mathcal{L}^2$ we define the trigonometric series

$$\langle f, g\rangle(w) := \sum_{\alpha\in\mathbb{Z}^s} (f, sh^\alpha g) e^{-i\alpha\cdot w}, \; w \in \mathbb{R}^s. \tag{2.31}$$

Let us estimate the $\ell^1(\mathbb{Z}^s)$ of the coefficients of $\langle f, g\rangle$,

$$\begin{aligned}
\sum_{\alpha\in\mathbb{Z}^s} |(f, sh^\alpha g)| &= \sum_{\alpha\in\mathbb{Z}^s} \left|\int_{\mathbb{R}^s} f(x)\overline{g(x-\alpha)}dx\right| \\
&\le \sum_{\alpha\in\mathbb{Z}^s}\sum_{\beta\in\mathbb{Z}^s} \int_{[0,1]^s+\beta} |f(x+\alpha)||g(x)|dx \\
&= \sum_{\beta\in\mathbb{Z}^s}\sum_{\alpha\in\mathbb{Z}^s} \int_{[0,1]^s} |f(x+\alpha+\beta)||g(x+\beta)|dx \\
&= \int_{[0,1]^s} f^0(x) g^0(x) dx \\
&\le |f|_2 \cdot |g|_2.
\end{aligned}$$

Hence $\langle f, g\rangle \in C([-\pi,\pi]^s)$.

Next, we express $\langle f, g\rangle$ in terms of the Fourier transform of f and g. Using the dominated convergence theorem, it follows for every $\xi \in \mathbb{R}^s$ that

$$\begin{aligned}
\langle f, g\rangle(w) &= \sum_{\alpha\in\mathbb{Z}^s} \int_{\mathbb{R}^s} f(x+\alpha)e^{-i\alpha\cdot\xi}\overline{g(x)}dx \\
&= \sum_{\alpha\in\mathbb{Z}^s}\sum_{\beta\in\mathbb{Z}^s} \int_{[0,1]^s} f(x+\alpha+\beta)e^{-i\alpha\cdot\xi}\overline{g(x+\beta)}dx \\
&= \int_{[0,1]^s} (Rf)(x)\overline{(Rg)(x)}dx
\end{aligned}$$

where R is the mapping defined by

$$(Rf)(x) = \sum_{\alpha\in\mathbb{Z}^s} f(x+\alpha)e^{-i(x+\alpha)\cdot\xi}.$$

Clearly $|(Rf)(x)| \le f^0(x)$ and so Rf is a one-periodic square integrable function on $[0,1]^s$ whose Fourier coefficients are given by

$$\begin{aligned}
\widehat{(Rf)}(\alpha) : &= \int_{[0,1]^s} (Rf)(x)e^{-2\pi i\alpha\cdot x}dx \\
&= \int_{\mathbb{R}^s} f(x)e^{-i(\xi+2\pi\alpha)\cdot x}dx \\
&= \hat{f}(\xi+2\pi\alpha), \; \alpha\in\mathbb{Z}^s.
\end{aligned}$$

Consequently, by Parseval's identity we have

$$\int_{[0,1]^s} (Rf)(x)\overline{(Rg)(x)}dx = \sum_{\alpha\in\mathbb{Z}^s} \hat{f}(\xi+2\pi\alpha)\hat{g}(\xi+2\pi\alpha).$$

Thus, we have demonstrated that

$$\langle f,g\rangle(w)=\sum_{\alpha\in\mathbb{Z}^s}\hat{f}(w+2\pi\alpha)\overline{\hat{g}(w+2\pi\alpha)},\quad w\in\mathbf{R}^s \tag{2.32}$$

from which it follows that $\|\langle f,g\rangle\|_2\le\|f\|_2\cdot\|g\|_2$.

Proposition 2.2 *Let $\phi\in\mathcal{L}^2$. Then ϕ satisfies (2.19) for some $m>0$ (equivalently, ϕ has ℓ^2-stable integer translates) if and only if either one of the following conditions hold.*

(i)

$$\langle\phi,\phi\rangle\ge m^2 \tag{2.33}$$

(ii) There exists a $d\in\ell^1(\mathbb{Z}^s)$ such that the function

$$g=[d,\phi] \tag{2.34}$$

satisfies

$$\langle g,\phi\rangle=1. \tag{2.35}$$

Before we prove this result we note that with it we can finish the proof of Theorem 2.1. To see this we evaluate $\langle\phi,\phi\rangle$ at zero and use (2.27) to get

$$m^2\le\langle\phi,\phi\rangle(0)=\sum_{\alpha\in\mathbb{Z}^s}|\hat{\phi}(2\pi\alpha)|^2=|\hat{\phi}(0)|^2.$$

Proof.

Pick any $\xi_0\in[-\pi,\pi]^s$ and set $m_0=\langle\phi,\phi\rangle(\xi_0)$. Since $\langle\phi,\phi\rangle$ is continuous, for every $\epsilon>0$ there is $\delta\in(0,\pi)$ such that

$$\langle\phi,\phi\rangle(\xi)\le m_0+\epsilon$$

whenever $|\xi-\xi_0|_\infty\le\delta$. Let h be the 2π-periodic function defined by

$$h(\xi)=\begin{cases}1, & |\xi-\xi_0|_\infty\le\delta\\ 0, & \delta<|\xi-\xi_0|_\infty\le\pi.\end{cases}$$

Therefore, there is a $b\in\ell^2(\mathbb{Z}^s)$ such that $h=\text{trig }b$. Set $f=[b,\phi]$ and observe that

$$\begin{aligned}m^2\|b\|_2^2\le\|f\|_2^2 &= (2\pi)^{-s}\|\hat{f}\|_2^2\\ &= (2\pi)^{-s}\|h\hat{\phi}\|_2^2\\ &= (2\pi)^{-s}\|h^2\langle\phi,\phi\rangle\|^2_{L_1([0,1]^s)}\\ &= (2\pi)^{-s}\int_{|\xi-\xi_0|_\infty\le\delta}|h(\xi)|^2\langle\phi,\phi\rangle(\xi)d\xi\\ &\le (m_0+\epsilon)(2\pi)^{-s}\int_{[-\pi,\pi]^s}|h(\xi)|^2d\xi\\ &= (m_0+\epsilon)\|b\|_2^2.\end{aligned}$$

Thus $m \le m_0 + \epsilon$ and since ϵ is arbitrary our first assertion (i) has been proved.

We now turn our attention to (i) implies (ii). By Wiener's lemma, cf. [5, p. 266] there is a $d \in \ell^1$ such that

$$\text{trig } d = \langle \phi, \phi \rangle^{-1}. \tag{2.36}$$

To prove (ii) we note the useful identity

$$\langle [b, f], g \rangle = (\text{trig } b)\langle f, g \rangle, \tag{2.37}$$

valid for $f, g \in \mathcal{L}^2$ and $b \in \ell^1(\mathbb{Z}^s)$. Appropriately specializing this formula gives

$$\langle g, \phi \rangle = \langle [d, \phi], \phi \rangle = (\text{trig } d)\langle \phi, \phi \rangle = 1.$$

To complete the equivalence we show finally that (ii) implies ϕ has ℓ^2-stable integer translates. To this end, we choose any $c \in \ell^2(\mathbb{Z}^s)$ and observe that

$$\begin{aligned} \|c\|_2 &= \|\text{trig } c\|_2 \\ &= \|(\text{trig } c)\langle \phi, g \rangle\|_2 \\ &= \|\langle [c, \phi], g \rangle\|_2 \le \|g\|_2 \|[c, \phi]\|_2 \end{aligned}$$

which by Proposition 2.1 is

$$\le |\phi|_2 \|d\|_2 \|[c, \phi]\|_2.$$

This inequality finishes the proof.

As an example of a function ϕ which satisfies the hypothesis of Theorem 2.1 we offer the cube spline. For its definition we require an $s \times n$ matrix X of rank s with integer entries. Then c_X is defined by the formula

$$\int_{[0,1]^n} f(Xt)dt = \int_{\mathbb{R}^s} f(x) c_X(x) dx \tag{2.38}$$

which is valid for all $f \in C(\mathbb{R}^s)$. Clearly, c_X is nonnegative and of compact support, in fact, it is zero off the set $Z(X) := X([0,1]^n)$. c_X is a piecewise polynomial which can be defined inductively in n. For $n = s$, we have

$$c_X(x) = \begin{cases} (\det X)^{-1}, & x \in Z(X) \\ 0, & \text{otherwise} \end{cases}$$

and

$$c_{[X|y]}(x) = \int_0^1 c_X(x - ty)dt, \; x \in \mathbb{R}^s.$$

Here $[X|y]$ stands for the block matrix whose first n columns are the columns of X and its last column is the vector y. The last formula provides an alternative recursive definition of c_X which defines c_X everywhere on $\mathbb{R}^s$.

Let's now check the hypothesis of Theorem 2.1. The cube spline satisfies a refinement equation (2.20) with no additional requirements on X. In fact, choosing $f(x) = e^{-iw\cdot x}$ gives

$$\hat{c}_X(w) = \int_{[0,1]^n} e^{-iw\cdot Xt}dt \tag{2.39}$$

$$= \Pi_{j=1}^{n}\left(\frac{1-e^{-ix^j\cdot w}}{ix^j\cdot w}\right). \tag{2.40}$$

Therefore,

$$2^s\frac{\hat{c}_X(2w)}{\hat{c}_X(w)} = 2^{s-n}\Pi_{j=1}^{n}(1+e^{-ix^j\cdot w})$$

$$: = (\text{trig } a)(w)$$

and

$$c_X = sc[a, c_X].$$

The ℓ^2-stability of the integer translates of c_X requires that X be unimodular. Recall that a matrix X is unimodular provided that every $s\times s$ submatrix of X has determinant $0, +1$ or -1. To show that c_X has ℓ^2-stable integer translates in this case, it suffices to demonstrate, in view of Proposition 2.2, that for all $\xi\in\mathbb{R}^s$ there is an $\alpha\in\mathbb{Z}^s$ with $\hat{c}_X(\xi+2\pi\alpha)\neq 0$. To find α, we first re-order the columns of X so that its first s columns are linearly independent. We consider the numbers $(x^1\cdot\xi)/2\pi,\ldots,(x^n\cdot\xi)/2\pi$ and suppose the first k of them are integers. If $k\geq s$ then $\xi = -2\pi\beta$ for some $\beta\in\mathbb{Z}^s$, since the minor of X consisting of its first s columns has determinant ±1. Moreover, $\hat{c}_X(\xi+2\pi\beta) = \hat{c}_X(0) = 1$; thus $\alpha=\beta$ will do in this case. When $k<s$, we choose $\beta\in\mathbb{Z}^s$ so that $x^i\cdot\beta = -(x^i\cdot\xi)/2\pi$, $i=1,\ldots,k$. Hence we conclude that

$$x^i\cdot(2\pi\beta+\xi) = \begin{cases} 0, & i=1,\ldots,k \\ \notin 2\pi\mathbb{Z}, & i=k+1,\ldots,n \end{cases}.$$

Consequently, it follows that $\hat{c}_X(\xi+2\pi\beta)\neq 0$ and so $\alpha := \beta$ is our choice in this case. Therefore c_X has ℓ^2-stable integer translates and by Theorem 2.1 c_X admits multiresolution.

We remark without proof that c_X has ℓ^2-stable integer translates if and only if X is unimodular.

3 Multivariate Wavelet Decomposition

The main result of this section is a rather general result which provides a wavelet decomposition by functions of compact support. The proof of the result below will occupy us throughout this section.

We find it convenient to use the set E of extreme points of $[0,1]^s$ as an index set. Thus $e\in E$ means each coordinate of e is either zero or one and clearly $\#E = 2^s$.

Theorem 3.1 *Let ϕ be a function of compact support in $L^2(\mathbb{R}^s)$ which admits multiresolution. Then there exist 2^s functions $\psi_e, e\in E$ in $L^2(\mathbb{R}^s)$ of compact support such that*

(i)

$$\psi_e \in V_1(\phi) = scV(\phi),\ \psi_0 = \phi. \tag{3.1}$$

(ii)

$$V_k(\psi_e) \perp V_k(\psi_{e'}),\ e \neq e',\ k \in \mathbb{Z}. \tag{3.2}$$

(iii) Let W_k be the orthogonal complement of V_k in V_{k+1}, denoted by $V_{k+1} = V_k \bigoplus W_k$. Then

$$W_k = \bigoplus_{e \in E \setminus \{0\}} V_k(\psi_e). \tag{3.3}$$

(iv)

$$\bigoplus_{k \in \mathbb{Z}^s} W_k = L^2(\mathbb{R}^s). \tag{3.4}$$

Before we get to the main details of the proof of this result we make some general comments. The first concerns the spaces $W_k, k \in \mathbb{Z}^s$. Obviously they are pairwise orthogonal, in fact, if $k < k'$ then

$$W_k \subset V_{k+1} \subset V_{k'} \perp W_{k'}.$$

By definition, $V_{k+1} = V_k \bigoplus W_k$, $k \in \mathbb{Z}$. From this fact and property (iii) of the multiresolution setup, (iv) of Theorem 3.1 follows. This is a standard fact and can be proved in the following way. Let f be any function which is orthogonal to all W_k, $k \in \mathbb{Z}^s$. Fix $\epsilon > 0$ and choose k_0 and a $g \in V_{k_0}$ such that $\|f - g_{k_0}\|_2 \leq \epsilon$. This is possible because of (iii) of our multiresolution setup. We write $g_{k_0} = g_{k_0-1} + v_{k_0}$ where $g_{k_0-1} \in V_{k_0}$ and $v_{k_0} \in W_{k_0}$. Observe that $\|f\|_2^2 \leq \epsilon^2 + 2(f, g_{k_0})$ and $\|g_{k_0-1}\|_2 \leq \|g_{k_0}\|_2$. Repeating this process to the residuals $g_{k_0-1}, g_{k_0-2}, \ldots$ we form successively a sequence $g_k \in V_k$ for $k \leq k_0$ such that $\|g_k\|_2 \leq \|g_{k_0}\|_2$ and $\|f\|^2 \leq \epsilon^2 + 2(f, g_k)$. The sequence $\{g_k : k \leq k_0\}$ contains a weakly convergent subsequence which, again in view of (iii) of the multiresolution setup must have a zero limit. Hence $\|f\|_2 \leq \epsilon$ and so since ϵ is arbitrary we conclude that $f = 0$. This proves (iv) of Theorem 3.1.

One additional comment about (ii) and (iii) of Theorem 3.1: it suffices to prove these results for $k = 0$, the remaining cases follow by a change of variable of integration. Let us now turn to the main details of the proof.

Proof.
For sake of illustration we begin with the univariate case, $s = 1$. Here we have several simplifying features which do not appear when $s > 1$. This case will also provide the inexperienced reader with some insight into the general case.

When $s = 1$, we need to choose one function ψ_1 in $V_1(\phi)$, the other, ψ_0, being set equal to ϕ. Here we can be quite concrete and choose, for instance,

$$\psi_1 = sc[b, \phi] \tag{3.5}$$

where

$$b_j := (-1)^j \mu_{j-1},\ j \in \mathbb{Z} \tag{3.6}$$

and

$$\mu_j := \int_{\mathbb{R}} \phi(2x+j)\overline{\phi}(x)dx,\ j \in \mathbb{Z}. \tag{3.7}$$

Obviously, μ_j is zero except for a finite number of values of j and so ψ_1 is clearly of compact support. Let us next show that $\psi_1 \in W_0 = V_1 \ominus V_0$. Using our definitions above we have

$$\begin{aligned}\int_{\mathbb{R}} \psi_1(x)\overline{\phi(x-j)}dx &= \sum_{n\in\mathbb{Z}}(-1)^n \mu_{n-1}\mu_{2j-n} \\ &= \sum_{m\in\mathbb{Z}} \mu_{2m-1}\mu_{2j-2m} \\ &\qquad - \sum_{m\in\mathbb{Z}} \mu_{2m}\mu_{2j-2m-1} \\ &= 0.\end{aligned}$$

Thus the heart of the matter reduces to showing that

$$W_0 = V(\psi_1). \tag{3.8}$$

Since ϕ admits multiresolution

$$\phi = sc[a,\phi] \tag{3.9}$$

for some $a \in \ell^2(\mathbb{Z}^s)$. More can be said about a since ϕ has ℓ^2-stable integer translates. According to Proposition 2.2, part (ii) there is a $d \in \ell^1(\mathbb{Z}^s)$ such that

$$g = [d,\phi] \tag{3.10}$$

satisfies

$$\langle g,\phi\rangle = 1. \tag{3.11}$$

Recall that this d was chosen so that

$$\text{trig } d = \langle \phi,\phi\rangle^{-1}.$$

Since in this case $\langle \phi,\phi\rangle$ is a positive trigonometric polynomial the sequence d has exponentially decaying components. Hence the function g also decays exponentially fast at infinity. Moreover, since

$$\begin{aligned}\langle sc^{-1}\phi, g\rangle &= \langle [a,\phi], g\rangle \\ &= \text{trig } a \cdot \langle \phi, g\rangle \\ &= \text{trig } a,\end{aligned}$$

we see that a also has exponentially decaying components.

To proceed further, we use the following notation. For every sequence $c = (c_j \cdot j \in \mathbb{Z})$ and $e \in E = \{0,1\}$ we set

$$(\text{trig } c^e)(w) = \sum_{j \in \mathbb{Z}} c_{2j+e} e^{-ijw}, \quad w \in \mathbb{R}$$

and $(c^e)_j := c_{2j+e}, j \in \mathbb{Z}$. Let us consider next the 2×2 matrix $M = (\text{trig } m^{e,e'})_{e,e' \in E}$ defined by

$$M := \begin{bmatrix} \text{trig } a^0 & \text{trig } a^1 \\ \text{trig } b^0 & \text{trig } b^1 \end{bmatrix}. \tag{3.12}$$

Clearly,

$$\det M = \text{trig } \{a^0 * b^1 - a^1 * b^0\}$$

and by a direct calculation this reduces to

$$\det M(w) = -\langle \phi, \phi \rangle(-w).$$

Hence $M(w)$ is invertible for all $w \in \boldsymbol{R}$ and so there exists a matrix

$$\begin{bmatrix} \text{trig } c^{0,0} & \text{trig } c^{0,1} \\ \text{trig } c^{1,0} & \text{trig } c^{1,1} \end{bmatrix} = M^{-1}, \tag{3.13}$$

where each biinfinite vector $c^{e,e}, e, e' \in E$ has components which decay exponentially fast.

Let us now observe that

$$\begin{aligned} \psi_0 = \phi &= sc[a, \phi] \\ &= [a^0, sc\phi] + [a^1, sc\ sh\phi] \\ &= \sum_{e \in E} a^e sc\ sh^e \phi. \end{aligned}$$

Similarly, we have

$$\psi_1 = sc[b, \phi] = \sum_{e \in E} [b^e, sc\ sh^e \phi].$$

Therefore, in view of our definition of M, we have in summary

$$\psi_e = \sum_{e' \in E} [m^{e,e'}, sc\ sh^{e'} \phi], \ e \in E, \tag{3.14}$$

where we used the fact that M has the form

$$M = (\text{trig } m^{e,e'})_{e,e' \in E}.$$

It follows that for each $e'' \in E$

$$\begin{aligned} &\sum_{e \in E} [c^{e'',e}, \psi_e] \\ &= \sum_{e \in E} \sum_{e' \in E} [c^{e'',e}, [m^{e,e'}, sc\ sh^{e'} \phi]] \\ &= \sum_{e' \in E} [\sum_{e \in E} c^{e'',e} * m^{e,e'}, sc\ sh^{e'} \phi] \\ &= \sum_{e' \in E} [\delta_{e'',e'}, sc\ sh^{e'} \phi] \\ &= sc\ sh^{e''} \phi. \end{aligned}$$

This proves $V_1 = V_0 \bigoplus W_0$ and also Theorem 3.1 in the case $s = 1$. We now turn to the general case.

We divide the proof into three main steps. Step 1 is based on the notion of extensibility of trigonometric series. The specific result we need is the following result. It allows us to construct the needed matrices which were easily constructed in the univariate case.

Proposition 3.1 *Suppose $F(w) := (f_1(w), \ldots, f_n(w))$ is a vector map of trigonometric polynomials from $\mathbb{R}^s$ into $\mathbb{C}^n \backslash \{0\}$ with $s < 2n-1$. Then there exists an $n \times n$ nonsingular matrix $M(w), w \in \mathbb{R}^s$, of trigonometric polynomials such that the first row of $M(w)$ is $F(w)$.*

For our subsequent needs it is important to note that the above proposition is also true if F is Hölder continuous map from the torus $T := [-\pi, \pi]^s$ into $\mathbb{C}^n \backslash \{0\}$, that is, for $x, y \in T$

$$|F(x) - F(y)| \leq \kappa |x - y|^\rho,\ \rho \in (0,1]. \tag{3.15}$$

The absolute value above denotes the euclidean norms on $\mathbb{R}^s, \mathbb{C}^s$ whichever is appropriate.

We'll leave the proof of Proposition 3.1 until later. Let us instead make immediate use of it and present, in the second step of the proof, a direct sum decomposition of V_1. We claim that there exist functions $\tilde{\psi}_e, e \in E$, in $L^2(\mathbb{R}^s)$ of compact support such that $\tilde{\psi}_0 = \phi$ and

$$V_1 = \sum_{e \in E} V(\tilde{\psi}_e). \tag{3.16}$$

Moreover, these functions (collectively) have ℓ^2-stable integer translates in the sense that there are positive constants $\tilde{m}, \tilde{K}$ such that

$$\tilde{m}^2 \sum_{e \in E} \|c^e\|^2 \leq \| \sum_{e \in E} [c^e, \tilde{\psi}_e] \|^2 \leq \tilde{K}^2 \sum_{e \in E} \|c^e\|^2. \tag{3.17}$$

We base their construction on Proposition 3.1 which requires us to identify an appropriate map F. First, as in the univariate case, we note that for some $a \in \ell^2(\mathbb{Z}^s)$

$$\phi = sc[a, \phi]. \tag{3.18}$$

The same argument used in the univariate case guarantees that the components of a decay exponentially fast. Next, we set

$$(\text{trig } a^e)(w) := \sum_{\beta \in E} a_{2\beta + e} e^{-i\beta \cdot w},\ w \in \mathbb{R}^s,$$

$((a^e)_\beta := a_{2\beta+e}, \beta \in \mathbb{Z}^s)$ and introduce the map

$$F := (\text{trig } a^e : e \in E). \tag{3.19}$$

Clearly F is a C^∞-map from the torus $T = [-\pi, \pi]^s$ into $\mathbb{C}^n$ where $n := 2^s$. Certainly $s < 2n - 1$. We shall show that F misses zero. To accomplish this we consider the sequences

$$\tau_\alpha := (\phi, sh^\alpha \phi),\ \alpha \in \mathbb{Z}^s, \tag{3.20}$$

$$\mu_\alpha := (sc\ sh^\alpha \phi, \phi), \alpha \in \mathbf{Z}^s. \tag{3.21}$$

These two sequences are related. The formula we need is obtained by substituting the refinement equation (3.18) into (3.20) and then simplifying to obtain

$$\tau_\alpha = \sum_{\beta \in \mathbf{Z}^s} a_\beta \mu_{2\alpha-\beta},\ \alpha \in \mathbf{Z}^s.$$

Letting

$$(\text{trig}\ \tilde{\mu}^e)(w) = \sum_{\beta \in \mathbf{Z}^s} \mu_{2\beta-e} e^{-i\beta \cdot w},\ w \in \mathbf{R}^s$$

$((\tilde{\mu}^e)_\beta := \mu_{\beta-2e},\ \beta \in \mathbf{Z}^s)$ it follows easily from the equation above that

$$\text{trig}\ \tau = \langle \phi, \phi \rangle = \sum_{e \in E} (\text{trig}\ a^e)(\text{trig}\ \tilde{\mu}^e) \tag{3.22}$$

and so Proposition 2.2 implies that $0 \notin$ range $F(T)$.

Therefore, by Proposition 3.1 there is a nonsingular $n \times n$ matrix $M(w), w \in \mathbf{R}^s$

$$M(w) = (\text{trig}\ m^{e,e'})_{e,e' \in E} \tag{3.23}$$

with

$$m^{0,e} = a^e, e \in E \tag{3.24}$$

and $m^{e,e'}, e \in E \backslash \{0\}, e' \in E$ being vectors of finite support. Using this matrix we define functions

$$\tilde{\psi}_e := \sum_{e' \in E} [m^{e,e'}, sc\ sh^{e'} \phi], e \in E. \tag{3.25}$$

Then, as in the univariate case, we check that

$$\begin{aligned} \tilde{\psi}_0 &= \sum_{e \in E} [\text{trig}\ a^e, sc\ sh^e \phi] \\ &= sc[a, \phi] = \phi. \end{aligned}$$

Moreover, there are sequences $c^{e,e'} e, e' \in E$ whose components decay exponentially fast such that

$$(\text{trig}\ c^{e,e'})_{e,e' \in E} = M^{-1}. \tag{3.26}$$

It then follows for any $e \in E$ that

$$sc\ sh^e \phi = \sum_{e' \in E} [c^{e,e'}, \tilde{\psi}_{e'}] \tag{3.27}$$

and hence

$$V^1 = \sum_{e \in E} V(\tilde{\psi}_e). \tag{3.28}$$

To obtain the stability estimate (3.17) we note the simple fact that for any collections of vectors $\{d^e\}_{e\in E}$ in $\ell^2(\mathbb{Z}^s)$

$$\sum_{e\in E}[d^e, sc\ sh^e\phi] = sc[d,\phi], \tag{3.29}$$

where $d = (d_\alpha : \alpha \in \mathbb{Z}^s)$ is defined by setting

$$d_{2\alpha+e} := (d^e)_\alpha,\ \alpha \in \mathbb{Z}^s. \tag{3.30}$$

Therefore by the ℓ^2-stability of the integer translates of ϕ we get

$$2^{-s}m\sum_{e\in E}\|d^e\|^2 \le \|\sum_{e\in E}[d^e, sc\ sh^e\phi]\|^2 \le 2^{-s}K\sum_{e\in E}\|d^e\|^2. \tag{3.31}$$

To get the bound (3.17) we merely choose

$$d^e = \sum_{e'\in E} m^{e,e'} * c^{e'},\ e \in E. \tag{3.32}$$

Since the matrix $M(w)$ is nonsingular for all $w \in T$ and has continuous elements, it follows that there are positive constants $\underline{\rho}$ and $\overline{\rho}$ such that

$$\underline{\rho}|\zeta|^2 \le |M(w)\zeta|^2 \le \overline{\rho}|\zeta|^2$$

for all $\zeta \in \mathbb{C}^s$ and $w \in T$.

Now, using (2.15), (3.32), and the formula

$$\sum_{e\in E}\|d^e\|_2^2 = \sum_{e\in E}\|\text{trig } d^e\|_2^2$$

we get

$$\underline{\rho}\sum_{e\in E}\|c^e\|_2^2 \le \sum_{e\in E}\|d^e\|_2^2 \le \overline{\rho}\sum_{e\in E}\|c^e\|^2.$$

Combining these bounds with (3.31) and (3.32) proves the stability inequality (3.17) with $\tilde{m}^2 = 2^{-s}m\underline{\rho}$ and $\tilde{K}^2 = 2^{-s}K\overline{\rho}$.

The final step of the proof is to apply a Gram-Schmidt orthogonalization procedure. In general terms we have the following situation in mind.

Define

$$V(\phi_1,\dots,\phi_n) := \{\sum_{j=1}^{n}[c^j,\phi_j] : c^j \in \ell^2(\mathbb{Z}^s), j = 1,\dots,n\}.$$

Proposition 3.2 *Suppose $\phi_1,\dots,\phi_n \in L^2(\mathbb{R}^s)$ are of compact support and have ℓ^2-stable integer translates. Then there exist $\psi_1,\dots,\psi_n \in L^2$ of compact support and an $n\times n$ lower triangular matrix $N = (n^{i,j})$ of biinfinite vectors with finite support such that*

$$\psi_i = \sum_{j=1}^{n}[n^{i,j},\phi_j] \tag{3.33}$$

and

$$\textit{trig}\ n^{i,i} > 0,\ i = 1,\dots,n. \tag{3.34}$$

Moreover

(i) $\psi_1 = \phi_1$ *(trig* $n^{1,1} = 1$)

(ii) $V(\phi_1, \dots, \phi_k) = V(\psi_1, \dots, \psi_k)$, $k = 2, \dots, n$.

(iii) $V(\psi_j)$, $j = 1, \dots, n$ *are mutually orthogonal subspaces of* $L^2(\mathbb{R}^s)$.

Proof.
The proof is by induction on n. The case $n = 1$ is trivial. Suppose it is true for n. Let $\phi_1, \dots, \phi_{n+1}$ in $L^2(\mathbb{R}^s)$ be of compact support with ℓ^2-stable integer translates. There exist $\psi_1, \dots, \psi_n$ which satisfy (i), (ii), (iii) and a lower triangular $n \times n$ matrix N which satisfies (3.33) and (3.34). Hence we have for $i, j = 1, \dots, n$

$$\langle \psi_i, \psi_j \rangle = \begin{cases} 0, & i \neq j \\ > 0, & i = j \end{cases} .$$

We define by Wiener's lemma $a_j \in \ell^1(\mathbb{Z}^s)$ by setting

$$\text{trig } a_j = \frac{\langle \phi_{n+1}, \psi_j \rangle}{\langle \psi_j, \psi_j \rangle}, \; j = 1, \dots, n.$$

Set

$$\text{trig } c = \pi_{j=1}^n \langle \psi_j, \psi_j \rangle,$$

so that trig $c > 0$. Also, $c = (c_\alpha : \alpha \in \mathbb{Z}^s)$ is of finite support.

Our next function is defined as

$$\begin{aligned} \psi_{n+1} &= [c, \phi_{n+1} - \sum_{j=1}^n [a_j, \psi_j]] \\ &= [c, \phi_{n+1}] - \sum_{j=1}^n [c * a_j, \psi_j]. \end{aligned}$$

Let us first observe that ψ_{n+1} is of compact support. This follows from the fact that

$$\begin{aligned} \text{trig } (c * a_j) &= \text{trig } c \cdot \text{trig } a_j \\ &= \langle \phi_{n+1}, \psi_j \rangle \pi_{k \neq j} \langle \psi_k, \psi_k \rangle, \end{aligned}$$

which is clearly a trigonometric polynomial. Next, to show that $V(\psi_j) \perp V(\psi_{n+1})$, $1 \leq j \leq n$ we

$$\begin{aligned} \langle \psi_{n+1}, \psi_j \rangle &= \langle [c, \phi_{n+1}] - \sum_{k=1}^n [c * a_k, \psi_k], \psi_j \rangle \\ &= \text{trig } c \cdot \langle \phi_{n+1}, \psi_j \rangle - \text{trig } (c * a_j) \langle \psi_j, \psi_j \rangle \\ &= \text{trig } c \{ \langle \phi_{n+1}, \psi_j \rangle - \text{trig } a_j \langle \psi_j, \psi_j \rangle \} \\ &= 0. \end{aligned}$$

We remark that by the definition of ψ_{n+1} and the induction hypothesis

$$V(\psi_1, \dots, \psi_{n+1}) \subset V(\phi_1, \dots, \phi_{n+1}).$$

For the reverse inclusion, we set trig $b := (\text{trig } c)^{-1}$ where $b \in \ell^1(\mathbb{Z}^s)$. Then

$$\phi_{n+1} = [b, \psi_{n+1}] + \sum_{j=1}^{n} [a_j, \psi_j].$$

Hence for the $n+1$-st stage (ii) follows directly.

Finally, we may write ψ_{n+1} in the form

$$\psi_{n+1} = \sum_{j=1}^{n+1} [n^{n+1,j}, \phi_j]$$

where each $n^{n+1,j}$ has finite support and trig $n^{n+1,n+1} = \text{trig } c > 0$. These sequences provide the $n+1$-row of a new lower triangular matrix whose upper left $n \times n$ block correspond to the matrix N.

It remains to prove that $\psi_1, \dots, \psi_{n+1}$ are stable. This is proven analogously to our argument that established the ℓ^2-stability of $\{\tilde{\psi}\}_{e \in E}$. We do not elaborate on the details. Thus we have proved Proposition 3.2.

Returning to Theorem 3.1 we see that by choosing $n = 2^s$ and $\{\phi_j\}_{1 \le j \le n} = \{\tilde{\psi}_e\}_{e \in E}$ in Proposition 3.2 the functions $\{\psi_e\}_{e \in E}$ provided by Proposition 3.2 have all the desired properties of Theorem 3.1.

Therefore we have reached the point in the proof of Theorem 3.1 where only the proof of Proposition 3.1 remains. Let us turn to this interesting matter and assume that F is a Hölder map from T into $\mathbb{C}^n \backslash \{0\}$. We begin with the fact that the normalized map

$$G := F/|F|$$

from T into $S_{\mathbb{C}}^{n-1}$ (the sphere in $\mathbb{C}^n$) is not onto. The proof uses measure theoretic reasoning to conclude that $G_0 := G|_{\mathbb{R}^{2n-1}}$ maps T into a set of measure zero in $B_{\mathbb{R}}^{2n-1}$, the unit ball in $\mathbb{R}^{2n-1}$ (here we view the range of G in real $2n$-dimensional euclidean spaces) see [2] for more details. Hence G_0 is not onto $B_{\mathbb{R}}^{2n-1}$ and so G is not onto $S_{\mathbb{C}}^{n-1}$. By a rotation of coordinates we may assume G omits the point $v = (1, 0, \dots, 0)$, designated the "north pole" of $S_{\mathbb{C}}^{n-1}$.

For every $x \in S_{\mathbb{C}}^{n-1} \backslash \{v\}$ we define the vector

$$y = \frac{v - x}{\sqrt{2(1 - Re x_1)}}, \quad x = (x_1, \dots, x_n)$$

and the complex number

$$\zeta = \frac{2(1 - Re x_1)}{1 - x_1}.$$

In terms of these quantities we introduce the $n \times n$ matrix

$$(Q_x)_{ij} = \delta_{ij} - \zeta y_i \overline{y}_j, \; i, j = 1, \dots n.$$

It easily follows that

$$Q_x x = v$$

and

$$|\det Q_x| = 1.$$

Clearly, also the elements of Q_x are continuous for $x \in S_{\mathbb{C}}^{n-1}\backslash\{v\}$.

We define

$$M^t(w) := |F(w)|Q^H_{G(w)},\ w \in T$$

so that the first row of M is F and M is nonsingular on T. We now approximate each row of M except its first row uniformly close on T by trigonometric polynomials to find a matrix close to M which is necessarily nonsingular and therefore satisfies all the requirements of Proposition 3.1.

4 Locally Finite Decomposition

The functions constructed in Theorem 3.1 have the property that

$$sc\ sh^e\phi = \sum_{e'\in E}[d^{e,e'}, \psi_{e'}],\ e \in E, \tag{4.1}$$

where $d^{e,e'}$ are sequences whose components decay exponentially fast. This is clear from the proof. These elements are obtained from (3.27) and also the matrix N of Proposition 3.2 applied to the collection of functions $\{\phi_j\}_{1\leq j\leq n} := \{\tilde{\psi}_e\}_{e\in E}$.

The representation (4.1) has some importance in practical data compression algorithms and so it is desirable that the vectors $d^{e,e'}$ have finite support. The next theorem gives a sufficient condition to achieve this improvement of Theorem 3.1.

Theorem 4.1 *Let $\phi \in L^2(\mathbb{R}^s)$ be of compact support, admit multiresolution and have orthogonal translates*

$$\int_{\mathbb{R}^s}\phi(x)\overline{\phi(x-\alpha)}dx = \delta_{0\alpha},\ \alpha \in \mathbb{Z}^s. \tag{4.2}$$

Then there exists 2^s functions $\psi_e, e \in E$ in $L^2(\mathbb{R}^s)$ of compact support such that

(i) $\psi_e \in V_1(\phi)$, $\psi_0 = \phi$

(ii) $V(\psi_e)\perp V(\phi)$, $e \in E\backslash\{0\}$

(iii) $sc\ sh^e\phi = \sum_{e'\in E}[d^{e,e'}, \psi_{e'}]$ where each $d^{e,e'}, e, e' \in E$ is a vector of finite support.

Of course, (ii) and (iii) imply that

$$W_k = \bigoplus_{e\in E\backslash\{0\}} V_k(\psi_e),\quad k \in \mathbb{Z},$$

and so

$$\bigoplus_{k\in\mathbb{Z}} W_k = L^2(\mathbb{R}^s).$$

Note that no claim is made about the pairwise orthogonality of the spaces $V(\psi_e), e \in E\backslash\{0\}$, as in Theorem 3.

Proof.

From (4.2) and (3.22) we have

$$1 = \sum_{e\in E}(\text{trig } a^e)(\text{trig } \tilde{\mu}^e).$$

This equation is valid over the torus. However, since all summands are finite trigonometric polynomials it persists over $\mathbb{C}^s$. Thus the family of trigonometric polynomials $\{(\text{trig } a^e)(w) : e \in E\}$ are zero free for $w \in \mathbb{C}^s$. By a theorem of Quillen and Suslin, cf. [3] there exists an $n \times n$ matrix, $n = 2^s$, of trigonometric polynomials

$$\tilde{M} = (\text{trig } \tilde{m}^{e,e'})_{e,e' \in E}$$

with

$$\text{trig } \tilde{m}^{0,e} = a^e, e \in E$$

and

$$\det \tilde{M}(w) = 1, \; w \in \mathbb{C}^s. \tag{4.3}$$

As in the proof of Theorem 3.1, we set

$$\tilde{\psi}_e = \sum_{e' \in E} [\tilde{m}^{e,e'}, sc \; sh^{e'} \phi].$$

But now, in view of (4.3),

$$M^{-1} = (\text{trig } \tilde{c}^{e,e'})_{e,e' \in E}$$

is also a matrix of trigonometric polynomials and as before

$$sc \; sh^e \phi = \sum_{e' \in E} [\tilde{c}^{e,e'}, \tilde{\psi}_{e'}], \quad e \in E.$$

Finally, we set

$$\text{trig } b_e := \langle \tilde{\psi}_e, \phi \rangle$$

and introduce the functions

$$\psi_e = \begin{cases} \tilde{\psi}_e - [b_e, \phi], & e \in E \backslash \{0\} \\ \phi, & e = 0 \end{cases}.$$

Then for $e \in E \backslash \{0\}$

$$\begin{aligned} \langle \psi_e, \phi \rangle &= \langle \tilde{\psi}_e, \phi \rangle - \text{trig } b_e \langle \phi, \phi \rangle \\ &= \langle \tilde{\psi}_e, \phi \rangle - \text{trig } b_e = 0. \end{aligned}$$

Moreover,

$$sc \; sh^e \phi = \sum_{e' \in E} [c^{e,e'}, \psi_{e'}]$$

where

$$c^{e,e'} := \begin{cases} \tilde{c}^{e,0} + \sum_{e'' \in E \backslash \{0\}} \tilde{c}^{e,e''} * b_{e''}, & e' = 0 \\ \tilde{c}^{e,e'}, & e' \in E \backslash \{0\} \end{cases}.$$

These functions have all the desired properties and the theorem is proved.

We end this section with a final observation from [4] on non-wavelet decomposition (not orthogonal). Here we seek a locally finite decomposition of V_1 without orthogonality. For this purpose, we recall that a function ϕ of compact support on $\mathbb{R}^s$ has algebraic linear independent integer translates if whenever $[c, \phi] = 0$ where $c = (c_\alpha : \alpha \in \mathbb{Z}^s)$ is any biinfinite vector implies that c is zero.

Theorem 4.2 *Let ϕ be a function of compact support defined on $\mathbb{R}^s$ with algebraic linear independent integer translates which satisfies a refinement equation $\phi = sc[a,\phi]$ for some finitely supported vector $a = (a_\alpha : \alpha \in \mathbb{Z}^s)$. Then there exist 2^s functions $\psi_e, e \in E$ in $L^2(\mathbb{R}^s)$ of compact support and vectors $m^{e,e'}, d^{e,e'}, e, e' \in E$ of finite support such that*

$$sc\ sh^e\phi = \sum_{e' \in E} [d^{e,e'}, \psi_{e'}], \quad e \in E,$$

$$\psi_e = \sum_{e' \in E} [m^{e,e'}, sc\ sh^{e'}\phi], \quad e \in E,$$

and $\psi_0 = \phi$.

Proof.
Let us first observe that the Laurent polynomials

$$a^e(z) := \sum_{\beta \in \mathbb{Z}^s} a_{2\beta+e} z^\beta, \ z = (z_1, \dots, z_s), \ e \in E,$$

have no common zeros on $(\mathbb{C}\backslash\{0\})^s$. To see this we note the identity

$$\sum_{\alpha \in \mathbb{Z}^s} z^\alpha \phi(x-\alpha) = \sum_{e \in E} a^e(z^{-1})(\sum_{\beta \in \mathbb{Z}^s} z^\beta \phi(2(x-\beta)-e)), \tag{4.4}$$

valid for $x \in \mathbb{R}^s$ and $z \in (\mathbb{C}\backslash\{0\})^s$. Thus by the Quillen-Suslin Theorem [3] we can find a matrix M of Laurent polynomial with determinant identically one whose first row consists of the polynomials $a^e(z), e \in \mathbb{Z}^s$.

We define functions of compact support in V_1 by setting

$$\psi_e = \sum_{e' \in E} [m^{e,e'}, sc\ sh^{e'}\phi], \quad e \in E.$$

Thus $\psi_0 = \phi$ and

$$sc\ sh^e\phi = \sum_{e' \in E} [d^{e,e'}, \psi_{e'}]$$

where $D = (d^{e,e'})_{e,e' \in E} := M^{-1}$ satisfies the conditions of the theorem.

We remark that the cube spline $c_X(\cdot)$ satisfies the hypothesis of Theorem 4.1 provided that X is unimodular.

5 References

1. R. Q. Jia and C. A. Micchelli, Using the refinement equation for the construction of pre-wavelets II: Powers of Two, to appear, Curves and Surfaces, P. J. Laurent, A. LeMéhauté, and L. L. Schumaker (eds.), Academic Press, New York, 1991.

2. R. Q. Jia and C. A. Micchelli, Using the refinement equation for the construction of pre-wavelets V: Extensibility of Trigonometric Series, February, 1991.

3. T. Y. Lam, Serre's Conjecture, Lecture Notes in Mathematics. Springer-Verlag, New York, 1978.

4. C. A. Micchelli, Using the refinement equation for the construction of pre-wavelets, Numerical Algorithms, 1 (1991), 75-116.

5. Rudin, W., Functional Analysis, McGraw Hill Book Company, New York, 1973.

USING THE REFINEMENT EQUATION FOR THE CONSTRUCTION OF PRE-WAVELETS VI: SHIFT INVARIANT SUBSPACES

CHARLES A. MICCHELLI[1]
IBM Research Center
P. O. Box 218
Yorktown Heights, New York 10598
e-mail address: CAM@YKTVMZ. bitnet

1 Introduction

This paper follows the format of our tutorial on multivariate wavelet decomposition, [6]. We demonstrate here that the methods employed by Jia and Micchelli [2] can be extended to subspaces of $L^2(\mathbb{R}^s)$ generated by a finite number of compactly supported functions. Specifically, we let $\Phi : \mathbb{R}^s \to \mathbb{C}^N$ be a vector of N functions $\Phi = (\phi_1, \ldots, \phi_N)$. As in our previous article in these proceedings we will be concerned with wavelet decomposition in $L^2(\mathbb{R}^s)$ built upon Φ and the basic operations of shift and scale. All the results in [2] pertain to the case $N = 1$. We now proceed to extend them to general N.

Our terminology here is the same as in [6]. However, for the present setting we require some additional notation. First, the linear space of N-tuples of elements from a Hilbert space X will be denoted X_N. We use the standard euclidean norm $|y|^2 = (y, y), y \in \mathbb{C}^N$ on the vector of norms $(\|x_1\|, \ldots, \|x_N\|)$ for the norm of $x = (x_1, \ldots, x_N)$ in X_N. This makes X_N into a Hilbert space. Typical examples that we have in mind are $L^2_N(\mathbb{R}^s)$ and $\ell^2_N(\mathbb{Z}^s)$. It is convenient to have $[c, \Phi]$ defined for two choices of c. First, if $c = (c_1, \ldots, c_N) \in \ell^2_N(\mathbb{Z}^s)$ then $[c, \Phi]$ signifies the scaler-valued function

$$\sum_{j=1}^{N} [c_j, \phi_j].$$

When c consists of an $N \times N$ matrix of elements in $\ell^2(\mathbb{Z}^s)$, $c = (c_{ij})_{i,j=1,\ldots,N}$, $c_{ij} \in \ell^2(\mathbb{Z}^s)$ we let $[c, \Phi]$ denote the vector-valued function

$$([c^1, \Phi], \ldots, [c^N, \Phi])$$

where c^k is the vector $(c_{k1}, \ldots, c_{kN})$, $k = 1, \ldots, N$. Let us also use $\ell^2_{N,N}(\mathbb{Z}^s)$ as the space of all $N \times N$ matrices whose elements are in $\ell^2(\mathbb{Z}^s)$.

We say $\Phi \in L^2_N(\mathbb{R}^s)$ has ℓ^2-stable integer translates if for some positive constants m, K

$$m\|c\|_2 \le \|[c, \Phi]\|_2 \le K\|c\|_2$$

[1]Partially supported by a DARPA grant and an SERC visiting fellowship to the University of Cambridge.

S. P. Singh (ed.), Approximation Theory, Spline Functions and Applications, 213–222.

for all $c \in \ell_N^2(\mathbb{Z}^s)$. In this case, we set

$$V(\Phi) = \{[c, \Phi] : c \in \ell_N^2(\mathbb{Z}^s)\},$$

which is a closed linear subspace of $L^2(\mathbb{R}^s)$. Moreover, as in the scalar case $N = 1$, we say Φ admits multiresolution if the scale of spaces

$$V_k(\Phi) = sc^k V(\Phi), \; k \in \mathbb{Z}$$

and Φ have the following properties

(i)

$$\Phi \text{ has } \ell^2 - \text{stable integer translates,} \tag{1.1}$$

(ii)

$$V_k(\Phi) \subset V_{k+1}(\Phi), \; k \in \mathbb{Z}, \tag{1.2}$$

(iii)

$$\cap_{k \in Z} V_k(\Phi) = 0, \; \overline{\cup_{k \in Z} V_k(\Phi)} = L^2(\mathbb{R}^s). \tag{1.3}$$

2 Multiresolution

Our first result gives a sufficient condition on Φ to admit multiresolution.

Theorem 2.1 *Let $\Phi \in \mathcal{L}_N^2$. Suppose there is a positive constant $m > 0$ such that*

$$m\|c\|_2 \leq \|[c, \Phi]\|_2, \; c \in \ell_N^2(\mathbb{Z}^s), \tag{2.1}$$

and $a = (a_\alpha : \alpha \in \mathbb{Z}^s)$ a biinfinite vector of $N \times N$ matrices whose elements are in $\ell^1(\mathbb{Z}^s)$ such that

$$\Phi = sc[a, \Phi], \tag{2.2}$$

then Φ admits multiresolution.

Proof.
All the properties (1.1), (1.2) and (1.3) follow easily from arguments used in [6] for the scaler case, $N = 1$, except for the density of the space

$$\cup_{k \in \mathbb{Z}} V_k(\Phi). \tag{2.3}$$

Even in the scalar case this required the most effort. As in the scalar case the proof proceeds by establishing some interesting consequences of our hypotheses (2.1) and (2.2) which are of some independent interest. In the ensuing discussion, we use trig a to denote the $N \times N$ matrix of trigonometric series

$$(\text{trig } a)(w) := \sum_{\alpha \in \mathbb{Z}^s} a_\alpha e^{-i\alpha \cdot w}. \tag{2.4}$$

From the refinement equation (2.2) we get for any $m \in \mathbb{Z}_+^s$ and $w \in \mathbb{R}^s$

$$\hat{\Phi}(w) = \Pi_{j=1}^{m}(2^{-s}a(w/2^j))\hat{\Phi}(w/2^m). \tag{2.5}$$

Let $\rho :=$ spectral radius of $2^{-s}a(0)$. We consider two cases:

Case 1. $\rho < 1$. In this case we know that for any $\rho' \in (\rho, 1)$ there is a norm $\|\cdot\|$ on $\mathbb{C}^N$ such that

$$\|2^{-s}a(0)x\| \leq \rho'\|x\|, \; x \in \mathbb{C}^N, \tag{2.6}$$

cf. [7, p. 384]. Hence by continuity, for every $w \in \mathbb{R}^s$ and $\varepsilon > 0$ there is a j_0 such that for $j \geq j_0$

$$\|2^{-s}a(w/2^j)x\| \leq (\rho' + \varepsilon)\|x\|, \; x \in \mathbb{C}^N$$

and

$$\|\hat{\Phi}(w/2^j)\| \leq \varepsilon + \|\hat{\Phi}(0)\|.$$

Thus from (2.5) with m replaced by $m + j_0$ we get

$$\|\hat{\Phi}(w)\| \leq k(\rho' + \varepsilon)^m$$

where

$$k := \Pi_{j=1}^{j_0}\|2^{-s}a(w/2^j)\|(\varepsilon + \|\hat{\Phi}(0)\|)$$

and so, letting $m \to \infty$ proves that $\hat{\Phi}(w) = 0$. This cannot happen in view of (2.1).

Case 2. $\rho \geq 1$. In this case we let $y \in \mathbb{C}^N \backslash \{0\}$ be an eigenvector of the adjoint of the matrix $2^{-s}a(0)$, viz.

$$(2^{-s}a(0))^* y = \rho y. \tag{2.7}$$

We will now show that the function

$$\phi := (y, \Phi) \tag{2.8}$$

has the property that

$$\hat{\phi}(2\pi\alpha) = 0, \; \alpha \in \mathbb{Z}^s\backslash\{0\}. \tag{2.9}$$

For this purpose, we choose $w = 2^{m+1}\pi\alpha$ in (2.5) to get

$$\hat{\Phi}(2^{m+1}\pi\alpha) = (2^{-s}a(0))^m\hat{\Phi}(2\pi\alpha).$$

Fix $\alpha \in \mathbb{Z}^s\backslash\{0\}$, then by the Riemann-Lebesgue lemma we have

$$\lim_{m\to\infty}(2^{-s}a(0))^m\hat{\Phi}(2\pi\alpha) = 0. \tag{2.10}$$

Taking the inner product of both sides of (2.10) with the vector y and using (2.7) proves that

$$\lim_{m\to\infty}\rho^m\hat{\phi}(2\pi\alpha) = 0 \tag{2.11}$$

from which (2.9) follows.

To proceed further we recall that in [1] it was proved that a $\Phi \in \mathcal{L}_N^2$ has ℓ^2-stable integer translates if and only if there does not exist a $\xi \in \mathbb{R}^s$ such that the vector

$$(\hat{\phi}_j(\xi + 2\pi\alpha))_{\alpha \in \mathbb{Z}^s},\ j = 1, \ldots, N$$

are linearly dependent. In view of (2.9), applying this fact to $\xi = 0$ proves that

$$\hat{\phi}(0) \neq 0. \tag{2.12}$$

Now, for the remaining part of the proof we employ the sequence of operators

$$T_k f = sc^k[ssc^{-k} f, \phi]/\hat{\phi}(0), \quad k \in \mathbb{Z}, \quad f \in C(\mathbb{R}^s).$$

Then $T_k f \in V_k(\Phi)$ and since we showed in the proof of Theorem 2.1 of [6] that $T_k f \to f$ weakly in $L^2(\mathbb{R}^s)$, if $f \in C_0(\mathbb{R}^s)$ the density of the subspace (2.3) follows.

3 Pre-Wavelet Decomposition

Our next theorem extends Theorem 3.1 of [6]. Much of the proof of the general case follows that for the scalar case $N = 1$.

Theorem 3.1 *Let $\Phi = (\phi_1, \ldots, \phi_N) \in L_N^2(\mathbb{R}^s)$ be of compact support and admit multiresolution. Then there exist $2^s N$ functions $\Psi_{e,j}, e \in E, 1 \leq j \leq N$ in $L^2(\mathbb{R}^s)$ of compact support such that*

(i) $\psi_{e,j} \in V_1(\Phi) = scV(\Phi)$, $\psi_{0,j} = \phi_j$, $1 \leq j \leq N$.

(ii) $V_k(\psi_{e,j}) \perp V_k(\psi_{e',j'})$, $e, e' \in E$, $1 \leq j, j' \leq N$ *if one of the following holds: either* $e = 0$ *and* $e' \neq 0$ *or* $e, e' \in E\backslash\{0\}$ *and either* $e \neq e'$ *or* $j \neq j'$.

(iii) Let $W_k := V_{k+1} \ominus V_k$ = *orthogonal complement of* V_k *in* V_{k+1}. *Then*

$$W_k = \bigoplus_{\substack{e \in E\backslash\{0\} \\ 1 \leq j \leq N}} V_k(\psi_{e,j}),\ k \in \mathbb{Z}.$$

(iv) $\oplus_{k \in \mathbb{Z}} W_k = L^2(\mathbb{R}^s)$.

Proof.
The proof depends on the construction of certain matrices of trigonometric series. First, we need the following lemma. We say that a family of $r \times n$ matrices $F(w)$ defined for $w \in T_s$, the s-torus is a Hölder matrix if each row of F is Hölder continuous in the sense of [6].

Lemma 3.2 *Let $F(w)$ be an $r \times n$ Hölder matrix, $1 \leq r < n$ defined on the s-torus with $2n > s + 2r - 1$. If $F(w)$ has rank r for each w in T_s then there exists an $n \times n$ nonsingular matrix $M(w), w \in T_s$ of continuous functions on T_s such that the first r row of M agree with the corresponding rows of F and the remaining rows are trigonometric polynomials.*

Proof.
The case $r = 1$ was proved in [3]. We prove the general result by induction on r. Suppose the result is true for r and F is an $(r+1)\times n$ Hölder matrix of rank $r+1$ with $2n > s+2r+1$. Let M_r be an $n \times n$ nonsingular matrix on T_s whose first r rows agree with the first r rows of F and the remaining rows are trigonometric polynomials. We define the $(r+1)\times n$ Hölder matrix $H := FM_r^{-1}$. Then $H_{ij} = \delta_{ij}, 1 \le i \le r, 1 \le j \le n$. Moreover, by the Cauchy-Binet formula all the $r+1 \times r+1$ minors of F are given by

$$\begin{aligned} F\begin{pmatrix} 1,\dots,r+1 \\ j_1,\dots,j_{r+1} \end{pmatrix} &= \sum_{1\le \ell_1<\dots<\ell_r\le n} H\begin{pmatrix} 1,\dots,r+1 \\ \ell_1,\dots,\ell_{n+1} \end{pmatrix} M_r\begin{pmatrix} \ell_1,\dots,\ell_{r+1} \\ j_1,\dots,j_{r+1} \end{pmatrix} \\ &= \sum_{\ell=r+1}^{n} H_{r+1,\ell} M_r\begin{pmatrix} 1,\dots,r,\ell \\ j_1,\dots,j_{r+1} \end{pmatrix}. \end{aligned}$$

Thus the Hölder vector map $K := (H_{r+1,r+1},\dots,H_{r+1,n})$ never vanishes on T_s. Since K has $m := n-r$ coordinates and $2m-1 = 2n-2r-1 > s$ there is an $n-r \times n-r$ nonsingular matrix W on T_s whose first row is K and remaining rows are trigonometric polynomials. We define the matrix

$$V_{ij} = \begin{cases} H_{ij}, & 1 \le i \le r+1, \quad 1 \le j \le n \\ 0, & r+1 < i \le n, \quad 1 \le j \le r \\ W_{i-r,j-r}, & r+1 < i \le n, \quad r+1 \le j \le n \end{cases}$$

and set $M_{r+1} := VM_r$. Since the first $r+1$ rows of V agree with the corresponding ones of H we see that the first $r+1$ rows of M_{r+1} are equal to the corresponding rows of F. Also, the remaining rows of V are trigonometric polynomials by the choice of W. Moreover, since W is nonsingular so too is V and in fact $\det M_{r+1} = \det W \cdot \det M_r$. Therefore, we have advanced the induction and proved the lemma.

To apply this result we note that since $V_0 \subseteq V_1$ there is a biinfinite vector of $N \times N$ matrices $a = (a_\alpha : \alpha \in \mathbb{Z}^s) \in \ell^2_{N,N}(\mathbb{Z}^s)$ such that

$$\Phi = sc[a, \Phi]. \tag{3.1}$$

Since Φ has compact support and ℓ^2-stable integer translates there is an N vector G of functions whose components are in V_0 and decay exponentially fast such that

$$\int_{\mathbb{R}^s} G(x) \otimes \Phi(x-\alpha)dx = \delta_{0\alpha} I_N, \tag{3.2}$$

see [1] or [2]. Here we use I_N for the $N \times N$ identity matrix and the tensor product notation

$$(u \otimes v)_{ij} := u_i \overline{v}_j, \quad i,j = 1,\dots,N$$

for $u = (u_1,\dots,u_N), v = (v_1,\dots,v_N) \in \mathbb{C}^N$. The case $N = 1$ of (3.2) is explained in some detail in [6]. Thus we get from (3.1) and (3.2)

$$a_\alpha = \int_{\mathbb{R}^s} G(x-\alpha) \otimes \Phi(x/2)dx, \tag{3.3}$$

so that the elements of a_α, decay exponentially fast for $\alpha \to \infty$, as well.

We set

$$(\text{trig } a^e)(w) := \sum_{\alpha \in \mathbb{Z}^s} a_{2\alpha+e} e^{-iw\cdot\alpha}, \tag{3.4}$$

and introduce the map

$$F = (\text{trig } a^e : e \in E).$$

We think of F as an $N \times n$ matrix, $n := 2^s N$ given in $N \times 2^s$ block form. Clearly, F is a C^∞ map on the torus T_s and surely $2n > s + 2N - 1$. Therefore to apply Lemma 3.1 it suffices to verify that F has rank N on T_s. To this end, we introduce two biinfinite sequences of $N \times N$ matrices

$$\tau_\alpha := \int_{\mathbb{R}^s} \Phi(x) \otimes \Phi(x - \alpha) dx, \ \alpha \in \mathbb{Z}^s, \tag{3.5}$$

and

$$\mu_\alpha := \int_{\mathbb{R}^s} \Phi(2x - \alpha) \otimes \Phi(x) dx, \ \alpha \in \mathbb{Z}^s. \tag{3.6}$$

Then it follows that

$$\tau_\alpha = \sum_{\beta \in \mathbb{Z}^s} a_\beta \mu_{2\alpha - \beta}, \tag{3.7}$$

and so

$$\text{trig } \tau = \sum_{e \in E} \text{trig } a^e \text{trig } \tilde{\mu}^e, \tag{3.8}$$

where

$$(\text{trig } \tilde{\mu}^e)(w) := \sum_{\beta \in \mathbb{Z}^s} \mu_{2\beta - e} e^{-i\beta\cdot w}, \ w \in \mathbb{R}^s. \tag{3.9}$$

By the Poisson summation formula we have

$$(\text{trig } \tau)(w) = \sum_{\alpha \in =z} \hat{\Phi}(w + 2\pi\alpha) \otimes \hat{\Phi}(w + 2\pi\alpha), \ w \in \mathbb{R}^s, \tag{3.10}$$

see also formula (2.32) of [6] with $f = \phi_i$ and $g = \phi_j, 1 \le i, j \le N$. Therefore, according to [1], trig τ is nonsingular on $\mathbb{R}^s$. Hence from (3.8) it follows that F has rank N. We are now in a position to apply Lemma 3.1 to F and conclude that there is an $n \times n$ nonsingular matrix $\tilde{M}(w), w \in T_s$ whose first N rows agree with those of $F(w)$ and the remaining are trigonometric polynomials.

We express $\tilde{M}$ in block form as

$$\tilde{M} = (\text{trig } \tilde{m}^{e,e'})_{e,e' \in E} \tag{3.11}$$

where each element $\tilde{m}^{e,e'}, e, e' \in E$ is a vector of $N \times N$ matrices. Hence by our choice of $\tilde{M}$ we have

$$\tilde{m}^{0,e} = a^e, \ e \in E. \tag{3.12}$$

Next, we introduce functions $\Theta_e \in L^2_N(\mathbf{R}^s)$ of compact support by setting

$$\Theta_e = \sum_{e' \in E} [\tilde{m}^{e,e'}, sc\ sh^{e'}\Phi]. \tag{3.13}$$

Then, as in the scalar case, we observe by (3.12) and (3.1) that

$$\begin{aligned} \Theta_0 &= \sum_{e \in E} [a^e, sc\ sh^e \Phi] \\ &= sc[a, \Phi] = \Phi. \end{aligned} \tag{3.14}$$

Moreover, there is an $n \times n$ matrix of trigonometric series $(\text{trig}\ c^{e,e'})_{e,e' \in E}$ where each $c^{e,e'} \in \ell^2_N(\mathbf{R}^s)$ with components which decay exponentially fast and

$$\tilde{M}^{-1} = (\text{trig}\ c^{e,e'})_{e,e' \in E}. \tag{3.15}$$

It follows that

$$sc\ sh^e \Phi = \sum_{e' \in E} [c^{e,e'}, \Theta_{e'}] \tag{3.16}$$

and therefore

$$V_1(\Phi) = \sum_{e \in E} V(\Theta_e). \tag{3.17}$$

For the next step of the proof we write $\Theta_e = (\Theta_{e,1}, \ldots, \Theta_{e,N})$ where each $\Theta_{e,j}, e \in E, 1 \le j \le N$ are scalar-valued functions of compact support in $L^2(\mathbf{R}^s)$. To finish the proof we need to subject these functions to a Gram Schmidt orthogonalization procedure. This important step is provided in [2] and explained in [6] for the case $N = 1$. There the first function was fixed in the orthogonalization procedure. In the present circumstances we apply this procedure to the coordinates of $\Theta_e, e \in E$ where the first N functions are chosen to be $\phi_1, \ldots, \phi_N$. This gives us $2^s N$ new functions, the first of which is ϕ_1. We replace the next $N-1$ functions by $\phi_2, \ldots, \phi_N$ and label all of them $\psi_{e,j}, e \in E, 1 \le j \le N$ where $\psi_{0,j} = \phi_j, 1 \le j \le N$. These functions satisfy all the properties of Theorem 3.1.

4 Locally Finite Decomposition

The focus of this section is similar to Section 4 of [6]. We wish to provide two applications of the Quillen-Suslin Theorem, [4], [8] to wavelet decomposition based upon a finite number of functions of compact support.

The ideas here parallel Theorem 3.1 of the previous section. However, our main tool in this section is the Quillen-Suslin theorem which replaces the elementary Lemma 3.2.

We begin with

Theorem 4.1 *Let $\Phi \in L^2_N(\mathbf{R}^s)$ be of compact support, admit multiresolution and have orthogonal translates, viz.*

$$\int_{\mathbf{R}^s} \Phi(x) \otimes \Phi(x - \alpha) dx = \delta_{0\alpha} I_N,\ \alpha \in \mathbf{Z}^s. \tag{4.1}$$

Then there exist 2^s functions $\triangle_e, e \in E$ in $L^2_N(\mathbf{R}^s)$ such that *(i)* $(\triangle_e)_j \in V_1(\Phi)$, $e \in E$, $j = 1, 2, \ldots, N, \triangle_0 = \Phi$

(ii) $W_k = V_{k+1} \ominus V_k = \bigoplus_{e \in E \setminus \{0\}} V_k(\triangle_e)$

(iii) *There exist* $N \times N$ *matrices* $d^{e,e'}, e, e' \in E \bigoplus_{e \in E \setminus \{0\}}$ *whose elements are biifinite finitely supported vectors such that*

$$sc\ sh^e \Phi = \sum_{e' \in E} [d^{e,e'}, \triangle_{e'}]$$

Proof.
According to (3.5), (3.8) and our hypothesis (4.1) the $N \times n, n = 2^s N$ matrix

$$F(w) = (\text{trig } a^e(w) : e \in E) \tag{4.2}$$

is of rank N for any $w \in \mathbb{C}^s$. Thus by the Quillen-Suslin Theorem [4], [8] there is an $n \times n$ matrix $\tilde{M}, w \in \mathbb{C}^s$ of trigonometric polynomials whose determinant is one and whose first r rows agree with the first r rows of F. As in the proof of Theorem 3.1 the functions

$$\ominus_e = \sum_{e \in E} [\tilde{m}^{e,e'}, sc\ sh^{e'} \Phi],$$

are of compact support with $\ominus_0 = \Phi$ but now

$$\tilde{M}^{-1} := (\text{trig } c^{e,e'})_{e,e' \in E}$$

where each $c^{e,e'}$ is an $N \times N$ matrix whose elements are biinfinite vectos of finite support. Thus we get a locally finite decomposition

$$sc\ sh^e \Phi = \sum_{e' \in E} [c^{e,e'}, \ominus_{e'}].$$

Paralleling the proof of Theorem 4.1 of [6] we introduce for $e \in E$ the $N \times N$ matrix of trigonometric polynomials

$$(\text{trig } b^e)(w) := \sum_{\alpha \in \mathbb{Z}^s} \left(\int_{\mathbb{R}^s} \ominus_e(x) \otimes \Phi(x - \alpha) dx \right) e^{-i\alpha \cdot w}$$

and define

$$\triangle_e = \begin{cases} \Phi, & e = 0 \\ \ominus_e - [b^e, \Phi], & e \in E \setminus \{0\} \end{cases}.$$

Note that b_e is an $N \times N$ matrix of biinfinite vectors of finite support. Also, for $e \in E \setminus \{0\}$ and $\alpha \in \mathbb{Z}^s$ we have

$$\begin{aligned} &\int_{\mathbb{R}^s} \triangle_e(x) \otimes \Phi(x - \alpha) dx \\ &= \int_{\mathbb{R}^s} \ominus_e(x) \otimes \Phi(x - \alpha) dx \\ &\quad - \sum_{\beta \in \mathbb{Z}^s} \int_{\mathbb{R}^s} (b^e)_\beta \Phi(x - \beta) \otimes \Phi(x - \alpha) dx \\ &= \int_{\mathbb{R}^s} \ominus_e(x) \otimes \Phi(x - \alpha) dx - (b^e)_\alpha = 0. \end{aligned}$$

Thus it is easy to see that these functions satisfy all the demands of Theorem 4.1.

Our last theorem concerns non-wavelets.

Theorem 4.2 *Let Φ be a function in $L^2_N(\mathbb{R}^s)$ of compact support with algebraic linear independent integer translates which satisfies a refinement equation $\Phi = sc[a, \Phi]$ for some $N \times N$ matrix $a = (a_\alpha : \alpha \in \mathbb{Z}^s)$ of biinfinite vectors where each element has finite support. Then there exists 2^s functions $\Delta_e, e \in E$ in $L^2_N(\mathbb{R}^s)$ of compact support and $N \times N$ matrices $m^{e,e'}, d^{e,e'}, e, e' \in E$ of elements which are biinfinite vectors of finite support such that*

$$\begin{aligned} sc\, sh^e \Phi &= \sum_{e' \in E} [d^{e,e'}, \Delta_{e'}], \\ \Delta_e &= \sum_{e' \in E} [m^{e,e'}, sc\, sh^{e'} \Phi], \end{aligned}$$

and $\Delta_0 = \Phi$.

Proof.
The essential thing to check is that the map F given by (4.2) has rank N for $w \in \mathbb{C}^s$. Thus we only have to point out that equation (4.5) of [6] extends immediately to the vector identity

$$\sum_{\alpha \in \mathbb{Z}^s} z^\alpha \Phi(x - \alpha) = \sum_{e \in E} a^e(z^{-1}) (\sum_{\beta \in \mathbb{Z}^s} z^\beta \Phi(2(x - \beta) - e)), \tag{4.3}$$

where $a^e(z)$ is the matrix of Laurent polynomials

$$a^e(z) = \sum_{\beta \in \mathbb{Z}^s} a_{2\beta+e} z^\beta.$$

Now, if F was not of rank N at some $z = e^{-iw}, w \in \mathbb{C}^s$, then there would be a $y \in \mathbb{C}^N$ such that $y a^e(z^{-1}) = 0$ for all $e \in E$. Consequently, from (4.3) it would follow that

$$\sum_{j=1}^{N} \sum_{\alpha \in \mathbb{Z}^s} z^\alpha y_j \phi_j(x - \alpha) = 0, \; x \in \mathbb{R}^s$$

contradicting the algebraic independence of the translates of Φ. Thus F has rank N on $\mathbb{C}^s$ and the rest follows as with Theorem 4.2 of [6].

5 References

1. R. Q. Jia and C. A. Micchelli, On the linear independence of integer translates of a finite number of functions, to appear in the Proceedings of the Edinburgh Mathematical Society.

2. R. Q. Jia and C. A. Micchelli, Using the refinement equation for the construction of pre-wavelets II: Powers of Two, to appear, Curves and Surfaces, P. J. Laurent, A. LeMéhauté, and L. L. Schumaker (eds.), Academic Press, New York, 1991.

3. R. Q. Jia and C. A. Micchelli, Using the refinement equation for the construction of pre-wavelets V: Extensibility of Trigonometric Series, February, 1991.

4. T. Y. Lam, Serre's Conjecture, Lecture Notes in Mathematics, #635, Springer-Verlag, New York, 1978.

5. C. A. Micchelli, Using the refinement equation for the construction of pre-wavelets, Numerical Algorithms, 1 (1991) 75-116.

6. C. A. Micchelli, A tutorial on multivariate wavelet decomposition, these Proceedings.

7. J. Stoer and R. Bulursch, Introduction to Numerical Analysis, Springer-Verlag, New York, Inc., 1980.

8. D. C. Youla and P. F. Pickel, The Quillen-Suslin Theorem and the structure of n-dimensional elementary polynomial matrices, IEEE Transactions on Circuits and Systems, Vol. Cas.-31, #6, (1984) 513-518.

ERROR ESTIMATES FOR NEAR-MINIMAX APPROXIMATIONS

G. M. PHILLIPS
The Mathematical Institute
University of St. Andrews
St. Andrews, Fife KY16 9SS
Scotland

ABSTRACT. We begin with a brief summary of the theory of minimax polynomial approximation and give a simple proof of Bernstein's well-known result in which the minimax error $E_n(f)$ is expressed in terms of the (n+1)th derivative of f. We define a near-minimax projection P from $C^{n+1}[-1, 1]$ to P_n as one where $\|f - Pf\|_\infty$ can similarly be expressed in terms of the (n+1)th derivative of f. We show that some of the most familiar polynomial approximations are particular examples of such projections and also give some general results concerning all such projections. Finally we demonstrate the advantages of presenting these error estimates with a divided difference in place of the derivative.

1. Minimax Approximation

Consider the approximation of members of the linear space C[-1, 1] by members of P_n, the subspace of polynomials of degree at most n. We say that $q^* \in P_n$ is a best approximation to $f \in C[-1, 1]$ with respect to a given norm if

$$\|f - q^*\| \le \|f - q\| \quad \forall\, q \in P_n. \tag{1.1}$$

A best approximation always exists and, at least for the most commonly used norms (including the maximum norm), is unique. (See, for example, Cheney [6], Davis [7], Rivlin [14], Powell [19], Singer [20], Watson [22].) Here we are concerned with the *maximum* norm, defined by

$$\|f\|_\infty = \max_{-1\le x\le 1} |f(x)|. \tag{1.2}$$

A necessary and sufficient condition for $q^* \in P_n$ to be a best approximation to $f \in C[-1, 1]$ with respect to the maximum norm is that there exist n+2 points x_i such that

$$-1 \le x_0 < x_1 < \dots < x_{n+1} \le 1,$$

S. P. Singh (ed.), Approximation Theory, Spline Functions and Applications, 223–241.

$$f(x_{i+1}) - q^*(x_{i+1}) = -(f(x_i) - q^*(x_i)), \quad 0 \le i \le n,$$
$$|f(x_i) - q^*(x_i)| = \|f - q^*\|_\infty, \quad 0 \le i \le n+1.$$

This characterizes best approximations with respect to the maximum norm. It is called the equioscillation property and (the unique) q^* is called the minimax approximation. (The points x_i on which $f - q^*$ equioscillates depend on f and are usually unknown. However, two of the equioscillation points are known in the following case : if $f^{(n+1)}$ is continuous and has constant sign on [-1, 1] it may be verified by an argument based on the repeated application of Rolle's theorem that $x_0 = -1$ and $x_{n+1} = 1$.) The simplest manifestation of the above characterization property of minimax approximation involves the Chebyshev polynomials, defined by

$$T_n(x) = \cos(n \cos^{-1}(x)), \quad -1 \le x \le 1.$$

Chebyshev showed that, for $f(x) = x^{n+1}$ on [-1, 1], the minimax approximation from P_n is

$$q^*(x) = x^{n+1} - \frac{1}{2^n} T_{n+1}(x).$$

It is clear from the oscillatory nature of T_{n+1} that in this case the error function,

$$f(x) - q^*(x) = \frac{1}{2^n} T_{n+1}(x),$$

equioscillates on the $n+2$ points where T_{n+1} attains its maximum modulus on [-1, 1]. Since the Chebyshev polynomial T_{n+1} is a cosine on [-1, 1], its maximum modulus on that interval is not greater than unity and it is easily verified that T_{n+1} assumes the value $(-1)^j$ at the $n+2$ points $x = \cos(\pi j/(n+1))$ for $0 \le j \le n+1$. These are called the extreme points. From Chebyshev's result above we see that the minimax approximation from P_n for x^{n+1} on [-1, 1] has maximum error $1/2^n$, that is

$$\inf_{q \in P_n} \|x^{n+1} - q\|_\infty = \frac{1}{2^n}.$$

It is of obvious interest to obtain general results concerning how well an element $f \in C[-1, 1]$ is approximated by its minimax polynomial and we define

$$E_n(f) = \inf_{q \in P_n} \|f - q\|_\infty . \tag{1.3}$$

Thus (1.3) is the minimax error. De La Vallée Poussin (see Cheney [6]) showed that if, for some $q \in P_n$, the error $f - q$ alternates in sign on $n+2$ consecutive points $t_0, t_1, \ldots, t_{n+1}$, then

$$\min_i |f(t_i) - q(t_i)| \le E_n(f) \le \|f - q\|_\infty . \tag{1.4}$$

In (1.4) the inequality on the right is a trivial consequence of the definition of $E_n(f)$ and the inequality on the left is not hard to establish by assuming the converse and establishing a contradiction. These inequalities are at the heart of the Remez algorithms for computing minimax approximations. In the Remez algorithms we repeatedly solve linear systems of equations of the form

$$f(\eta_i) - q(\eta_i) \;=\; (-1)^i e, \quad 0 \le i \le n+1, \tag{1.5}$$

where $q \in P_n$. It is easy to demonstrate that the equations (1.5) are nonsingular if the η_i are distinct, We solve (1.5) to determine $q \in P_n$, which involves n+1 of the n+2 unknowns, and the real number e. It is the precise procedure adopted for changing the point set $\{\eta_0, \eta_1, \ldots, \eta_{n+1}\}$ after solving the linear equations in each iteration that distinguishes different versions of the algorithm. In the simplest version we change only one point in each iteration, introducing a point where $\|f - q\|_\infty$ is attained and deleting one of the neighbouring points from the point set $\{\eta_0, \eta_1, \ldots, \eta_{n+1}\}$ in such a way that $f - q$ still oscillates in sign on the new point set. Note that the left inequality in (1.4) implies that $|e|$ from (1.5) is a lower bound for $E_n(f)$. Thus on each iteration of the algorithm we have bounds for $E_n(f)$ given by

$$|e| \;\le\; E_n(f) \;\le\; \|f - q\|_\infty. \tag{1.6}$$

At each iteration of a Remez algorithm the inequalities in (1.6) become sharper.

Jackson showed that, for $f \in C[-1, 1]$,

$$E_n(f) \;\le\; \left(1 + \frac{\pi^2}{2}\right) \omega\left(\frac{1}{n}\right) \tag{1.7}$$

where $\omega(\delta)$ is the modulus of continuity of f, defined by

$$\omega(\delta) \;=\; \sup |f(x_1) - f(x_2)|,$$

the supremum being taken over all $x_1, x_2 \in [-1, 1]$ such that $|x_1 - x_2| \le \delta$. Jackson obtained other estimates of $E_n(f)$ corresponding to different orders of smoothness of f. (See, for example, Cheney [6], Davis [7], Rivlin [14].) However, we are more interested here in the following estimate of $E_n(f)$, which is due to Bernstein [1]: if $f \in C^{n+1}[-1, 1]$ then there exists $\xi \in (-1, 1)$ such that

$$E_n(f) \;=\; \frac{|f^{(n+1)}(\xi)|}{2^n\,(n+1)!}. \tag{1.8}$$

It is easily deduced from the equioscillation property that the minimax approximation for f interpolates f at certain n+1 points in [-1, 1]. Phillips [15] obtained the following simple proof of (1.8) based on this interpolation property.

Let $\xi_1, \xi_2, ..., \xi_{n+1}$ denote the zeros of T_{n+1} and let $q \in P_n$ denote the polynomial which interpolates f on these points. We now use the well-known expression for the error of interpolation. For any $f \in C^{n+1}[-1, 1]$ there exists $\xi_x \in (-1, 1)$ such that

$$f(x) - q(x) = (x-\xi_1)...(x-\xi_{n+1}).\frac{f^{(n+1)}(\xi_x)}{(n+1)!}. \tag{1.9}$$

We will derive lower and upper bounds for $E_n(f)$. Since, from Chebyshev's result quoted above,

$$\|(x-\xi_1)...(x-\xi_{n+1})\|_\infty = \frac{1}{2^n} \tag{1.10}$$

and, by definition,

$$E_n(f) \le \|f - q\|_\infty$$

we deduce from (1.9) the upper bound

$$E_n(f) \le \frac{1}{2^n} \cdot \max_{-1 \le x \le 1} |f^{(n+1)}(x)| \, / \, (n+1)! . \tag{1.11}$$

Second, we deduce from the characterizing property that the minimax approximation $q^* \in P_n$ interpolates f at certain points, say $t_1, t_2, ..., t_{n+1}$, in $(-1, 1)$. Thus there is a number $\eta_x \in (-1, 1)$ such that

$$f(x) - q^*(x) = (x-t_1)...(x-t_{n+1}).\frac{f^{(n+1)}(\eta_x)}{(n+1)!}. \tag{1.12}$$

Hence we obtain

$$E_n(f) \ge \|(x-t_1)...(x-t_{n+1})\|_\infty \min_{-1 \le x \le 1} |f^{(n+1)}(x)| \, / \, (n+1)! .$$

and from

$$\|(x-t_1)...(x-t_{n+1})\|_\infty \ge E_n(x^{n+1}) = \frac{1}{2^n}$$

we derive the lower bound

$$E_n(f) \ge \frac{1}{2^n} \cdot \min_{-1 \le x \le 1} |f^{(n+1)}(x)| \, / \, (n+1)! . \tag{1.13}$$

Finally we apply the mean value theorem to deduce from (1.11) and (1.13) that if $f \in C^{n+1}[-1, 1]$ there exists $\xi \in (-1, 1)$ such that (1.8) holds.

We note that , if f is not a member of P_n , (1.8) is simplest when f(x) is x^{n+1}. This suggests re-writing Bernstein's estimate of $E_n(f)$ in the form

$$E_n(f) \quad = \quad \frac{|f^{(n+1)}(\xi)|}{(n+1)!} E_n(x^{n+1}), \tag{1.14}$$

for some $\xi \in (-1, 1)$.

Phillips [16] has generalized (1.14), showing that it is still valid when $E_n(f)$ is re-defined, replacing the infinity norm in (1.3) by any of the p-norms, given by

$$\|f\|_p \quad = \quad \left(\int_{-1}^{1} | f(x) |^p \, dx \right)^{1/p}, \tag{1.15}$$

for $p \geq 1$. On letting $p \to \infty$ in (1.15) we obtain the maximum norm. This explains why the latter is sometimes called the infinity norm. The best approximation for $f \in C[-1, 1]$ with respect to any of the p-norms interpolates f on n+1 points of the interval [-1, 1], as we saw for the maximum norm. (For the case p = 1 we need to impose the constraint on f that f-q has at most n+1 zeros in [-1, 1], for any $q \in P_n$.) The above interpolatory property is used to establish (1.14) when

$$E_n(f) \quad = \quad \inf_{q \in P_n} \quad \|f - q\|_p \; . \tag{1.16}$$

For the 1-norm, $E_n(f) = 1/2^n$ and

$$E_n(f) \quad = \quad 2^{n+1} \left\{ \frac{2}{2n+3} \right\}^{1/2} \Big/ \binom{2n+2}{n+1}$$

for the 2-norm. For the general p-norm, not so much is known about $E_n(f)$. However, Phillips [15] has shown that, for $1 < p < \infty$,

$$1/2^{n+1-1/p} \quad < \quad E_n(f) \quad < \quad 1/2^{n-1/p} \, . \tag{1.17}$$

This generalizes inequalities for $E_n(f)$ in the case of the 2-norm given in Timan [20].

Kimchi and Richter-Dyn [13] have further generalized (1.14) to all *monotone* norms, that is, to any norm for which

$$|f(x)| \geq |g(x)| \;\; \text{for } -1 \leq x \leq 1 \quad \Rightarrow \quad \|f\| \geq \|g\| \, . \tag{1.18}$$

This includes all the p-norms.

2. Near-minimax Approximations

Although minimax polynomials may be satisfactorily computed by a Remez algorithm, the mapping which sends $f \in C[-1, 1]$ into its minimax polynomial $q^* \in P_n$ is nonlinear. It is natural to seek *linear* mappings which are in some sense close to the nonlinear minimax mapping. Such a linear mapping P will send $f \in C[-1, 1]$ into $Pf \in P_n$. Since we will want $Pf = f$ for all $f \in P_n$, this entails $P^2 = P$ and thus P is a *projection* from $C[-1, 1]$ to P_n. In view of Bernstein's estimate (1.8) for the minimax error, we will say that such a projection P is *near-minimax* if for each $f \in C^{n+1}[-1, 1]$ there is a ξ such that

$$\|f - Pf\|_\infty = \frac{|f^{(n+1)}(\xi)|}{2^n (n+1)!}. \tag{2.1}$$

It turns out that those polynomial approximations which have long been recognised (almost as part of numerical folk-lore) as being close to minimax approximations are near-minimax approximations in the sense we have just defined.

The simplest of these is the polynomial which interpolates f on the zeros of the Chebyshev polynomial T_{n+1}. Denoting this by Pf we note that, from the uniqueness of interpolation, $Pf = f$ if $f \in P_n$ and thus P is a projection. We have already seen in (1.11) that, for this projection,

$$\|f - Pf\|_\infty \leq \frac{1}{2^n} \cdot \max_{-1 \leq x \leq 1} |f^{(n+1)}(x)| \,/\, (n+1)! \,. \tag{2.2}$$

We derive a lower bound in a similar way to that obtained for $E_n(f)$: we write

$$f(x) - Pf(x) = (x-\xi_1)\ldots(x-\xi_{n+1}).\frac{f^{(n+1)}(\xi_x)}{(n+1)!},$$

where $\xi_1, \xi_2,\ldots, \xi_{n+1}$ denote the zeros of T_{n+1}. We deduce that

$$\|f - Pf\|_\infty \geq \|(x-\xi_1)\ldots(x-\xi_{n+1})\|_\infty \cdot \min_{-1 \leq x \leq 1} |f^{(n+1)}(x)| \,/\, (n+1)!$$

and therefore

$$\|f - Pf\|_\infty \geq \frac{1}{2^n} \cdot \min_{-1 \leq x \leq 1} |f^{(n+1)}(x)| \,/\, (n+1)! \,. \tag{2.3}$$

From (2.2), (2.3) and the continuity of $f^{(n+1)}$ it follows that Pf satisfies (2.1) and so is a near-minimax projection.

Second, let us define Pf as the polynomial q which satisfies the Remez-type equations (1.5), where we choose the η_i as the extreme points of T_{n+1}. If $f \in P_n$ we see that $Pf = f$, with e=0 in (1.5), and so P is a projection. There is an easier way to obtain this approximating polynomial q than by solving the linear equations (1.5). We begin by constructing the polynomial $q_{n+1} \in P_{n+1}$ which interpolates f on $H = \{\eta_0, \eta_1, ..., \eta_{n+1}\}$. Using the divided difference form of the error of interpolation, we have

$$f(x) - q_{n+1}(x) = (x-\eta_0)...(x-\eta_{n+1}) \, . \, f[x, \eta_0, ..., \eta_{n+1}] \, .$$

Let us write q_{n+1} in the Newton divided difference form

$$q_{n+1}(x) = f(\eta_0) + (x-\eta_0) f[\eta_0, \eta_1] + (x-\eta_0)(x-\eta_1) f[\eta_0, \eta_1, \eta_2]$$

$$+ ...+ (x-\eta_0)...(x-\eta_n) f[\eta_0, ..., \eta_{n+1}] \, .$$

We now "economize" q_{n+1}, by expressing it as a linear combination of the Chebyshev polynomials and removing the term in T_{n+1} to give, say,

$$q(x) = q_{n+1}(x) - \frac{1}{2^n} f[\eta_0, ..., \eta_{n+1}] \, T_{n+1}(x) \, , \tag{2.4}$$

where $q \in P_n$. Since T_{n+1} takes the values +1 and -1 alternately on the η_j, we see that the polynomial q defined by (2.4) satisfies the Remez-type equations (1.5), with

$$|e| = \frac{1}{2^n} \, |f[\eta_0, ..., \eta_{n+1}]| \, . \tag{2.5}$$

(We can use the fact that $T_{n+1}(1) = 1$ to fix the sign of e.) From the de La Vallée Poussin inequalities (1.4), together with (1.5) and (2.4),we have

$$\|f - Pf\|_\infty \geq |f(\eta_j) - Pf(\eta_j)| = \frac{1}{2^n} \, |f[\eta_0, ..., \eta_{n+1}]| \, .$$

Since for $f \in C^{n+1}[-1, 1]$ there is a number $\eta \in (-1, 1)$ such that

$$f[\eta_0, ..., \eta_{n+1}] = f^{(n+1)}(\eta) \, / \, (n+1)! \, ,$$

we have the lower bound

$$\|f - Pf\|_\infty \geq \frac{1}{2^n} \, . \, \min_{-1 \leq x \leq 1} |f^{(n+1)}(x)| \, / \, (n+1)! \, . \tag{2.6}$$

We remark that we could have derived (2.6) otherwise by deducing that this projection interpolates f at certain n+1 points in (-1, 1) and repeating the argument used in deriving the lower bound (1.13) for $E_n(f)$. (We note that this shows that the lower bound (2.6) holds for any projection which interpolates f at n+1 points of (-1, 1) or, equivalently, such that f - Pf oscillates in sign on n+2 consecutive points of [-1, 1].) For the projection whose error equioscillates on the extreme points of T_{n+1}, Phillips and Taylor [18] expressed the error as (see (2.4) and above)

$$f(x) - Pf(x) = (x-\eta_0)...(x-\eta_{n+1}) . f[x, \eta_0,..., \eta_{n+1}] + \frac{1}{2^n} f[\eta_0,..., \eta_{n+1}] T_{n+1}(x),$$

whence they obtained an upper bound for $\|f - Pf\|_\infty$ of the form (2.2), so verifying that this projection is near-minimax. (The upper bound is much harder to establish than the above lower bound.) Brutman [3] has studied similar projections which are not near-minimax, and alternating *trigonometrical* polynomials are considered in Brutman [4]. The third commonly used approximation which we wish to discuss is the partial Chebyshev series for f, truncated after the term involving T_n. This is

$$Pf(x) = \sum_{r=0}^{n}{}' a_r T_r(x), \tag{2.7}$$

where the dash denotes a summation whose first term is halved and the Chebyshev coefficient a_r is given by

$$a_r = \frac{2}{\pi} \int_{-1}^{1} (1-x^2)^{-1/2} f(x) T_r(x) dx . \tag{2.8}$$

Recall that the Chebyshev polynomials are orthogonal on [-1, 1] with respect to the weight function $(1-x^2)^{-1/2}$, that is

$$\int_{-1}^{1} (1-x^2)^{-1/2} T_r(x) T_s(x) dx = 0, \quad r \neq s. \tag{2.9}$$

If $f \in P_n$ then since $\{T_0,..., T_n\}$ is a basis for P_n we can write

$$f(x) = \sum_{r=0}^{n}{}' b_r T_r(x) \tag{2.10}$$

for some choice of b_r. Let us substitute (2.10) in (2.8) and from the orthogonality of the Chebyshev polynomials twe deduce that $a_r = b_r$. We have shown that Pf = f for all $f \in P_n$ and thus the truncated Chebyshev series operator is a projection. We now show that this projection interpolates f on n+1 points of (-1, 1). First we note that

$$\int_{-1}^{1} (1-x^2)^{-1/2} (f(x) - Pf(x)) T_r(x) dx = 0, \quad 0 \leq r \leq n. \tag{2.11}$$

This follows from (2.8) and the orthogonality of the Chebyshev polynomials. In particular, with r = 0 in (2.11), we see that f - Pf must have at least one zero in (-1, 1). Suppose f - Pf has exactly k zeros in (-1, 1), at $t_1,\dots, t_k$. Let us write

$$\pi(x) = (x - t_1)\dots(x - t_k) = \sum_{r=0}^{k}{}' c_r T_r(x) , \tag{2.12}$$

for some choice of c_r. Then, since f - Pf and π have the same zeros in (-1, 1), it follows that

$$\int_{-1}^{1} (1-x^2)^{-1/2}(f(x) - Pf(x))\pi(x)dx \neq 0 . \tag{2.13}$$

However, (2.11 and (2.12) show that we require $k > n$ so that the integral in (2.13) will be non-zero. This confirms that Pf interpolates f at n+1 points in (-1, 1) and so $\|f - Pf\|_\infty$ has a lower bound of the form (2.6). In a very fine paper, Brass [2] shows that $\|f - Pf\|_\infty$ also has an upper bound of the form (2.2) and thus the truncated Chebyshev series projection is near-minimax. The approach used by Brass allows him to obtain results not only for the truncated Chebyshev series but also for a class of orthogonal series which includes the other two near-minimax projections discussed above. Let us now examine the orthogonality relations concerning these two projections. The interpolation near-minimax projection can be expressed as

$$Pf(x) = \sum_{r=0}^{n}{}' a_r T_r(x), \tag{2.14}$$

where

$$a_r = \frac{2}{n+1} \sum_{j=1}^{n+1} f(\xi_j)T_r(\xi_j) \tag{2.15}$$

and the ξ_j are the zeros of T_{n+1}. This is a consequence of the orthogonality property

$$\sum_{j=1}^{n+1} T_r(\xi_j)T_s(\xi_j) = 0 \quad \text{for } r \neq s \text{ and } 0 \leq r,s \leq n. \tag{2.16}$$

The projection whose error equioscillates on the extreme points of T_{n+1} may likewise be written as

$$Pf(x) = \sum_{r=0}^{n}{}' a_r T_r(x), \tag{2.17}$$

where

$$a_r = \frac{2}{n+1} \sum_{j=0}^{n+1}{}'' f(\eta_j) T_r(\eta_j) . \tag{2.18}$$

In (2.18) the double dash modifying the summation signifies that the first and last terms are halved.

An alternative proof to that of Brass that the Chebyshev series projection is near-minimax is given by Elliott, Paget, Phillips and Taylor [9]. Concerning the Chebyshev series projection we note that, due to the orthogonality property (2.9),

$$\int_{-1}^{1} (1-x^2)^{-1/2} (x^{n+1} - q(x))^2 \, dx \tag{2.19}$$

is minimized over $q \in P_n$ by choosing q so that

$$x^{n+1} - q(x) = \frac{1}{2^n} T_{n+1}(x).$$

Let us now write

$$\|f\|_C = \left\{ \int_{-1}^{1} (1-x^2)^{-1/2} (f(x))^2 \, dx \right\}^{1/2} . \tag{2.20}$$

This is a weighted 2-norm, with weight function $(1-x^2)^{-1/2}$. Then, using the above result concerning (2.19), we can obtain upper and lower bounds for $\|f\|_C$ as we did above for $E_n(f)$ and the errors of the near-minimax projections. We thus find that, for the Chebyshev series projection,

$$\|f - Pf\|_C = \left\{\frac{\pi}{2}\right\}^{1/2} \cdot \frac{|f^{(n+1)}(\xi)|}{2^n (n+1)!} , \tag{2.21}$$

for some $\xi \in (-1, 1)$, assuming that $f \in C^{n+1}[-1, 1]$. The constant on the right of (2.21) follows from

$$\int_{-1}^{1} (1-x^2)^{-1/2} (T_{n+1}(x))^2 \, dx = \frac{\pi}{2} . \tag{2.22}$$

In this connection, it is also worth noting (see Elliott [8]) that the Chebyshev coefficient satisfies

$$a_{n+1} = \frac{|f^{(n+1)}(\eta)|}{2^n (n+1)!} , \tag{2.23}$$

for $\eta \in (-1, 1)$, provided that $f \in C^{n+1}[-1, 1]$. This may be compared with (see Rivlin [14])

$$|a_{n+1}| \leq \frac{4}{\pi} E_n(f) . \tag{2.24}$$

We conclude this section by mentioning a general result on near-minimax projections.

DEFINITION. A semi-norm E_n on $C^{n+1}[-1, 1]$ is said to satisfy Property B of order n if

$$|f^{(n+1)}(x)| \leq g^{(n+1)}(x), \quad -1 \leq x \leq 1, \tag{2.25}$$

implies that $E_n(f) \leq E_n(g)$. □

This definition was proposed by Holland, Phillips and Taylor [12] who prove the following result.

THEOREM 1. If E_n is a semi-norm on $C^{n+1}[-1, 1]$ which satisfies Property B of order n then, for each $f \in C^{n+1}[-1, 1]$ there exists $\xi \in (-1, 1)$ such that

$$E_n(f) = \frac{|f^{(n+1)}(\xi)|}{(n+1)!} E_n(x^{n+1}) . \quad \square \tag{2.26}$$

This generalizes Bernstein's result (1.8) above.

Chalmers, Phillips and Taylor [5] derived further results concerning near-minimax projections; for example, they show that a necessary condition for a projection from $C^{n+1}[-1, 1]$ to P_n to be near-minimax is that $P(T_{n+1}) = 0$.

3. Improved Error Estimates

Let us return to the lower and upper bounds for $E_n(f)$ from which Bernstein's estimate (1.8) was derived above. The upper bound is

$$E_n(f) \leq \frac{1}{2^n} \cdot \max_{-1 \leq x \leq 1} |f^{(n+1)}(x)|/(n+1)! \tag{3.1}$$

and the lower bound is similar, with 'min' in place of 'max'. Elliott and Taylor [11] obtained the following lower and upper bounds for $E_n(f)$ in terms of divided differences:

$$2^{-n} |f[\eta_0,...,\eta_{n+1}]| \leq E_n(f) \leq 2^{-n} |f[\xi, \xi_1, \xi_2,..., \xi_{n+1}]| . \tag{3.2}$$

In (3.2) the η_j and the ξ_j are respectively the extreme points and zeros of T_{n+1} and ξ is some point in (-1, 1). For any y_j in [-1, 1] and any $f \in C^{n+1}[-1, 1]$ there is an η in (-1, 1) such that

$$f[y_0,\ldots,y_{n+1}] = f^{(n+1)}(\eta)/(n+1)! \tag{3.3}$$

(see, for example, Phillips and Taylor [17]) and we see that for $f \in C^{n+1}[-1, 1]$ the bounds in (3.2) are in general sharper than those above involving derivatives. Moreover, the bounds in (3.2) have the added advantage that they are valid for all $f \in C^1[-1, 1]$, whereas those in (1.11) and (1.13) are valid only for $f \in C^{n+1}[-1, 1]$. (Note that it is because, in the right-hand inequality in (3.2), ξ may coincide with one of the Chebyshev zeros ξ_j that the bounds in (3.2) are valid for $f \in C^1[-1, 1]$ rather than for $f \in C[-1, 1]$.)

We now discuss bounds analogous to those of (3.2) for $\|f - Pf\|_\infty$, where P is one of the near-minimax projections discussed in section 2. From (3.2) and the de La Vallée Poussin inequalities (1.4) , we have immediately the lower bound

$$2^{-n}\,|f[\eta_0,\ldots,\eta_{n+1}]| \le E_n(f) \le \|f - Pf\|_\infty, \tag{3.4}$$

valid for *any* projection P from C[-1, 1] to P_n . Note also that these lower bounds hold for all $f \in C[-1, 1]$.

Following the approach of Elliott and Taylor [11] it is easily verified that, for the projection P obtained by interpolating on the Chebyshev zeros, the error $\|f - Pf\|_\infty$ satisfies the same inequalities as does $E_n(f)$ in (3.2).

Let $\{\zeta_1^+,\ldots,\zeta_n^+\}$, $\{\zeta_1^-,\ldots,\zeta_n^-\}$ denote the zeros of the polynomials $(T_{n+1}(x) \pm T_n(x)) / (x \pm 1)$ respectively. (Since, for all n, $T_n(1) = 1$, it follows that $x-1$ is a factor of $T_{n+1}(x)-T_n(x)$ and thus $(T_{n+1}(x)-T_n(x))/(x-1) \in P_n$. We deduce from $T_n(-1) = -1$ that $(T_{n+1}(x)+T_n(x))/(x+1) \in P_n$ also.) Elliott and Taylor [11] obtained the following upper bound for the truncated Chebyshev series projection:

$$\|f - Pf\|_\infty \le 2^{-n} \max\{M^+, M^-\}, \tag{3.5}$$

where

$$M^+ = \max|f[\xi, \eta, \zeta_1^+,\ldots, \zeta_n^+]|, \quad M^- = \max|f[\xi, \eta, \zeta_1^-,\ldots, \zeta_n^-]|,$$

the latter two maxima being taken over all $\xi, \eta \in (-1, 1)$. Since it is possible that three points may coincide in each of the two divided differences, the upper bound (3.5) holds only for $f \in C^2[-1, 1]$.

In the remainder of this section we describe work carried out by Elliott and Phillips [10]. First we need to label individual members of the four sets of points of points named above. We label the zeros of T_{n+1} in the order

$$\xi_i = \cos \frac{(2i-1)\pi}{2(n+1)}, \quad 1 \le i \le n+1, \tag{3.6}$$

so that $-1 < \xi_{n+1} < \xi_n < ... < \xi_1 < 1$, and label the extreme points of T_{n+1} as

$$\eta_i = \cos \frac{i\pi}{n+1}, \quad 0 \le i \le n+1, \tag{3.7}$$

so that $-1 = \eta_{n+1} < \eta_n < ... < \eta_1 < \eta_0 = 1$. From (3.6) and (3.7) we note that

$$\eta_{i+1} < \xi_i < \eta_i, \quad 0 \le i \le n. \tag{3.8}$$

The zeros of the polynomials $(T_{n+1}(x) \pm T_n(x)) / (x \pm 1)$ are given respectively by

$$\zeta_i^+ = \cos \frac{(2i-1)\pi}{2n+1}, \quad 1 \le i \le n, \tag{3.9}$$

and

$$\zeta_i^- = \cos \frac{2i\pi}{2n+1}, \quad 1 \le i \le n. \tag{3.10}$$

By working through the details of the analysis given by Phillips and Taylor [18] and resisting the impulse to convert divided differences into derivatives, using (3.3), we may derive the following upper bound for the Chebyshev equioscillation projection:

$$\|f - Pf\|_\infty \le 2^{-n} \max\{M_1, M_2\}, \tag{3.11}$$

where

$$M_1 = \max |f[\xi,\eta_1,...,\eta_n,\eta_{n+1}]|, \quad M_2 = \max |f[\xi,\eta_0,\eta_1,...,\eta_n]|,$$

the latter two maxima being taken over all $\xi \in (-1, 1)$. This bound for the Chebyshev equioscillation projection is valid for all $f \in C^1[-1, 1]$.

We shall define a function $f \in C^{n+2}[-1, 1]$ to be monotonic if $f^{(n+2)}$ has constant sign on $[-1, 1]$ and, without loss of generality, we shall assume that $f^{(n+2)}(x) \ge 0$ for all $x \in [-1, 1]$. (Our non-standard use of the word 'monotonic' should not cause any confusion.) We state and prove the following property of divided differences.

THEOREM 2. Suppose that $f \in C^{n+2}[-1, 1]$, $f^{(n+2)}(x) \geq 0$ on $[-1, 1]$ and that $-1 \leq x_i \leq y_i \leq 1$ for $0 \leq i \leq n+1$. Then

$$f[x_0, x_1,..., x_{n+1}] \leq f[y_0, y_1,..., y_{n+1}] . \tag{3.12}$$

PROOF. First let us compare $f[x_0, x_1,..., x_{n+1}]$ with $f[y_0, x_1,..., x_{n+1}]$. We have

$$f[y_0, x_1,..., x_{n+1}] - f[x_0, x_1,..., x_{n+1}]$$
$$= (y_0 - x_0) f[x_0, y_0, x_1,..., x_{n+1}],$$

using the recurrence relation for divided differences, and from this and (3.3) we see that

$$f[y_0, x_1,..., x_{n+1}] - f[x_0, x_1,..., x_{n+1}]$$
$$= (y_0 - x_0) f^{(n+2)}(\xi) / (n+2)! \tag{3.13}$$

for some $\xi \in (-1, 1)$. It follows that

$$f[y_0, x_1,..., x_{n+1}] - f[x_0, x_1,..., x_{n+1}] \geq 0.$$

We now write

$$f[y_0, y_1,..., y_{n+1}] - f[x_0, x_1,..., x_{n+1}]$$
$$= f[y_0, x_1,..., x_{n+1}] - f[x_0, x_1,..., x_{n+1}]$$
$$+ f[y_0, y_1,..., x_{n+1}] - f[y_0, x_1,..., x_{n+1}]$$
$$+ ...$$
$$+ f[y_0, y_1,..., y_{n+1}] - f[y_0, y_1,...,y_n, x_{n+1}] . \tag{3.14}$$

Since $f[x_0, x_1,..., x_{n+1}]$ is a symmetric function of its n+2 arguments, each pair of terms in the above sum may be combined as in (3.13) and shown to be non-negative. This completes the proof. □

Consider the evaluation of $2^{-n} | f[\eta_0,...,\eta_{n+1}] |$ in (3.4) as a lower bound for both $E_n(f)$ and $\|f - Pf\|_\infty$, where P is any projection from C[-1, 1] to P_n. This bound requires a single evaluation of a divided difference. On the other hand, the evaluation of the upper bounds in (3.2) and (3.11), for $E_n(f)$ and the error of the Chebyshev equioscillation projection respectively, require finding the maximum of a function of one variable, while the evaluation of the upper bound (3.5) for the error of the Chebyshev series projection requires finding the maximum of a function of two variables. In what follows we will assume that $f \in C^{n+2}[-1, 1]$ and $f^{(n+2)}(x) \geq 0$ for all $x \in [-1, 1]$ and see how, in this case, the evaluation of these upper bounds is greatly simplified. We will also find it of interest to observe that, for this class of functions, the various upper bounds can be arranged in order.

Let us apply Theorem 2 first to (3.2) to obtain, for $E_n(f)$ and the projection involving interpolation on the Chebyshev zeros,

$$E_n(f) \le \|f - Pf\|_\infty \le 2^{-n} f[1, \xi_1, \xi_2,\dots, \xi_{n+1}], \tag{3.15}$$

the divided difference being non-negative, since $f^{(n+2)}(x) \ge 0$. For the Chebyshev equioscillation projection under the same conditions we deduce from (3.11) that

$$\|f - Pf\|_\infty \le 2^{-n} f[1, 1, \eta_1,\dots, \eta_n]. \tag{3.16}$$

In this case the divided difference has to be evaluated at two coincident points, which requires evaluating the first derivative of f. (We will pursue this below.) Finally, for the Chebyshev series projection, since $\zeta_i^+ > \zeta_i^-$ for $1 \le i \le n$, we deduce from (3.5) that

$$\|f - Pf\|_\infty \le 2^{-n} f[1, 1, \zeta_1^+,\dots,\zeta_n^+] . \tag{3.17}$$

We can order the upper bounds given in (3.15), (3.16) and (3.17). From (3.6), (3.7) and (3.9) it is easily verified that

$$\xi_{i+1} < \eta_i < \zeta_i^+, \quad 1 \le i \le n,$$

so that under the conditions of Theorem 2 we have

$$f[1,\xi_1,\xi_2,\dots,\xi_{n+1}] \le f[1,1,\eta_1,\dots,\eta_n] \le f[1,1,\zeta_1^+,\dots,\zeta_n^+] . \tag{3.18}$$

Let $\{y_0,\dots, y_{n+1}\}$ and $\{x_0,\dots, x_{n+1}\}$ be ordered pair-wise as in the statement of Theorem 2. Then, on expressing each pair of terms in the sum on the right of (3.14) as in (3.13), we obtain

$$f[y_0, y_1,\dots, y_{n+1}] - f[x_0, x_1,\dots, x_{n+1}]$$

$$\le \frac{\|f^{(n+2)}\|_\infty}{(n+2)!} \sum_{i=0}^{n+1} (y_i - x_i) . \tag{3.19}$$

Let us apply this result to the two sets of points $\{\eta_0,\dots,\eta_{n+1}\}$,which gives the lower bound, and $\{1, 1, \zeta_1^+,\dots,\zeta_n^+\}$ which from (3.18) is the largest of the upper bounds. We readily verify that

$$\sum_{i=0}^{n+1} \eta_i = 0 \quad \text{and} \quad \sum_{i=1}^{n} \zeta_i = \pm\frac{1}{2},$$

where the plus and minus signs correspond to $\zeta_i = \zeta_i^+$ and $\zeta_i = \zeta_i^-$ respectively.

It then follows from (3.19) that

$$0 \leq f[1, 1, \zeta_1^+,\ldots,\zeta_n^+] - f[\eta_0,\eta_1,\ldots,\eta_{n+1}] \leq \frac{5}{2}\,\frac{\|f^{(n+2)}\|_\infty}{(n+2)!}. \tag{3.20}$$

Under the conditions of Theorem 2 we also have

$$\max_{-1\leq x\leq 1} |f^{(n+1)}(x)| \,/\, (n+1)! \;-\; \min_{-1\leq x\leq 1} |f^{n+1}(x)| \,/\, (n+1)!$$

$$= f[1, 1,\ldots, 1, 1] - f[-1, -1,\ldots, -1, -1]$$

$$\leq 2(n+2)\,\frac{\|f^{(n+2)}\|_\infty}{(n+2)!} \tag{3.21}$$

from (3.19). The comparison of (3.20) and (3.21) gives a measure of the "gain" achieved by using divided differences instead of derivatives in the above bounds.

We conclude by giving simple formulas for the divided differences which we require for the bounds given above in (3.4), (3.15), (3.16) and (3.17). When all the arguments are distinct, we may evaluate divided differences from repeated application of he recurrence relation

$$f[x_0, x_1,\ldots, x_{k+1}] = \frac{f[x_1,x_2,\ldots,x_{k+1}] - f[x_0,x_1,\ldots,x_k]}{x_{k+1} - x_0}. \tag{3.22}$$

When we have a repeated parameter we define

$$f[x_0, x_0, x_1,\ldots, x_k] = \lim_{h\to 0} f[x_0+h, x_0, x_1,\ldots, x_k]$$

$$= \lim_{h\to 0} \frac{1}{h}\{f[x_0+h, x_1,\ldots, x_k] - f[x_0, x_1,\ldots, x_k]\}$$

$$= \left(\frac{d}{dx} f[x, x_1,\ldots, x_k]\right)_{x=x_0}. \tag{3.23}$$

The symmetric form for divided differences is

$$f[x_0, x_1,\ldots, x_k] = \sum_{i=0}^{k} f(x_i) \Big/ \prod_{j\neq i} (x_i - x_j). \tag{3.24}$$

From this and (3.23) we readily obtain

$$f[x_0, x_0, x_1,\ldots, x_k] = g[x_0, x_1,\ldots, x_k] - (\pi'(x_0)f(x_0) - \pi(x_0)f'(x_0))/(\pi(x_0))^2 \tag{3.25}$$

where

$$g(x) = f(x)/(x - x_0)^2 \quad \text{and} \quad \pi(x) = \prod_{j=1}^{k} (x - x_j). \tag{3.26}$$

Then, using (3.24), (3.25) and (3.26), we derive

$$f[\eta_0,\eta_1,\ldots,\eta_{n+1}] = \frac{2^n}{(n+1)} \sum_{j=0}^{n+1}{}'' (-1)^j f(\eta_j) . \tag{3.27}$$

Similarly we obtain

$$f[1,\xi_1,\xi_2,\ldots,\xi_{n+1}] = 2^n \Big\{ f(1) + \sum_{j=1}^{n+1} \alpha_j f(\xi_j) \Big\} \tag{3.28}$$

where

$$\alpha_j = \frac{(-1)^j}{n+1} \cdot [(1+\xi_j)/(1-\xi_j)]^{1/2} ,$$

$$f[1,1,\eta_1,\ldots,\eta_n] = \frac{2^n}{n+1} \Big\{ \sum_{j=1}^{n} \beta_j f(\eta_j) - \frac{1}{3} n(n+2) f(1) + f'(1) \Big\} \tag{3.29}$$

where $\beta_j = (-1)^{j-1} \cdot \dfrac{1+\eta_j}{1-\eta_j}$

and

$$f[1,\ 1,\zeta_1,\ldots,\zeta_n] = \frac{2^{n+1/2}}{2n+1} \sum_{j=1}^{n} \gamma_j f(\zeta_j)$$

$$- \ 2^n \{n(n+1)f(1) - f'(1)\} , \tag{3.30}$$

where $\gamma_j = (-1)^{j-1} (1+\zeta_j)/(1-\zeta_j)^{3/2}$ and, for simplicity we have written ζ_j in place of ζ_j^+. There is a similar expression to (3.30) for the case where $\zeta_j = \zeta_j^-$.

The expressions (3.27) - (3.30) are satisfactory only for small values of n . As n is increased there is a liklihood of severe cancellation with consequent loss of significant digits. For particular choices of the function f, Elliott and Phillips [10] illustrate how we can estimate divided differences by using a method based on contour integration. This follows from the relation

$$f[y_0, y_1, \ldots, y_{n+1}] = \frac{1}{2\pi i} \int \frac{f(z)\, dz}{\pi(z)} , \tag{3.31}$$

where

$$\pi(z) = \prod_{j=0}^{n+1} (z - y_j) ,$$

the integral being taken around some closed contour which contains all the y_j, the function f being analytic on and within the closed contour.

References

1. Bernstein, S.N. (1926) Lecons sur les propriétés extrémales et la meilleure approximation des fonctions analytiques d'une variable réele, Gauthier-Villars, Paris.
2. Brass, H. (1984) 'Error estimates for least squares approximation by polynomials', J. Approx. Theory 41, 345-349.
3. Brutman, L. (1983) 'Generalized alternating polynomials, some properties and numerical applications', Mathematics Publication Series, University of Haifa, Israel.
4. Brutman, L. (1987) 'Alternating trigonometric polynomials', J. Approx. Theory 49, 64-74.
5. Chalmers, B.L., Phillips, G.M. and Taylor, P.J. (1988) 'Polynomial approximation using projections whose kernels contain the Chebyshev polynomials', J. Approx. Theory 53, 321-334.
6. Cheney, E.W. (1966) Introduction to Approximation Theory, McGraw-Hill, New York.
7. Davis, P.J. (1963) Interpolation and Approximation, Blaisdell, New York.
8. Elliott, David (1964) 'The evaluation and estimation of the coefficients in the Chebyshev series expansion of a function. Math. of Comp. 18, 274-2 .
9. Elliott, David, Paget, D.F., Phillips, G.M. and Taylor, P.J. (1987) 'Error of truncated Chebyshev series and other near minimax polynomial approximations', J. Approx. Theory 50, 49-57.
10. Elliott , David and Phillips, George M. (*in the press*) 'Improved error bounds for near-minimax approximations', BIT.

11. Elliott , David and Taylor, Peter J. (*in the press*) 'Polynomial approximation errors for functions of low order continuity', Constructive Approximation.

12. Holland, A.S.B., Phillips, G.M. and Taylor, P.J. (1980) 'Generalization of a theorem of Bernstein', in E.W. Cheney (ed.), Approximation Theory III, Academic Press, New York.

13. Kimchi, E. and Richter-Dyn, N. (1978) 'Restricted range approximation of k-convex functions in monotone norms', SIAM J. Numer. Anal. 15, 1030-1038.

14. Rivlin, T.J. (1969) An Introduction to the Approximation of Functions, Blaisdell, Waltham, Mass.

15. Phillips, G.M. (1968) 'Estimate of the maximum error in best polynomial approximations', Comp. J. 11, 110-111.

16. Phillips, G.M. (1970) 'Error estimates for best polynomial approximations', in A. Talbot (ed.), Approximation Theory, Academic Press, London.

17. Phillips, G.M. and Taylor, P.J. (1973) Theory and Applications of Numerical Analysis, Academic Press, London.

18. Phillips, G.M. and Taylor, P.J. (1982) 'Polynomial approximation using equioscillation on the extreme points of the Chebyshev polynomials', J. Approx. Theory 36, 257-264.

19. Powell, M.J.D. (1981) Approximation Theory and Methods, Cambridge University Press, Cambridge.

20. Singer, I. (1970) Best Approximation in Normed Linear Spaces by Elements of Linear Subspaces, Springer-Verlag, Berlin (translated by Radu Georgescu from the Romanian edition of 1966).

21. Timan , A.F. (1963) Theory of Approximation of Functions of a Real Variable, Pergamon, Oxford (translated by J. Berry from the Russian edition of 1966).

22. Watson, G.A. (1980) Approximation Theory and Numerical Methods, Wiley, Chichester.

DIFFERENT METRICS AND LOCATION PROBLEMS

E. CASINI and P.L. PAPINI

Department of Mathematics
Piazza Porta S. Donato, 5
I-40127 BOLOGNA

Department of Mathematics
Piazza Porta S. Donato, 5
I-40127 BOLOGNA

ABSTRACT. Many problems in operations research and in economics reduce to the finding of one or more points minimizing some function of the distance. The euclidean distance is most commonly used: such a metric is often well suited to modeling problems, but - above all - it is simple and behaves most regularly with respect to mathematical properties. Anyhow, some problems are better modeled (at least from some points of view) by other metrics; e.g., by the rectilinear one. Therefore, it is useful to have estimates concerning the difference among solutions based on different choices of the distance function. In this note we give, hopefully, a contribution in this direction.

1. Introduction And Notations

Many problems arising in economics and in operations research (for example, location problems in the continuous case) are described by using -among other theoretical tools- different types of distance. The euclidean distance, which is used in many cases and has nice properties, is often adequate to provide an analytic formulation of the problem considered. However, in some cases, such distance is not the best one (or at least, it is not the only one) which can give a suitable description of a specific problem. For this reason, it is important to know the relations existing among the various distances which are used more frequently. In fact, in some cases the use of a simple distance (e.g. the euclidean or the rectilinear one), to approximate other distances which could better fit the requirements of the problem involved, could be tempting.

Here we examine the relations among the euclidean distance and other distances of type d_p, with respect to some of the most common parameters (means, medians and so on). Among other things, this paper sharpens and enlarges some of the arguments contained in a paper by Krarup and Pruzan [8]; in particular, we show that some of the rather optimistic conclusions indicated there are not completely justified by the theory.

1991 Mathematics Subject Classification: primary 90B85; secondary 41A28.

Key words and phrases: mean values; metrics.

S. P. Singh (ed.), Approximation Theory, Spline Functions and Applications, 243–253.

For the sake of simplicity we discuss only the two-dimensional case (R^2), but similar arguments carry over to n-dimensional spaces. Also, we limit our discussion to metrics like

$$d_p(x,y) = \sqrt[p]{|x_1-y_1|^p+|x_2-y_2|^p} \quad (x=(x_1,x_2);\; y=(y_1,y_2))\;; \quad 1\leq p<\infty \tag{1}$$

and (with standand notation)

$$d_\infty(x,y) = \max\ (|x_1-y_1|, |x_2-y_2|)\,. \tag{1'}$$

All these distances can be defined by norms; thus we shall also write $\|x\|_p$ instead of $d_p(\theta,x)$.

For $r>0$, we denote by $\mathbf{B_p(r)}$ and $\mathbf{S_p(r)}$ - respectively - the set $\{x\in R^2;\ \|x\|\leq r\}$ and its boundary. Also, we shall simply write $\mathbf{B_p}$ and $\mathbf{S_p}$ instead of $B_p(1)$ and $S_p(1)$.

Recall that for x, y fixed in R^2, $d_p(x,y)$ as a function of $p\in[1,+\infty]$ is non increasing. Also, for any pair (x,y), the following relations hold among the different metrics:

$$d_p(x,y) \leq d_q(x,y)\cdot 2^{1/p-1/q} \quad \text{for } p\leq q\,. \tag{2}$$

This follows, by setting $x-y=z$, from the equality $\sup\limits_{x\ y\in R^2} \dfrac{d_p(x,y)}{d_q(x,y)} = \sup\limits_{x\in R^2} \dfrac{d_p(x,y)}{d_q(x,y)}$ taking into account that for $p<q$, the maximum of $\|z\|_p$ over $S_q(r)$ is attained for $z=(z_1,z_2)$ with $|z_1|=|z_2| = \dfrac{r}{2^{1/q}}$ (see e.g. [2], § 3).

Inequality (2) is sharp in the following sense: equality in it holds when the two components of x-y have the same modulus.

We recall that metrics not of type "d_p" have been used in the literature: see e.g. [2], [10], § 10, [15] and [16]; see also [8] for some related discussions. Some of these are obtained by combining metrics of d_p type. It is not difficult to see how our results can be extended to more general situations.

2. Some Mean Values

Assume we "approximate" a metric by another one in evaluating some quantities. It is more interesting to know a "mean" of the error done rather than the maximum possible error (for some special case). We discuss in full a few estimates, mainly in case the approximating metric is the euclidean one (d_2). We limit ourselves to write the results, when the steps needed to obtain them reduce to computing simple integrals.

2.1. MEANS RELATIVE TO THE MEASURE ON S_p.

Given $\rho>0$ and $q\in[1,+\infty]$, consider the set $S_q(\rho)$; on the sphere we shall consider the usual measure, that will be denoted by $dS(x)$; also, we set

$$\mathrm{meas}(S_q(\rho)) = \int_{S_q(\rho)} dS(x) .$$

Now set, for $1\leq p, q\leq \infty$:

$$\mu_p^q = \frac{1}{\mathrm{meas}(S_q(\rho))} \int_{S_q(\rho)} \|x\|_p \, dS(x) \tag{3}$$

These numbers, which are independent of ρ, represent the mean value of $\|\cdot\|_p$ over S_q; also, they depend continuously on p and q.

The first quadrant of S_q, $1\leq q<\infty$, can be parametrized e.g. by $x=(\cos\theta)^{2/q}$, $y=(\sin\theta)^{2/q}$, $0\leq\theta\leq\frac{\Pi}{2}$; the first quadrant of S_∞, by $x=1$, $y=t$, and by $x=t$, $y=1$ $(0\leq t\leq 1)$.

Now set $f_q(\theta) = \frac{2}{q}\left((\cos\theta)^{4/q\,-2}\sin^2\theta + (\sin\theta)^{4/q\,-2}\cos^2\theta\right)^{1/2}$ $(1\leq q<\infty)$; note that $f_1(\theta)=2\sqrt{2}\cos\theta\sin\theta$; $f_2(\theta)=1$.

For simmetry reasons we have

$$\mu_p^q = \frac{4}{\mathrm{meas}(S_q)} \int_0^{\Pi/2} \left((\cos\theta)^{2p/q} + (\sin\theta)^{2p/q} \right)^{1/p} f_q(\theta)\, d\theta \qquad 1\leq p<\infty \tag{3'}$$

$$\mu_\infty^q = \frac{4}{\mathrm{meas}(S_q)} \left(\int_0^{\Pi/4} (\cos\theta)^{2/q} f_q(\theta)\, d\theta + \int_{\Pi/4}^{\Pi/2} (\sin\theta)^{2/q} f_q(\theta)\, d\theta \right) =$$

$$= \frac{8}{\mathrm{meas}(S_q)} \int_0^{\Pi/4} (\cos\theta)^{2/q} f_q(\theta)\, d\theta \qquad 1\leq q<\infty . \tag{3''}$$

We have also $\text{meas}(S_q) = 4 \int_0^{\Pi/2} f_q(\theta)\, d\theta$ (for $1 \le q < \infty$) and $S_\infty = 8$. Moreover

$$\mu_p^\infty = \frac{1}{8} \int_{S_\infty} \|x\|_p \, dS(x) = \int_0^1 (1+t^p)^{1/p} \, dt \qquad (1 \le p < \infty) . \tag{3'''}$$

Evaluating the above integrals is not trivial in general; but in some cases this can be done easily. For example:

$$\mu_\infty^2 = \frac{4}{\Pi} \int_0^{\Pi/4} \cos\theta \, d\theta = \frac{2\sqrt{2}}{\Pi} \cong 0.9.$$

$$\mu_\infty^1 = \frac{4}{4\sqrt{2}}\, 2 \int_0^{\Pi/4} 2\sqrt{2}\, \cos^3\theta \sin\theta \, d\theta = 4 \int_0^{\Pi/4} \cos^3\theta \sin\theta \, d\theta = -\left(\frac{\sqrt{2}}{2}\right)^4 + 1 = \frac{3}{4}$$

$$\mu_1^\infty = \frac{1}{8} \int_0^1 (1+t) \, dt = \frac{3}{2}$$

$$\mu_1^2 = \frac{4}{2\Pi} \int_0^{\Pi/2} (\cos\theta + \sin\theta) \, d\theta = \frac{4}{\Pi} \cong 1.273 \qquad \text{(cf. [7] p. 643 and [3] p.162)}$$

and with some computation we obtain

$$\mu_2^\infty = \int_0^1 (1+t^2)^{1/2} \, dt = \frac{\sqrt{2}+\log(1+\sqrt{2})}{2} \cong 1.148 \quad \text{(see [3] p.158)} .$$

Another way to parametrize the first quadrant of the unit sphere is the following one: $x(t)=t$; $y(t)=(1-t^p)^{1/p}$ $(0 \le t \le 1)$. So we reobtain for example easily:

$$\mu_\infty^1 = 2 \int_{1/2}^1 t \, dt \quad \left(= \frac{3}{4}\right)$$

and

$$\mu_2^1 = \int_0^1 \sqrt{1-2t+2t^2}\,dt = 2\int_0^1 \frac{1}{2\sqrt{2}}\sqrt{1+x^2}\,dx \cong 0.8116.$$

Note that in general μ_p^q is not the reciprocal of μ_q^p. This can be expressed in the following way: the substitution of a distance with another one is not independent on the order of the substitution.

The discussion in [8, § 3.1] concerns mainly the number μ_1^2; the estimate 1.26 obtained there (and in [3]) is not far from the theoretical value for such number.

Other interesting parameters are the following ones, which give a sort of median for spheres. Fixed q, find a number r_p^q (depending on p) such that :

$$\text{meas}\,\{\, x\in S_q;\ \|x\|_p \geq r_p^q \} = \frac{1}{2}\,\text{meas}\,(S_q) \quad (p\neq q)\,. \tag{4}$$

These numbers are called in the literature "Levy's means", and have been widely studied also in the context of functional analysis; estimates of these means, depending on the norm and the dimension of the space, have been given: see e.g. [11], and the references there.

Now we shall give estimates for these Levy's means; for $q\in[1,\infty)$ fixed, we obtain r_∞^q from r_p^q by passing to the limit for $p\to\infty$. By a simmetry reasoning we have:

$$r_p^q = \sqrt[p]{(\cos\tfrac{\Pi}{8})^{2p/q}+(\sin\tfrac{\Pi}{8})^{2p/q}} \quad \text{for } 1\leq p<\infty, \text{ thus } r_1^2 = \cos\tfrac{\Pi}{8} + \sin\tfrac{\Pi}{8} \cong 1.307;$$

$$r_\infty^2 = \cos\tfrac{\Pi}{8} \cong 0.924\,;$$

$$r_p^1 = \sqrt[p]{(\tfrac{3}{4})^p + (\tfrac{1}{4})^p} \quad \text{for } 1\leq p<\infty\,, \text{ thus } r_2^1 = \frac{\sqrt{10}}{4} \cong 0.79\,;\quad r_\infty^1 = \frac{3}{4}\,;$$

$$r_p^\infty = \|(1,\tfrac{1}{2})\|_p\,, \text{ thus } r_1^\infty = \frac{3}{2}\ ;\quad r_2^\infty = \frac{\sqrt{5}}{2} \cong 1.118\,.$$

2.2. MEANS RELATIVE TO THE MEASURE ON B_p.

Other relations of interest are the following ones. Fixed q, find a number ρ_p^q (depending on p) such that $\text{meas}(B_p(\rho_p^q)) = \text{meas}\,(B_q)$. Thus the following relations must be satisfied $(\text{meas}(B_\infty)=4)$:

$$\int_0^{\rho_p^q} dx \int_0^{((\rho_p^q)^p - x^p)^{1/p}} dy = \frac{1}{4}\,\text{meas}(B_q) = \int_0^1 dx \int_0^{(1-x^q)^{1/q}} dy \qquad \text{for } 1 \le p < \infty$$

and $\displaystyle\int_0^{\rho_\infty^q} dx \int_0^{\rho_\infty^q} dy = (\rho_\infty^q)^2 = \frac{1}{4}\,\text{meas}(B_q)$.

In particular, we obtain for example:

for $q=2$, $\rho_1^2 = \sqrt{\frac{\Pi}{2}} \cong 1.253$; $\qquad \rho_\infty^2 = \frac{\sqrt{\Pi}}{2} \cong 0.886$;

for $q=1$, $\rho_2^1 = \sqrt{\frac{2}{\Pi}} \cong 0.798$; $\qquad \rho_\infty^1 = \frac{\sqrt{2}}{2} \cong 0.707$;

for $q=\infty$, $\rho_1^\infty = \sqrt{2} \cong 1.414$; $\qquad \rho_2^\infty = \frac{2}{\sqrt{\Pi}} \cong 1.128$.

Of course, we have (for any pair p, q) $\rho_p^q \cdot \rho_q^p = 1$.

The following table resumes some of the results indicated for $p, q = 1, 2, \infty$.

p \ q	1	2	∞	
	1	1.273	1.5	μ_p^q
1		1.307	1.5	r_p^q
	1	1.253	1.414	ρ_p^q
	0.8116	1	1.148	μ_p^q
2	0.79		1.118	r_p^q
	0.798	1	1.128	ρ_p^q
	0.750	0.9	1	μ_p^q
∞	0.750	0.924		r_p^q
	0.707	0.886	1	ρ_p^q

Now we shall indicate a possible use of some of the relations indicated before. Suppose we choose "random" a point in B_p, in the sense that we choose random the two random coordinates x, y. For the expected value of the distance between such point and the origin we obtain (by using a simple probability reasoning) the value 2/3; in fact, for p=1, 2, ∞ we have to calculate -respectively- the following integrals, see [3], p.153:

$$2\int_0^1 dx \int_0^{1-x} (x+y)\, dy \; ; \quad \frac{1}{\Pi} 2\Pi \int_0^1 \rho^2\, d\rho \; ; \quad \int_0^1 dx \int_{-x}^{x} x\, dy \, .$$

We denote by dx the measure on the ball; in these considerations the probability used is the usual measure on B_p divided by meas(B_p).

Assume now we want estimate the expected value of the distance d_q between a point in B_p and the origin; i.e.,

$$\frac{1}{\mathrm{meas}(B_q)} \int_{B_q} \|x\|_p \, dx \; ;$$

we have, for $p, q \in \{1, 2, \infty\}$:

$$\frac{1}{\mathrm{meas}(B_q)} \int_{B_q} \|x\|_p \, dx = \frac{2}{3} \mu_p^q$$

For example, let $p=\infty$ (i.e., we consider a square); we want to estimate the expected value of the euclidean distance between the center and a random point. Such value is $\frac{1}{3} | \sqrt{2}+\log(1+\sqrt{2}) | \cong 0.765$ (see [3], p.157) and can be obtained immediately as the product $\frac{2}{3} \mu_2^\infty$ (for these estimates and related ones see also [4], or [12], p.357).

The expected value of d_1 for a random point in B_2 is $\frac{2}{3} \mu_1^2 = \frac{8}{3\Pi} \cong 0.849$ (see [9] p. 588). In a similar way estimates for other values of p and q can be obtained.

Consider now the mean value of the distance d_2 between two points taken in B_2 random. The exact value is $\frac{128}{45\Pi} \cong 0.9054$ (for the calculation see [3], p.154 or [9], p.590).

Also: the expected value of the distance d_1 between two random points in B_2 is $\frac{128}{45\Pi} \mu_1^2 = \frac{512}{45\Pi^2} \cong 1.153$ (see [3], p.163 or [9], p. 589)).

It is not difficult to prove (see [3], p. 164) that the expected value of the d_1 distance between two random points (random coordinates) in B_∞ is $\frac{4}{3}$; this implies that an estimate of expected value for the d_2 distance is $\frac{4}{3}\mu_2^1 \cong 1.082$ (compare with the rather complicated computations done in [3], p. 159; see also [13]). Since in general $\mu_p^q \neq \mu_q^p$, it is not immediate to transform estimates concerning pairs of "random" points; also, the meaning of "random" should be clearly checked.
For more complicated situations see also [13] and [14].

3. Centers And Medians

Given a subset A of the plane, the following numbers can be useful when dealing with location problems:

$$\delta_p(A) = \sup\{d_p(x,y);\ x \in A;\ y \in A\} \tag{5}$$

$$r_p(A) = \inf_{x \in R^2} \sup\{d_p(x,y);\ y \in A\}\ . \tag{6}$$

For example: we want to find in the plane an otimal location for a service center serving all the points in A. If no capability restriction is done, then the optimal location is (for p given) a point C_p such that

$$r_p(A) = \sup\{d(C_p,y);\ y \in A\}.$$

Recall that for any A, we have $r_2(A) \leq \frac{1}{\sqrt{3}}\ \delta_2(A)$ (Jung's inequality). In general, we have $r_p(A) \leq \frac{2}{3}\ \delta_p(A)$ for any $p \geq 1$, while it is possible to substitute $\frac{2}{3}$ by better coefficients, depending on p (see [1] for a general discussion of this topic).
Any compact set has at least one center, which is also unique for $p \in (1,\infty)$, but not in general for $p=1$ or $p=\infty$. For example, if $A=\{(0,1), (0,-1)\}$ and $p=\infty$, then all the points in the segment joining (-1,0) and (1,0) are "centers". A similar remark applies for $p=1$, by considering the set $\{(-1,-1), (1,1)\}$.

Also, note that r_p depends continuously on p ($p \in [1,\infty]$). The same is true for C_p if $p \in (1,\infty)$; if C_1 (C_∞) is unique, then we have continuity also for $p \to 1$ (respectively: for $p \to \infty$): se e.g. [5]; see also [6], § 33, for general properties of centers.

Recall that r_p (p arbitrary) has the same value for a set A and for its closed convex hull. For any A, inequalities similar to those in (2) hold for δ_p and δ_q, or r_p and r_q. In fact, they trivially hold also for these quantities; better inequalities are not possible in general, as we can see by considering a set A which is not a singleton and which is contained in one of the bisecting lines.

In other cases, the optimal location for a service center is the solution of a "minisum" problem (searching for a "median"); namely, given p, if $A=\{P_1,P_2,...,P_n\}$, then a point M_p minimizing $\sum_{i=1}^{n} d_p(M_p,P_i)$ must be found.

Estimates concerning the difference among values of quantities depending on the distance (and not simply values of the distance) are sometimes complicated. Nevertheless, some of the remarks already done lead to the conclusion that statements in [8, § 3.1] and resumed in works concerning experimental data analysis are somewhat optimistic. In particular, the following is said in [8]: solutions of "minisum" problems are relatively insensitive to distance measure,... with a difference between euclidean and not euclidean around 0.01: the cases considered here for "median" problems with respect to p-norms held differences around 0.5 % on the average and less than 2.6 % as a maximum.
Apart from sets contained on a line (as the previous indicated one), the following simple example, concerning a three-point-set, shows how location problems can be seriously affected by the distance used.

Let $A=\{(2,0),(0,1),(0,-1)\}$. We have:
$r_1(A)=\frac{3}{2}$, and $C_1=(\frac{1}{2},0)$ is the unique center of A with respect to d_1;

$r_2(A)=\frac{5}{4}$, and $C_2=(\frac{3}{4},0)$ is the unique center of A with respect to d_2;

$r_\infty(A)=1$, and $C_\infty=(1,0)$ is the unique center of A with respect to d_∞.

Therefore we obtain $\frac{|r_2(A)-r_\infty(A)|}{r_2(A)}=\frac{20}{100}$; $\frac{|r_2(A)-r_\infty(A)|}{r_\infty(A)}=\frac{25}{100}$;

$$d(C_\infty, C_2) = \frac{1}{4} = \frac{d(C_\infty, C_2)}{r_\infty(A)} .$$

$$\frac{|r_2(A)-r_1(A)|}{r_2(A)}=\frac{1}{5}; \quad \frac{|r_2(A)-r_1(A)|}{r_1(A)}=\frac{1}{6}=\frac{d(C_2,C_1)}{r_1(A)}; \quad d(C_2,C_1)=\frac{1}{4} .$$

It is easily seen that there exists for A a unique median M_p with respect to the p-norm, for $p=1, 2, \infty$. In fact, for d_1, d_2, d_∞ we have respectively $M_1=(0,0)$; $M_2=(\frac{3}{4}, 0)$; $M_\infty=(1,0)$. Thus $d(M_2,M_1)=$ \F(3,4) ; $d(M_2,M_\infty)=$ \F(1,4) ; $d(M_1,M_\infty)=1$. Also,

$$s_1=\sum_{i=1}^{3} d_1(M_1,A_i)=4; \quad s_2=\sum_{i=1}^{3} d_2(M_2,A_i)=\frac{15}{4}; \quad s_\infty=\sum_{i=1}^{3} d_\infty(M_\infty,A_i)=3 .$$

We write d to indicate the fact that the values of d_1, d_2, d_∞ coincide.

Suppose we want to know how an estimate obtained e.g. by d_2 as an approximant for d_1 fits. For A the same set as above, we indicate the values concerning some quantities which are interesting in this context :

$$\frac{d(M_2,M_1)}{s_2}=\frac{1}{5}; \qquad \frac{|s_1-s_2|}{s_2}=\frac{1}{16}; \qquad \frac{d(M_2,M_1)}{s_1}=\frac{3}{16}$$

and by setting $d_p(i,j)=d_p(P_i,P_j)$

$$\sum_{i\,j=1}^{3}\left(\frac{d_2(i,j)-d_1(i,j)}{\sqrt{d_1(i,j)}}\right)^2=2\left(\frac{3-\sqrt{5}}{\sqrt{3}}\right)^2=\frac{28-12\sqrt{5}}{3} \cong 0.389.$$

$$E_{i,j}\left(\frac{d_2(i,j)-d_1(i,j)}{\sqrt{d_1(i,j)}}\right)=\frac{2(3-\sqrt{5})}{3\sqrt{3}} \cong 0.294 .$$

Of course, the differences become in general smaller when we consider a set containing more points, in they are scattered enough (but not when we have groups of points located more or less in the same position). An estimate of the size of these differences is given by the different values of the Jung's constant, for different metrics.

4. Some Final Remarks

Here we limited ourselves to coordinate some remarks concerning different metrics. Of course, in special cases other estimates could me more interesting. For example, "partial" means (i.e., concerning pairs of points along special directions) could be executed. Note that while d_p is sensitive to directions, d_2 is not: see e.g. [7] for d_1 and [2] for d_p, $p\neq 1$.

Our remarks could also be extended to other cases: e.g., to estimates of "variance" like

$$V_2 = (4/\Pi) \int_0^{\Pi/4} (\cos\theta - \mu_2^{\infty})^2 \, d\theta \,. \tag{7}$$

We did not consider such estimates here: no important theoretical tool is needed to extend our remarks, but the size of computations (in many cases) may rapidly increase. Anyway, in our opinion, simple general remarks could often help avoiding tedious statistics, giving at the same time sharper results. Hopefully, the examples discussed here should support this opinion.

ACKNOWLEDGEMENTS. The authors are indebted to R. Durier and C. Michelot for many helpful remarks concerning the paper, and to several other mathematicians who sent them their papers, often not appearing in the most common mathematical journals.

Also, the authors acknowledge support given by G.N.A.F.A.-C.N.R. and by the National Research Group "Functional Analysis" (M.U.R.S.T.) .

References

[1] Amir, D. 'On Jung's constant and related constants in normed linear spaces', Pacific J. Math. 118 (1985), 1-15.

[2] Brimberg, J. and Love, R.F. 'Estimating travel distances by the weighted ℓ_p norm', Naval Res. Logist. Quart., to appear.

[3] Eilon, S., Watson-Gandy, C.D.T., and Christofides, N. Distribution management: mathematical modelling and practical analysis, Griffin, London 1971.

[4] Erlenkotter, D. 'The general optimal market area model', preprint (Los Angeles, May 1987).

[5] Freimer, M. and Yu, P.L. 'Some new results on compromise solutions for group decision problems', Management Sci. 22 (1976), 688-693.

[6] Holmes, R.B. A course on optimization and best approximation, Lecture Notes in Mathematics 257, Springer Verlag, Berlin 1972.

[7] Huriot, J.M. and Perreur, J. 'Modèles de localisation et distance rectilinéaire', Revue d'Econ. Polit. 83 (1973), 640-662.

[8] Krarup, J. and Pruzan, P.M. 'The impact of distance on location problems', European J. Oper. Res. 4 (1980), 256-269.

[9] Lew, J.S., Frauenthal, J.C. and Keyfitz, N. 'On the average distances in a circular disc', SIAM Rev. 20 (1978), 584-592.

[10] Love, R.F., Morris, J.G. and Wesolowski, G.O. Facilities location: Models and Methods, North Holland, New York 1988.

[11] Martinelli, S. and Struppa, D. 'Sulla distanza media dall'origine dei convessi simmetrici di R^n', Atti Accad. Sci. Torino Cl. Sci. Fis. Mat. Natur. 118 (1984), 149-156.

[12] Newell, G.F. 'Scheduling, location, transportation, and continuum mechanics; some simple approximations to optimization problems', SIAM J. Appl. Math. 25 (1973), 346-360.

[13] Pfiefer, R.E. 'Minimum average distance between points in a rectangle', Amer. Math. Monthly 96 (1989), 64-65, Problem E 3217.

[14] Stone, R.E. 'Some average distance results', Transportation Sci. 25 (1991), 83-91.

[15] Ward, J.E. and Wendell, R.E. 'Using block norms for location modeling', Oper. Res. 28 (1980), 836-844.

[16] Witzgall, C. 'Optimal location of a central facility: Mathematical models and concepts', National Bureau of Standards, Report 8388 (1964), Washington D.C..

ON THE EFFECTIVENESS OF SOME INVERSION METHODS FOR NOISY FOURIER SERIES

L. DE MICHELE, M. DI NATALE, D. ROUX
Dipartimento di Matematica
Università degli Studi
Via C.Saldini,50
20133 Milano
Italy

ABSTRACT. A recent very general method for inversion of noisy Fourier series (given in [2]) employs suitable approximate units. In this paper, for some classical kernels, we give the estimates of the parameters involved in the method and we test it numerically.

1. Let us consider the ill-posed problem of reconstructing a periodic L^p ($1 \leq p < +\infty$, $p \neq 2$) function or a continuous periodic function knowing only an approximation of its Fourier coefficients. Recently, by using suitable approximate units, we gave a very general method which solves the problem for continuous or L^p periodic functions ([2],[3]). The method gives both a good norm and pointwise approximation, the last one at every Lebesgue point of f. Also *a priori* estimates of these approximations are given for particular classes of functions.

In this paper, for some classical kernels, we give the estimates of the parameters involved in the method and we test it numerically.

2. We confine ourselves to the one-dimensional case. Let T be the one-dimensional torus (identified with $\left[-\frac{1}{2}, \frac{1}{2}\right)$).

If $f \in L^1(T)$, let us denote by $\hat{f} = \{\hat{f}_n\}_{-\infty}^{+\infty} \in l^\infty$ the sequence of its Fourier coefficients and for every $\lambda = \{\lambda_n\}_{-\infty}^{+\infty} \in l^\infty$ let us set

$$\delta_0 = |\hat{f}_0 - \lambda_0|, \qquad \delta^* = \sup_{n \neq 0} |\hat{f}_n - \lambda_n| \qquad \delta = \max(\delta_0, \delta^*).$$

If $G \in L^1(\mathbb{R})$ and $\hat{G}(0) = 1$, for every $\lambda \in l^\infty$ and $\sigma \geq 0$ we give the following formal definition

$$R_\sigma \lambda \approx \sum_{-\infty}^{+\infty} \hat{G}(\sigma n) \lambda_n e^{2\pi i n t}, \qquad t \in T.$$

With some more conditions on G ([3], Th. 1 and 3), $R_\sigma \lambda$ gives "a good", pointwise or norm, approximation of f providing δ sufficiently small and σ chosen in a suitable way.

S. P. Singh (ed.), Approximation Theory, Spline Functions and Applications, 255–267.

Let us now consider the following classes of $L^1(T)$ functions. If $E \subset T$ we say that a function $f \in L^1(T)$ belongs to the Lebesgue class $K\text{Leb}\,(\alpha, E)$ if for every $t \in E$ and for every r, $0 < r \leq 1/2$

$$\int_{|x-t|\leq r} |f^*(x) - f(t)|\, dx < K r^{1+\alpha}$$

where f^* is the periodic continuation of f in $\mathbb{R}$.

If f satisfies

$$|f^*(x) - f(t)| \leq c|x-t|^\alpha$$

locally at t, then obviously $f \in K\text{Leb}(\alpha; \{t\})$ for some $K \geq c$, but in general it is not so easy to compute the value of $\inf K$ such that $f \in K\text{Leb}(\alpha; \{t\})$ (see [5]).

However it is easy to see that the following proposition holds.

PROPOSITION 2.1. If for every $t_1, t_2 \in \left(-\frac{1}{2}, \frac{1}{2}\right)$

$$|f(t_1) - f(t_2)| \leq c|t_1 - t_2| \tag{2.1}$$

then for every $t \in T$, $|t| \neq 1/2$, $f \in K(t)\text{Leb}(1, \{t\})$, where

$$K(t) = c(1 - 4t^2)^{-1}. \tag{2.2}$$

If, moreover, (2.1) holds for every $t_1, t_2 \in T$, then $f \in c\text{Leb}(1, T)$.

To proove (2.2) one can carry on the same calculations of [5], par. 4, function $f(t) = t$. The second part of the proposition is obvious.

For functions of the classes $K\text{Leb}(\alpha, E)$ we can give an *a priori* estimate of the pointwise approximation of $f(t)$ by $R_\sigma\lambda(t)$ $(t \in E)$ which is uniform on E. In order to do this let $M(x) = \text{Sup ess}_{|y|\geq|x|}\, |G(y)|$. Then the following theorem holds.

THEOREM 1 *If $\{\hat{G}(\sigma n)\}_{-\infty}^{+\infty} \in l^1$ for every $\sigma > 0$ and*

$$\int_0^{+\infty} x^\alpha M(x) dx = c_\alpha < +\infty,$$

then for every $f \in K\text{Leb}(\alpha, E)$, $\sigma > 0$ and $t \in E$ we have

$$|f(t) - R_\sigma\lambda(t)| \leq K((1+\alpha)c_\alpha + d_\alpha)\sigma^\alpha + \delta_0 + \|\{\hat{G}(\sigma n)\}_{n\neq 0}\|_1 \delta^*$$

where

$$d_\alpha = d_\alpha(\sigma) = 2^{-\alpha}\left(\sigma^{-(\alpha+1)} M\left(\frac{1}{2\sigma}\right) + \sigma^{-\alpha}\int_{1/(2\sigma)}^{+\infty} M(x)\, dx\right).$$

COROLLARY. *If, moreover, $\hat{G} \in L^1(\mathbb{R})$, for every $\sigma_0 > 0$ there exist C_α and D (depending only on σ_0 and G) such that if $0 < \sigma \leq \sigma_0$*

$$|f(t) - R_\sigma\lambda(t)| \leq K C_\alpha \sigma^\alpha + D\delta^*\sigma^{-1} + \delta_0 \tag{2.3}$$

for every $t \in E$.

Indeed, since $\lim\limits_{\sigma\to 0} d_\alpha(\sigma) = 0$ we can assume

$$C_\alpha = (1+\alpha)c_\alpha + \sup_{0<\sigma\leq\sigma_0} d_\alpha(\sigma)$$

and since

$$\left\| \{\hat{G}(\sigma n)\}_{n\neq 0} \right\|_1 = \sigma^{-1} \int_{-\infty}^{+\infty} |\hat{G}(x)|\, dx \qquad (1 + o(1))$$

we can assume

$$D = \sup_{0<\sigma<\sigma_0} \sigma \left\| \{\hat{G}(\sigma n)\}_{n\neq 0} \right\|_1$$

Remarks.

1) The statement of the theorem is given in a form more convenient for the applications than [3], Th. 4, but the proof is substantially the same.

2) If $|\hat{G}(x)|$ is decreasing with respect to $|x|$ we have

$$D = \|\hat{G}\|_1 \qquad \forall\, \sigma_0 > 0.$$

If moreover $\hat{G} \geq 0$ then

$$D = G(0) \qquad \forall\, \sigma_0 > 0. \tag{2.4}$$

3) The right hand side of (2.3) assumes the minimum value for

$$\sigma = \sigma^* = \sigma^*(\delta^*) = \left(\frac{D\delta^*}{K\alpha C_\alpha} \right)^{1/(1+\alpha)}. \tag{2.5}$$

Then this value σ^* of σ is an *a priori* choice of the regularizing parameter $\sigma = \sigma(\delta)$ for the uniform approximation of f on E which fits all functions $f \in K\mathrm{Leb}(\alpha, E)$ and all sequences $\lambda \in l^\infty$ such that $\sup_{n\neq 0} |\hat{f}_n - \lambda_n| \leq \delta^*$.

3. For making the method effective, chosen a function G satisfying the hypothesis we have to evaluate C_α and D. In this section we carry out this evaluation for the classical Poisson and Gauss kernels and for one of the kernels of the family studied in [1]. The functions G connected with these kernels satisfy the hypotheses of the theorem, with $\hat{G}$ positive even L^1 function, monotonically decreasing for $x > 0$. Then (2.3) holds with D given by (2.4).

In order to compute C_α let us set

$$H_\alpha(\sigma) = \sigma^{-(\alpha+1)} M\left(\tfrac{1}{2\sigma}\right)$$
$$I_\alpha(\sigma) = \sigma^{-\alpha} \textstyle\int_{1/(2\sigma)}^{+\infty} M(x)\, dx.$$

1) Poisson kernel.

We have

$$G(x) = \pi e^{-2\pi|x|} = M(x), \qquad \hat{G}(x) = (1 + x^2)^{-1}.$$

Then

$$D = \pi;$$
$$c_\alpha = 2^{-\alpha-1}\pi^{-\alpha} \textstyle\int_0^{+\infty} t^\alpha e^{-t}\, dt$$
$$= 2^{-(\alpha+1)}\alpha\pi^{-\alpha}\Gamma(\alpha);$$
$$H_\alpha(\sigma) = \pi\sigma^{-(\alpha+1)} e^{-\pi/\sigma};$$
$$I_\alpha(\sigma) = 2^{-1}\sigma^{-\alpha} e^{-\pi/\sigma}.$$

The functions $H_\alpha(\sigma)$ and $I_\alpha(\sigma)$ get the maximum value respectively at $\sigma = \frac{\pi}{1+\alpha}$ and $\sigma = \frac{\pi}{\alpha}$. Then for every σ, $0 < \sigma \leq \sigma_0 \leq \frac{\pi}{1+\alpha}$ we have

$$d_\alpha(\sigma) \leq 2^{-\alpha}\sigma_0^{-(\alpha+1)}(\pi + 2^{-1}\sigma_0)e^{-\pi/\sigma_0}.$$

2) Gauss kernel.
We have

$$G(x) = e^{-\pi x^2} = M(x) = \hat{G}(x).$$

Then

$$\begin{aligned}
D &= 1; \\
c_\alpha &= 2^{-1}\pi^{-(\alpha+1)/2} \int_0^{+\infty} t^{(\alpha-1)/2}e^{-t}\,dt \\
&= 2^{-1}\pi^{-(\alpha+1)/2}\Gamma\left(\tfrac{\alpha+1}{2}\right); \\
H_\alpha(\sigma) &= \sigma^{-(\alpha+1)}e^{-\pi/(4\sigma^2)}; \\
I_\alpha(\sigma) &= \sigma^{-\alpha} \int_{1/(2\sigma)}^{+\infty} e^{-\pi x^2}\,dx \\
&\leq 2\sigma^{1-\alpha} \int_{1/(2\sigma)}^{+\infty} x e^{-\pi x^2}\,dx \\
&\leq \pi^{-1}\sigma^{1-\alpha}e^{-\pi/(4\sigma^2)}.
\end{aligned}$$

Since $H_\alpha(\sigma)$ gets its maximum value at $\sigma = \left(\frac{\pi}{2(1+\alpha)}\right)^{1/2}$ and the majorant of $I_\alpha(\sigma)$ monotonically increases in $(0, +\infty)$, then for every $0 < \sigma \leq \sigma_0 \leq \left(\frac{\pi}{2(1+\alpha)}\right)^{1/2}$ we have

$$d_\alpha(\sigma) \leq 2^{-\alpha}\sigma_0^{-(\alpha+1)}(1 + \pi^{-1}\sigma_0^2)e^{-\pi/(4\sigma_0^2)}.$$

3) Let us now consider the function

$$G(x) = \pi e^{-\sqrt{2}\pi|x|}\sin\left(\frac{\pi}{4} + \sqrt{2}\pi|x|\right).$$

The interest in this function is due to the fact that the related kernel allows the approximation of f with a good control on the size of the second derivative of the approximant (see [4]).

We have

$$\begin{aligned}
M(x) &\leq \pi e^{-\sqrt{2}\pi|x|}; \\
\hat{G}(x) &= (1 + x^4)^{-1}; \\
D &= \pi/\sqrt{2}; \\
c_\alpha &\leq 2^{-(\alpha+1)/2}\pi^{-\alpha} \int_0^{+\infty} t^\alpha e^{-t}\,dt \\
&= 2^{-(\alpha+1)/2}\pi^{-\alpha}\alpha\Gamma(\alpha); \\
H_\alpha(\sigma) &\leq \pi\sigma^{-(\alpha+1)}e^{-\sqrt{2}\pi/(2\sigma)}; \\
I_\alpha(\sigma) &\leq \tfrac{\sqrt{2}}{2}\sigma^{-\alpha}e^{-\sqrt{2}\pi/(2\sigma)}.
\end{aligned}$$

In the last two inequalities, the right hand side gets its maximum value, respectively, at $\sigma = \frac{\sqrt{2}}{2}\frac{\pi}{1+\alpha}$ and $\sigma = \frac{\sqrt{2}}{2}\frac{\pi}{\alpha}$. Then for every $0 < \sigma \leq \sigma_0 \leq \frac{\sqrt{2}}{2}\frac{\pi}{1+\alpha}$ we have

$$d_\alpha(\sigma) \leq 2^{-\alpha}\sigma_0^{-(\alpha+1)}\left(\pi + \frac{\sqrt{2}}{2}\sigma_0\right)e^{-\sqrt{2}\pi/(2\sigma_0)}.$$

4. In order to compare the behaviour of the considered kernels in the approximation of functions of the classes $K\text{Leb}(\alpha, E)$, we confine ourselves to the case $\alpha = 1$, which is the most interesting case in the applications.

In this case, (2.5) becomes

$$\sigma^* = \left(\frac{D\delta^*}{KC_1}\right)^{1/2}$$

and (2.3) gives, for every $t \in E$

$$|f(t) - R_{\sigma^*}\lambda(t)| \leq 2(C_1 D K \delta^*)^{1/2} + \delta_0 \tag{4.1}$$

provided $\sigma^* \leq \sigma_0$, that is δ^* sufficiently small.

For kernels 1), 2), 3) above considered we have respectively for sufficiently small σ_0 (e.g. $\sigma_0 \leq 3/4$)

1) $$C_1(\sigma_0) = (2\pi)^{-1} + (2\sigma_0{}^2)^{-1}(\pi + 2^{-1}\sigma_0)e^{-\pi/\sigma_0}$$

2) $$C_1(\sigma_0) = \pi^{-1} + (2\sigma_0{}^2)^{-1}(1 + \pi^{-1}\sigma_0{}^2)e^{-\pi/(4\sigma_0{}^2)}$$

3) $$C_1(\sigma_0) = \pi^{-1} + (2\sigma_0{}^2)^{-1}(\pi + 2^{-1/2}\sigma_0)e^{-\sqrt{2}\pi/(2\sigma_0)}.$$

Since the values of D are respectively π, 1, $\pi/\sqrt{2}$, the estimates of the approximation given by (4.1) for all functions of the class $K\text{Leb}(1, E)$ are essentially of the same order whatever is the kernel we use.

Then it is interesting to compare the effective approximations obtained for particular functions of the class by using different kernels. To this aim we performed a numerical experience briefly described in the following.

We chose some subsets E of T, some values of K and some functions $f \in K\text{Leb}(1, E)$ of different behaviour (e.g. $C^{(0)}(T)$, $C^{(0)}$ $(-1/2, 1/2)$,..., monotonous, oscillating,... $L^1(T)$ but not $L^2(T)$...).

We inserted a noise in $\{\hat{f}_n\}_{-\infty}^{+\infty}$ by putting $\hat{f}_n = 0$ if $|n| \geq m$ (e.g. $m = 50$) and by adding to $\hat{f}_n$ ($|n| \leq m$, $n \neq 0$) a random error using the standard normal distribution $N(0, 1)$, in such a way to obtain a sufficiently small δ^* (e.g. $\delta^* = 0.01$).

We calculated $\|f - R_\sigma\lambda\|_{L^\infty(E)}$ for $\sigma = 0$ and for $\sigma = \sigma_i{}^*$ ($i = 1, 2, 3$) (the values of σ^* for the three above kernels) by evaluating $|f(t) - R_\sigma\lambda(t)|$ on a finite set of equidistributed points $t \in E$. Moreover we plotted the corresponding graphs of $|f(t) - R_\sigma\lambda(t)|$ in E.

As a comment to the numerical experiments we can say that:

1) The choice of $\sigma = \sigma_i{}^*$ gives an effective improvement in the approximation on E with respect to $\sigma = 0$ (i.e. the noisy Fourier polynomial).

2) The graphics show that the effectiveness of the methods relies on the smoothness of the approximations. Indeed for every kernel we have a good control of the first derivative. Hence the approximating functions $R_{\sigma^*}\lambda$ graphically fit the function f much better than the noisy Fourier polynomial.

3) Comparing the behaviour of the three considered kernels, Poisson and Gauss kernels seem to approximate f better than the third one even though the errors are not too different.

In tables 1, 2 and figures 1,...,6 we present some qualitative but significant results of the numerical experience in the case $K = 5$, $\delta^* = 0.01$, intervals $E = [-0.35, 0.35]$ (see table 1) and $E = [0.15, 0.35]$ (see table 2 and figures 1,...,6), for the following test functions of the class 5Leb(1, E) (see Prop. 1 for $f_1, \ldots, f_5$ and [5] for f_6):

$$f_1 = 5t^2, \qquad f_2 = 5|t|, \qquad f_3 = 1.5e^t, \qquad f_4 = 0.5(e^t + 0.1\sin 10\pi t),$$

$$f_5 = 0.5(e^t + 0.1\sin 30t), \qquad f_6 = 0.4(1 - 4t^2)^{-1/2}.$$

Table 1. $E = [-0.35, 0.35]$; $E_0 = \|f - R_0\lambda\|_\infty$; $E_i = \|f - R_{\sigma_i^*}\lambda\|_\infty$, $i = 1,2,3$.

	f_1	f_2	f_3	f_4	f_5	f_6
E_0	1.1 E−1	1.7 E−1	1.4 E−1	1.4 E−1	1.9 E−1	1.2 E−1
E_1	1.8 E−2	1.5 E−1	3.9 E−2	4.0 E−2	5.2 E−2	3.1 E−2
E_2	1.3 E−2	1.1 E−1	4.6 E−2	3.5 E−2	4.9 E−2	1.6 E−2
E_3	1.6 E−2	4.9 E−2	5.8 E−2	3.7 E−2	5.2 E−2	2.6 E−2

Table 2. $E = [0.15, 0.35]$; $E_0 = \|f - R_0\lambda\|_\infty$; $E_i = \|f - R_{\sigma_i^*}\lambda\|_\infty$, $i = 1,2,3$.

	f_1	f_2	f_3	f_4	f_5	f_6
E_0	8.8 E−2	1.0 E−1	1.4 E−1	1.2 E−1	1.5 E−1	1.2 E−1
E_1	1.8 E−2	1.7 E−2	2.5 E−2	3.5 E−2	4.6 E−2	2.7 E−2
E_2	1.4 E−2	2.0 E−2	2.5 E−2	3.0 E−2	4.3 E−2	1.2 E−2
E_3	1.7 E−2	3.4 E−2	3.2 E−2	1.9 E−2	4.8 E−2	1.6 E−2

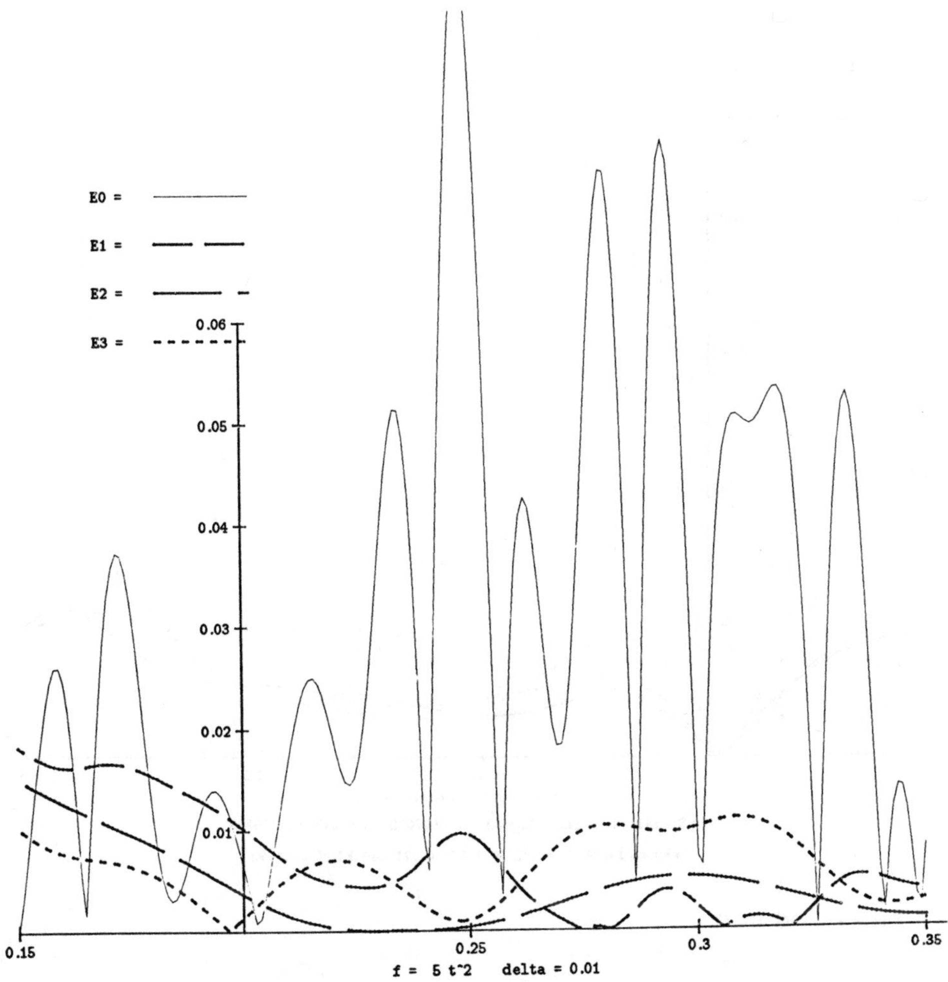

Figure 1. Function f1.

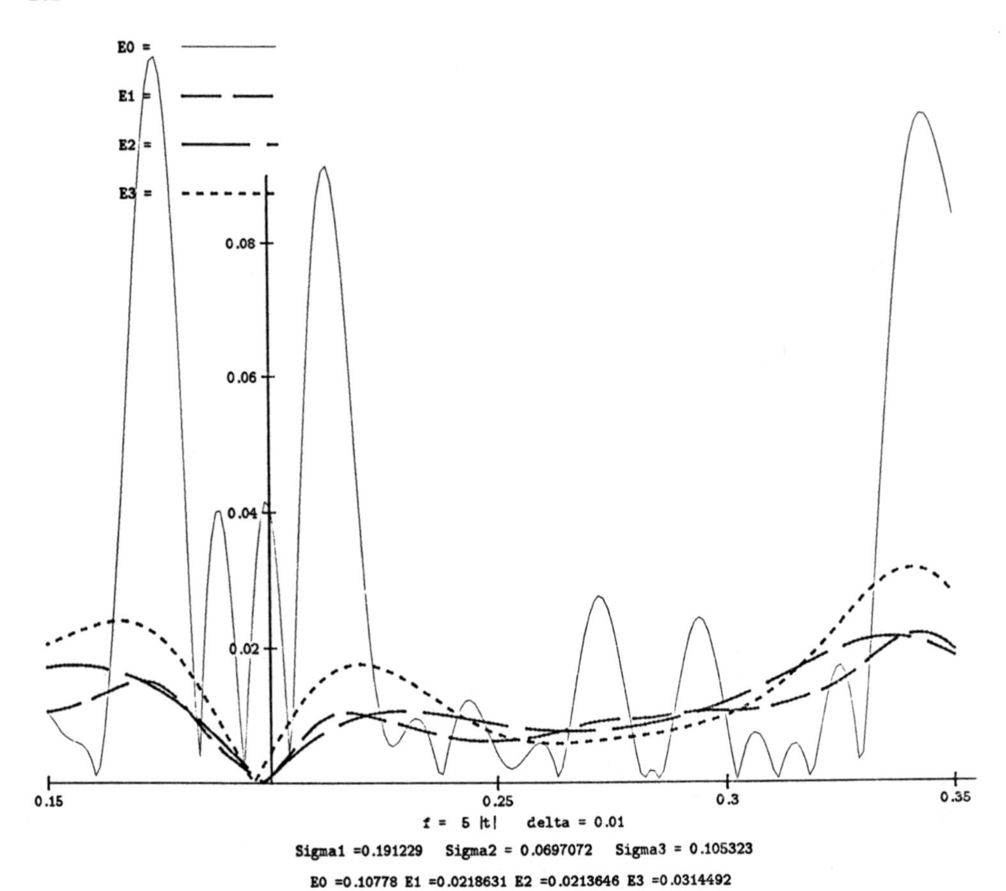

Figure 2. Function f2.

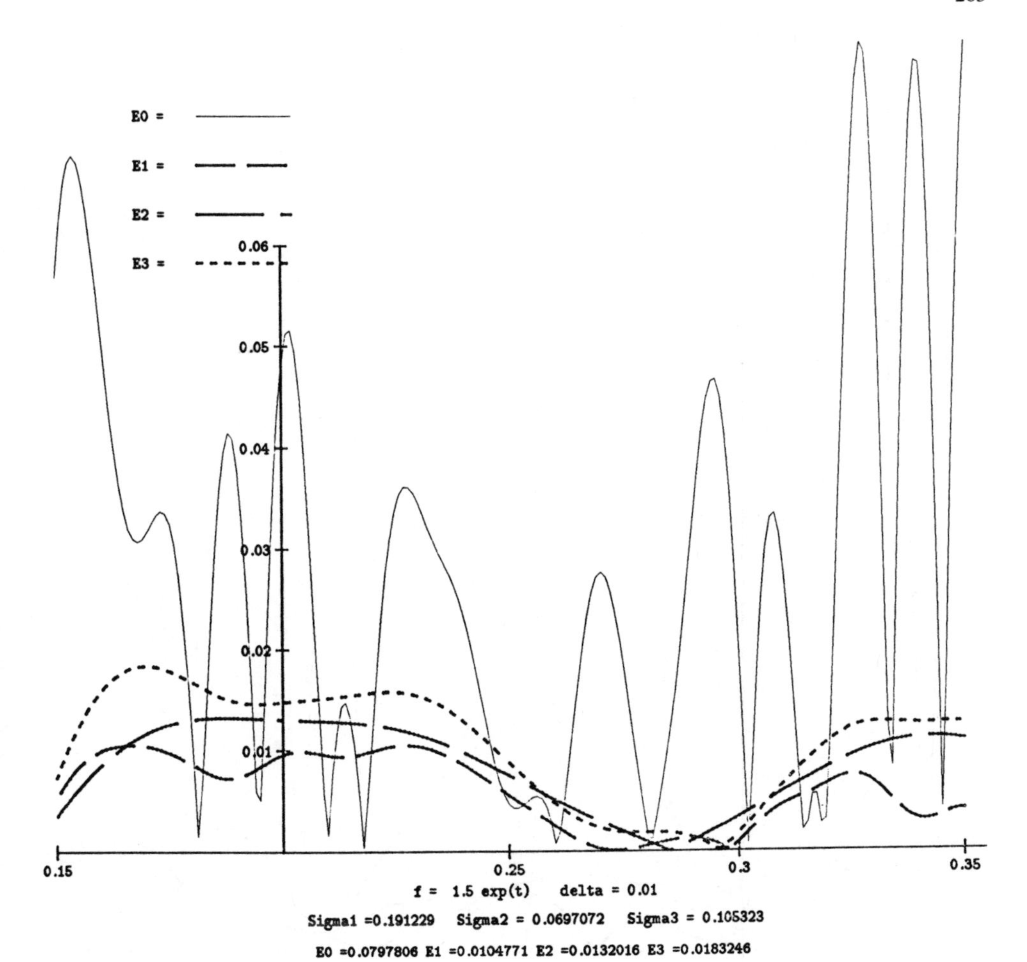

Figure 3. Function f3.

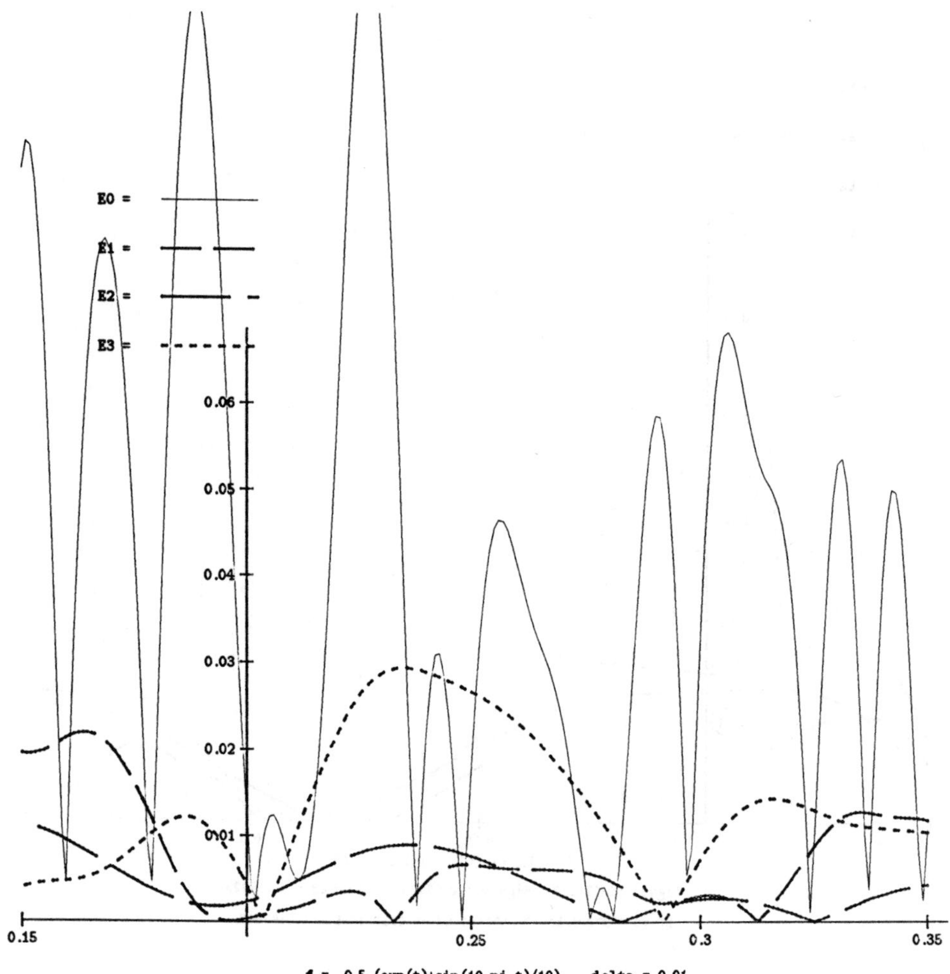

f = 0.5 (exp(t)+sin(10 pi t)/10) delta = 0.01

Sigma1 =0.191229 Sigma2 = 0.0697072 Sigma3 = 0.105323

E0 =0.114288 E1 =0.0217992 E2 =0.0113603 E3 =0.0293805

Figure 4. Function f4.

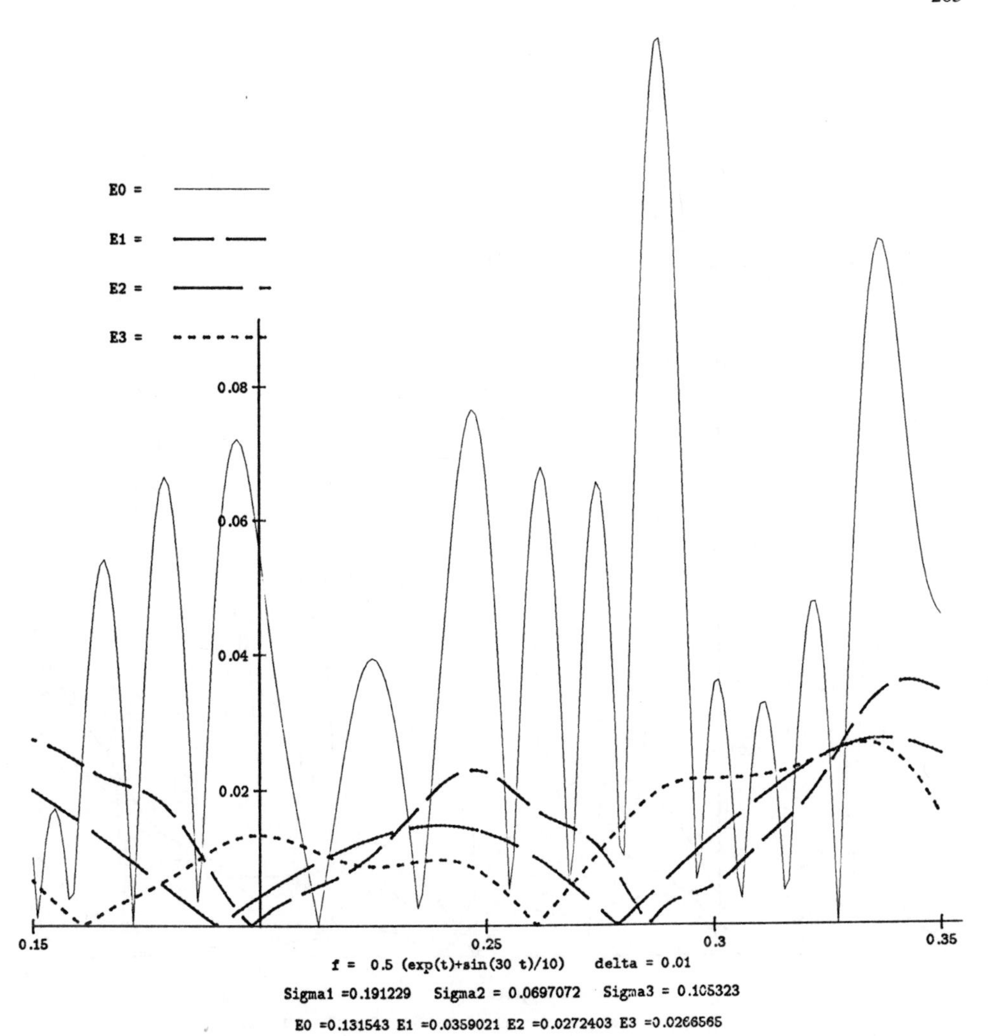

Figure 5. Function f5.

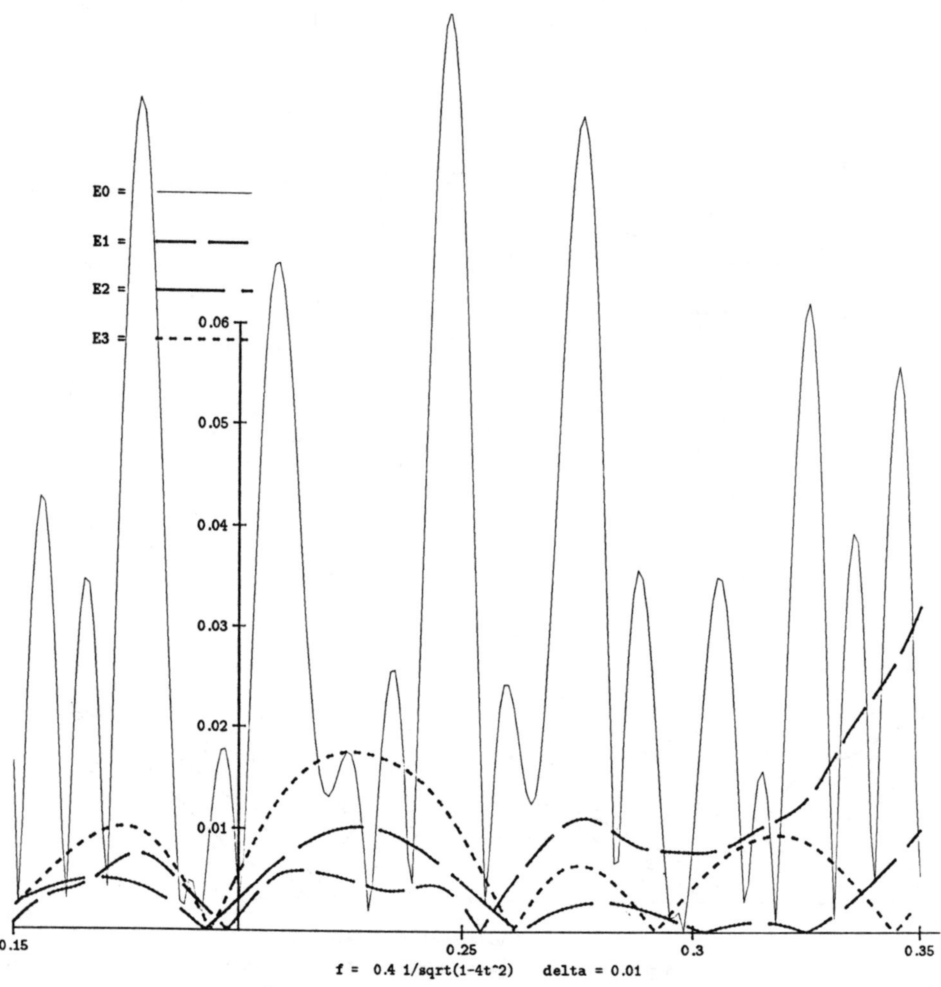

Figure 6. Function f6.

References

[1] L. De Michele, M. Di Natale, D. Roux, *Sharp properties of a regularization method for inversion of Fourier series*, Proceedings of the Meeting "Trends in Functional Analysis and Approximation Theory", Atti Sem. Mat. Fis. Univ. Modena, 39 (1991), 253–267.

[2] L. De Michele, M. Di Natale, D. Roux, *Inverse Fourier Transform*, Atti Accad. Naz. Lincei Rend. Cl. Sci. Fis. Mat. Natur. (9) 1 (1990), 305–308.

[3] L. De Michele, M. Di Natale, D. Roux, *Inversion methods for Fourier Transform*, Rend. Accad. Naz. Sci. XL Mem. Mat. (1991).

[4] L. Gotusso, D. Roux, *Remarks on a smoothing problem*, Atti Accad. Sci. Torino Cl. Sci. Fis. Mat. Natur., 118 (1984), 136–142.

[5] L. Gotusso, D. Roux, P. Zanzi, *Sul calcolo di una costante relativa ad un metodo di inversione delle serie di Fourier*, Ist. Lomb. Accad. Sci. Lett. Rend. A, 124 (1990), 85–98.

Work supported by Italian M.U.R.S.T.

Acknowledgement.

We thank Miss Paola Gemelli for her precious contribution in realizing by *Mathematica* the graphic part of this work.

A GENERALIZATION OF N-WIDTHS

ASUMAN G. AKSOY
Department of Mathematics
Claremont McKenna College
Claremont, CA. 91711
U.S.A.

Abstract

This paper is a study of the n-widths defined by Kolmogorov. In section I we give definitions of n-widths of a set in a Banach space and n-widths of an operator acting between Banach spaces. Several important well known results about this concepts are also included in section I. In section II, we introduce a refined concept of an approximation scheme with respect to which a refined concept of n-widths can be defined. Theorems about generalized n-widths illustrate the fact that this is a genuine generalization. We finish by the question of finding concept of n-widths in the context of Orlicz modular spaces.

I. N-Widths of a Set

Let X be a normed linear space and X_n be its n-dimensional subspace of X, for each $x \in X$ the distance, $d(x;X_n)$ of X_n to x is defined by:

$$d(x;X_n) = \text{Inf}\ \{\| x - y \| : y \in X_n\}.$$

If there is a $y^* \in X_n$ for which $d(x;Xn) = \| x - y^* \|$ holds then y^* is the best approximation to x from X_n. More than 100 years ago Weierstrass proved that given a continuous function f(x) on [a,b] and $\varepsilon > 0$, there exists a polynomial p(x) such that $\|f(x) - p(x)\| < \varepsilon$. Which tells us that $d(f; P_n) \to 0$ as $n \to \infty$ for each n, where $P_n = \text{span}(1, x^1, \cdots, x^n)$.

Now let us suppose instead of a single element x, we are given a subset A of X, then how well n-dimensional subspace X_n

S. P. Singh (ed.), Approximation Theory, Spline Functions and Applications, 269–278.

of X approximate the subset A? To answer this question one looks at the deviation of A from X_n, namely:

$$d(A; X_n) = \text{Sup} \{ d(a, X_n) : a \in A \}$$

Thus, $d(A; X_n)$ measures the extent to which the "worst element" of A can be approximated from X_n. If we take this process one step further by allowing n-dimensional subspaces X_n vary within X, then the question is how well one can approximate A by n-dimensional subspaces of X? The answer to this question was first given by Kolmogorov.

Definition: Let X be a normed linear space and A a subset of X, the n-th width or n-diameter (or Kolmogorov n-th diameter) of A in X is:

$$d_n(A;X) = \text{Inf}\{ d(A;X_n) : X_n \text{ is n-dimensional subspace of } X\}$$

Thus $d_n(A;X) = \inf_{X_n} \sup_{a \in A} \inf_{x \in X_n} \| a - x \|$.

We often drop X and write $d_n(A)$.
A subspace X_n of X of dimension at most n, for which $d_n(A;X) = d(A;X_n)$ is called the optimal subspace for $d_n(A;X)$.

Besides defining the concept of n-widths, Kolmogorov also computed $d_n(A;X)$ for some particular cases. For example, he showed that [13]

$$d_o(A ; L_2) = \infty, \text{ and}$$

$$d_{2n-1}(A ; L_2) = d_{2n}(A ; L_2) = n^{-k}$$

where $L_2 = L_2[0;2\pi]$ square integrable functions on $[0;2\pi]$, and

$$A = \{ f : f \in W_2^{(k)}, \quad \| f^{(k)} \| \leq 1 \}$$

and $W_2^{(k)}$ is the space of 2π periodic, real valued, (k-1) times differentiable functions whose (k-1) st derivative is absolutely continuous and whose kth derivative is in L_2.

In general it is impossible to obtain $d_n(A ; X)$ for all A and X although there is a considerable effort devoted to calculate $d_n(A;X)$ for specific choices of A and X [See 13]. A usual method of calculation is to find an upper bound by calculating $d_n(A;X_n)$ for a "reasonable" choice of X_n, and then to show that the quantity obtained is infact the lower bound as well. It is also important to determine asymptotic behavior of $d_n(A;X)$ as $n \to \infty$. In many cases very simple n-dimensional subspaces may approximate A in an asymptotically optimal manner.

N-widths of integral operators and n- widths of Soboloev spaces can be found in [13]. Let D be a fixed nxn matrix and the set A is

$$A = \{ Dx : \| x \|_{l_p^n} \leq 1 \} \subset l_q^n \quad \text{where } p, q \in [1,\infty]$$

Very little are known about $d_n(A;l_q^n)$ unless $p=q=2$ or $p=q=\infty$ and D is totally positive. Therefore one usually considers the case that D is a diagonal matrix. Following is such a result the proof of which can be found in [13] :

Let $D = \text{diag}\{a_1, a_2, \cdots, a_m\}$ be an mxm real diagonal matrix,assume that $a_1 \geq a_2 \geq \cdots \geq a_m > 0$. Given $1 \leq q \leq p \leq \infty$. Let $1/r = 1/q - 1/p$. Then

$$d_n(D_p;\ l_q^m) = \Big(\sum_{k=n+1}^{m} a_k^r\Big)^{1/r}, \text{ where } D_p = \{ Dx : \| x \|_p \leq 1 \}$$

It can be easily seen that the n-width $d_n(A;X)$ can also be written as

$$d_n(A\ ;\ X) = \operatorname*{Inf}_{X_n} \inf \{ \varepsilon > 0 \ :\ A \subset \varepsilon U_X + X_n\}$$

where U_X is the unit ball of X. This definition allows us the following generalization.

Let A, B be non-empty subsets of a normed linear space X. Assume that B absorbs A then n- width of A with respect to B, $d_n(A, B; X)$, is defined by

$$d_n(A, B; X) = \operatorname*{Inf}_{X_n} \inf \{ \varepsilon > 0 : A \subset \varepsilon B + X_n\}.$$

This definition is used in the concept of diametral dimension of nuclear spaces [3, 12].

The basic properties of n-widths can be found in [9,10,12,13]. It is easy to show that if X be a normed linear space and A be a closed subset of X, then

A is compact if and only if $d_n(A) \downarrow 0$ and A is bounded.

N-Widths of an Operator

Let $T : X \longrightarrow Y$ be an operator between two normed linear spaces. The n-width of T:

$$d_n(T) = d_n(T(U_X); Y) = \operatorname{Inf}\{ r > 0 : T(U_X) \subset r\,U_Y + Y_n \}.$$

It is known that

$$T \text{ is compact if and only if } d_n(T) \downarrow 0 .$$

<u>Notation:</u> Let F(X,Y) and K(X,Y) denote the closed subsets of L(X,Y) consist of finite rank and compact operators respectively. F(X,Y) is a subset of K(X,Y) and need not equal K(X,Y). The n-th approximation number $a_n(T)$ of $T \in L(X,Y)$ for n $= 0,1,2,\cdots$ defined as

$$a_n(T) = \operatorname{Inf}\{ \| T - A \| : A \in F_n(X,Y) \}$$

where $F_n(X,Y)$ is the collection of all mappings whose range is at most n- dimensional. It is known that

$$T \in F(X,Y) \text{ if and only if } \lim_{n \to \infty} a_n(T) = 0$$

so, $a_n(T)$ provides a measure how well T can be approximated by finite mappings whose range is at most n-dimensional. Algebraic and analytic properties of $a_n(T)$ can be found in [9,12]. The following theorem [5] gives the relationship between the n-widths and the approximation numbers:

<u>Theorem:</u> For any $T \in L(X,Y)$, the following inequality is valid:

$$d_n(T) \leqslant a_n(T) < (\sqrt{n} + 1)\, d_n(T).$$

The best value p(n) for which $a_n(T) < p(n)\, d_n(T)$ is not known. But p(n) can not be replaced by a constant independent of n. There are spaces for which

$$\lim_n d_n(T) = 0 \quad \text{and} \quad \lim_n a_n(T) \neq 0.$$

It should be noted that if $T: H \to H$ is a compact operator on a Hilbert space H, then one can define $(d_n(T))$ as the sequence of eigenvalues of the positive operator $|T| = (TT^*)^{1/2}$. In this case:

i) $a_n(T) = d_n(T)$

ii) $\prod_{i=1}^{n} |\lambda_i(T)| \leq \prod_{i=1}^{n} d_i(T)$ (H.Weyl Inequality, 1949) [14]

where $(\lambda_i(T))$ is an eigenvalue sequence [6]. The last inequality can be viewed as relating the eigenvalues of T to those of $|T|$.

From (ii) it may be deduced that for all $n \in N$ and all $p \in (0,\infty)$,

$$\sum_{i=1}^{n} |\lambda_i(T)|^p \leq \sum_{i=1}^{n} a_i^p(T)$$

which implies that if $(a_i(T)) \in l_p$ then $(\lambda_i(T)) \in l_p$. This result can be used to obtain information about the distribution of eigenvalues of certain non-self-adjoint elliptic problems [see chapter XII of 4]. Although Weyl's inequality was given in Hilbert space setting, a simple proof of it in the context of Banach spaces can be found in [4].

II. Generalized N-Widths

Let X be a Banach space and $(A_n)_{n \in N}$ be a sequence of subsets of X satisfying the following conditions:

1) $(0) = A_0 \subset A_1 \subset A_2 \subset \cdots \subset X$

2) $\lambda A_n \subset A_n$ for all scalars λ and $n = 1,2,\cdots$

3) $A_n + A_m \subset A_{n+m}$ for $m,n = 1,2,\cdots$

then (X,A_n) is called an <u>approximation scheme.</u> The use of an approximation scheme on a Banach space and its use in approximation theory can be found in Butzer and Scherer [2] and in Pietsch [11].For example one can consider $X = l_p$ with $p > 1$ and A_n to be the set of all scalar sequences (a_m) such that $a_m = 0$ when $m > n$ or $X = L_p [0,1]$ $2 \leq p \leq \infty$ and $A_n = L_{p+1/n} [0,1]$.

Instead of looking at subset of X with the above properties, if we consider $Q = Q_n(X)$ a <u>family</u> of subsets of X with the same properties (replace A_n by Q_n in above 1,2,3) then it is possible to define a refined notion of approximation scheme. For example, for a given Banach space X, Q_n will be the set of all n-dimensional subspaces or for a given Banach space E, consider $X = L(E)$ and Q_n will be the set of all n-nuclear maps on E.

This refined approximation scheme allows us to define n- width $d_n(A ; Q)$ with respect to this approximation scheme as follows:

<u>Definitions:</u> 1) Let U_X be the closed unit ball of X and D be a bounded subset of X. Then the <u>generalized n-th width</u> of D with respect to U_X is defined by:

$$d_n(D;Q) = \text{Inf} \{ r > 0 : D \subset r U_X + A \quad A \in Q_n(X) \}.$$

The generalized n-th width $d_n(T;Q)$ of $T \in L(X)$ is defined as $d_n(T(U_X); Q)$. From the stated definition it follows that $(d_n(T;Q))$ is non- increasing sequence of non- negative numbers and

$$\| T \| = d_o(T Q) \geq d_1(T;Q) \geq \cdots \geq d_n(T;Q) \geq \cdots$$

Notice that if one choses Q_n to be the at most n- dimensional subspaces of X, then $d_n(T;Q)$ coincides with the usual definition of $d_n(T)$.

2) A bounded set D of X is said to be <u>Q- compact set</u> if $\lim_n d_n(D;Q) = 0$ and $T \in L(X)$ is said to be <u>Q-compact operator</u> if $\lim_n d_n(T;Q) = 0$. That is $T(Ux)$ is a Q- compact set.

We assume that each $A_n \in Q_n (n \in N)$ is separable, then it is immediate from the definitions that Q-compact sets are separable and Q-compact maps have separable range.

Q-Compactness Does Not Imply Compactness

We show that in $L_p[0,1]$, $2 \leq p \leq \infty$, with suitably defined approximation scheme, one can find a Q-compact map which is not compact.

Let $[r_n]$ be the space spanned by Radamacher functions and R_p be the closure of $[r_n]$ in L_p [0, 1]. Define an approximation scheme A_n on $L_p[0,1]$ as $A = L_{p+1/n}$. $L_{p+1/n} \subset L_{p+1/n+1}$ gives us $A_n \subset A_{n+1}$ for $n = 1, 2, \cdots$ and it is easily seen that $A_n + A_m \subset A_{n+m}$ for $n, m = 1,2, \cdots$ and that $\lambda A_n \subset A_n$.

Next we observe the existence of a projection

$$P : L_p[0,1] \dashrightarrow R_p \quad \text{for } p \geq 2 .$$

In fact $P = j \circ P_2 \circ i$ where i, j are isomorphisms shown in the diagram below and P_2 is the orthogonal projection.

$$L_p \xrightarrow{i} L_2 \xrightarrow{P_2} R_2 \xrightarrow{j} R_p .$$

<u>Theorem:</u> For $p \geq 2$ the projection $P: L_p[0, 1] \dashrightarrow R_p$ is Q-compact but not compact.

It is easy to show that $P(U_{L_p}) \subset L_{p+1/n}$ thus $d_n(P;Q) \dashrightarrow 0$. To see P is not a compact operator observe that $\dim R_p = \infty$ and $I - P$ is a projection with kernel R_p, thus $I - P$ is not a Fredholm operator so, P can not be a compact operator. For details of the proof of the above theorem see [1].

<u>Definitions:</u> 1) A sequence $(x_{n,k})_k \subset A_n$ is said to be order c_o-sequence if followings hold:

i) For every $n \in N$, there exists an $A_n \in Q_n$ and $(x_{n,k})_k \subset A_n$.

ii) $\| x_{n,k} \| \dashrightarrow 0$ as $n \dashrightarrow \infty$ uniformly in k.

2) Suppose $(x_{n,k})_k$ is an order- c_o-sequence in X. Then the set S_m associated with $(x_{n,k})_k$ is:

$$S_m = \{\sum_{n=1}^{m} \lambda_n x_{n,k(n)} \quad : \quad \sum_{n=1}^{m} |\lambda_n| \leq 1 \}.$$

where $x_{1,k(1)} \in A_1$, $x_{2,k(2)} \in A_2$, $\cdots$, $x_{m,k(m)} \in A_m$.

Clearly $S_m \subset A_1 + A_2 + \cdots + A_m \in Q_{m^2}$.So if Q_n is n-dimensional, S_n is at most n^2-dimensional.

For a bounded set D in X, we define the <u>ball measure of non-Q-compactness</u> $\alpha(D;Q)$ of D by

$$\alpha(D, Q) = \inf\{r > 0 : \text{order-}c_o\text{-sequence } (x_{n,k})_k \text{ and associated } S_n \text{ such that } D \subset \bigcup_{x \in S_n} B(x, r) \text{ for some } n\}.$$

Following are the several results about Q- compact sets and Q- compact maps. The proofs of all are presented in [1].

<u>Theorems:</u> 1) Suppose (X, Q_n) is an approximation scheme with sets A_n Q_n assumed to be solid (i.e., $|\lambda|$ A_n A_n for $|\lambda| \leq 1$). Then a bounded set D of X is Q- compact if and only if there exists an order co- sequence $(x_{n,k})_k$ A_n such that

$$D \subset \{\sum_{n=1}^{\infty} \lambda_n x_{n,k(n)} \quad : \quad x_{n,k(n)} \in (x_{n,k}), \ \sum_{n=1}^{\infty} |\lambda_n| \leq 1\}.$$

This theorem can be considered an analogue of the Dieudonne-Schwartz lemma on compact sets in terms of standard Kolmogorov diameter. Again if one choses Q_n to be at most n- dimensional subspaces of X, one can show that Q- compactness of a bounded subset D coincides with the usual definition of compactness of D.

2) The uniform limit of Q-Compact maps is Q- compact and an ideal of Q- compact maps is equal to its surjective hull.

3) Given (X,Q_n), assume that each $A_n \in Q_n$ is a vector subspaces of X. Then, a bounded set D of X is Q-compact if and

only if $D \subset T(U_E)$ for a suitable Banach space E and a Q- compact map T on E into X.

4) Let X be a Banach space with approximation scheme Q_n and let D be a bounded subset of X; then

$$\alpha(D;Q) = \lim_{n \to \infty} d_n(D;Q)$$

Theorem (4) defines the ball measure of non-Q-compactness as a limit of generalized n-widths.

We finish by posing the following question: Suppose Orlicz function space L^{Ψ} is given (for definitions see [7]). If L^{Ψ} is considered with the norm $\| \ \|_{\psi}$. It is well known that $(L^{\Psi}, \| \ \|_{\psi})$ is a Banach space [8]. Therefore n-widths $d_n(A)$ of a norm, bounded set A can be defined as usual. On the other hand it is more natural to consider L^{Ψ} with its Orlicz modular ρ where

$$\rho(f) = \int_X \psi(f(x)) \, dx$$

after all $\| f \|_{\psi} = \text{Inf} \{ \lambda > 0 : \rho(f/\lambda) \leq 1 \}$, defined in terms of this modular. Can one define an n-width of a modular bounded set A, say $d_n(A,\rho)$, such that $d_n(A,\rho) = d_n(A)$ and can this $d_n(A,\rho)$ be related with measures of non-compactness?

References

1. Aksoy, A. G. (1991) "Q-compact maps and Q- compact sets," *Math.Japon.* 36 No.1, 1-7.

2. Butzer, P. L., and Scherer, K. (1968) *Approximationsprozesse und Interpolationsmethoden* , Mannheim/ Zurich.

3. Dubinsky, E. (1979) *The Structure of Nuclear Frechet Spaces* Springer-Verlag, Lecture Notes in Math. No.720.

4. Edmunds D. E., and Evans W. D. (1987) *Spectral Theory and Differential Operators* Oxford Science Publications.

5. Hutton, C. V., Morrell, J. S., and Retherford, J. R. (1976) "Diagonal operators, Approximation numbers, and Kolmogorov diameter," *Journal of Approximation Thoery* , 16, 48-80.

6. Konig, H. (1986) *Eigenvalue Distribution of Compact Operators*, Birkhäuser Verlag, OT 16, Basel.

7. Krasnosel'skii, M. A., Rutickii, Ya. B. (1961) *Convex functions Orlicz spaces* , P. Noordhoff Ltd/ Groningen .

8. Musielak, J. (1983) *Orlicz spaces and modular spaces*, Springer-Verlag, Lect. Notes in Math., No: 1034.

9. Pietsch, A. (1980) *Operator Ideals* , North Holland, Amsterdam.

10. ________ (1987) *Eigenvalues and s- numbers* , Cambridge University Press.

11. ________ (1981) " Approximation spaces," *Journal of Approximation Theory*, 32, 115-134.

12. ________ (1972) *Nuclear Locally Convex Spaces*, Springer-Verlag, New York.

13. Pinkus, A. (1985) *N-widths in Approximation theory*, Springer-Verlag,A Series of Modern Surveys in Mathematics.

14. Weyl, H. (1949) "Inequalities between the two kinds eigenvalues of a linear transformation." Proc. Nat. Acad. Sci. USA 35, 408-411.

The Equivalence of the Usual and Quotient Topologies for $C^\infty(E)$ when $E \subset \mathbb{R}^n$ is Whitney p-Regular

L. P. Bos
Department of Mathematics
University of Calgary
Calgary, Alberta
Canada T2N1N4

P. D. Milman
Department of Mathematics
University of Toronto
Toronto, Ontario
Canada M5S1A1

Abstract

We show that when $E \subset \mathbb{R}^n$ is a compact Whitney p-regular domain then the usual and quotient topologies for $C^\infty(E)$ are equivalent with a special estimate of the continuity constants. It follows that in the equivalence of Markov and Sobolev type inequalities given in [2], the quotient norm may be replaced by the usual norm in case E is Whitney p-regular.

§1. Introduction

Suppose that $E \subset \mathbb{R}^n$ is compact and $C^\infty-$determining in the sense that if $f \in C^\infty(\mathbb{R}^n)$ and $f|_E = 0$ then all derivatives of f also restrict to 0 on E. In [2] we show that E admits a Markov type inequality on the derivatives of polynomials iff E also admits a Sobolev type inequality on the intermediate derivatives of C^∞ functions. To make these notions more precise, for $\alpha = (\alpha_1, \cdots, \alpha_n) \in \mathbb{Z}_+^n$, a multiindex, and $x \in \mathbb{R}^n$ let $|\alpha| := \alpha_1 + \cdots + \alpha_n$, $\alpha! := \alpha_1! \cdots \alpha_n!$, $\binom{|\alpha|}{\alpha} := |\alpha|!/\alpha!$, $x^\alpha := x_1^{\alpha_1} \cdots x_n^{\alpha_n}$ and $D^\alpha f(x) = \frac{\partial^{|\alpha|}}{\partial x_1^{\alpha_1} \cdots \partial x_n^{\alpha_n}} f(x)$. Further, let

$$C^\infty(E) := \{f : E \to \mathbb{R} \mid \exists g \in C^\infty(\mathbb{R}^n) \quad and \quad g|_E = f\},$$

$$\|f\|_E := \max_{x \in E} |f(x)|,$$

$$|f|_{E,j} := \sum_{|\alpha| = j} \|D^\alpha f\|_E$$

and

$$\|f\|_{E,m} := \sum_{|\alpha| \le m} \|D^\alpha f\|_E.$$

1.1 Definition. *We say that E admits a <u>global</u> <u>Markov</u> <u>inequality</u> if there are constants $M > 0$ and $r \ge 1$, depending only on E, such that*

$$\|D^\alpha Q\|_E \le (M(deg(Q))^r)^{|\alpha|} \|Q\|_E$$

for all polynomials Q.

S. P. Singh (ed.), Approximation Theory, Spline Functions and Applications, 279–292.

A local version of Markov's inequality, introduced in [2], is the following.

1.2 Definition. *We say that E admits a* <u>*local Markov inequality*</u> *if there are constants $C > 0$ and $\rho \geq 1$, depending only on E, such that*

$$|D^\alpha Q(x_0)| \leq (C\epsilon^{-\rho})^{|\alpha|}\|Q\|_{E\cap B_\epsilon(x_0)}$$

for all $x_0 \in E$, polynomials Q and $0 < \epsilon \leq 1/deg(Q)$. Here $B_\epsilon(x_0)$ is the ball of radius ϵ centred at x_0.

Moreover, suppose that $K \subset \mathbb{R}^n$ is a fixed compact cube that contains E in its interior, then for $f \in C^\infty(E)$, let

$$|||f|||_m := \inf\{\|g\|_{K,m} \mid g \in C^\infty(\mathbb{R}^n) \quad and \quad g|_E = f\};$$

i.e., the norm defining the quotient topology of $C^\infty(E)$.

1.3 Definition. *We say that E admits a* <u>*Sobolev type inequality*</u> *(in the quotient topology) if there are constants $c_1, c_2, c_3 > 0$ and $r \geq 1$, depending only on E and K such that*

$$|f|_{E,j} \leq c(j,m)|||f|||_0^{1-rj/m}|||f|||_m^{rj/m}$$

for all $f \in C^\infty(E)$, $0 \leq j \leq m/r$ and where

$$c(j,m) = c_1(c_2 m)^{c_3 j}.$$

In [2] we have introduced the above version of a Sobolev inequality and have shown that E admits a Markov inequality (1.1) iff it admits a local Markov inequality (1.2) iff it admits a Sobolev inequality (1.3) (actually much more is true but we refer the reader to [2] for the details). It is the purpose of this note to show that if E satisfies a mild regularity condition, (1.4) below, then the quotient topology in (1.3), and indeed in the statement of our main theorem in [2], may be replaced by the usual topology. This is essentially the statement of Theorem 4.2 of [2], given there without proof.

1.4 Definition. *Suppose that p is a positive integer and that $E \subset \mathbb{R}^n$ is compact. We will say that E is Whitney p-regular if there is a positive constant, $C_p \leq 1$, such that for all $x, y \in E$ there is a rectifiable path $\sigma \subset E$, of length $|\sigma|$, connecting x and y with the property that $|x - y| \geq C_p^p|\sigma|^p$.*

Our claim follows easily from the equivalence of norms given below.

1.5 Theorem. *Suppose that E is a Whitney p-regular compact domain and that $K \supset E$ is a fixed cube which contains E in its interior. Then there are constants $c_1, c_2, c_3 > 0$, depending only on E and K such that*

$$|||f|||_m \leq c_1(c_2 m)^{c_3 m}\|f\|_{E,pm}$$

for all $f \in C^\infty(E)$ and $m = 0, 1, 2, \cdots$.

§2. The Equivalence of the Norms

The proof of Theorem 1.5 depends strongly on a reworking of Whitney's extension theorem[4] while keeping more careful notice of the continuity constants that arise. We will follow the presentations of Bierstone[1] and Tougeron[3]. We begin with the definition of a Whitney field. Throughout $E \subset \mathbb{R}^n$ will be a compact domain. Let

$$J^m(E) := \{(F^\alpha)_{|\alpha| \leq m} \mid F^\alpha \in C(E)\}$$

denote the vector space of jets of order m on E. Then for $F \in J^m(E)$,

$$\|F\|_{E,m} := \sup_{\substack{x \in E \\ |\alpha| \leq m}} |F^\alpha(x)|,$$

$$T_a^m F(x) := \sum_{|\alpha| \leq m} \frac{F^\alpha(a)}{\alpha!}(x-a)^\alpha$$

is the Taylor polynomial of F of degree m at $a \in E$ and

$$(R_a^m F)^\alpha(x) := F^\alpha(x) - \sum_{|k| \leq m-|\alpha|} \frac{F^{\alpha+k}(a)}{k!}(x-a)^k$$

is the remainder jet. The set of Whitney fields on E is defined as

$$\mathcal{E}^m(E) := \{F \in J^m(E) \mid (R_x^m F)^k(y) = o(|x-y|^{m-|k|}) \ \forall |k| \leq m\}.$$

$\mathcal{E}^m(E)$ is equipped with the norm

$$\|F\|_m^E := \|F\|_{E,m} + \sup_{\substack{x,y \in E \\ x \neq y \\ |k| \leq m}} \frac{|(R_x^m F)^k(y)|}{|x-y|^{m-|k|}}.$$

2.1 Theorem. *(Whitney extension theorem with constants) Suppose that K is a compact cube that contains E in its interior. There exists a continuous linear mapping*

$$W_m : \mathcal{E}^m(E) \to C^m(K)$$

such that $D^k W_m(F)(x) = F^k(x)$ for $x \in E$, $|k| \leq m$. Moreover, there are constants $c_1, c_2, c_3 > 0$ such that

$$\|W_m(F)\|_{K,m} \leq c_1 (c_2 m)^{c_3 m} \|F\|_m^E.$$

The proof of Theorem 2.1 depends on the construction of a Whitney partition of unity with control of the growth of the derivatives of the partition functions.

2.2 Lemma. *(Whitney partition of unity with constants) There exists a countable family of functions $\Phi_L \in C^\infty(\mathbb{R}^n - E)$, $L \in I$, such that:*
(1) $\{supp(\Phi_L)\}_{L\in I}$ is locally finite: in fact each $x \in \mathbb{R}^n$ belongs to at most 6^n of the $supp(\Phi_L)$.
(2) $\Phi_L \geq 0$ $\forall L \in I$ and $\sum_{L\in I} \Phi_L(x) = 1$ $\forall x \in \mathbb{R}^n - E$.
(3) $2 \cdot dist(supp(\Phi_L), E) \geq diam(supp(\Phi_L))$ $\forall L \in I$.
(4) There are positive constants c_1 and c_2 depending only on n such that

$$|D^\alpha \Phi_L(x)| \leq (c_1|\alpha|)^{c_2|\alpha|}(1 + \frac{1}{dist(x, E)^{|\alpha|}})$$

for all α, $L \in I$ and $x \in \mathbb{R}^n - E$.

Proof: (of lemma) The construction is the standard one; we only establish the estimate on the continuity constant, (4). Again, our presentation is based on those of [1] and [3]. To begin, decompose $\mathbb{R}^n - E$ into a collection of closed cubes as follows. For each non-zero integer p, subdivide $\mathbb{R}^n$ into the closed cubes formed by the hyperplanes $x_i = j/2^p$, $1 \leq i \leq n$, $j \in \mathbb{Z}$. Each such cube will have side length $1/2^p$. Let Σ_p be the set of these cubes. Now, let $S_0 \subset \Sigma_0$ be those cubes, L, for which $dist(L, E) \geq diam(L) = \sqrt{n}$ and inductively let $S_p \subset \Sigma_p$ consist of those cubes, L, not in any cube of $S_0, \cdots, S_{p-1}$ such that $dist(L, E) \geq diam(L) = \sqrt{n}/2^p$. Then $I := \cup_p S_p$ is a subdivision of $\mathbb{R}^n - E$. Further, suppose that $L, M \in I$ are two cubes which touch. Then $L \in S_{p_1}$ and $M \in S_{p_2}$ for some p_1 and p_2 where we may assume that $p_1 \geq p_2$. But then, since the cubes intersect,

$$\begin{aligned} dist(M, E) &\leq dist(L, E) + diam(L) \\ &\leq 2\, diam(L) \\ &= 2\sqrt{n}/2^{p_1} \\ &= \sqrt{n}/2^{p_1 - 1}. \end{aligned}$$

Hence if $p_1 - 1 > p_2$, $dist(M, E) < diam(M)$ contradicting the construction of I. Thus cubes in I from S_p can only touch cubes from S_{p-1}, S_p or S_{p+1}. Continuing the construction, let $\psi \in C^\infty(\mathbb{R}^n)$ be a cutoff function with the properties that:
(a) $0 \leq \psi \leq 1$
(b) $\psi(x) = 1$ on $[-1/2, 1/2]^n$
(c) $\psi(x) = 0$ outside $(-3/4, 3/4)^n$
(d) there are constants $a_1, a_2 > 0$ such that $\forall x \in \mathbb{R}^n$

$$|D^\alpha \psi(x)| \leq (a_1|\alpha|)^{a_2|\alpha|}.$$

The existence of such a ψ is guaranteed by our Lemma 3.15 of [2].

For each $L \in I$ let x_L denote the centre of L and λ_L the length of its sides. We set

$$\psi_L(x) := \psi((x - x_L)/\lambda_L)$$

and

$$\Phi_L(x) := \frac{\psi_L(x)}{\sum_{M \in I} \psi_M(x)} \quad .$$

For each $x \in \mathbb{R}^n - E$ the above sum is actually finite as any $x \in \mathbb{R}^n - E$ is in at most 6^n of the $supp(\psi_L)$. To see this first note that for a fixed cube, L, the union of L together with all cubes which touch L must contain the support of ψ_L since any cube which touches L has diameter at least half of L. Thus if $supp(\psi_L)$ overlaps some cube $M \in I$, M and L must touch. Since any $x \in \mathbb{R}^n - E$ must lie in some cube $M \in I$, $card\{L \in I \mid x \in supp(\psi_L)\}$ is bounded above by the number of cubes in I which touch M. But M may touch only cubes in I with either the same, twice or half the diameter of M. The number of neighbouring cubes of the same size is bounded by 3^n and each of these may be at most split into 2^n cubes of half diameter, giving a maximum of 6^n.

We have thus shown property (1). Property (2) is immediate. To show (3) consider $L \subset I$, then

$$\begin{aligned} dist(supp(\Phi_L), E) &\geq dist(L, E) - (1/4)diam(L) \\ &\geq (3/4)diam(L) \\ &\geq (1/2)diam(supp(\Phi_L)). \end{aligned}$$

To establish (4), we have for $x \in \mathbb{R}^n - E$,

$$\begin{aligned} |D^\beta \psi_L(x)| &= \frac{1}{\lambda_L^{|\beta|}} |D^\beta \psi(\frac{x - x_L}{\lambda_L})| \\ &\leq \frac{1}{\lambda_L^{|\beta|}} (a_1 |\beta|)^{a_2 |\beta|} \end{aligned}$$

by property (d) above. Now if $x \in supp(\psi_L)$, x must lie in L itself or a cube which touches L. Thus if $x \in supp(\psi_M)$ also, M must either touch L or a neighbour of L. Hence $\lambda_M \geq (1/4)\lambda_L$. Then by Lemma 2.7 below, using the fact that $\sum_{M \in I} \psi_M(x) \geq 1$ and that at most 6^n of the $\psi_M(x)$ are non-zero,

$$\max_{x \in supp(\psi_L)} \left| D^\gamma \frac{1}{\sum_{M \in I} \psi_M(x)} \right| \leq \frac{4^{|\gamma|}}{\lambda_L^{|\gamma|}} (b_1 |\gamma|)^{b_2 |\gamma|}.$$

From Leibnitz's rule it then follows that

$$|D^\alpha \Phi_L(x)| \leq b_1 (b_2 |\alpha|)^{b_3 |\alpha|} \frac{1}{\lambda_L^{|\alpha|}}$$

for appropriately adjusted b_1, b_2 and b_3. Now if $L \in S_0$, then $\lambda_L = 1$ so that

$$|D^\alpha \Phi_L(x)| \leq b_1 (b_2 |\alpha|)^{b_3 |\alpha|}.$$

Otherwise, suppose that $L \in S_p$ with $p \geq 1$. Let L' be a cube of Σ_{p-1} which contains L. By construction of the S_p, $dist(L', E) < diam(L') = 2\,diam(L)$, so that for all $y \in L$,

$$\begin{aligned} dist(y, E) &\leq dist(L', E) + diam(L') \\ &\leq 2\,diam(L') \\ &= 4\,diam(L). \end{aligned}$$

Further, for all $x \in supp(\Phi_L)$ and $y \in L$ the point closest to x,

$$\begin{aligned} dist(x, E) &\leq dist(x, y) + dist(y, E) \\ &\leq (1/4)diam(L) + 4\,diam(L) \\ &= (17/4)diam(L) \\ &= (17/4)\sqrt{n}\lambda_L. \end{aligned}$$

Therefore,

$$|D^\alpha \Phi_L(x)| \leq b_1(b_2|\alpha|)^{b_3|\alpha|}\left(1 + \frac{(17\sqrt{n})^{|\alpha|}}{4^{|\alpha|}dist(x, E)^{|\alpha|}}\right)$$

from which the result follows. ■

Proof: (of Theorem 2.1) Let $\{\Phi_L\}$ be the Whitney partition of unity of Lemma 2.2. For each $L \in I$, choose $a_L \in E$ such that

$$dist(supp(\Phi_L), E) = dist(supp(\Phi_L), a_L).$$

For each $F \in \mathcal{E}^m(E)$ define a function $f = W_m(F)$ on $\mathbb{R}^n$ by

$$\begin{aligned} f(x) &= F^0(x) \quad x \in E \\ f(x) &= \sum_{L \in I} \Phi_L(x) T^m_{a_L} F(x) \quad x \notin E. \end{aligned}$$

We will show that the mapping W_m satisfies the required properties. Clearly W_m is linear and $W_m(F) \in C^\infty(\mathbb{R}^n - E)$.

Now, by a modulus of continuity is meant a continuous increasing function $\alpha : [0, \infty) \to [0, \infty)$ such that $\alpha(0) = 0$ and $\alpha(t)$ is concave down. In Tougeron[3] it is shown that there is a modulus of continuity α such that

$$|(R^m_a F)^k(x)| \leq \alpha(|x - a|) \cdot |x - a|^{m-|k|} \tag{2.3}$$

for all $x, a \in E$ and $|k| \leq m$. Moreover,

$$\|F\|^E_m = \|F\|_{E,m} + \alpha(diam(E)).$$

Now write

$$f^k(x) = F^k(x) \quad x \in E$$
$$f^k(x) = D^k f(x) \quad x \notin E.$$

We claim that there exist constants $b_1, b_2, b_3 > 0$ such that

$$|f^k(x) - D^k T_a^m F(x)| \leq b_1 (b_2 m)^{b_3 m} \alpha(|x-a|) \cdot |x-a|^{m-|k|} \tag{2.4}$$

for all $|k| \leq m$, $a \in E$ and $x \in K$. To see this first note that if $x \in E$ then (2.4) holds by (2.3) for any $b_1, b_2, b_3 \geq 1$. Suppose that $x \in K - E$. Then

$$f(x) - T_a^m F(x) = \sum_{L \in I} \Phi_L(x)(T_{a_L}^m F(x) - T_a^m F(x))$$

and so by Leibnitz's formula

$$f^k(x) - D^k T_a^m F(x) = \sum_{\ell \leq k} \binom{k}{\ell} S_\ell(x)$$

where

$$S_\ell(x) := \sum_{L \in I} D^\ell \Phi_L(x) D^{k-\ell}(T_{a_L}^m F(x) - T_a^m F(x)).$$

To bound $S_\ell(x)$ we first bound

$$D^j(T_{a_L}^m F(x) - T_a^m F(x)).$$

Now, expanding in a Taylor series about a_L,

$$\begin{aligned} T_{a_L}^m F(x) - T_a^m F(x) &= \sum_{|k| \leq m} \frac{(x-a_L)^k}{k!} D^k(T_{a_L}^m F(x) - T_a^m F(x))(a_L) \\ &= \sum_{|k| \leq m} \frac{(x-a_L)^k}{k!} (R_a^m F)^k(a_L), \end{aligned}$$

so that

$$D^j(T_{a_L}^m F(x) - T_a^m F(x)) = \sum_{|k| \leq m-|j|} \frac{(x-a_L)^k}{k!} (R_a^m F)^{j+k}(a_L).$$

Hence

$$|D^j(T_{a_L}^m F(x) - T_a^m F(x))| \leq \sum_{|k| \leq m-|j|} \frac{|x-a_L|^{|k|}}{k!} \alpha(|a_L - a|)\, |a_L - a|^{m-|j|-|k|}.$$

Now if $|x - a_L| \leq |x - a|$ then

$$\begin{aligned} |a_L - a|^{m-|j|-|k|} &\leq (|x - a_L| + |x - a|)^{m-|j|-|k|} \\ &\leq 2^{m-|j|-|k|}|x - a|^{m-|j|-|k|} \end{aligned}$$

so that

$$\begin{aligned} |D^j(T^m_{a_L} F(x) - T^m_a F(x))| &\leq \sum_{|k| \leq m-|j|} \frac{|x - a|^{m-|j|}}{k!} 2^{m-|j|-|k|} \alpha(|a_L - a|) \\ &\leq 2^{m-|j|}|x - a|^{m-|j|}\alpha(|a_L - a|) \sum_{|k| \leq m-|j|} \frac{1}{k!} \\ &\leq 2^{m-|j|} e^n |x - a|^{m-|j|}\alpha(|a_L - a|). \end{aligned}$$

On the other hand, if $|x - a| \leq |x - a_L|$ then $|a_L - a| \leq 2\,|x - a_L|$ so that

$$|D^j(T^m_{a_L} F(x) - T^m_a F(x))| \leq 2^{m-|j|} e^n |x - a_L|^{m-|j|}\alpha(|a_L - a|).$$

In general we have

$$|D^j(T^m_{a_L} F(x) - T^m_a F(x))| \leq 2^{m-|j|} e^n \{|x - a_L|^{m-|j|} + |x - a|^{m-|j|}\}\alpha(|a_L - a|). \quad (2.5)$$

We now proceed to bound $S_0(x)$. Note that if $x \in supp(\Phi_L)$, then

$$\begin{aligned} |x - a_L| &\leq diam(supp(\Phi_L)) + dist(supp(\Phi_L), E) \\ &\leq 3\, dist(supp(\Phi_L), E) \end{aligned}$$

by Lemma 2.2, property (3). But $x \in supp(\Phi_L)$ and $a \in E$ so

$$|x - a_L| \leq 3\,|x - a|$$

and hence

$$|a - a_L| \leq |x - a| + |a_L - a| \leq 4\,|x - a|$$

and

$$\alpha(|a_L - a|) \leq 4\,\alpha(|x - a|)$$

since α is a modulus of continuity. Therefore by (2.5)

$$\begin{aligned} |S_0(x)| &\leq \sum_{L \in I} \Phi_L(x)|D^k(T^m_{a_L} F(x) - T^m_a F(x))| \\ &\leq \sum_{L \in I} \Phi_L(x) 2^{m-|k|} e^n \alpha(|a_L - a|)\{|x - a_L|^{m-|k|} + |x - a|^{m-|k|}\} \\ &\leq 6^n 2^{m-|k|} e^n\, 4\,\alpha(|x - a|)\{3^{m-|k|}|x - a|^{m-|k|} + |x - a|^{m-|k|}\} \\ &\leq (6e)^n 2^{m-|k|}\{3^{m-|k|} + 1\}\, 4\,\alpha(|x - a|)|x - a|^{m-|k|}. \end{aligned}$$

For $\ell \neq 0$ note that $\sum_{L\in I} D^\ell \Phi_L(x) = 0$ so that

$$S_\ell(x) = \sum_{L\in I} D^\ell \Phi_L(x) D^{k-\ell}(T^m_{a_L} F(x) - T^m_a F(x))$$
$$= \sum_{L\in I} D^\ell \Phi_L(x) D^{k-\ell}(T^m_{a_L} F(x) - T^m_b F(x))$$

for any $b \in E$. Choose $b \in E$ so that $|x-b| = dist(x,E)$. Then, as before, $|x-a_L| \leq 3|x-b| = 3\,dist(x,E)$ and hence

$$|b-a_L| \leq |x-b| + |x-a_L| \leq 4\,dist(x,E)$$

and

$$\alpha(|b-a_L|) \leq 4\,\alpha(dist(x,E)).$$

Thus by (2.5) and Lemma 2.2

$$|S_\ell(x)| \leq (c_1|\ell|)^{c_2|\ell|}\left(1 + \frac{1}{dist(x,E)^{|\ell|}}\right) 6^n 2^{m-|k|+|\ell|} e^n$$
$$\times (3^{m-|k|+|\ell|}+1)\,4\,\alpha(dist(x,E))(dist(x,E))^{m-|k|+|\ell|}$$
$$\leq (c_1|\ell|)^{c_2|\ell|}((dist(x,E))^{|\ell|}+1)(6e)^n 2^{m-|k|+|\ell|}$$
$$\times 4\,(3^{m-|k|+|\ell|}+1)\alpha(dist(x,E))dist(x,E)^{m-|k|}.$$

Our claim (2.4) now follows as for $a \in E$, $|x-a| \geq dist(x,E)$. To complete the proof of the theorem, let (j) denote the multiindex whose jth component is 1 and the others 0. For $a \in E$ and $|k| < m$,

$$|f^k(x) - f^k(a) - \sum_{j=1}^n (x_j - a_j) f^{k+(j)}(a)| \leq |f^k(x) - D^k T^m_a F(x)|$$
$$+ |D^k T^m_a F(x) - D^k T^m_a F(a) - \sum_{j=1}^n (x_j - a_j) D^{k+(j)} T^m_a F(a)|.$$

The first term on the right hand side is $o(|x-a|)$ by (2.4) and the second is $o(|x-a|)$ since $T^m_a F(x)$ is a polynomial. Hence f^k is continuously differentiable on K and in fact, $D^{(j)} f^k = f^{k+(j)}$. Consequently $W_m(F) = f \in C^m(K)$ and $D^k f|_E = F^k$ for all $|k| \leq m$. It remains to show the bound on $\|W_m(F)\|_{K,m}$. To that end, suppose that $x \in K$ and that $a \in E$ is such that $dist(x,E) = |x-a|$. By (2.4)

$$|D^k f(x)| \leq |D^k T^m_a F(x)| + b_1(b_2 m)^{b_3 m} \alpha(|x-a|)|x-a|^{m-|k|}$$
$$\leq |D^k T^m_a F(x)| + b_1(b_2 m)^{b_3 m} \alpha(\lambda)\lambda^{m-|k|}$$

where we have set $\lambda := \sup_{x\in K} dist(x,E)$. Hence,

$$\begin{aligned}
&|D^k f(x)| \\
&\leq |\sum_{|\ell|\leq m-|k|} F^{k+\ell}(a)(x-a)^\ell/\ell!| + b_1(b_2 m)^{b_3 m}\lambda^{m-|k|}\alpha(\lambda) \\
&\leq \left(\sum_{|\ell|\leq m-|k|} \frac{\lambda^{|\ell|}}{\ell!}\right) \|F\|_{E,m} + b_1(b_2 m)^{b_3 m}\lambda^{m-|k|}\frac{\lambda}{diam(E)}\alpha(diam(E)) \\
&\leq e^{n\lambda}\|F\|_{E,m} + b_1(b_2 m)^{b_3 m}\lambda^{m-|k|}\frac{\lambda}{diam(E)}(\|F\|^E_m - \|F\|_{E,m}).
\end{aligned}$$

The result now follows from the fact that $\|F\|_{E,m} \leq \|F\|^E_m$. ■

2.6 Lemma. *Suppose that* $g, h \in C^\infty(\mathbb{R}^n)$ *are such that* $h = \log(g)$ *and* $g = \exp(h)$. *Let*

$$a_\beta := \frac{\frac{1}{\beta!}D^\beta g(x)}{g(x)},$$

$$A_m := \max_{1\leq|\beta|\leq m} |a_\beta|^{1/|\beta|},$$

$$b_\alpha := \frac{1}{\alpha!}D^\alpha h(x)$$

and

$$B_m := \max_{1\leq|\alpha|\leq m} |b_\alpha|^{1/|\alpha|}.$$

Then there are constants $a_1, a_2 > 0$, *independent of* g *and* h, *such that*

$$A_m \leq (a_1 m)^{a_2} B_m$$

and

$$B_m \leq (a_1 m)^{a_2} A_m.$$

Proof: Expanding in formal power series we have

$$\begin{aligned}
w := \sum_{|\alpha|\geq 1} b_\alpha y^\alpha &= \log(g(x+y)) - \log(g(x)) \\
&= \log(g(x+y)/g(x)) \\
&= \log(1 + \sum_{|\beta|\geq 1} a_\beta y^\beta) \\
&= z - \frac{z^2}{2} + \frac{z^3}{3} + \cdots
\end{aligned}$$

where $z := \sum_{|\beta|\geq 1} a_\beta y^\beta$. Comparing coefficients we see that

$$|b_\alpha| \leq |a_\alpha| + \frac{1}{2} \sum_{\gamma^{(1)}+\gamma^{(2)}=\alpha} |a_{\gamma^{(1)}}|\,|a_{\gamma^{(2)}}| + \frac{1}{3} \sum_{\gamma^{(1)}+\gamma^{(2)}+\gamma^{(3)}=\alpha} |a_{\gamma^{(1)}}|\,|a_{\gamma^{(2)}}|\,|a_{\gamma^{(3)}}|$$
$$+ \cdots + \frac{1}{|\alpha|} \sum_{\gamma^{(1)}+\cdots+\gamma^{|\alpha|}=\alpha} |a_{\gamma^{(1)}}| \cdots |a_{\gamma^{(|\alpha|)}}|$$

with all $|\gamma^{(j)}| \geq 1$. Hence,

$$\begin{aligned}
|b_\alpha| &\leq \sum_{j=1}^{|\alpha|} \frac{1}{j} \sum_{\gamma^{(1)}+\cdots+\gamma^{(j)}=\alpha} A_m^{|\gamma^{(1)}|+\cdots+|\gamma^{(j)}|} \\
&= A_m^{|\alpha|} \sum_{j=1}^{|\alpha|} \frac{1}{j} \sum_{\gamma^{(1)}+\cdots+\gamma^{(j)}=\alpha} 1 \\
&= A_m^{|\alpha|} \sum_{j=1}^{|\alpha|} \frac{1}{j} \prod_{k=1}^{n} \binom{\alpha_k + j - 1}{j-1}.
\end{aligned}$$

This last equality follows from the fact that $\gamma^{(1)} + \cdots + \gamma^{(j)} = \alpha$ iff $\gamma_k^{(1)} + \cdots + \gamma_k^{(j)} = \alpha_k$ for all components $1 \leq k \leq n$. Therefore, for $|\alpha| \leq m$,

$$\begin{aligned}
|b_\alpha| &\leq A_m^{|\alpha|} |\alpha| \binom{2|\alpha| - 1}{|\alpha| - 1}^n \\
&\leq A_m^{|\alpha|} |\alpha| (|\alpha| + 1)^{n(|\alpha|-1)}
\end{aligned}$$

and one direction follows.

Conversely, since

$$1 + \sum_{|\beta|\geq 1} a_\beta y^\beta = e^w = 1 + w + w^2/2! + \cdots,$$

$$|a_\beta| \leq \sum_{j=1}^{|\beta|} \frac{1}{j!} \sum_{\gamma^{(1)}+\cdots+\gamma^{(j)}=\beta} |b_{\gamma^{(1)}}| \cdots |b_{\gamma^{(j)}}|$$

and the result follows by a similar estimate to the one above. ∎

2.7 Lemma. *Suppose that $g \in C^\infty(\mathbb{R}^n)$ is such that $g(x) \geq 1$ and $|D^\beta g(x)| \leq a_1(a_2|\beta|)^{a_3|\beta|}$ for some constants $a_1, a_2, a_3 > 0$. Then there are other constants $b_1, b_2 > 0$ such that*

$$\left| D^\alpha \frac{1}{g(x)} \right| \leq (b_1|\alpha|)^{b_2|\alpha|}.$$

Proof: Writing $1/g(x) = \exp(-\log(g(x))$ the estimate follows immediately from Lemma 2.6. ∎

2.8 Theorem. *Suppose that $E \subset \mathbb{R}^n$ is a compact p-regular domain. Then there are constants $a_1, a_2 > 0$ such that for all $F \in \mathcal{E}^{mp}(E)$,*

$$\|F\|_m^E \leq a_1 a_2^m \|F\|_{E,mp}.$$

Proof: Again we follow Bierstone[2]. Suppose that $g \in C^q(\mathbb{R}^n)$ for some $q \geq 1$. If $x, y \in \mathbb{R}^n$ then by the Mean Value Theorem,

$$|g(y) - g(x)| \leq \sqrt{n}|x - y| \sup_{\substack{\eta \in [x,y] \\ |\ell|=1}} |D^\ell g(\eta)|.$$

Hence if σ is a piecewise linear path joining x and y of length $|\sigma|$,

$$|g(y) - g(x)| \leq \sqrt{n}|\sigma| \sup_{\substack{\eta \in \sigma \\ |\ell|=1}} |D^\ell g(\eta)|.$$

By passing to the limit, we see that this holds for any rectifiable path σ connecting x and y. Now suppose that g is $q-1$ flat at x. Then by iterating the above inequality we obtain

$$|g(y)| \leq n^{q/2}|\sigma|^q \sup_{\substack{\eta \in \sigma \\ |\ell|=q}} |D^\ell g(\eta)|. \tag{2.9}$$

Now for $F \in \mathcal{E}^m(E)$ and $|k| \leq m$ take $g = D^k(W_m(F) - T_x^m F)$ (where W_m is the extension map of Theorem 2.1) and apply (2.9) to g with $x, y \in E$ and $q = m - |k|$. It follows that

$$\begin{aligned}
|(R_x^m F)^k(y)| &\leq n^{(m-|k|)/2}\delta(x,y)^{m-|k|} \sup_{\substack{\eta \in E \\ |\ell|=m-|k|}} |D^{k+\ell}(W_m(F) - T_x^m F)(\eta)| \\
&= n^{(m-|k|)/2}\delta(x,y)^{m-|k|} \sup_{\substack{\eta \in E \\ |\ell|=m}} |F^\ell(\eta) - F^\ell(x)| \qquad (2.10) \\
&\leq 2n^{(m-|k|)/2}\delta(x,y)^{m-|k|}\|F\|_{E,m}
\end{aligned}$$

where $\delta(x,y)$ denotes the least distance from x to y through E.

Now suppose that $F \in \mathcal{E}^{mp}(E)$, then

$$\begin{aligned}
|(R_x^m F)^k(y)| &= |F^k(y) - (T_x^m F)^k(y)| \\
&\leq |F^k(y) - (T_x^{mp} F)^k(y)| + |(T_x^{mp} F)^k(y) - (T_x^m F)^k(y)| \\
&= |(R_x^{mp} F)^k(y)| + |(T_x^{mp} F)^k(y) - (T_x^m F)^k(y)|.
\end{aligned}$$

But

$$
\begin{aligned}
&|(T_x^{mp}F)^k(y) - (T_x^m F)^k(y)| \\
&= \left| \sum_{|\ell|\le mp-|k|} F^{k+\ell}(x)\frac{(y-x)^\ell}{\ell!} - \sum_{|\ell|\le m-|k|} F^{k+\ell}(x)\frac{(y-x)^\ell}{\ell!} \right| \\
&= \left| \sum_{m-|k|<|\ell|\le mp-|k|} F^{k+\ell}(x)\frac{(y-x)^\ell}{\ell!} \right| \\
&\le \|F\|_{E,mp} \sum_{m-|k|<|\ell|\le mp-|k|} \frac{|y-x|^{|\ell|}}{\ell!} \\
&\le \|F\|_{E,mp}|y-x|^{m-|k|+1} \sum_{|\ell|\le mp-m-1} \frac{|y-x|^{|\ell|}}{\ell!} \\
&\le \|F\|_{E,mp}|y-x|^{m-|k|+1} \sum_{|\ell|\le mp-m-1} \frac{(diam(E))^{|\ell|}}{\ell!} \\
&\le e^{n\, diam(E)}\|F\|_{E,mp}|y-x|^{m-|k|+1}.
\end{aligned}
$$

Therefore,

$$
|(R_x^m F)^k(y)| \le |(R_x^{mp}F)^k(y)| + e^{n\, diam(E)}\|F\|_{E,mp}|y-x|^{m-|k|+1|}. \tag{2.11}
$$

But by (2.10) (with m replaced by mp)

$$
\begin{aligned}
|(R_x^{mp}F)^k(y)| &\le 2\,n^{(mp-|k|)/2}\delta(x,y)^{mp-|k|}\|F\|_{E,mp} \\
&\le 2\,n^{(mp-|k|)/2}(\frac{1}{C_p}|x-y|^{1/p})^{mp-|k|}\|F\|_{E,mp} \\
&= 2\,n^{(mp-|k|)/2}(1/C_p)^{mp-|k|}|x-y|^{m-|k|/p}\|F\|_{E,mp} \\
&\le 2\,n^{mp/2}(1/C_p)^{mp}(diam(E))^{|k|(1-1/p)}|x-y|^{m-|k|}\|F\|_{E,mp}
\end{aligned}
$$

where we have used the fact that $C_p \le 1$. Combining this with (2.11) we see that

$$
|(R_x^m F)^k(y)| \le b_1\, b_2^m \|F\|_{E,mp}|y-x|^{m-|k|}
$$

for some constants $b_1, b_2 > 0$ and the result follows. ■

We are now ready to prove our main Theorem 1.5.

Proof: (of Theorem 1.5) Since $f \in C^\infty(E)$, $f \in \mathcal{E}^m(E)$ for all $m \ge 0$. Let $g = W_m(f)$, the Whitney extension of f to K. We may approximate g arbitrarily closely on K by C^∞ functions. Hence

$$
\begin{aligned}
|||f|||_m &\le \|W_m(f)\|_{K,m} \\
&\le c_1(c_2 m)^{c_3 m}\|f\|_m^E \\
&\le c_1(c_2 m)^{c_3 m} a_1\, a_2^m \|f\|_{E,mp}
\end{aligned}
$$

and the result follows for suitably adjusted c_1, c_2 and c_3. ■

§References

1. Bierstone, E. (1980) 'Differentiable Functions', Bol. Soc. Bras. Mat., vol. 11, no. 2, 139-190.
2. Bos, L. and Milman P. (1991) 'The equivalence of Markov and Sobolev type inequalities on compact subsets in $\mathbb{R}^n$', preprint.
3. Tougeron, J. (1972) Idéaux de fonctions différentiables, Springer, Berlin.
4. Whitney, H. (1934) 'Analytic extensions of differentiable functions defined in closed sets', Trans. Amer. Math. Soc. 36, 369-387.

KOROVKIN THEOREMS FOR VECTOR-VALUED CONTINUOUS FUNCTIONS

MICHELE CAMPITI
Department of Mathematics
University of Bari
Traversa 200 Via Re David, 4
70125 Bari
Italy

ABSTRACT. In this paper we consider some Korovkin type results in the space of continuous functions with values in a fixed locally convex space; we give some conditions which generalize in a natural way those well-known for continuous real-valued functions.

1. Preliminaries and notation

Let X be a compact Hausdorff topological space and E a locally convex Hausdorff space. We denote by $\mathscr{C}(X,E)$ the space of all continuous functions from X in E and we consider on $\mathscr{C}(X,E)$ the topology of the uniform convergence. Our interest will be devoted to the convergence of nets of suitable continuous linear operators from $\mathscr{C}(X,E)$ in itself to a continuous linear operator.

More precisely, given a class $\mathscr{M}$ of continuous linear operators from $\mathscr{C}(X,E)$ in itself and an operator L in $\mathscr{M}$, we try to find conditions on a subset H of $\mathscr{C}(X,E)$ which ensure the validity of the following property

if $(L_\iota)_{\iota\in I}$ *is an equicontinuous net of linear operators in* $\mathscr{M}$ *such that* $(L_\iota(h))_{\iota\in I}$ *converges to* $L(h)$ *for each* $h\in H$, *then* $(L_\iota(f))_{\iota\in I}$ *converges to* $L(f)$ *for each* $f\in \mathscr{C}(X,E)$. (1.1)

A subset H of $\mathscr{C}(X,E)$ satisfying condition (1.1) will be called an **L-Korovkin set with respect to the class $\mathscr{M}$ of linear continuous operators.**

Obviously, we can assume that H is a subspace of $\mathscr{C}(X,E)$ without loss of generality and in this case we shall speak of L-Korovkin subspaces rather than L-Korovkin subsets of $\mathscr{C}(X,E)$.

Moreover, if L is the identity operator we shall use the expression Korovkin set (or subspace) instead of L-Korovkin set (or subspace).

In the case where $E=\mathbb{R}$, $\mathscr{M}$ is the class of all positive linear operators and L is the identity operator, the subspaces H of $\mathscr{C}(X,\mathbb{R})$

S. P. Singh (ed.), Approximation Theory, Spline Functions and Applications, 293–302.

satisfying (1.1) have been studied by many authors and have completely characterized in different ways. An interesting characterization for our purposes is that of Berens and Lorentz [1] which states that a subspace H of $\mathscr{C}(X,\mathbb{R})$ is a Korovkin subspace with respect to positive operators if and only if it satisfies the following condition

$$\text{for each } f \in \mathscr{C}(X,\mathbb{R}),\ x_0 \in X \text{ and } \varepsilon > 0, \text{ there exist}$$
$$h,k \in H \text{ such that } h \leq f \leq k \text{ and } k(x_0) - h(x_0) < \varepsilon. \tag{1.2}$$

In the same contest, if L is an arbitrary positive linear operator, Ferguson and Rusk [6] have given the following similar characterization of L-Korovkin subspaces of $\mathscr{C}(X,\mathbb{R})$ with respect to positive operators

$$\text{for each } f \in \mathscr{C}(X,\mathbb{R}),\ x_0 \in X \text{ and } \varepsilon > 0, \text{ there exist}$$
$$h,k \in H \text{ such that } h \leq f \leq k \text{ and } L(k)(x_0) - L(h)(x_0) < \varepsilon. \tag{1.3}$$

The preceding conditions will be at the center of our investigation; in the subsequent Sections we shall be interested in extending them to larger context.

If E is a Banach lattice, this problem was studied in the same papers of Berens and Lorentz [1] and Ferguson and Rusk [6], while in the case where E has not an order structure, there has been not a similar development. In fact, the first problem we meet with is what class of linear operators is the natural substitute of the class of positive operators in the real case.

Some authors have considered this problem and in the following Sections we give some details of their arguments.

We conclude this Section with some notation which will be useful in the sequel. Let E be a locally convex Hausdorff space and let $y \in E$ and $\varphi: X \to \mathbb{R}$ a continuous real function; we shall denote by $\varphi \cdot y$ the continuous function from X in E defined by setting $(\varphi \cdot y)(x) = \varphi(x) \cdot y$ for each $x \in X$.

Moreover, if $\varphi_1,\ldots,\varphi_n \in \mathscr{C}(X,E)$ and $\varphi \in \mathscr{C}(X,E)$, we shall write $\varphi \in co(\varphi_1,\ldots,\varphi_n)$ to indicate $\varphi(x) \in co(\varphi_1(x),\ldots,\varphi_n(x))$ for each $x \in X$, where $co(\varphi_1(x),\ldots,\varphi_n(x))$ denotes the convex hull of the set $\{\varphi_1(x),\ldots,\varphi_n(x)\}$.

2. Quasi-positive operators and convexity-monotone operators

In this Section we consider two classes of linear operators which both extend monotone operators in the real case. The first one is the class of quasi-positive operators in the setting of Banach spaces [7], and the second is that of convexity-monotone operators in the setting of locally convex spaces [5]; we shall see that if E is a Banach space, convexity-monotone operators are quasi-positive and if $E=\mathbb{R}$ the two classes coincide and are strictly related to monotone linear operators.

First, we consider the case where E is a Banach space. In this setting, Nishishiraho [7] has introduced the following Definition.

Definition 2.1. *Let* E *be a Banach space and* $L:\mathscr{C}(X,E)\to\mathscr{C}(X,E)$ *be a linear continuous operator. We shall say that* L *is quasi-positive if, for each* $\varphi,\psi \in \mathscr{C}(X,\mathbb{R})$ *such that* $|\varphi(x)| \le \psi(x)$ *for each* $x \in X$, *we have*

$$\|L(\varphi\cdot y)(x)\| \le \|L(\psi\cdot y)(x)\| \text{ for each } x \in X \text{ and } y \in E. \tag{2.1}$$

Let $\mathscr{P}$ be the class of all quasi-positive linear operators from $\mathscr{C}(X,E)$ in itself. In [7], Nishishiraho has established some results concerning with the convergence of nets of quasi-positive operators satisfying suitable conditions. The main theorem is the following [7, Theorem 1].

Theorem 2.2. *Let* E *be a Banach space,* $(L_\alpha)_{\alpha\in I}$ *a net of quasi-positive linear operators from* $\mathscr{C}(X,E)$ *in itself and* $L:\mathscr{C}(X,E)\to\mathscr{C}(X,E)$ *a quasi-positive linear operator.*

Moreover, let $\varphi:X\times X\to\mathbb{R}$ *be a non-negative continuous function which is strictly positive in the complement of the diagonal* $\Delta = \{(x,x) \mid x\in X\}$ *of* X.

If

$$\lim_{\alpha\in I}\ \sup_{x\in X} \|L_\alpha(\varphi(\cdot,x)\cdot y)(x)\| = 0$$

for each $y \in E$ *and if there exists a strictly positive continuous function* $u \in \mathscr{C}(X,\mathbb{R})$ *such that*

$$\lim_{\alpha\in I} L_\alpha(u\cdot y) = L(u\cdot y)$$

for each $y \in E$, *then*

$$\lim_{\alpha\in I} L_\alpha(f) = L(f)$$

for each $f \in \mathscr{C}(X,E)$.

The preceding result has not a strict connection with conditions (1.2) and (1.3), but it can be applied in various circumstances.

For example, in [7, Corollaries 3 and 4] it is shown that if $L:\mathscr{C}(X,E)\to\mathscr{C}(X,E)$ is a quasi-positive linear operator, the following subsets of $\mathscr{C}(X,E)$ are L-Korovkin sets with respect to the class $\mathscr{P}$.

1. Let $m\ge 1$ and $v_1,\dots,v_m,w_1,\dots,w_m:X\to\mathbb{R}$ be continuous functions such that the map $\varphi:X\times X\to\mathbb{R}$ defined by setting

$$\varphi(s,t) = \sum_{i=1}^{m} v_i(s)w_i(t)$$

for each $(s,t) \in X \times X$ is non-negative and $\varphi = 0$ exactly on $\Delta = \{(t,t) \mid t \in X\}$.

Then, the set

$$H = \{v_i \cdot y \mid i=1,\dots,m,\ y \in E\}$$

is an L-Korovkin sets with respect to the class $\mathcal{P}$.

2. If S is a finite subset of $\mathcal{C}(X,\mathbb{R})$ which separates the points of X (i.e. for each $s,t \in X$, there exists $h \in S$ such that $h(s) \neq h(t)$), then the set H consisting of all the functions

$$1 \cdot y \quad (y \in E), \qquad h \cdot y \quad (y \in E,\ h \in S), \qquad (\sum_{h \in S} h^2) \cdot y \quad (y \in E)$$

is an L-Korovkin sets with respect to the class $\mathcal{P}$.

Other applications of Theorem 2.2 can be given concerning particular examples of approximating operators, but we shall return later on these questions in a more detailed form.

At this point, we consider another class of linear continuous operators which will be compared with that of quasi-positive operators already introduced; this class has been implicitly introduced in [3] and successively studied in more details in [5]; we refer to these papers for the proofs of the related results.

We return to the general context where E is a locally convex Hausdorff space and we fix a base $\mathfrak{B}$ of convex open neighborhoods at the origin of E.

For simplicity, we restrict our arguments to the case where the compact Hausdorff topological space X is also connected.

Definition 2.3. *Let* X *be a connected compact Hausdorff topological space,* E *be a locally convex Hausdorff space and* $L: \mathcal{C}(X,E) \to \mathcal{C}(X,E)$ *be a continuous linear operator from* $\mathcal{C}(X,E)$ *in itself.*

We say that L *is convexity-monotone if it satisfies the following condition*

$$\text{if } \varphi_1, \varphi_2 \in \mathcal{C}(X,E) \text{ have disjoint graphs and if } \varphi \in co(\varphi_1,\varphi_2), \text{ then } L(\varphi) \in co(L(\varphi_1), L(\varphi_2)). \tag{2.2}$$

Moreover, we shall denote by $\mathcal{M}$ *the class of all convexity-monotone operators from* $\mathcal{C}(X,E)$ *in itself.*

The requirement on $\varphi_1, \varphi_2 \in \mathcal{C}(X,E)$ to have disjoint graphs is necessary to make the class $\mathcal{M}$ sufficiently large; otherwise, we observe that non zero monotone operators in the case $E = \mathbb{R}$ may not satisfy (2.2) for each $\varphi_1, \varphi_2 \in \mathcal{C}(X,E)$ (for example, consider $X = [-1,1]$ and the functions $\varphi_1(x) = -x$, $\varphi_2(x) = x$ and $\varphi(x) = x^2$ for each $x \in X$).

In the next Proposition we give some characterizations of convexity-monotone operators. The proof is based on straightforward arguments (cf. [5, Proposition 2.4]).

Proposition 2.4. *Let* X *be a connected compact Hausdorff topological space,* E *be a locally convex Hausdorff space and* $L:\mathscr{C}(X,E)\to\mathscr{C}(X,E)$ *be a continuous linear operator from* $\mathscr{C}(X,E)$ *in itself.*
Then, the following statements are equivalent

a) L *is convexity-monotone;*

b) *if* $\varphi_1,\ldots,\varphi_n \in \mathscr{C}(X,E)$ *are affinely independent at each point of* X [1] *and if* $\varphi \in co(\varphi_1,\ldots,\varphi_n)$, *then* $L(\varphi) \in co(L(\varphi_1),\ldots,L(\varphi_n))$;

c) *if* $\psi \in \mathscr{C}(X,E)$ *satisfies* $\psi(x) \neq 0$ *for each* $x \in X$ *and if* $\varphi \in co(0,\psi)$, *then* $L(\varphi) \in co(0,L(\psi))$.

In the case where E is a Banach space, it is shown in [5, Proposition 2.6] that the class $\mathscr{M}$ is contained in $\mathscr{P}$. Moreover, if we consider the special case $E = \mathbb{R}$, we have the following interesting relation of both quasi-positive and convexity-monotone operators with monotone operators (cf. [3, Proposition 1.3] and [5, Proposition 2.7]).

Proposition 2.5. *Let* X *be a connected compact Hausdorff topological space and* $L:\mathscr{C}(X,\mathbb{R})\to\mathscr{C}(X,\mathbb{R})$ *be a continuous linear operator from* $\mathscr{C}(X,\mathbb{R})$ *in itself.*
Then, the following statements are equivalent

a) L *is convexity-monotone;*

b) L *is quasi-positive;*

c) *there exist two closed subsets* X^+ *and* X^- *of* X *such that* $X = X^+ \cup X^-$ *and*

$$\varphi \in \mathscr{C}(X,\mathbb{R}),\ \varphi \geq 0 \Rightarrow L(\varphi) \geq 0 \text{ on } X^+ \text{ and } L(\varphi) \leq 0 \text{ on } X^-. \quad (2.3)$$

Moreover, if L *satisfies* a) *or equivalently* b) *or* c), *we can take*

$$X^+ = \{x \in X \mid L(1)(x) \geq 0\}, \qquad X^- = \{x \in X \mid L(1)(x) \leq 0\}.$$

Observe that condition (2.3) is equivalent to the following

$$\varphi,\psi \in \mathscr{C}(X,\mathbb{R}),\ \varphi \leq \psi \Rightarrow L(\varphi) \leq L(\psi) \text{ on } X^+, \quad L(\varphi) \leq L(\psi) \text{ on } X^-.$$

Moreover, the implication c) ⇒ b) in Proposition 2.5 also holds if X is not connected, but the implication b) ⇒ c) does not hold in general (cf. [3], Remark 1.4).

By the preceding Proposition, we have that each monotone linear

---

[1] i.e., for each $x \in X$, the family $\{\varphi_1(x),\ldots,\varphi_n(x)\}$ is affinely independent.

operator $L: \mathscr{C}(X,\mathbb{R}) \to \mathscr{C}(X,\mathbb{R})$ is convexity-monotone. However, in general there exist continuous linear operators $L: \mathscr{C}(X,\mathbb{R}) \to \mathscr{C}(X,\mathbb{R})$ which are convexity-monotone but not monotone; an example is furnished by the operator $L: \mathscr{C}([a,b],\mathbb{R}) \to \mathscr{C}([a,b],\mathbb{R})$ $(a,b \in \mathbb{R},\ a<b)$ defined by putting, for each $\varphi \in \mathscr{C}([a,b],\mathbb{R})$ and $x \in [a,b]$,

$$L(\varphi)(x) = \int_{x_0}^{x} \varphi(t)\, dt,$$

where x_0 is a fixed element in the open interval (a,b).

At this point, we shall consider some Korovkin-type results with respect to the class $\mathscr{M}$ of convexity-monotone operators; we shall see that the restriction from quasi-positive to convexity-monotone operators is useful in order to formulate conditions similar to the initial ones (1.2) and (1.3) in the real case. The following main Theorem has been proved in [3] and [5, Theorem 3.2]; in [3], the proof involves another important property of convexity-monotone operators, which is the possibility of extending them to linear monotone operators between cones of set-valued continuous functions (for more details, see [3, Proposition 1.5] and [5]).

Theorem 2.6. *Let* X *be a connected compact Hausdorff topological space,* E *a locally convex Hausdorff space and* $\mathscr{V}$ *a base of open convex neighborhoods at* 0 *in* E.

If $L: \mathscr{C}(X,E) \to \mathscr{C}(X,E)$ *is a convexity-monotone linear operator from* $\mathscr{C}(X,E)$ *in itself, and if* H *is a subset of* $\mathscr{C}(X,E)$ *satisfying the following condition*

for each $\varphi \in \mathscr{C}(X,E)$, $x_0 \in X$ *and* $V \in \mathscr{V}$, *there exist* $\varphi_1,\dots,\varphi_n \in H$ *which are affinely independent at each point of* X *and satisfy* (2.4)

$$\varphi \in co(\varphi_1,\dots,\varphi_n),$$

$$co(L(\varphi_1)(x_0),\dots,L(\varphi_n)(x_0)) \subset L(\varphi)(x_0) + V,$$

then H *is an* L*-Korovkin set with respect to the class* $\mathscr{M}$.

In particular, if H *is a subset of* $\mathscr{C}(X,E)$ *satisfying the following condition*

for each $\varphi \in \mathscr{C}(X,E)$, $x_0 \in X$ *and* $V \in \mathscr{V}$, *there exist* $\varphi_1,\dots,\varphi_n \in H$ *which are affinely independent at each point of* X *and satisfy* (2.5)

$$\varphi \in co(\varphi_1,\dots,\varphi_n),$$

$$co(\varphi_1(x_0),\dots,\varphi_n(x_0)) \subset \varphi(x_0) + V,$$

then H *is a Korovkin set with respect to the class* $\mathscr{M}$.

In the case where L is the identity operator, we can give the following Corollary (cf. [5, Corollary 3.3]).

Corollary 2.7. *Let* X *be a connected compact Hausdorff topological space and* E *a locally convex Hausdorff space.*

If Γ is a Korovkin set in $\mathscr{C}(X,\mathbb{R})$, *then the set*

$$H = \{\varphi \cdot y \mid \varphi \in \Gamma, y \in E\}$$

is a Korovkin set in $\mathscr{C}(X,E)$.

The preceding Corollary furnishes many examples of Korovkin sets in $\mathscr{C}(X,E)$ by considering those well-known in the real case; the preceding Example 2. is also a consequence of Corollary 2.7.

We point out the analogy of (2.4) with (1.3) and (2.5) with (1.2). We have to observe that the classical definition of L-Korovkin set involves equicontinuous nets of monotone linear operators rather than convexity-monotone linear operators; it follows that an L-Korovkin set with respect to the class $\mathscr{M}$ (cf. (1.1)) is always an L-Korovkin in the classical sense. Therefore, if we state Theorem 2.6 in the case $E = \mathbb{R}$, we obtain a more general version of the results of Berens-Lorentz and Ferguson-Rusk, since the class $\mathscr{M}$ of convexity-monotone linear operators contains that of positive operators (cf. Proposition 2.5, c)). As in [3, Corollary 1.12], we can state condition (2.4) and (2.5) only for $n=2$ and obtain the following corollary.

Corollary 2.8. *Let* X *be a connected compact Hausdorff topological space and* $L:\mathscr{C}(X,\mathbb{R})\to\mathscr{C}(X,\mathbb{R})$ *be a convexity-monotone linear operator from* $\mathscr{C}(X,\mathbb{R})$ *in itself.*

If a subset H *of* $\mathscr{C}(X,\mathbb{R})$ *satisfies the following condition*

$$\text{for each } f\in\mathscr{C}(X,\mathbb{R}),\ x_0\in X \text{ and } \varepsilon>0, \text{ there exist } h,k \in H \text{ such that } h \le f \le k \text{ and } |L(h)(x_0) - L(k)(x_0)| \le \varepsilon, \tag{2.6}$$

then H *is an* L*-Korovkin set in* $\mathscr{C}(X,\mathbb{R})$ *with respect to the class* $\mathscr{M}$ *of convexity-monotone operators.*

Moreover, if a subset H *of* $\mathscr{C}(X,\mathbb{R})$ *satisfies the following condition*

$$\text{for each } f\in\mathscr{C}(X,\mathbb{R}),\ x_0\in X \text{ and } \varepsilon>0, \text{ there exist } h,k \in H \text{ such that } h \le f \le k \text{ and } k(x_0) - h(x_0) \le \varepsilon, \tag{2.7}$$

then H *is a Korovkin set in* $\mathscr{C}(X,\mathbb{R})$ *with respect to the class* $\mathscr{M}$ *of convexity-monotone operators.*

Another interesting case is when $E = \mathbb{R}^p$ $(p\ge 1)$; in this case, conditions (2.4) and (2.5) can be stated for $n = p+1$ (cf. [3, Corollary 1.12] and [5]).

In the particular case where L is the identity operator from $\mathscr{C}(X,\mathbb{R}^p)$ in itself, some examples of Korovkin sets with respect to the class $\mathscr{M}$

can be derived by the well-known examples of Korovkin sets in $\mathscr{C}(X,\mathbb{R})$. Indeed, we state the following Proposition [3, Corollary 1.13] which indicates a possible way to realize this.

Corollary 2.9. *Let* X *be a connected compact Hausdorff topological space and* H *be a subset of* $\mathscr{C}(X,\mathbb{R})$ *satisfying condition* (1.2).

Then, the set

$$H_p = \{h \in \mathscr{C}(X,\mathbb{R}^p) \mid \textit{there exists } j=1,\ldots,p \textit{ such that } \\ pr_j \circ h \in H \textit{ and } pr_i \circ h = 0 \textit{ for each } i=1,\ldots,p,\ i \neq j\} \tag{2.8}$$

(where pr_i *denotes the* i*-projection from* $\mathbb{R}^p$ *in* $\mathbb{R}$*) is a Korovkin set in* $\mathscr{C}(X,\mathbb{R}^p)$ *with respect to the class* $\mathcal{M}$ *of convexity-monotone linear operators.*

Condition (2.8) allows us to give many examples of Korovkin sets in $\mathscr{C}(X,\mathbb{R}^p)$ with respect to the class $\mathcal{M}$ of convexity-monotone linear operators. In particular, from a Korovkin set in $\mathscr{C}(X,\mathbb{R})$ consisting of m elements, by (2.8) we obtain a Korovkin set in $\mathscr{C}(X,\mathbb{R}^p)$ with respect to the class $\mathcal{M}$ consisting exactly of $m \cdot p$ elements; if X is the compact real interval [0,1], we can take m = 3 to obtain a Korovkin set in $\mathscr{C}(X,\mathbb{R}^p)$ with respect to the class $\mathcal{M}$ consisting of 3p elements.

We remind to [3] and [5] for other general applications of Theorem 2.6; here we turn our interest to some particular sequences of quasi-positive linear operators between spaces of vector-valued continuous functions.

3. Approximation processes for vector-valued continuous functions

In this last Section, we briefly describe two simple approximation processes for the identity operator.

The operators under consideration generalize Bernstein polynomials on the standard simplex and on the hypercube of $\mathbb{R}^p$; the proof of the results that are stated here can be found in [7] (see also [3]).

1. Consider $p \geq 1$ and let $X = X^p$ be the standard simplex in $\mathbb{R}^p$:

$$X^p = \{(x_1,\ldots,x_p) \in \mathbb{R}^p \mid x_i \geq 0 \text{ for each } i=1,\ldots,p \text{ and } \sum_{i=1}^{p} x_i \leq 1\}.$$

Moreover, let E be a locally convex Hausdorff space; for each $n \in \mathbb{N}$, the n-th Bernstein operator $B_n : \mathscr{C}(X,E) \to \mathscr{C}(X,E)$ is defined by setting, for each $\varphi \in \mathscr{C}(X,E)$ and $(x_1,\ldots,x_p) \in X^p$,

$$B_n(\varphi)(x_1,\ldots,x_p) = \sum_{\substack{h_1,\ldots,h_p\in\mathbb{N}\\ h_1+\ldots+h_p\leq n}} \frac{n!}{h_1!\ldots h_p!(n-h_1-\ldots-h_p)!}\, x_1^{h_1}\cdots x_p^{h_p}\Big(1-\sum_{i=1}^{n} x_i\Big)^{n-h_1-\ldots-h_p}\, \varphi\Big(\frac{h_1}{n},\ldots,\frac{h_p}{n}\Big). \tag{3.1}$$

It is easy to show that every B_n is quasi-positive. By using the arguments of Nishishiraho [7] or, alternatively, by extending these operators to set-valued Bernstein operators as in [3] (cf. also [2]), we have that the sequence $(B_n(f))_{n\in\mathbb{N}}$ converges to f for each $f\in\mathscr{C}(X,E)$.

2. Consider $p\geq 1$ and let $X=[0,1]^p$ be the hypercube of $\mathbb{R}^p$. In this case, for each $n\in\mathbb{N}$, the n-th Bernstein operator $B_n:\mathscr{C}(X,E)\to\mathscr{C}(X,E)$ is defined by setting, for each $\varphi\in\mathscr{C}(X,E)$ and $(x_1,\ldots,x_p)\in X$,

$$B_n(\varphi)(x_1,\ldots,x_p) = \sum_{h_1,\ldots,h_p=0}^{n} \binom{n}{h_1}\cdots\binom{n}{h_p}\, x_1^{h_1}(1-x_1)^{h_1}\cdots x_p^{h_p}(1-x_p)^{h_p}\, \varphi\Big(\frac{h_1}{n},\ldots,\frac{h_p}{n}\Big). \tag{3.2}$$

Also in this case, for every $n\in\mathbb{N}$, B_n is a quasi-positive linear operator and therefore, by using the above arguments, we have the convergence to f of the sequence $(B_n(f))_{n\in\mathbb{N}}$ for each $f\in\mathscr{C}(X,E)$.

References

[1] H. Berens and G. G. Lorentz, *Geometric theory of Korovkin sets,* J. Approx. Theory 15 (1975), no. 3, 161-189.

[2] M. Campiti, *Approximation of set-valued continuous functions in Fréchet spaces,* I, to appear in Anal. Numér. Théor. Approx. 20 (1991), no. 1.

[3] M. Campiti, *Approximation of set-valued continuous functions in Fréchet spaces,* II, to appear in Anal. Numér. Théor. Approx. 20 (1991), no. 2.

[4] M. Campiti, *Convergence of linear monotone operators between cones of set-valued continuous functions,* to appear.

[5] M. Campiti, *Convexity-monotone operators in Korovkin theory,* to appear.

[6] L. B. O. Ferguson and M. D. Rusk, *Korovkin sets for an operator on a space of continuous functions,* Pacific J. Math. 65 (1976), no. 2, 337-345.

[7] T. Nishishiraho, *Convergence of quasi-positive linear operators,* Proc. Internat. Conf. "Trends in Functional Analysis and Approximation Theory", Acquafredda di Maratea, 1989, to appear in Atti Sem. Mat. Fis. Univ. Modena.

ON MODIFIED BOJANIC-SHISHA OPERATORS

A.S. CAVARETTA
Dept. of Math. & Comp. Sci.
Kent State University
Kent, Ohio 44242
U.S.A.
e-mail: cavaretta@mcs.kent.edu

and

SHUN SHENG GUO
Dept. of Mathematics
Hebei Normal University
Shijiazhuang, Hebei 050016
People's Republic of China

ABSTRACT. Rates of convergence of certain approximation methods are given for functions whose smoothness is measured in terms of the second modulus of continuity.

1. Introduction

Recently Chui, He and Hsu introduced a class of general summation-integral operators (cf. [5], [6]). We want to introduce an operator of this kind generalizing the Bojanic-Shisha operators, which are as follows (cf. [1]).

$$K_n(f,x) = \frac{2}{m_n+2} \sum_{k=1}^{m_n+2} f(t_{k,n}) \Phi_n(t_{k,n} - x) \tag{1.1}$$

where Φ_n is a nonnegative cosine polynomial of degree $\le m_n$ and

$$t_{k,n} = \frac{2k\pi}{m_n+2} \qquad k = 1, 2, \cdots, m_n+2.$$

We shall consider here modified operator

$$L_n(f,x) = \frac{2}{2m_n+1} \sum_{k=0}^{2m_n} \Phi_n(t_{k,n} - x) \frac{1}{\pi} \int_0^{2\pi} f(t) \Phi_n(t_{k,n} - t) dt. \tag{1.2}$$

where $t_{k,n} = 2k\pi/(2m_n+1)(k = 0, 1, \cdots, 2m_n)$ and $m_n = \ell$ or $\ell + \frac{1}{2}$, $\ell \in \mathbb{N}$.

$$\Phi_n(x) = \frac{1}{2} + \sum_{k=1}^{[m_n]} p_{k,n} \cos kx.$$

which satisfies the conditions: $\Phi_n(x) \ge 0$ and $p_{1,n} \to 1 (n \to \infty)$.

In this paper we will instigate direct and inverse theorems in $L^p_{2\pi} (1 \le p \le \infty)$ for the operator (1.2).

S. P. Singh (ed.), Approximation Theory, Spline Functions and Applications, 303–310.

We need use some notations:

$$\omega_2(f,t)_p = \sup_{0<|h|<t} (\int_0^{2\pi} |f(x+h)-2f(x)+f(x-h)|^p dx)^{1/p}$$

$$K_2(f,t^2)_p = \inf_{g\in L_p^2} \{\|f-g\|_p + t^2\|g''\|_p\}$$

here $L_p^2 =: \{f|f, f'' \in L_{2\pi}^p\}$. It is known that

$$\omega_2(f,t)_p \asymp K_2(f,t^2). \tag{1.3}$$

Now we can state main results.

THEOREM. If $f \in L_{2\pi}^p$ and there exists $M > 0$.

$$\int_0^{2\pi} |\Phi_n''(x)|dx \leq M(1-\rho_{1,n}^2)^{-1} \tag{1.4}$$

then for operator (1.2), we have

$$(1)\|L_n''(f,x)\|_p = O((1-p_{1,n}^2)^{-1+\alpha}) \Leftrightarrow \omega_2(f,t)_p = O(t^{2\alpha}) \quad (\alpha \leq 1) \tag{1.5}$$

$$(2)\|L_n f - f\|_p = O(1(1-p_{1,n}^2)^{\alpha}) \Leftrightarrow \omega_2(f,t)_p = O(t^{2\alpha}). \quad (\alpha < 1). \tag{1.6}$$

2. Lemmas

To prove (1.5) and (1.6) we need some Lemmas.

LEMMA 1. ([2]) Let x_0 be a point and $x_k = x_0 + \frac{2k\pi}{n}(k = 0, \cdots, n-1)$. If T is a trigonometric polynomial of degree $\leq n-1$, then

$$\frac{2}{n}\sum_{k=0}^{n-1} T(x_k) = \frac{1}{\pi}\int_0^{2\pi} T(x)dx.$$

LEMMA 2. The operator (1.2) can be expressed as follows

$$L_n(f,x) = \frac{1}{\pi^2}\int_0^{2\pi}\int_0^{2\pi} f(t)\Phi_n(u-x)\Phi_n(u-t)dudt. \tag{2.1}$$

Proof. From the definition of (1.2) and Lemma 1, it is easy to get (2.1).
LEMMA 3. We have

$$L_n(1,x) = 1,$$

$$L_n(\sin(t-x),x) = 0, \tag{2.2}$$

$$L_n(\sin^2\tfrac{t-x}{2},x) = \tfrac{1}{2}(1-p_{1,n}^2).$$

Proof. We only prove last one. Using (2.1) we have

$$\begin{aligned} L_n(\sin^2 \tfrac{t-x}{2}, x) &= \tfrac{1}{2} - \tfrac{1}{2\pi^2}\int_0^{2\pi}\int_0^{2\pi} \cos(t-x)\Phi_n(u-x)\Phi_n(u-t)dudt \\ &= \tfrac{1}{2} - \tfrac{1}{2\pi^2}\int_0^{2\pi}\int_0^{2\pi} (\cos(t-u)\cos(u-x) + \sin(t-u) \\ &\quad \sin(u-x))\Phi_n(u-x)\Phi_n(u-t)dudt \\ &= \tfrac{1}{2}(1-p_{1,n}^2). \end{aligned}$$

LEMMA 4. If $f \in L_p^2$, then

$$\|L_n''(f,x)\|_p \le \|f''\|_p. \tag{2.3}$$

Proof. By (2.1) we have

$$\begin{aligned} L_n''(f,x) &= \tfrac{1}{\pi^2}\int_0^{2\pi}\int_0^{2\pi} \Phi_n''(x-u)\Phi_n(u-t)f(t)dtdu \\ &= \tfrac{1}{\pi^2}\int_0^{2\pi}\int_0^{2\pi} \Phi_n''(x-u)f(u-t)\Phi_n(t)dtdu \\ &= \tfrac{1}{\pi^2}\int_0^{2\pi}\int_0^{2\pi} \Phi_n(x-u)f''(u-t)\Phi_n(t)dtdu \\ &= \tfrac{1}{\pi^2}\int_0^{2\pi}\int_0^{2\pi} \Phi_n(u)f''(x-u-t)\Phi_n(t)dtdu. \end{aligned} \tag{2.4}$$

Notice $\frac{1}{\pi}\int_0^{2\pi} \Phi_n(t)dt = 1$. So when $p = \infty$, we have

$$\|L_n''(f,x)\|_\infty \le \|f''\|_\infty.$$

When $1 \le p < \infty$ by Minkowski's inequality, we have

$$\|L_n''(f,x)\|_p = \left(\int_0^{2\pi} \left|\frac{1}{\pi^2}\int_0^{2\pi}\int_0^{2\pi} \Phi_n(u)f''(x-u-t)\Phi_n(t)dudt\right|^p dx\right)^{1/p}$$

$$\le \frac{1}{\pi^2}\int_0^{2\pi}\int_0^{2\pi} \Phi_n(u)\Phi_n(t)\left(\int_0^{2\pi} |f''(x-u-t)|^p dx\right)^{1/p} dudt = \|f''\|_p.$$

LEMMA 5. Denote $\beta_n^2 = \frac{1}{2}(1-p_{1,n}^2)$. If Φ_n satisfies condition (1.4) we have for $f \in L_{2\pi}^p (1 \le p \le \infty)$

$$\|L_n''(f,x)\|_p \le M\beta_n^{-2}\|f\|_p \tag{2.5}$$

where M is a constant independent of n and f.

Proof. Applying the similar procedure as used in the (2.4), we get

$$L_n''(f,x) = \frac{1}{\pi^2}\int_0^{2\pi}\int_0^{2\pi} \Phi_n''(u) f(x-u-t)\Phi_n(t)dudt.$$

Then using the same procedure as used in the proof of Lemma 4, we arrive at (2.5).

3. Main Results

THEOREM 1. If $f \in L_{2\pi}^p$ and Φ_n satisfies condition (1.4), then there exists $M > 0$ such that

$$\|L_n''(f,x)\|_p \le M\beta_n^{-2}\omega_2(f,\beta_n)_p. \tag{3.1}$$

Proof. Using Lemma 4 and 5 for every $g \in L_p^2$ we have

$$\begin{aligned}\|L_n''(f,x)\|_p &\le \|L_n''(f,g,x)\|_p + \|L_n''(g,x)\|_p \\ &\le M_1\beta_n^{-2}(\|f-g\|_p + \|g''\|_p) \\ &\le M_2\beta_n^{-2}(\|f-g\|_p + \beta_n^2\|g''\|_p)\end{aligned}$$

hence

$$\|L_n''(f,x)\|_p \le M_2\beta_n^{-2}K_2(f,\beta_n^2).$$

By (1.3) we can obtain (3.1).

THEOREM 2. If $f \in L_{2\pi}^p$, we have

$$\|L_n(f,x) - f(x)\|_p \le (1+\pi^2)\omega_2(f,\beta_n)_p. \tag{3.2}$$

Proof. Since f, Φ_n are periodic functions and Φ_n is even, using some technique as used in the proof of Lemma 4, we have

$$\begin{aligned}&\|L_n(f,x) - f(x)\|_p \\ &= \tfrac{1}{2\pi^2}\|\int_0^{2\pi}(\int_{u-\pi}^{u+\pi}\Phi_n(u)\Phi_n(t)[f(x+(t-u)) + f(x-(t-u)) - 2f(x)]dt)du\|_p \\ &\le \omega_2(f,\delta)_p\tfrac{1}{2\pi^2}\int_0^{2\pi}(\int_{u-\pi}^{u+\pi}(1+\frac{|t-u|}{\delta})^2\Phi_n(u)\Phi_n(t)dt)du \\ &\le \omega_2(f,\delta)_p\tfrac{1}{\pi^2}\int_0^{2\pi}(\int_{u-\pi}^{u+\pi}(1+\frac{|t-u|}{\delta^2})^2\Phi_n(u)\Phi_n(t)dt)du \\ &\le \omega_2(f,\delta)_p(1+\tfrac{1}{\delta^2}\int_0^{2\pi}\int_{u-\pi}^{u+\pi}\sin^2\frac{t-u}{2}\Phi_n(u)\Phi_n(t)dtdu).\end{aligned} \tag{3.3}$$

Here we have used the inequality $t \le \pi \sin \frac{t}{2}$ $(t \in [0,\pi])$. Notice that

$$\frac{1}{\pi^2}\int_0^{2\pi}\int_{u-\pi}^{u+\pi} \sin^2\frac{t-u}{2}\Phi_n(u)\Phi_n(t)dtdu$$

$$= \frac{1}{\pi^2}\int_0^{2\pi}\int_0^{2\pi} \sin^2\frac{t-u}{2}\Phi_n(u)\Phi_n(t)dtdu = \beta_n^2.$$

Hence choosing $\delta = \beta_n$, from (3.3) we get (3.2).

THEOREM 3. If $f \in L_{2\pi}^p$ and $\alpha \le 1$, and Φ_n satisfies (1.4), then

$$\|L_n''(f,x)\|_p = O(\beta_n^{2(\alpha-1)}) \Rightarrow \omega_2(f,t)_p = O(t^{2\alpha}), \tag{3.4}$$

provided $\lim_{n\to\infty} \dfrac{\beta_n}{\beta_{n-1}} > c > 0$ for some constant c.

Remark: This last condition is satisfied in all interesting cases.

Proof. to prove (3.4) we only need show that for any k

$$\|L_n''(f,x)\|_p = O(\beta_n^{2(\alpha-1)}) \Rightarrow \omega_2(L_kf,t) \le Bt^{2\alpha} \tag{3.5}$$

where B is a constant independently of k.

In fact, from

$$\begin{aligned}\omega_2(f,x)_p &\le \omega_2(f-L_kf,t)_p + \omega_2(L_kf,t)_p \\ &\le c\|f-L_kf\|_p + Bt^{2\alpha}.\end{aligned}$$

Let $k \to \infty$, then we get $\omega_2(f,x)_p \le Bt^{2\alpha}$. Now we will show (3.5). From (1.3) and definition of $K_2(f,t^2)_p$, we know

$$K_2(L_kf,t^2)_p \le \|L_kf - L_nL_kf\|_p + t^2\|L_n''(L_kf,x)\|_p \tag{3.6}$$

By Theorem 2 we have

$$\begin{aligned}\|L_kf - L_nL_kf\|_p &\le (1+\pi^2)\omega_2(L_kf,\beta_n)_p \\ &\le AK_2(L_kf,\beta_n^2)_p.\end{aligned} \tag{3.7}$$

By (2.1) it is easy to show $L_nL_kf = L_kL_nf$. Hence using Lemma 4, one has

$$\|\frac{d^2}{dx^2}L_n(L_kf,x)\|_p = \|\frac{d^2}{dx^2}L_k(L_nf,x)\| \le \|L_n''f\|_p \le L\beta_n^{2(\alpha-1)}, \tag{3.8}$$

for some constant $L > 0$. Combining (3.6)-(3.8) we have

$$K_2(L_kf,t^2)_p \le AK_2(L_kf,{\beta_n}^2)_p + L\beta_n^{2(\alpha-1)}t^2. \tag{3.9}$$

Choose $R \ge 1$ such that $A/R^{2\alpha} = q < 1$ and without losing generality we can choose n such that $\beta_n \le \frac{t}{R} < \beta_{n-1}(n \ge 2)$, then

$$\beta_n^{2(\alpha-1)}t^2 \le ct^{2\alpha}R^2.$$

Recall (3.9), we have

$$K_2(L_kf,t^2)_p \le AK_2(L_kf,\frac{t^2}{R^2})_p + Lt^{2\alpha}R^2.$$

Applying above inequality again and again to m times, then

$$\begin{aligned} K_2(L_kf,t^2)_p &\le A^mK_2(L_kf,\tfrac{t^2}{R^{2m}})_p \\ &+ 2Lt^{2\alpha}R^2(1+\tfrac{A}{R^{2\alpha}}+\cdots+(\tfrac{A}{R^{2\alpha}})^{m-1}) \\ &= A^mK_2(L_kf,\tfrac{t^2}{R^{2m}})_p + 2Lt^{2\alpha}R^2(1-q^m)/(1-q). \end{aligned}$$

By the definition of $K_2(L_kf,\cdot)$ and Lemma 5, we have

$$\begin{aligned} A^mK_2(L_kf,\tfrac{t^2}{R^{2m}})_p &\le A^m\tfrac{t^2}{R^{2m}}\|L_k''f\|_p \\ &\le A^m\tfrac{t^2}{R^{2m}}M\cdot\beta_k^{-2}\|f\|_p \end{aligned}$$

But k is fixed and $A/R^2 \le q < 1$, so that

$$\lim_{m\to\infty} A^mK_2(L_kf,\frac{t^2}{R^{2m}})_p = 0.$$

Hence

$$K_2(L_kf,t^2)_p \le \frac{2L}{1-q}R^2t^{2\alpha} = Bt^{2\alpha}.$$

This is (3.5), then Theorem 3 is proved.

THEOREM 4. If $f \in L^p_{2\pi}, \alpha < 1$ and Φ_n satisfies (1.4), then

$$\|L_nf - f\|_p = O(\beta_n^{2\alpha}) \Rightarrow \omega_2(f,t)_p = O(t^{2\alpha}). \tag{3.10}$$

Proof. First

$$K_2(f,t^2)_p \le \|L_nf - f\|_p + t^2\|L_n''(f)\|_p.$$

By Theorem 1, we have

$$\|L_n''(f)\}_p \le M\beta_n^{-2}\omega_2(f,\beta_n)_p \le M_1\beta_n^{-2}K_2(f,\beta_n^2)_p.$$

Notice

$$\|L_nf - f\|_p \le A_1\beta_n^{2\alpha},$$

therefore

$$\begin{aligned} K_2(f,t^2)_p &\le A_1\beta_n^{2\alpha} + M_1\beta_n^{-2}t^2K_2(f,\beta_n^2)_p \\ &\le A_2(\beta_n^{2\alpha} + \beta_n^{-2}t^2K_2(f,\beta_n^2)_p). \end{aligned}$$

For $0 < \delta < 1$, choose n such that $\beta_n \leq \delta < 2\beta_n$, then for $\alpha < 1$ we have

$$K_2(f,t^2)_p \leq A_3(\delta^{2\alpha} + (\frac{t}{\delta})^2 K_2(f,\delta^2)_p).$$

According to Berens-Lorentz's Lemma (see [3]) for $\alpha < 1$, we get

$$K_2(f,t^2)_p \leq c_1 t^{2\alpha}.$$

Now from Theorem 1-4, we obtain (1.5) and (1.6).

4. Remarks

We introduced in [4] the following three operators

$$F_n(f,x) = \sum_{k=0}^{2n} V_{nk}(x)\frac{2n+1}{2\pi}\int_0^{2\pi} V_{nk}(t)f(t)dt \tag{4.1}$$

$$G_n(f,x) = \sum_{k=0}^{4n-1} J_{nk}(x)\frac{2\pi}{\pi}\int_0^{2\pi} J_{nk}(t)f(t)dt \tag{4.2}$$

$$H_n(f,x) = \sum_{k=0}^{2n-1} K_{nk}(x)\frac{n}{\pi}\int_0^{2\pi} K_{nk}(t)f(t)dt \tag{4.3}$$

where $V_{nk}(x)$, $J_{nk}(x)$ and $K_{nk}(x)$ are defined respectively by the following.

$$\begin{aligned} V_{nk}(x) &= \tfrac{2^{2n}(n!)^2}{(2n+1)!}\cos^{2n}\tfrac{1}{2}(x - \tfrac{2k\pi}{2n+1}), \\ J_{nk}(x) &= \tfrac{3}{4n^2(2n^2+1)}(\tfrac{\sin\frac{n}{2}(x-\frac{k\pi}{2n})}{\sin\frac{1}{2}(x-\frac{k\pi}{2n})})^4, \\ K_{nk}(x) &= \tfrac{1}{n}\sin^2\tfrac{\pi}{n}(\tfrac{\cos\frac{n}{2}(x-\frac{k\pi}{n})}{\cos(x-\frac{k\pi}{n})-\cos\frac{\pi}{n}})^2. \end{aligned}$$

These operators are the special case of operator (1.2). For Operator (4.1), we have to set $\Phi_n(t) = \frac{2^{2n-1}(n!)^2}{(2n)!}\cos^{2n}\frac{t}{2}$ in (1.2), then $L_n(f,x) = F_n(f;x)$. In particular, we find

$$F_n(\sin^2\frac{t-x}{2}, x) = \frac{1}{2}(1 - P_{1,n}^2) = \frac{1}{n+1}(\frac{2n+1}{2n+2}) < \frac{1}{n}$$

(this case we can take $\beta_n = 1/\sqrt{n}$.) and

$$\begin{aligned} \tfrac{1}{\pi}\int_0^{2\pi} |\Phi_n''(t)|dt &\leq \tfrac{1}{\pi}\int_0^{2\pi} \frac{2^{2n-1}(n!)^2}{(2n)!}[n^2\cos^{2n-2}(\frac{t}{2})\sin^2\frac{t}{2} + \frac{n}{2}\cos^{2n}\frac{t}{2}]dt \\ &= \tfrac{n^2}{2n-1} + \tfrac{n}{2} < 2n. \end{aligned}$$

Hence condition (1.4) is satisfied for (4.1). For operator (4.2) we can obtain $G_n(\sin^2\frac{t-x}{2}, x) \leq \frac{3}{n^2}$, and condition (1.4) can easily be deduced from Bernstein's inequality. The operator $H_n(f,x)$ can be treated entirely similarly as $G_n(f,x)$. From (1.5) and (1.6) we can obtain corresponding results for operators (4.1), (4.2) and (4.3) respectively.

References

[1] R. Bojanic and O. Shisha, Approximation of continuous, periodic functions by discrete linear operators, J. Approx. Th. 11, 231-235.

[2] R.A. DeVore, The approximation of continuous by positive linear operators, Springer-Verlag, New York, 1972.

[3] H. Berens and G.G. Lorentz, Inverse theorems for Bernstein polynomials, Indiana Univ. Math. J. 21(1972), 693-708.

[4] S. S. Guo and L.C. Hsu, Inverse theorems for a certain class of operators, Demonstratio Math. XXI 3(1988), 745-760.

[5] C.K. Chui, T.X. He and L.C. Hsu, On a general class of multivariate linear smoothing operators, J. Approx. Th. 55(1988), 35-48.

[6] C.K. Chui, T.X. He and L.C. Hsu, Asymptotic properties of positive summation-integral operators, J. Approx. Th. 55(1988), 49-60.

A PROPERTY OF ZEROS AND COTES NUMBERS OF HERMITE AND LAGUERRE ORTHOGONAL POLYNOMIALS

FRANCESCO COSTABILE
IBM Department of Mathematics
University of Calabria
87036 Arcavacata of Rende, Italy

A property of localization of the zeros of three consecutive orthogonal polynomials of Hermite and Laguerre is proved. Then, a monotonic property of Cotes Numbers for abscissas of Hermite and Laguerre is, also proved.

1 Hermite's Orthogonal Polynomials

Let $x_{i,n} i = 1, \ldots, [n/2]$ be the positive zeros of $H_n(x)$, Hermite orthogonal polynomial of degree n, enumerated in decreasing order:

$$x_{1,n} > x_{2,n} > \ldots > x_{[n/2],n} > 0. \tag{1}$$

Then we have the following

Theorem 1.1 *For the zeros of $H_{n-1}(x), H_n(x), H_{n+1}(x)$ we have*

$$\ldots x_{i-1,n} > x_{i-1,n-1} > x_{i,n+1} > x_{i,n} \ldots .$$

Proof.
For the function

$$g(x) = H_n(x) H_{n-1}(x)$$

we have

$$\begin{cases} g(x) > 0 & \text{for} \quad x_{i,n} < x < x_{i-1,n-1} \\ g(x) < 0 & \text{for} \quad x_{i,n-1} < x < x_{i,n} \end{cases} \tag{2}$$

S. P. Singh (ed.), Approximation Theory, Spline Functions and Applications, 311–316.

But by [2. pp 106]

$$\begin{cases} H_{n+1}(x) & = & 2xH_n(x) - H_n'(x) \\ H_n'(x) & = & 2xH_{n-1}(x) \end{cases} \tag{3}$$

we have

$$g(x_{i,n+1}) = \frac{[H_n'(x_{i,n+1})]^2}{4x_{i,n+1}} > 0$$

and the thesis follows by (2).

Now let

$$\lambda_{\nu,n} = \pi^{\frac{1}{2}} 2^{n+1} n! [H_n'(x_{\nu,n})]^{-2}, \quad \nu = 1, \dots, [n/2]$$

be the Cotes numbers for the abscissas of Hermite polynomials, then we have

Theorem 1.2 *The sequence* $\{\lambda_{\nu,n}\} n \in N$ *is decreasing.*

Proof.

We consider the function

$$q(x) = 2(n+1)[H_{n+1}(x)]^2 + [H_{n+1}'(x)]^2 \tag{4}$$

which has $q(x)$ increasing for $x > 0$. Indeed we have

$$q'(x) = 4(n+1)H_{n+1}'(x)H_{n+1}(x) + 2H_{n+1}''(x)H_{n+1}'(x)$$

and by (3)

$$q'(x) = 16x(n+1)[H_n(x)]^2 > 0 \quad \text{for } x > 0.$$

By (4) we have

$$q(x_{i,n+1}) \quad = \quad [H_{n+1}'(x_{i,n+1})]^2$$

$$q(x_{i,n}) \quad = \quad 2(n+1)[H_{n+1}(x_{i,n})]^2$$

and by (3)

$$H_{n+1}(x_{i,n+1}) = -H_n'(x_{i,n})$$

for which

$$q(x_{i,n}) = 2(n+1)[H_n'(x_{i,n})]^2.$$

But

$$x_{i,n} < x_{i,n+1} \Rightarrow q(x_{i,n}) < q(x_{i,n+1}) \Leftrightarrow 2(n+1)[H_n'(x_{i,n})]^2 < H_{n+1}'(x_{i,n+1})]^2$$

by which follows the thesis.

2 Laguerre's Orthogonal Polynomials

Let $L_n^{(\alpha)}(x), \alpha > -1$ be the orthogonal polynomials of Laguerre of degree n and $x_{i,n}$ $i = 1, \dots, n$ the zeros enumerated in decreasing order.

Theorem 2.1 *For the zeros of $L_{n-1}^{(\alpha)}(x), L_n^{(\alpha)}(x), L_{n+1}^{(\alpha)}(x)$, we have:*

if

$$x_{i-1,n-1} < 2n - \alpha + 1$$

then

$$\dots x_{i-1,n} > x_{i,n+1} > x_{i-1,n-1} > x_{i,n} \dots \tag{5}$$

and if

$$x_{i-1,n} > 2n - \alpha + 1$$

then

$$\dots x_{i-1,n} > x_{i-1,n-1} > x_{i,n+1} > x_{i,n} \dots \tag{6}$$

Proof.
We consider the function

$$g(x) = L_n^{(\alpha)}(x) L_{n+1}^{(\alpha)}(x)$$

it has

$$\begin{cases} g(x) > 0 & \text{for} \quad x_{i,n} < x < x_{i,n+1} \\ g(x) < 0 & \text{for} \quad x_{i,n+1} < x < x_{i-1,n} \end{cases} . \tag{7}$$

By (2) it has

$$L_{n+1}^{(\alpha)}(x) = \frac{1}{n+1} \{(-x + 2n - \alpha + 1) L_n^{(\alpha)}(x) - (n + \alpha) L_{n-1}^{(\alpha)}(x)\} \tag{8}$$

and then

$$g(x_{i-1,n-1}) \geq 0 \quad \text{if} \quad x_{i-1,n-1} \leq 2n - \alpha + 1$$

$$g(x_{i-1,n-1}) \leq 0 \quad \text{if} \quad x_{i-1,n-1} \geq 2n - \alpha + 1$$

by which the thesis for (7).

Let $\alpha_{i,n}$ $i = 1, \dots, n$ be the zeros of $L_n^{(\alpha)\prime}(x)$, enumerated in decreasing order, we have

Theorem 2.2 *For the zeros of* $L_n^{(\alpha)}(x), L_n^{(\alpha)'}(x), L_{n+1}^{(\alpha)}(x), L_{n+1}^{(\alpha)'}(x)$ *it has*

if

$$x_{i,n+1} < n+1+\alpha \tag{9}$$

then

$$\dots \alpha_{i-1,n+1} > x_{i,n+1} > \alpha_{i-1,n} > \alpha_{i,n+1} > x_{i,n} > \alpha_{i,n} > \dots$$

if

$$x_{i,n+1} > n+1+\alpha \tag{10}$$

then

$$\dots > \alpha_{i-1,n+1} > \alpha_{i-1,n} > x_{i,n+1} > \alpha_{i,n+1} > x_{i,n} > \alpha_{i,n} > \dots$$

Proof.

The first, third and fifth inequalities are well known, therefore we are proving the second and the fourth. We consider the function

$$g(x) = L_n^{(\alpha)'}(x) L_{n+1}^{(\alpha)'}(x)$$

and we have

$$g(x) \le 0 \quad \text{for} \quad \alpha_{i,n} \le x \le \alpha_{i,n+1}$$

$$g(x) \ge 0 \quad \text{for} \quad \alpha_{i,n+1} \le x \le \alpha_{i-1,n}.$$

Observing that (2)

$$L_{n+1}^{(\alpha)'}(x) = \frac{1}{x}\{(n+1)L_{n+1}^{(\alpha)}(x) - (n+1+\alpha)L_n^{(\alpha)}(x)\} \tag{11}$$

and applying (9) we have

$$g(x_{i,n}) = [L_n^{(\alpha)'}(x_{i,n})]^2 > 0$$

by which the fourth inequality.

Now for (11) we have

$$g(x_{i,n+1}) = -\frac{1}{x_{i,n+1}}(n+1+\alpha)L_n^{(\alpha)'}(x_{i,n+1})L_n^{(\alpha)}(x_{i,n+1}) \tag{12}$$

$$L_n^{(\alpha)'}(x_{i,n+1}) = \frac{1}{x_{i,n+1}} n L_n^{(\alpha)}(x_{i,n+1}) - \frac{1}{x_{i,n+1}}(n+\alpha)L_{n-1}^{(\alpha)}(x_{i,n+1})$$

and substituting in (12) we have

$$g(x_{i,n+1}) = -\frac{1}{x_{i,n+1}^2} n(n+1+\alpha)[L_n^{(\alpha)}(x_{i,n+1})]^2 + \frac{1}{x_{i,n+1}^2}(n+\alpha)(n+1+\alpha) \quad (13)$$

$$\cdot L_n^{(\alpha)}(x_{i,n+1}) L_{n-1}^{(\alpha)}(x_{i,n+1})$$

applying, also (9) finally we have

$$g(x_{i,n+1}) = -\frac{1}{x_{i,n+1}^2}(n+1+\alpha)[L_n^{(\alpha)}(x_{i,n+1})]^2[-x_{i,n+1} + n + 1 + \alpha]$$

by which the thesis.

Let

$$B_{i,n} = \frac{\Gamma(n+\alpha+1)}{\Gamma(n+1)} \frac{1}{x_{i,n}} [L_n^{(\alpha)\prime}(x_{i,n})]^{-2}$$

be the number of Cotes for abscissas of Laguerre polynomials then we have

Theorem 2.3 *The sequence* $\{B_{i,n}\}$ *is decreasing for* $-1 < \alpha \leq 0$ *and* $x_{i,n} > \alpha + 1/2$.

Proof.

First we observe that

$$\frac{\Gamma(n+\alpha+1)}{\Gamma(n+1)} \text{ is decreasing for } -1 < \alpha \leq 0. \quad (14)$$

We consider the function

$$q(x) = (n+1)[L_{n+1}^{(\alpha)}(x)]^2 + x[L_{n+1}^{(\alpha)\prime}(x)]^2. \quad (15)$$

It is increasing for $x > \alpha + 1/2$. We have

$$q(x_{i,n+1}) = x_{i,n+1}[L_{n+1}^{(\alpha)\prime}(x_{i,n+1})]^2 \quad (16)$$

$$q(x_{i,n}) = (n+1)[L_{n+1}^{(\alpha)}(x_{i,n})]^2 + x_{i,n}[L_{n+1}^{(\alpha)\prime}(x_{i,n})]^2. \quad (17)$$

But it is, also,

$$(n+1)^2[L_{n+1}^{(\alpha)}(x_{i,n+1})]^2 = (n+\alpha)^2[L_{n-1}^{(\alpha)}(x_{i,n})]^2$$

and

$$(n+\alpha)^2[L_{n-1}^{(\alpha)}(x_{i,n})]^2 = x_{i,n}^2[L_n^{(\alpha)\prime}(x_{i,n})]^2$$

and substituting in (17) we have

$$q(x_{i,n}) = \frac{x_{i,n}^2}{n+1}[L_{n+1}^{(\alpha)\prime}(x_{i,n})]^2 + x_{i,n}[L_{n+1}^{(\alpha)\prime}(x_{i,n})]^2$$

and also

$$[L_{n+1}^{(\alpha)'}(x_{i,n})]^2 = \frac{1}{x_{i,n}^2}(n+1)^2[L_{n+1}^{(\alpha)}(x_{i,n})]^2 = [L_n^{(\alpha)'}(x_{i,n})]^2$$

by which

$$q(x_{i,n}) = [L_n^{(\alpha)'}(x_{i,n})]^2 \left(\frac{x_{i,n}}{n+1} + 1\right) x_{i,n}.$$

Therefore for $x > \alpha + 1/2$

$$x_{i,n} < x_{i,n+1} \Rightarrow q(x_{i,n}) < q(x_{i,n+1})$$

by which the thesis.

Note.

For ultraspherical polynomials, $P_n^{(\lambda)}(x)$, we have in [1] similar results on the separation of zeros and the decrease for the Cotes numbers when $0 < \lambda \leq 1/2$.

References

1. F. Costabile, "Un teorema di separazione degli zeri dei polimoni untrasferice e relative applicazioni," *B.U.M.I.*, 5, 13A (1976) 651-659.

2. G. Szegö, "Orthogonal polynomials," *Amer. Math. Soc. Col. Pub.*, 23, New York (1959).

HERMITE-FEJÉR AND HERMITE INTERPOLATION[1]

G. CRISCUOLO, B. DELLA VECCHIA, G. MASTROIANNI
Istituto per Applicazioni della Matematica, C.N.R.
Via P. Castellino 111, 80131 Napoli, Italy

Dipartimento di Matematica, Università della Basilicata
Via N.Sauro 85, 85100 Potenza, Italy

ABSTRACT. The authors consider two procedures of Hermite and Hermite-Fejér interpolation based on the zeros of Jacobi polynomials plus additional nodes and prove that such procedures can always well approximate a function and its derivatives simultaneously.

1. Introduction.

The problem of uniform and weighted L^p convergence of Hermite-Fejér and Hermite interpolating polynomials on the zeros of Jacobi polynomials was studied by many authors and the convergence conditions consist in some restrictions for the Jacobi parameters (see subsections 1.1 and 1.2).
In the present paper we want to extend the convergence results for Hermite-Fejér and Hermite interpolation to other nodes matrices. In particular in Sections 2 and 3 we show two new good classes of nodes matrices realizing convergence of Hermite-Fejér and Hermite interpolation in uniform and weighted L^p norm and in some cases also improvements of previous results.

1.1. HERMITE-FEJÉR OPERATOR

Given a continuous function f on $[-1,1]$ and a matrix V, we denote by $H_m(V;f)$ the corresponding Hermite-Fejér polynomial of degree $2m-1$ interpolating the function f on the nodes of V.
Although invented about 75 years ago, Hermite-Fejér interpolation is still quite a timely object of research in approximation theory.

[1]This material is based upon work supported by the Italian Research Council (all the authors) and by the Ministero della Università e della Ricerca Scientifica e Tecnologica (the third author).

S. P. Singh (ed.), Approximation Theory, Spline Functions and Applications, 317–331.

For a comprehensive bibliography on Hermite-Fejér interpolation, the interested reader can consult the paper by Gonska and Knoop [13].(See also [18]).
Here we want to recall some known results.
The Hermite-Fejér interpolating polynomials were introduced in 1916 in [10] by Fejér, who called them "step parabolas". He proved that $H_m(V;f) \to f$, uniformly in $[-1,1]$, for every continuous function on $[-1,1]$, if the given matrix V is formed by the zeros of Chebyshev polynomials of first kind.
A complete treatment for the zeros of Jacobi polynomials was then given by Szegö [33]. He proved that, if V is a Jacobi matrix, then $H_n(V;f) \to f$ uniformly in $[-1,1]$ for every continuous function on $[-1,1]$, if and only if the Jacobi parameters are less than 0. In such a case the corresponding interpolation matrix V is called strongly ρ-normal and the Hermite-Fejér operator is positive.
If the Jacobi parameters are equal 0, i.e. the Legendre case, then Fejér proved that $H_m(f)$ is not converging to f, uniformly in $[-1,1]$, for every continuous function on $[-1,1]$. In [9] Egervary and Turan modified the Legendre matrix by adding the endpoints ± 1, i.e. they considered a quasi Hermite-Fejér interpolation process.
Other results about quasi Hermite-Fejér interpolation can be found in [26,27,29,30].
The convergence conditions for Hermite-Fejér interpolation, when the Jacobi parameters are greater than 0, are a little more complicated and they involve structural properties of f and some peculiar integral conditions (see also [29]).
Other convergence results can be found in the book, recently appeared, by P. Vértesi and J. Szabados [32]. For example, Vértesi proved that, if X is a matrix formed by the zeros $x_{i,n}, i = 1, \ldots, n$, of the $n-$th Jacobi polynomial with Jacobi parameters $\alpha, \beta \geq -1$, then we have, for $x \in [-1+\epsilon, 1]$, $\epsilon > 0$,

$$|H_n(X;f;x) - f(x)| = O(1)\{\omega(f;|x - x_{j,n}|)+$$

$$+n^2(\theta - \theta_{j,n})^2 \sum_{i=1}^{n} [\omega(f; \frac{i\sqrt{1-x^2}}{n}) + \omega(f; \frac{i^2|x|}{n^2})]i^{2\gamma-1}\},$$

where $\gamma = max\{\alpha, -1/2\}$ and $x_{j,n} = \cos\theta_{j,n}$ is, as usual, the nearest root to $x = \cos\theta$.
Hermite-Fejér interpolation on matrices formed by roots of generalized Jacobi Polynomials was also investigated and convergenge results were obtained [22].
The weighted L^p convergence of Hermite-Fejér interpolation has been also object of research by some authors. (See e.g. [21]).
Finally the extension to Hermite-Fejér interpolating polynomial of higher order has been studied and some convergence theorems can be found in [25,36-38].

1.2. HERMITE OPERATOR

We denote by $\mathcal{H}_m(V;f)$ the corresponding Hermite polynomial of degree $2m-1$

interpolating the function f and its derivative f' on the nodes of V, i.e.

$$\mathcal{H}_m(V;f;x_i)=f(x_i),\quad \mathcal{H}'_m(V;f;x_i)=f'(x_i), i=1,\ldots,m.$$

The problem of convergence of Hermite interpolating polynomials on the zeros of Jacobi polynomials was studied by several authors; among the others we mention [11,12,19,23,28].
A complete theorem on the convergence of the derivatives of Hermite polynomials on the zeros of Jacobi polynomials was also given by Neckermann and Runk in [20].
Recently Xu [39], working on a more general weight

$$w(x)=\phi(x)(1-x)^{\alpha}(1+x)^{\beta},\quad \alpha,\beta>-1,\phi>0,\phi'\in \mathrm{Lip}1,$$

obtained the estimate, $\forall f\in C^1([-1,1])$,

$$||\mathcal{H}_n(f)-f||\leq \text{ const } E_n(f')\frac{\log m}{m},\quad \gamma\leq -1/2,$$

$$||\mathcal{H}_n(f)-f||\leq \text{ const } n^{2\gamma}E_n(f'),\quad \gamma>-1/2,$$

where $\gamma=max\{\alpha,\beta\}$ and $E_n(g)=\min_{P\in\mathcal{P}_n}||g-P||,\quad g\in C([-1,1])$, with $\mathcal{P}_n$ the set of algebraic polynomials of degree at most n.
Moreover Xu investigated about the weighted L^p convergence of $\mathcal{H}_n(f)$ and he proved that, if $u\in L^p$ is a Jacobi weight, then [39]

$$||(H_n(f)-f)u||_p^1\leq \text{ const } E_n(f'),$$

if and only if $w(x)(1-x^2)u(x)\in L^p$, where $||f||^1=max\{||f||,||f'||\}$.

From all these papers it follows that the convergence conditions for Hermite-Fejér and Hermite interpolation consist in some restrictions for the Jacobi parameters, namely they must be less than 0.

Here we are going to show other two classes of nodes matrices realizing Hermite-Fejér and Hermite interpolation.

2. Hermite-Fejér and Hermite interpolation at the Jacobi zeros plus additional points

Let

$$w(x)=w^{(\gamma,\delta)}(x)=(1-x)^{\gamma}(1+x)^{\delta},\quad \gamma,\delta>-1,-1\leq x\leq 1, \tag{2.1}$$

be a Jacobi weight and let $\{p_m^{(\alpha,\beta)}\}$ be the corresponding system of orthonormal polynomials on $[-1,1]$. If $x_{i,m}=x_{i,m}(w)$, $i=1,2,\ldots,m$ are the zeros of $p_m^{(\alpha,\beta)}$, we define the infinite triangular matrix of nodes X, whose m-th line is:

$$-1=y_{1,m}<\ldots<y_{s,m}<x_{1,m}<\ldots<x_{m,m}<z_{1,m}<\ldots<z_{r,m}=1,$$

$$y_{j,m} = -1 + \frac{j-1}{s}(1 + x_{1,m}), \quad j = 1, \ldots, s, \tag{2.2}$$

$$z_{i,m} = x_{m,m} + \frac{i}{r}(1 - x_{m,m}), \quad i = 1, \ldots, r.$$

Additional points were used recently in many different contexts for Lagrange interpolation in [1,19,24,31] and for Hermite-Fejér and Hermite interpolation in [4-7,9,14–16].

In this Section we are going to give some convergence theorems about Hermite-Fejér and Hermite interpolation based on the nodes of X.

2.1. HERMITE FEJÉR INTERPOLATION

If f is a given continuous function on $[-1,1]$, we denote by $H_m(w;f)$ the Hermite-Fejér polynomial interpolating the function f on the zeros $x_{k,m}$ of $p_m^{(\alpha,\beta)}$, i.e.

$$H_m(w; f; x_{k,m}) = f(x_{k,m}), \quad k = 1, \ldots, m,$$

$$H'_m(w; f; x_{k,m}) = 0, \quad k = 1, \ldots, m.$$

Then, we denote by $H_{m,r,s}(w;f)$ the Hermite-Fejér polynomial of degree $2m + r + s - 1$ interpolating f on the points $y_{i,m}, i = 1, \ldots, s, \quad x_{k,m}, k = 1, \ldots, m, \quad z_{i,m}, i = 1, \ldots, r$, i.e.

$$\begin{aligned} H_{m,r,s}(w; f; x_{k,m}) &= f(x_{k,m}), \quad k = 1, \ldots, m, \\ H_{m,r,s}(w; f; y_{k,m}) &= f(y_{k,m}), \quad k = 1, \ldots, r, \\ H_{m,r,s}(w; f; z_{k,m}) &= f(z_{k,m}), \quad k = 1, \ldots, s, \\ H'_{m,r,s}(w; f; x_{k,m}) &= 0, \quad k = 1, \ldots, m. \end{aligned} \tag{2.3}$$

We complete the definition by putting $H_{m,0,0}(w;f) = H_m(w;f)$.
The polynomial $H_{m,r,s}(w;f)$ is not strictly an Hermite-Fejér interpolating polynomial, since we do not require conditions for the first derivative on additional points $y_{j,m}, j = 1, \ldots, s$ and $z_{i,m}, i = 1, \ldots, r$, but nevertheless we will call it *Hermite-Fejér interpolating polynomial.*

Denoted by $\|\ \|_\Delta$ the supremum norm on the set Δ and setting $\|\ \|_{[-1,1]} = \|\cdot\|$, we can state the following theorems proved in [5].

Theorem 2.1. *Let $H_{m,r,s}(w;f)$ be the Hermite-Fejér polynomial defined by (2.3). If the parameters α, β verify the conditions*

$$r = \alpha + 1/2, \quad s = \beta + 1/2,$$

then, for all $f \in C([-1,1])$ *we have*

$$\|f - H_{m,r,s}(w; f)\| \leq \text{ const } \sum_{i=1}^{m} \frac{1}{i^2}\omega(g; \frac{i}{m}), \tag{2.4}$$

with $g = f(\cos)$ *and for some constant independent of* f *and* m.

Remark 1. Vértesi in [34] generalized a previous result by Bojanic [2]; by using the zeros of Jacobi polynomials, he obtained the estimate

$$\|f - H_m(w; f)\|_{[-a,a]} \leq \text{const} \sum_{i=1}^{m} \frac{1}{i^2}\omega(f; \frac{i}{m}), \tag{2.5}$$

where $\alpha, \beta \leq -1/2$ and $|a| < 1$.
The estimate (2.4) is better than (2.5); in fact, if $f(x) = (1-x)^\gamma \psi(x)$, with $0 < \gamma < 1$, and $\psi(x)$ a good function (for example $\psi \in C^2([-1,1])$), then from (2.5), it follows that the error goes to 0 as $O(m^{-\gamma})$. On the contrary, by (2.4), we have that the error goes to 0 as $O(m^{-2\gamma})$, if $\gamma < 1/2$ and $O(\frac{\log m}{m})$, if $\gamma \geq 1/2$.
Moreover the estimate (2.4) is verified by infinitely many values of the parameters α, β, in particular, when $\alpha = \beta = -1/2$.

2.2. HERMITE INTERPOLATION

Analogously we denote by $\mathcal{H}_m(w; f)$ the Hermite polynomial $\mathcal{H}_m(w; f)$ interpolating the differentiable function f on the Jacobi zeros $x_{k,m}$, $k = 1, \ldots, m$, i.e.

$$\mathcal{H}_m^{(i)}(w; f; x_{k,m}) = f^{(i)}(x_{k,m}), \quad i = 0, 1, \quad k = 1, \ldots, m.$$

Now we consider the Hermite polynomial $\mathcal{H}_{m,r,s}(w; f)$ based on the nodes of X and defined by:

$$\begin{aligned} \mathcal{H}_{m,r,s}^{(i)}(w; f; x_{k,m}) &= f^{(i)}(x_{k,m}), \quad i = 0, 1, \quad k = 1, \ldots, m, \\ \mathcal{H}_{m,r,s}(w; f; y_{j,m}) &= f(y_{j,m}), \quad j = 1, \ldots, s, \\ \mathcal{H}_{m,r,s}(w; f; z_{i,m}) &= f(z_{i,m}), \quad i = 1, \ldots, r. \end{aligned} \tag{2.6}$$

We complete the definition by putting $\mathcal{H}_{m,0,0}(w; f) = \mathcal{H}_m(w; f)$.
Now we can state the following

Theorem 2.2. [5] *Let* $\mathcal{H}_{m,r,s}(w; f)$ *be the polynomial defined by (2.6) and let* $f \in C^q([-1,1])$, $q \geq 1$ *and let* ℓ *be a nonnegative integer, with* $\ell \leq q$. *Then, if the parameters* α, β *verify*

$$\frac{\ell}{2}+\alpha+\frac{1}{2}\leq r<\frac{\ell}{2}+\alpha+\frac{3}{2}, \qquad (2.7)$$
$$\frac{\ell}{2}+\beta+\frac{1}{2}\leq s<\frac{\ell}{2}+\beta+\frac{3}{2},$$

then, for $|x|\leq 1$ and $h=0,\ldots,\ell$,

$$|\,[f(x)-\mathcal{H}_{m,r,s}(w;f;x)]^{(h)}\,|\leq \text{const}\left[\frac{\sqrt{1-x^2}}{m}+\frac{1}{m^2}\right]^{\ell-h}E_{m-q}(f^{(q)})\frac{\log m}{m^{q-\ell}}, \quad (2.8)$$

with some constant independent of f and m.

Remark 1. For the Hermite polynomial interpolating f and f' only on the zeros of $p_m^{(\alpha,\beta)}, \alpha,\beta>0$, the following estimate

$$||f-\mathcal{H}_m(w;f)||\leq \text{const}\ m^{2\gamma}\omega(f';1/m), \quad \gamma=max\{\alpha,\beta\}, f'\in C([-1,1])$$

holds [39]. This estimate is worst than

$$||f-\mathcal{H}_m(w;f)||\leq \text{const}\ E_{m-1}(f')\frac{\log m}{m},$$

coming from (2.8), when $q=1$, $\ell=h=0$ and therefore for $r=[\alpha+3/2]$ and $s=[\beta+3/2]$.
Then Theorem 2.2 assures us that, chosen as interpolation nodes, the zeros of Jacobi polynomial $p_m^{(\alpha,\beta)}$, with parameters $\alpha,\beta>-1$, we can determine, by (2.7), the number r of the nodes to be added near $+1$ and the number s of the nodes to be added near -1 such that (2.8) holds.
The estimate (2.8) seems optimal; indeed P. Vértesi proved that, for every nodes matrix V, there exists a function $f\in C^1([-1,1])$, such that

$$||\mathcal{H}_m(V;f)-f||\geq \text{const}\ \omega(f';\frac{1}{m})\frac{\log m}{m}.$$

We also remark that in (2.8), when $h=0$, the factor "$\log m$" can be replaced by "$\log(m\sqrt{1-x^2}+1)+2$".

Now, letting : $||f||_h=\max_{0\leq i\leq h}||f^{(i)}||$, $f^{(0)}=f, h\geq 0$, from the previous theorem it follows:

Corollary 2.3. *Let $f\in C^{(q)}([-1,1])$ and $1\leq h\leq q$. Then there exist integers r and s, such that*

$$||f-\mathcal{H}_{m,r,s}(w;f)||_h\leq \text{const}\ \frac{E_{m-q}(f^{(q)})}{m^{q-h}}\log m.$$

Moreover the integers r and s are defined by (2.7).
From the previous corollary it follows the useful inequality

$$||\mathcal{H}_{m,r,s}(w)||_h = \sup_{||f||_h=1} ||\mathcal{H}_{m,r,s}(w;f)||_h \leq \text{ const } \log m,$$

which holds in the hypotheses of the corollary.
Remark 2. We observe that by Theorem 2.2 we can solve also the following problem: assume that $f \in C^q([-1,1])$ and that $f(1)$, $f'(1)$, ..., $f^{(r-1)}(1)$, $f(-1)$, $f'(-1)$, ..., $f^{(s-1)}(-1)$ are known. Choose m nodes $t_1,\ldots,t_m \in (-1,1)$ and construct the Hermite polynomial P_m defined by

$$P_m^{(j)}(t_i) = f^{(j)}(t_i), \quad i = 1,\ldots,m, \quad j = 0,1,$$

$$P_m^{(j)}(-1) = f^{(j)}(-1), \quad j = 0,\ldots,s-1,$$

$$P_m^{(j)}(1) = f^{(j)}(1), \quad j = 0,\ldots,r-1,$$

such that

$$|\,[f(x) - \mathcal{H}_{m,r,s}(w;f;x)]^{(h)}\,| \leq \text{const}\,[\frac{\sqrt{1-x^2}}{m}]^{\ell-h} E_{m-q}(f^{(q)})\frac{\log m}{m^{q-\ell}}. \tag{2.9}$$

Theorem 2.2 assures us that such a polynomial exists and coincides with $\mathcal{H}_{m,r,s}(w;f;x) =: \overline{\mathcal{H}}_{m,r,s}(w;f;x)$, where the node -1 has multeplicity s and the node $+1$ has multeplicity r, with $r,s \leq q+1$ and with the parameters α,β,r,s satisfying (2.7).
In addition, since (2.7) can we rewritten as

$$r - \frac{\ell}{2} - \frac{3}{2} < \alpha \leq r - \frac{\ell}{2} - \frac{1}{2}, \quad r > \frac{\ell-1}{2}, \tag{2.10}$$

$$s - \frac{\ell}{2} - \frac{3}{2} < \beta \leq s - \frac{\ell}{2} - \frac{1}{2}, \quad s > \frac{\ell-1}{2},$$

it is possible to choose $t_1,\ldots,t_m$ in infinitely many ways, that is the nodes t_i are the zeros of any Jacobi polynomial $p_m^{(\alpha,\beta)}$, with α,β satisfying (2.10).
For example assume to know $f(\pm 1)$ and $f'(\pm 1)$ and to want to approximate the function $f \in C^q([-1,1])$ and its first derivative. Then, by (2.10), we know that the required polynomial is $\overline{\mathcal{H}}_{m,2,2}(w;f)$, with $0 < \alpha,\beta \leq 1$ and the corresponding error estimate is

$$|\,[f(x) - \overline{\mathcal{H}}_{m,2,2}(w;f;x)]^{(h)}\,| \leq \text{const}\,[\frac{\sqrt{1-x^2}}{m}]^{1-h}\omega(f^{(q)};\frac{1}{m})\frac{\log m}{m^{q-1}}, h = 0,1.$$

For the polynomial $\overline{\mathcal{H}}_{m,r,s}(w;f)$ a stronger estimate can be obtained, when $f^{(q)} \in \mathrm{Lip}_M\lambda, 0 < \lambda \leq 1$.

Theorem 2.4. [5] *Let $\overline{\mathcal{H}}_{m,r,s}(w;f)$ be the polynomial introduced above and let $f \in C^q([-1,1])$, $q \geq 1$, $f^{(q)} \in Lip_M\lambda$, $0 < \lambda \leq 1$ and let ℓ be a nonnegative integer, with $\ell \leq \min\{r,s\}$. If the parameters α, β verify*

$$-\frac{\ell+\lambda}{2} + \alpha - \frac{1}{2} < r \leq -\frac{\ell+\lambda}{2} + \alpha - \frac{3}{2}, \tag{2.11}$$

$$-\frac{\ell+\lambda}{2} + \beta - \frac{1}{2} < s \leq -\frac{\ell+\lambda}{2} + \beta - \frac{3}{2},$$

then, for $|x| \leq 1$ and $h = 0,\ldots,\ell$,

$$\frac{\mid [f(x) - \overline{\mathcal{H}}_{m,r,s}(w;f;x)]^{(h)} \mid}{(\sqrt{1-x^2})^{\ell-h+\lambda}} \leq \mathrm{const}\frac{\log m}{m^{q-h+\lambda}}, \tag{2.12}$$

for some constant independent of f and m.

Theorems 2.1–2.4 are the analogous ones of results established in [14] for the Lagrange interpolation.

Now we give some weighted L^p convergence results for the polynomial $\mathcal{H}_{m,r,s}(w;f)$ defined by (2.6).
We recall that $f \in L^p$ or $f \in (L\log^+ L)^p$, $0 < p \leq \infty$, if and only if

$$||f||_p = \{\int_{-1}^{1} |f(x)|^p dx\}^{1/p} < \infty$$

or respectively

$$||f\log^+|f|||_p = \{\int_{-1}^{1} [|f(x)|\log^+|f(x)|]^p dx\}^{1/p} < \infty.$$

We also recall that $\bar{u}$ is a generalized Jacobi weight ($\bar{u} \in GJ$), if

$$\bar{u}(x) = \phi(x)v^{(\gamma,\delta)}(x) = \phi(x)(1-x)^\gamma(1+x)^\delta, \quad \gamma,\delta > -1,$$

with ϕ nonnegative and $\phi^{\pm 1} \in L^\infty$.
Then we have

Theorem 2.5. [6] *Let w be the weight function defined by (2.2) and let*

$u \in (L \log^+ L)^p$, *with* $0 < p < \infty$.
If

$$wv^{(1-r,1-s)} \in L^1, \qquad uw^{-1}v^{(r-1/2,s-1/2)} \in L^p, \tag{2.13}$$

where r *and* s *are nonnegative integers, then for every function* $f \in C^1([-1,1])$

$$\lim_{m\to\infty} \|[f - H_{m,r,s}(w;f)]u\|_p = 0. \tag{2.14}$$

Furthermore

Theorem 2.6. [6] *Let* $f \in C^q$, *with* $q \geq 1$. *Let* w *be the weight function defined by* (2.2) . *Assume* $u \in$GJ *and* $0 < p < \infty$.
If

$$wv^{(\frac{q+1}{2}-r,\frac{q+1}{2}-s)} \in L^1, \qquad u \in L^p, \qquad uw^{-1}v^{(r-\frac{\ell+1}{2},s-\frac{\ell+1}{2})} \in L^p, \tag{2.15}$$

where r, s, ℓ *are nonnegative integers, with* $\ell \leq q$, *then*

$$\left\|[f - H_{m,r,s}(w;f)]^{(\ell)}u\right\|_p \leq \text{const}\ \frac{E_{m-q}(f^{(q)})}{m^{q-\ell}}, \tag{2.16}$$

with some constant independent of f *and* $m \geq 4q+5$.

For the proofs see [6].
We note that, if we interpolate f and f' only on the zeros of $p_m^{(\alpha,\beta)}$ and $H_m(w;f)$ is the corresponding Hermite interpolating polynomial, (i.e. $r = s = 0$), the above estimate, with $q = \ell = 1$, gives us back a result stated by Xu in [39].
To complete the previous results, we remark that generally speaking the polynomial $H_m(w;f)$ interpolating f at the zeros of $p_m^{(\alpha,\beta)}, \alpha,\beta > 0$, does not converge to f. Theorem 2.5 assures that, when the hypotheses (2.13) are satisfied, then we can obtain an interpolating polynomial by adding knots near the endpoints ± 1 which realizes the L^p convergence to the given function f. In addition Theorem 2.6 garantees the simultaneous L^p approximation of the function and of its derivatives by Hermite interpolating polynomial.
If in particular the weight u is defined by

$$u(x) = v^{(\gamma,\delta)}(x), \tag{2.17}$$

then we can give an explicit expression to the hypotheses (2.15).
For instance, Theorem 2.6 implies

Corollary 2.7 . *Let* $f \in C^q$, *with* $q \geq 1$. *Let* w *and* $u \in L^p$ *be defined by* (2.2) *and*

(2.17) *respectively. Let* $0 \le \ell \le q$ *and* $0 < p < \infty$. *Then, there exist two integers* r *and* s, *defined by*

$$\frac{\ell}{2} + \alpha - \gamma - \frac{1}{p} + \frac{1}{2} < r < \frac{q}{2} + \alpha + \frac{3}{2}$$

and

$$\frac{\ell}{2} + \beta - \delta - \frac{1}{p} + \frac{1}{2} < s < \frac{q}{2} + \beta + \frac{3}{2}$$

such that

$$\left\{ \int_{-1}^{1} |f^{(\ell)}(x) - H^{(\ell)}_{m,r,s}(w; f; x)|^p u^p(x) dx \right\}^{1/p} \le \text{const } \frac{E_{m-q}(f^{(q)})}{m^{q-\ell}},$$

with some constant independent of f *and* $m \ge 4q + 5$.

An useful consequence of Corollary 2.7 is the following

Corollary 2.8. *Let* $f \in C^q$, *with* $q \ge 1$ *and let* $0 \le \ell \le q$. *Let* w *and* $u \in L^p$ *be defined by (2.2) and (2.17), respectively. Then, there exist two integers* r *and* s *defined by*

$$\frac{\ell}{2} + \alpha - \gamma - \frac{1}{2} < r < \frac{q}{2} + \alpha + 3/2$$

and

$$\frac{\ell}{2} + \beta - \delta - \frac{1}{2} < s < \frac{q}{2} + \beta + 3/2$$

such that

$$\int_{-1}^{1} |f^{(\ell)}(x) - H^{(\ell)}_{m,r,s}(w; f; x)| u(x) dx \le \text{const } \frac{E_{m-q}(f^{(q)})}{m^{q-\ell}},$$

with some constant independent of f *and* $m \ge 4q + 5$.

The last corollary has interesting applications in quadrature processes, when we want to approximate integrals of the type $\int_{-1}^{1} f^{(q)}(x)u(x)dx$, $1 \le q < \infty$, $u \in GJ$, by an interpolatory product rule obtained by replacing f by an Hermite interpolating polynomial.

We remark that all the above theorems hold if $w^{(\alpha,\beta)}$ is replaced by a generalized smooth Jacobi weight, i.e. $w^{(\alpha,\beta)}(x) = \phi(x)(1-x)^\alpha(1+x)^\beta, \alpha, \beta > -1$ and $0 < \phi \in C^1([-1,1]), \phi' \in \text{Lip}_M \lambda, 0 < \lambda \le 1$.
Finally we observe that analogous results on weighted L^p convergence of Lagrange interpolation were obtained in [15].

3. Hermite-Fejér and Hermite interpolation on extended matrices

Recently in [8] the two sequences of orthogonal polynomials $\{p_m^{(\alpha,\beta)}\}_{m=0}^{\infty}$ and $\{p_m^{(\alpha+1,\beta+1)}\}_{m=0}^{\infty}$ were considered. In the same paper it was proved that the zeros of the polynomial $q_{2m+1}(x) = p_m^{(\alpha+1,\beta+1)} p_{m+1}^{(\alpha,\beta)}$ are simple and they have an arcsin distribution. This property allows us to introduce so called *extended interpolation matrix* Y, having as nodes the zeros of q_{2m+1}, i.e.

$$Y = \{t_{i,2m+1}, i = 1, \ldots, 2m+1 | t_{i,2m+1} \text{zeros of } q_{2m+1}\}.$$

However the extended matrices generally are not good matrices for Hermite-Fejér and Hermite interpolation, in the sense that the corresponding Lebesgue constants in uniform norm are greater than $O(\log m)$.
Therefore we are going to transform the above extended matrix Y by the same procedure given in Section 2, i.e. by adding points near ± 1.

So, we consider the matrix $\overline{Y}$, whose $(2m+1)$–rst line is:

$$-1 = y_{1,m} < \ldots < y_{s,m} < t_{1,2m+1} < \ldots < t_{2m+1,2m+1} < z_{1,m} < \ldots < z_{r,m} = 1,$$

$$y_j = y_{j,m} = -1 + \frac{j-1}{s}(1 + t_{1,2m+1}), \quad j = 1, \ldots, s, \tag{3.1}$$

$$z_i = z_{i,m} = t_{2m+1,2m+1} + \frac{i}{r}(1 - t_{2m+1,2m+1}), \quad i = 1, \ldots, r$$

with r and s fixed nonnegative integers.

3.1. HERMITE FEJÉR INTERPOLATION

Generalizing a procedure showed in [27], given a bounded function f on $[-1,1]$, we denote by $H_m(Y; f)$ the corresponding Hermite-Fejér interpolating polynomial on the zeros of q_{2m+1} and we denote by $H_{m,r,s}(\overline{Y}; f)$ the Hermite-Fejér interpolating polynomial of degree $4m + r + s + 1$ based on the nodes (3.1) and defined by

$$\begin{aligned} H_{m,r,s}(\overline{Y}; f; t_{k,2m+1}) &= f(t_{k,2m+1}), \quad k = 1, \ldots, 2m+1, \\ H'_{m,r,s}(\overline{Y}; f; t_{k,2m+1}) &= 0, \quad k = 1, \ldots, 2m+1, \\ H_{m,r,s}(\overline{Y}; f; y_{j,m}) &= f(y_{j,m}), \quad j = 1, \ldots, s \\ H_{m,r,s}(\overline{Y}; f; z_{i,m}) &= f(z_{i,m}), \quad i = 1, \ldots, r. \end{aligned} \tag{3.2}$$

We complete the definition by putting $H_{m,0,0}(\overline{Y}; f) = H_m(Y; f)$.

For the polynomial $H_{m,r,s}(\overline{Y}; f)$ we can prove convergence theorems analogous to those ones showed above in subsection 2.1.
For example we have [4]

Theorem 3.1 *Let $f \in C([-1,1])$. If for some integers r, s the exponents $\alpha, \beta > -1$ of the weight w satisfy the conditions*

$$r = 2\alpha + 2, \quad s = 2\beta + 2$$

then

$$||f - H_{m,r,s}(\overline{Y}, f)|| \leq \text{const} \sum_{i=1}^{m} \frac{1}{i^2}\omega(g; \frac{i}{m}), \tag{3.3}$$

with $g = f(\cos)$ and some constant independent of f and m.
We observe that, if we interpolate only on the zeros of q_{2m+1} and $H_m(Y; f)$ is the Hermite-Fejér interpolating polynomial, then it is easy to prove that the Lebesgue constants verify

$$||H_m(Y)|| \leq m^{4\tau+4}, \quad \tau = max\{\alpha, \beta\}.$$

We also remark that in the particular case $\alpha = \beta = -1/2$, that is $r = s = 1$, the polynomial $H_{m,r,s}(\overline{Y}; f)$ reduces to quasi Hermite-Fejér interpolating polynomial on the zeros of U_{2m+1}.

3.2. HERMITE INTERPOLATION

Analogously, if f is a given differentiable function on $[-1,1]$, we denote by $\mathcal{H}_m(\overline{Y}; f)$ the corresponding Hermite interpolating polynomial on the zeros of q_{2m+1}.
Similarly we denote by $\mathcal{H}_{m,r,s}(\overline{Y}; f)$ the Hermite interpolating polynomial of degree $4m + r + s + 1$ based on the nodes (3.1) and defined by

$$\mathcal{H}^{(i)}_{m,r,s}(\overline{Y}; f; t_{k,2m+1}) = f^{(i)}(t_{k,2m+1}), \quad i = 0, 1, \quad k = 1, \dots, 2m+1,$$

$$\mathcal{H}_{m,r,s}(\overline{Y}; f; y_{j,m}) = f(y_{j,m}), \quad j = 1, \dots, s,$$

$$\mathcal{H}_{m,r,s}(\overline{Y}; f; z_{i,m}) = f(z_{i,m}), \quad i = 1, \dots, r.$$

We complete the definition by putting $\mathcal{H}_{m,0,0}(\overline{Y}; f) = \mathcal{H}_m(Y; f)$.

Also for the polynomial $\mathcal{H}_{m,r,s}(\overline{Y}; f)$ we can show convergence results analogous to those ones given above in subsection 2.2.
For instance we recall the following results proved in [4,7].

Theorem 3.2. *Let $f \in C^q([-1,1])$, $q \geq 1$ and $f^{(q)} \in Lip_M\lambda$, $0 < \lambda \leq 1$. Assume that the values $f^{(i)}(-1), i = 0, \dots, s-1$, and $f^{(j)}(1), j = 1, \dots, r-1$, with $\dfrac{q+\lambda}{2} < r, s \leq q+1$ are known and let $y_{j,m} = -1, j = 1, \dots, s$ and $z_{j,m} = 1$, $j = 1, \dots, r$. Then there exist infinitely many Jacobi weights $w^{(\alpha,\beta)}$, with exponents α, β satisfying*

$$-\frac{q+\lambda}{4}+\frac{r}{2}-\frac{3}{2}<\alpha\leq-\frac{q+\lambda}{4}+\frac{r}{2}-1,$$

$$-\frac{q+\lambda}{4}+\frac{s}{2}-\frac{3}{2}<\beta\leq-\frac{q+\lambda}{4}+\frac{s}{2}-1,$$

such that the Hermite polynomial $\overline{\mathcal{H}}_{m,r,s}(\overline{Y};f)$ *defined by*

$$\overline{\mathcal{H}}^{(i)}_{m,r,s}(\overline{Y};f;t_{k,2m+1})=f^{(i)}(t_{k,2m+1}),\quad i=0,1,k=1,\dots,2m+1,$$

$$\overline{\mathcal{H}}^{(i)}_{m,r,s}(\overline{Y};f;-1)=f^{(i)}(-1),\quad i=0,\dots,s-1,$$

$$\overline{\mathcal{H}}^{(i)}_{m,r,s}(\overline{Y};f;1)=f^{(i)}(1),\quad i=0,\dots,r-1,$$

satisfies, for $|x|\leq 1$ *and* $h=0,\dots,\min\{r,s\}$,

$$|\,[f(x)-\overline{\mathcal{H}}_{m,r,s}(\bar{Y};f;x)]^{(h)}\,|\leq \text{const}\,[\frac{\sqrt{1-x^2}}{m}]^{q+\lambda-h}\log m,$$

for some constant independent of f *and* m.

Theorem 3.3. *Let* $f\in C^q$, *with* $q\geq 1$ *and let* $0\leq\ell\leq q$. *Let* w *and* u *be defined by* (2.2) *and* (2.17) *respectively. Then, there exist two integers* r *and* s *defined by*

$$\frac{\ell}{2}+2\alpha-\gamma+1<r<\frac{q}{2}+2\alpha+3$$

and

$$\frac{\ell}{2}+2\beta-\delta+1<s<\frac{q}{2}+2\beta+3$$

such that

$$\int_{-1}^{1}|f^{(\ell)}(x)-H^{(\ell)}_{m,r,s}(\bar{Y};f;x)|u(x)dx\leq\text{const}\frac{E_{m-q}(f^{(q)})}{m^{q-\ell}},$$

with some constant independent of f *and* $m\geq 4q+5$.

References

[1] Balasz, K. and Kilgore, T. (1990) 'Simultaneous approximation of derivatives', Jour. Approx. Theory 60, 231-244.

[2] Bojanic, R. (1971) 'A note on the precision of interpolation by Hermite-Fejér polynomials', Proceedings of the Conference on Constructive Theory of Functions, (Budapest), 69-76.

[3] Bojanic, R., Varma, A.K. and Vértesi P. 'Necessary and sufficient conditions for uniform convergence of quasi Hermite- and extended Hermite-Fejér interpolation', Studia Sci. Math. Hungar. (to appear).

[4] Criscuolo, G., Della Vecchia, B. and Mastroianni, G. (1991) 'Approximation by Extended Hermite-Fejér and Hermite interpolation', to appear on Proceedings of Conference on Approximation Theory, Keckemet, Hungary.

[5] Criscuolo, G., Della Vecchia, B. and Mastroianni, G. (1990) 'Approximation by Hermite-Fejér and Hermite Interpolation', IAM Tec. Rep. n. 74/90, submitted.

[6] Criscuolo, G., Della Vecchia, B. and Mastroianni, G. (1990) 'Hermite interpolation and mean convergence of its derivatives', IAM Tec. Rep. n. 76/90, submitted.

[7] Criscuolo, G., Della Vecchia, B. and Mastroianni, G. (1990) 'Extended Hermite interpolation on Jacobi zeros and mean convergence of its derivatives', IAM Tec. Rep. n. 77/90.

[8] Criscuolo, G., Mastroianni, G. and Occorsio, D. (1990) 'Convergence of extended Lagrange interpolation', Math. Comp. 55, 197-210.

[9] Egervary, E. and Turan, P. (1958) 'Notes on interpolation, V', Acta Math. Acad. Sci. Hungar. 9, 259-267.

[10] Fejér, L. (1916) 'Uber Interpolation', Gottinger Nachrichten, 66-91.

[11] Freud, G. (1968) 'Uber eine Klasse Lagrangescher Interpolations verfahren', Studia Sci. Math. Hungar. 3, 249-255.

[12] Freud, G. (1969) 'Ein Beitrag zur Theorie des Lagrangeschen Interpolations verfahren', Studia Sci. Math. Hungar. 4, 374-384.

[13] Gonska, H.H. and Knoop, H.-B. (1987) 'On Hermite-Fejér interpolation: A Bibliography (1914-1987)', Dept. of Mathematics and Computer Science, Technical Report, Drexel University, Philadelphia, 1-52.

[14] Mastroianni, G. (1990) 'Uniform convergence of derivatives of Lagrange interpolation', submitted.

[15] Mastroianni, G. and Nevai, P. (1991) 'Mean Convergence of Derivatives of Lagrange Interpolation', Jour. of Comp. and Appl. Math. 34, 385-396.

[16] Mastroianni, G. and Vértesi, P. (1990) 'Simultaneous pointwise approximation of Lagrange Interpolation', submitted.

[17] Mate, A. and Nevai, P. (1988) 'Necessary conditions for mean convergence of Hermite-Fejér interpolation', preprint.

[18] Mills, T.M. (1980) 'Some techniques in approximation theory', Math. Sci. 5, 105-120.

[19] Muneer, Y.E. (1987) 'On Lagrange and Hermite interpolation', Acta Math. Hungar. 49, 293-305.

[20] Neckermann, L. and Runk, P.O. (1985) 'Uber Approximationseigenschaften differenzierter Hermitescher Interpolationspolynome mit Jacobischen Abszissen', Acta Sci. Math. 48, 361-373.

[21] Nevai, P. and Vértesi, P. (1985) 'Mean convergence of Hermite-Fejér interpolation', J. Math. Anal. and Appl., 105, 26-58.
[22] Nevai, P. and Vértesi, P. (1989) 'Convergence of Hermite-Fejér interpolation at zeros of generalized Jacobi polynomials', Acta Sci. Math., 53, 77-98.
[23] Pottinger, P. (1978) 'On the approximation of functions and their derivatives by Hermite polynomials', J. Approx. Theory 23, 267-273.
[24] Runk, P. and Vértesi, P. (1990) 'Some good point systems for derivatives of Lagrange interpolatory operators', Acta Math. Sci. Hungar. 56, 337-342.
[25] Sakai, R. (1991) 'Hermite-Fejér interpolation prescribing higher order of derivatives', J. Approx. Theory (to appear).
[26] Schonage, A. (1972) 'Zur Konvergenz der Stufenpolynome uber den Nullstellen der Legendre-Polynome', in: Linear Operators and Approximation, ISNM, Vol. 20, Birkhauser (Basel), 448-451.
[27] Srivastava, K. B. (1986) 'Some results in the theory of interpolation using the Legendre polynomial and its derivative', J. Approx. Theory 47, 1-16.
[28] Steinhaus, B. (1987) 'On the C^1−norm of the Hermite interpolation operator', J. Approx. Theory 50, 160-166.
[29] Szabados, J. (1972) 'On Hermite-Fejér interpolation for the Jacobi abscissas', Acta Math. Acad. Sci. Hungar. 23, 449-464.
[30] Szabados, J. (1973) 'On the convergence of the Hermite-Fejér interpolation based on the roots of Legendre polynomials', Acta Sci. Math. Szeged 34, 367-370.
[31] Szabados, J. (1987) 'On the convergence of the derivatives of projection operators', Analysis 7, 349-357.
[32] Szabados, J. and Vértesi, P. (1988) 'Interpolation of functions', World Scientific, Budapest.
[33] Szegö, G. (1939) 'Orthogonal Polynomials', Amer. Math. Soc. no. 23, Providence, Rhode Island.
[34] Vértesi, P. O. H. (1973) 'Notes on the Hermite-Fejér interpolation based on the Jacobi abscissas', Acta Math. Acad. Sci. Hungar., 24, 233-239.
[35] Vértesi, P. (1982) 'Hermite-Fejér type interpolations, IV (Convergence criteria for Jacobi abscissas)', Acta Math. Acad. Sci. Hungar. 39, 83-93.
[36] Vértesi, P. (1989) 'Hermite-Fejér interpolation of higher order I', Acta Math. Hungar. 54, 135-152.
[37] Vértesi, P. 'Hermite and Hermite-Fejér interpolation of higher order II', Acta Math. Hungar. (to appear).
[38] Vértesi, P. and Xu, Y. 'Truncated Hermite interpolation polynomials', Studia Sci. Hungar. (to appear).
[39] Xu, Y. (1989) 'Norm of the Hermite interpolation operator', Approx. Theory IV, vol. II, (ed. by C.K. Chui, L.L. Schumaker, J.D. Ward), 683-686.

NEW RESULTS ON LAGRANGE INTERPOLATION

G. CRISCUOLO and G. MASTROIANNI
Istituto per Applicazioni della Matematica, C.N.R.
Via P.Castellino 111 - 80131 Napoli, Italy
Dipartimento di Matematica, Università della Basilicata
Via N.Sauro 85 - 85100 Potenza, Italy

ABSTRACT. Uniform convergence of Lagrange interpolation at zeros of Jacobi polynomials or at the zeros of product Jacobi polynomials, as well as at the zeros of generalized smooth Jacobi polynomials is investigated. We show that by a simple procedure it is always possible to transform the matrices of these zeros into matrices such that the corresponding Lagrange interpolating polynomial well approximate a given function and its derivatives simultaneously.

1. Introduction

Let $X = \{x_{m,k},\ k = 1, \ldots, m,\ m \in \mathbb{N}\} \subseteq [-1, 1]$ be an infinite triangular matrix of nodes and f a continuous function. We denote by $\mathcal{L}_m(X; f)$ the Lagrange interpolation polynomial of degree $m - 1$

$$\mathcal{L}_m(X; f; x) = \sum_{k=1}^{m} f(x_{m,k})\ell_{m,k}(X; x),$$

where $\ell_{m,k}(X)$ are the fundamental polynomials of Lagrange interpolation.
If the elements of X are the zeros of the orthogonal polynomials $p_m(w)$, $m \in \mathbb{N}$, with respect to a given weight w, then we will use the other notation $\mathcal{L}_m(w; f)$.
Together with the Lagrange interpolation polynomial we define the Lebesgue constants $\|\mathcal{L}_m(X)\|$ in the supremum norm, as

$$\|\mathcal{L}_m(X)\| := \sup_{\|f\| \leq 1} \|\mathcal{L}_m(X; f)\| = \max_{-1 \leq x \leq 1} \sum_{k=1}^{m} |\ell_{m,k}(X; x)|.$$

The Lebesgue constants are the crucial point in the convergence to continuous functions of the Lagrange interpolation polynomials.
On the other hand, it is

$$\|\mathcal{L}_m(X; f)\| \leq (1 + \|\mathcal{L}_m(X)\|)E_{m-1}(f).$$

Here $E_{m-1}(f)$ denotes the error of the best uniform approximation, i.e.

$$E_{m-1}(f) := \inf_{p \in \mathbb{P}_{m-1}} \|f - p\|,$$

where f is a continuous function and $\mathbb{P}_{m-1}$ denotes the set of the polynomials of degree $\leq m - 1$.
L. Fejér [6] proved that for the Chebyshev matrix the Lebesgue constant has order $\log m$ and, on the other hand, G. Faber [7] and S. Bernstein [1] showed that the order $\log m$ cannot improved,

S. P. Singh (ed.), Approximation Theory, Spline Functions and Applications, 333–340.

because for any matrix X, it is $\|\mathcal{L}_m(X)\| \geq$ const $\log m$. Then, it is natural give the following definition: X is a *good* matrix when $\|\mathcal{L}_m(X)\| = \mathcal{O}(\log m)$. For instance, the zeros of the Jacobi polynomials $p_m^{\alpha,\beta}$ with $\alpha, \beta \leq -1/2$ are *good* nodes, whereas the zeros of $p_m^{\alpha,\beta}$ with $\max(\alpha,\beta) > -1/2$ are *bad* nodes.

Good matrices are of interest in approximation theory and in numerical methods based on Lagrange interpolation process. Since there exist many *bad* matrices with easily computable nodes, it is natural to consider the following problem "to transform *bad* matrices into *good* ones". In the meaning that we would find a procedure transforming a given matrix X such that $\|\mathcal{L}_m(X)\| >$ const $\log m$ into another matrix X^* with $\|\mathcal{L}_m(X^*)\| = \mathcal{O}(\log m)$.

The solution of the problem for every matrix X is very hard. Here, we give the solution in the following three cases: *i)* X is a Jacobi matrix, *ii)* X is an extended Jacobi matrix, *iii)* X is a generalized smooth Jacobi matrix.

2. Interpolation by Jacobi Matrices

Let $\{x_{m,k}^{\alpha,\beta}\}_{i=1}^{m}$ be the zeros of the Jacobi polynomials $p_m^{\alpha,\beta}$ and

$$X = \{x_{m,k}^{\alpha,\beta},\ \alpha,\beta > -1,\ k = 1,\ldots,m,\ m \in \mathbb{N}\},$$

the corresponding Jacobi matrix. Then, we denote by $\mathcal{L}_m(w^{\alpha,\beta}; f)$ the Lagrange polynomial interpolating the function f at the nodes $x_{m,k}^{\alpha,\beta}$, $k = 1,\ldots,m$. If $\max(\alpha,\beta) \geq -1/2$, then X is a *bad* matrix; but we can transform it into a *good* matrix by the following procedure. Chosen the mth line $x_{m,1}^{\alpha,\beta},\ldots,x_{m,m}^{\alpha,\beta}$ of the basic matrix X, we add $s(\geq 0)$ nodes

$$y_j = y_{m,j} = -1 + (j-1)\frac{1 + x_{m,1}^{\alpha,\beta}}{s}, \qquad j = 1,\ldots,s,$$

between -1 and $x_{m,1}^{\alpha,\beta}$, and $r(\geq 0)$ nodes

$$z_j = z_{m,j} = 1 + (j-1)\frac{1 - x_{m,m}^{\alpha,\beta}}{r}, \qquad j = 1,\ldots,r,$$

between $x_{m,1}^{\alpha,\beta}$ and 1. So, the matrix X is replaced by the matrix X^* having as mth line

$$-1 = y_1 < \cdots < y_s < x_{m,1}^{\alpha,\beta} < \cdots < x_{m,m}^{\alpha,\beta} < z_1 < \cdots < z_r = 1. \tag{2.1}$$

Now, we denote by $\mathcal{L}_{m,r,s}(w^{\alpha,\beta}; f)$ the Lagrange polynomial interpolating the function f at the nodes of (2.1). Obviously the matrix X^* depends on the parameters $r, s \in \mathbb{N}$ and on $\alpha,\beta > -1$. The following theorem provides to determine the previous parameters in order to X^* is a *good* matrix.

Theorem 2.1. *Let $f \in C^q([-1,1])$, $q \geq 0$. Let h be an integer with $0 \leq h \leq q$. If*

$$\frac{h+\alpha}{2} + \frac{1}{4} \leq r \leq \frac{h+\alpha}{2} + \frac{5}{4}, \tag{2.2}$$

$$\frac{h+\beta}{2} + \frac{1}{4} \leq s \leq \frac{h+\beta}{2} + \frac{5}{4}, \tag{2.3}$$

then, for $m \geq 4q+5$

$$\left|f^{(i)}(x) - \mathcal{L}_{m,r,s}^{(i)}(w^{\alpha,\beta}; f; x)\right|$$

$$\leq \text{const} \left[\frac{\sqrt{1-x^2}}{m} + \frac{1}{m^2}\right]^{h-i} \frac{E_{m-q}(f^{(q)})}{m^{q-h}} \log m, \quad |x| \leq 1,\ i = 0,1,\ldots,h, \tag{2.4}$$

with some constant independent of f, x and m.

The proof of this theorem can be found in [8] and the idea of adding nodes near the end points ± 1 is dued to Szabados [11] and Runck and Vértesi [10]. The estimate (2.4) is more careful than those obtained in [10] and [11].
Setting

$$\|f\|_h = \max_{0\le i\le h} \|f^{(i)}\|, \qquad f^{(0)} = f, \quad h \ge 0,$$

from Theorem 2.1 the next corollary follows.

Corollary 2.2. *Let $f \in C^q([-1,1])$, $q \ge 0$. Let h be an integer with $0 \le h \le q$. For any exponents $\alpha, \beta > -1$ of the weight $w^{\alpha,\beta}$ there exist the nonnegative integers r and s such that*

$$\left\| f - \mathcal{L}_{m,r,s}(w^{\alpha,\beta}; f) \right\|_h \le \text{const } E_{m-q}(f^{(q)}) \frac{\log m}{m^{q-h}}. \tag{2.5}$$

Moreover, the integers r and s are defined by (2.2) and (2.3).

From (2.5) we deduce the useful estimate

$$\left\| \mathcal{L}_{m,r,s}(w^{\alpha,\beta}) \right\|_h := \sup_{\|f\|_h \le 1} \left\| \mathcal{L}_{m,r,s}(w^{\alpha,\beta}; f) \right\|_h \le \text{const } \log m, \quad h \ge 0, \tag{2.6}$$

which holds when r and s satisfy the conditions (2.2) and (2.3), i.e. $r = [\frac{h+\alpha}{2} + \frac{5}{4}]$, $s = [\frac{h+\beta}{2} + \frac{5}{4}]$. Sumarizing, any Jacobi matrix X can be transformed into a matrix X^* such that (2.6) holds. We remark that in the introduction we have only requested (2.6) with $h = 0$.
As an example, assume $h = 0$ in (2.6) and consider the systems of interpolation nodes

$$X_1 = \left\{ -\cos \frac{k}{m+1}\pi, \quad k = 1, \dots, m, \quad m \in \mathbb{N} \right\},$$

$$X_2 = \left\{ -\cos \frac{2k}{2m+1}\pi, \quad k = 1, \dots, m, \quad m \in \mathbb{N} \right\},$$

$$X_3 = \left\{ -\cos \frac{2k-1}{2m+1}\pi, \quad k = 1, \dots, m, \quad m \in \mathbb{N} \right\},$$

corresponding to the Jacobi weights

$$w^{1/2,1/2}(x) = \sqrt{1-x^2}, \quad w^{1/2,-1/2}(x) = \frac{\sqrt{1-x}}{\sqrt{1+x}}, \quad w^{-1/2,1/2}(x) = \frac{\sqrt{1+x}}{\sqrt{1-x}},$$

respectively. The Lebesgue constants corresponding to these systems have order m. On the other hand, in view of (2.2) and (2.3) with $h = 0$, we obtain $r = s = 1$ for the first, $s = 0$, $r = 1$ for the second, and finally $s = 1$, $r = 0$ for the last. So, the matrices $X_1^* = X_1 \cup \{\pm 1\}$, $X_2^* = X_2 \cup \{1\}$, $X_3^* = X_3 \cup \{-1\}$ have Lebesgue constants with order $\log m$. Note that X_1^* is the case of well-known Clenshaw's abscissas.

Before we have added s nodes between -1 and the first node $x_{m,1}^{\alpha,\beta}$ of the basic matrix and r nodes between $x_{m,m}^{\alpha,\beta}$ and 1. Now, instead of taking $s + r$ nodes, we add the point -1 with multiplicity s and the point 1 with multiplicity r. In other words, we consider the other problem of finding a polynomial $\widehat{L}_{m,r,s}$ of degree $m + r + s - 1$ satisfing

$$\widehat{L}_{m,r,s}(w^{\alpha,\beta}; f; x_{m,k}^{\alpha,\beta}) = f(x_{m,k}^{\alpha,\beta}), \qquad k = 1, \dots, m,$$

$$\widehat{L}_{m,r,s}^{(j)}(w^{\alpha,\beta}; f; -1) = f^{(j)}(-1), \qquad j = 0, 1, \dots, s-1,$$

$$\widehat{L}_{m,r,s}^{(j)}(w^{\alpha,\beta}; f; 1) = f^{(j)}(1), \qquad j = 0, 1, \dots, r-1,$$

which well approximate f and its derivatives. Obviously, in this case we have a mixed Lagrange-Hermite interpolation process (Lagrange on the starting points and Hermite on the additional nodes ± 1). Further, we must assume $f \in C^q$ with $q \geq r-1$, $q \geq s-1$. For this polynomial we can state the following theorem.

Theorem 2.3. *Let $f \in C^q([-1,1])$, $q \geq 0$, $f^{(q)} \in \mathrm{Lip}\lambda$, $0 < \lambda \leq 1$ with $q \geq r-1$, $q \geq s-1$. If*

$$\frac{q+\alpha+\lambda}{2}+\frac{1}{4} \leq r \leq \frac{q+\alpha+\lambda}{2}+\frac{5}{4},$$

$$\frac{q+\beta+\lambda}{2}+\frac{1}{4} \leq s \leq \frac{q+\beta+\lambda}{2}+\frac{5}{4},$$

then, for $m \geq 4q+5$

$$\left|f^{(i)}(x) - \widehat{L}^{(i)}_{m,r,s}(w^{\alpha,\beta}; f; x)\right|$$
$$\leq \mathrm{const} \left[\frac{\sqrt{1-x^2}}{m}\right]^{q-i+\lambda} \log m, \quad |x| \leq 1, \; i = 0,1,\ldots,\min(r,s), \tag{2.7}$$

with some constant independent of f, x and m.

For the proof see [8]. Remark that the estimate (2.7) is of Gopengauz-Telyakovskii type.

3. Interpolation by Extended Matrices

Let $\{p_m(w)\}_{m=0}^{\infty}$ and $\{p_n(u)\}_{m=0}^{\infty}$ be two sequences of orthogonal polynomials on the interval $[-1,1]$ corresponding to the weights w and u. Then the polynomial $Q_{m+n}(w,u) = p_m(w)p_n(u)$ of degree $m+n$ has zeros belonging to $(-1,1)$, but these are not generally distinct. In the favourable case when the polynomial $Q_{m+n}(w,u)$ has $m+n$ simple zeros, we can define the infinite matrix of nodes

$$X = \{\text{ zeros of } Q_{m+n}(w,u), \; m,n \in \mathbb{N}\},$$

which we term an extended matrix.
The extended matrices turn out to be very useful for solving numerically singular integral equations by collocation methods. Indeed, the integrals which appear in the equations are often dealt with two different interpolation process (one for the quadrature and one for the collocation). In order to avoid divergence and numerical cancellation it is necessary that the collocation points and the quadrature knots are not only distinct, but sufficiently separated. Therefore, the extended matrices, whose construction is generally difficult because of the density of the zeros of the orthogonal polynomials, are very useful when we are able to evaluate the distance between two consecutive knots.
We would give some examples of extended matrices. For any weight w, let $w^*(x) = (1-x^2)w(x)$ and let $Q_{2m+1}(w,w^*) = p_{m+1}(w)p_m(w^*)$. Denote by $\chi_{2m+1,k} = \cos\theta_{2m+1,k} = \cos\theta_k$, $k = 1,\ldots,2m+1$, $0 < \theta_k < \pi$, the zeros of $Q_{2m+1}(w,w^*)$. Further, let $\theta_{2m+1,0} = \theta_0 = \pi$, $\theta_{2m+1,2m+2} = \theta_{2m+2} = 0$. As regarding to the distribution of the zeros of $Q_{2m+1}(w,w^*)$, we recall the following result [4].

Theorem 3.1. *For any weight w we have*

$$\theta_k \neq \theta_j, \qquad k \neq j, \; j,k = 0,1,\ldots,2m+2, \; m \in \mathbb{N}. \tag{3.1}$$

If

$$w(x) = \varphi(x)(1-x)^{\alpha}(1+x)^{\beta} \prod_{i=1}^{M} |x-t_i|^{\gamma_i} \in \mathrm{GSJ}, \tag{3.2}$$

with $\alpha, \beta, \gamma_i > -1$, $i = 1, \ldots, M$, $-1 < t_1 < \cdots < t_M < 1$, $0 < \varphi \in \mathrm{Lip}\lambda$, *then*

$$\theta_k - \theta_{k+1} \sim m^{-1}, \qquad k = 0, 1, \ldots, 2m+1, \; m \in \mathbb{N}. \tag{3.3}$$

We would remark that the inequality (3.1) was already known only for the classical Jacobi weights. Further, the inequality (3.3) allows us to say that in the case w is a generalized smooth Jacobi weight, i.e. w is defined by (3.2), the zeros $\chi_{2m+1,k}$ have a distribution similar to that of the Chebyshev zeros.
The same authors of [4] have constructed many other extended matrices. Recall, for instance, that the same results of the previous theorem hold also for the zeros of $Q_{2m}(w_1, w_2) = p_m(w_1)p_m(w_2)$ with $w_1(x) = (1-x)w(x)$ and $w_2(x) = (1+x)w(x)$ [5].

The extended matrices are not *good* interpolation matrices. For example, for the matrix $X = \{\text{zeros of } T_{m+1}U_m,\ m \in \mathbb{N}\}$ with T_m and U_m Chebyshev polynomials of first and second kind respectively, we have $\|\mathcal{L}_{2m+1}(X)\| \geq \text{const } m$. Nevertheless, for the matrix $X = \{\text{ zeros of } p^{\alpha,\beta}_{m+1}p^{\alpha+1,\beta+1}_m,\ m \in \mathbb{N}\}$, by the previous procedure of adding some nodes we can obtain good results. Denoted by $\chi_{2m+1,1}, \chi_{2m+1,2}, \ldots, \chi_{2m+1,2m+1}$ the zeros of $p^{\alpha,\beta}_{m+1}p^{\alpha+1,\beta+1}_m$, we transform the previous matrix X into the other X^* having as mth line

$$-1 \leq y_1 < \ldots < y_s < \chi_{2m+1,1} < \cdots < \chi_{2m+1,2m+1} < z_1 < \ldots < z_r \leq 1, \tag{3.4}$$

with

$$y_j = y_{m,j} = -1 + (j-1)\frac{1+\chi_{2m+1,1}}{s}, \qquad j = 1, \ldots, s, \quad s \geq 0,$$

$$z_j = z_{m,j} = 1 + (j-1)\frac{1-\chi_{2m+1,2m+1}}{r}, \qquad j = 1, \ldots, r, \quad r \geq 0.$$

So, if we denote by $\mathcal{L}_{2m+1,r,s}(w^{\alpha,\beta}, w^{\alpha+1,\beta+1}; f)$ the Lagrange polynomial interpolating the function f at the points of (3.4), then we can state the following theorem (see [3]).

Theorem 3.2. *Let* $f \in C^q([-1,1])$, $q \geq 0$. *Let* h *be an integer with* $0 \leq h \leq q$.
If

$$\frac{h}{2} + \alpha + 1 \leq r \leq \frac{h}{2} + \alpha + 2, \tag{3.5}$$

$$\frac{h}{2} + \beta + 1 \leq s \leq \frac{h}{2} + \beta + 2, \tag{3.6}$$

then, for $m \geq 4q + 5$

$$\left|f^{(i)}(x) - \mathcal{L}^{(i)}_{2m+1,r,s}(w^{\alpha,\beta}, w^{\alpha+1,\beta+1}; f; x)\right|$$
$$\leq \text{const}\left[\frac{\sqrt{1-x^2}}{m} + \frac{1}{m^2}\right]^{h-i} \frac{E_{m-q}(f^{(q)})}{m^{q-h}} \log m, \quad |x| \leq 1, \; i = 0, 1, \ldots, h, \tag{3.7}$$

with some constant independent of f, x *and* m.

The estimate (3.7) is very similar to (2.4); that is, by adding a suitable number of nodes near ± 1, the Lagrange interpolating polynomial with respect to the extended matrix $X = \{\text{ zeros of } p^{\alpha,\beta}_{m+1}p^{\alpha,\beta}_m,\ m \in \mathbb{N}\}$ can have the same behaviour of that corresponding to a classical Jacobi matrix. In the case of the previous extended matrix the integers r and s must satisfy (3.5) and (3.6) instead of (2.2) and (2.3). Moreover, for these extended matrices can be also established results analogous to those of Theorem 2.3. Similar results have been proved for the other extended matrix $X = \{\text{ zeros of } p^{\alpha+1,\beta}_m p^{\alpha,\beta+1}_m,\ m \in \mathbb{N}\}$ (see [3]).
Nevertheless, the tecnique of adding points near ± 1 does note give always good result: it is not

generally possible to transform *bad* matrices into *good* ones by add of nodes. For example, if $X = \{$ zeros of $T_m T_{m+1},\ m \in \mathbb{N}\}$, then $||\mathcal{L}_m(X)|| = \mathcal{O}(m)$, and adding nodes the order of the Lebesgue constants does not improve. This holds for the more general matrix $X = \{$ zeros of $p_m^{\alpha,\beta} p_{m+1}^{\alpha,\beta},\ m \in \mathbb{N}\}$.

4. Interpolation by Generalized Smooth Jacobi matrices

We recall that a *generalized smooth Jacobi* weight w ∈GSG is defined by

$$w(x) = \varphi(x)(1-x)^{\alpha}(1+x)^{\beta} \prod_{i=1}^{M} |x - t_i|^{\gamma_i} \in \text{GSJ}, \qquad x \in [-1,1], \tag{4.1}$$

where $\alpha, \beta, \gamma_i > -1,\ i = 1, \dots, M,\ -1 < t_1 < \dots < t_M < 1$ and $0 < \varphi \in \text{Lip}\lambda$. Now, let $\{p_m(w)\}_{m=0}^{\infty}$ be the system of the orthogonal polynomials corresponding to the weight function w ∈GSJ. We consider the matrix X having as elements the zeros $x_{m,k} = x_{m,k}(w),\ k = 1, \dots, m$ of $p_m(w),\ m \in \mathbb{N}$.
The behaviour of the Lebesgue constants for the weight defined by (4.1) depends besides upon α and β, also upon γ_i. Indeed, it is known that for $\alpha, \beta \leq -1/2$ and $\gamma_i \leq 0,\ i = 1, \dots, M$ it is $||\mathcal{L}_m(w)|| = \mathcal{O}(\log m)$. Nevertheless, if $\gamma_j > 0$ for some index $j \in \{1, \dots, M\}$, then the Lebesgue constant has order higher than $\log m$ (see [9]).
To simplify the study, we cosider the weight

$$w(x) = \frac{|x-t|^{\gamma}}{\sqrt{1-x^2}}, \qquad x \in [-1,1],\ -1 < t < 1, \quad \gamma > -1. \tag{4.2}$$

Obviously, the weight w defined by (4.2) is a particular case of that defined by (4.1). For a such weight we have $||\mathcal{L}_m(w)|| = \mathcal{O}(\log m)$ if $\gamma \leq 0$ and $||\mathcal{L}_m(w)|| = \mathcal{O}(m^{\gamma/2})$ if $\gamma > 0$. Moreover, in this last case, the addition of nodes near the end points ± 1 is useless.
Before, in the case of classical Jacobi weights, when $\max(\alpha, \beta) > -1/2$, we added nodes near ± 1. Now, it is natural to add points near t. Thus, we consider the basic matrix X having as elements the zeros $x_{m,k}$ of $p_m(w)$. For a suitable index m_0 we have $x_{m,i} \leq t \leq x_{m,i+1},\ m \geq m_0$. So, we add the new nodes

$$\tau_j = \tau_{m,j} = \begin{cases} t + j\frac{x_{m,i+1}-t}{\rho+1} & \text{if } t - x_{m,i} \leq x_{m,i+1} - t \\ x_{m,i} + j\frac{t - x_{m,i}}{\rho+1} & \text{if } t - x_{m,i} > x_{m,i+1} - t \end{cases}, \qquad j = 1, \dots, \rho, \quad \rho \geq 0,$$

between $x_{m,i}$ and $x_{m,i+1}$, obtaining the new matrix X^* whose elements of mth line are

$$x_{m,1} < \dots < x_{m,i} < \tau_1 < \dots < \tau_{\rho} < x_{m,i+1} < \dots < x_{m,m}. \tag{4.3}$$

Then, let $\mathcal{L}_{m,\rho}$ the Lagrange polynomial interpolating f at the nodes of (4.3). Now, the problem is to find the number ρ of nodes that we must add so that the starting matrix becames *good*. The answer is given by the following theorem.

Theorem 4.1. *Let w ∈GSJ be defined by* (4.2). *For any exponent γ of the weight w, there exists the integer ρ defined by*

$$\frac{\gamma}{2} \leq \rho \leq \frac{\gamma}{2} + 1,$$

such that for $m \geq 4q + 5$ and $m \geq m_0$

$$||\mathcal{L}_{m,\rho}(w)|| = \mathcal{O}(\log m).$$

This is a particular case of a more general result proved in [2], stating that
"Any bad matrix corresponding to a weight $w \in GSJ$ defined by (4.1) can be transformed into a good matrix."
By Theorem 4.1, we deduce immediately the following error estimate for the Lagrange interpolation

$$\|f - \mathcal{L}_{m,\rho}(w)\| \leq \text{const } E_{m-1}(f) \text{ log} m, \qquad f \in C^0([-1,1]),$$

with $m \geq 4q+5$ and $m \geq m_0$.
We can have also results about the simultaneous approximation. But, in this case it is necassary to add nodes also near ± 1. Let $\mathcal{L}_{m,r,s,\rho}(w)$ be the Lagrange polynomial interpolating at the points $x_{m,k}$, $k = 1, \dots, m$ and at the additional nodes

$$y_j = y_{m,j} = -1 + (j-1)\frac{1 + x_{m,1}}{s}, \qquad j = 1, \dots, s, \quad s \geq 0,$$

$$z_j = z_{m,j} = 1 + (j-1)\frac{1 - x_{m,m}}{r}, \qquad j = 1, \dots, r, \quad r \geq 0.$$

$$\tau_j = \tau_{m,j} = x_{m,i} + j\frac{x_{m,i+1} - x_{m,i}}{\rho + 1}, \qquad j = 1, \dots, \rho, \quad \rho \geq 0.$$

Then, we can state the following theorem.

Theorem 4.2. *Let $w \in GSJ$ be defined by (4.2). Let $f \in C^q([-1,1])$, $q \geq 0$. Let h be an integer with $0 \leq h \leq q$.*
If

$$\frac{h}{2} \leq r \leq \frac{h}{2} + 1,$$

$$\frac{h}{2} \leq s \leq \frac{h}{2} + 1,$$

$$\frac{\gamma}{2} \leq \rho \leq \frac{\gamma}{2} + 1,$$

then, for $m \geq 4q+5$ and $m \geq m_0$

$$\left| f^{(i)}(x) - \mathcal{L}^{(i)}_{m,r,s,\rho}(w; f; x) \right|$$

$$\leq \text{const} \left[\frac{\sqrt{1-x^2}}{m} + \frac{1}{m^2} \right]^{h-i} \frac{E_{m-q}(f^{(q)})}{m^{q-h}} \text{log} m, \quad |x| \leq 1, \; i = 0, 1, \dots, h,$$

with some constant independent of f, x and m.

Remark that the number ρ of nodes that we must add near t does not depend to the order of derivative that we would approximate.
Further, also Theorem 4.2 is a particular case of a result proved in [2] for the more general weight w defined by (4.1).

References

[1] Bernstein, S. (1918) 'Quelques remarques sur l'interpolation', Math. Ann., 79, 1–12.

[2] Criscuolo, G. and Mastroianni, G. 'Lagrange interpolation on generalized Jacobi zeros with additional nodes', submitted.

[3] Criscuolo, G. and Mastroianni, G. and Vértesi, P. 'Pointwise simultaneous convergence of the extended Lagrange interpolation with additional knots', submitted.

[4] Criscuolo, G. and Mastroianni, G. and Occorsio, D. (1990) 'Convergence of extended Lagrange interpolation', Math. Comp., 55, 197–212.
[5] Criscuolo, G. and Mastroianni, G. and Occorsio, D. 'Uniform convergence of derivatives of the extended Lagrange interpolation', to appear on Numer. Math.
[6] Fejér, L. (1930) 'Die Abschätzungen eines Polynoms in einem Intervalle', Math. Z., 32, 426–457.
[7] Faber, G. (1914) 'Über die interpolatorische Darstellung stetiger Funktionen', Jahresber. der deutschen Math. Verein., 23, 190–210.
[8] Mastroianni, G. 'Uniform convergence of Lagrange interpolation', submitted.
[9] Nevai, P. (1979) Orthogonal Polynomials, Mem. Amer. Math. Soc., 213, Amer. Mathematical Soc., Providence, RI.
[10] Runck, P.O. and Vértesi, P. (1990) 'Some good point systems for derivatives of Lagrange interpolatory operators', Acta Math. Hung., 56, 337–342.
[11] Szabados, J. 'On the convergence of the derivatives of projection operators', to appear on Analysis.

AMBIGUOUS LOCI IN BEST APPROXIMATION THEORY

F.S. DE BLASI
Department of Mathematics
University of Roma II
00133 - Roma, Italy

J. MYJAK
Department of Mathematics
University of L'Aquila
67100 - L'Aquila, Italy

ABSTRACT. Given a nonempty closed bounded subset X of a Banach space E, we study the ambiguous locus $A_e(X)$ (resp. $A_u(X)$, $A_{wp}(X)$) of X, i.e. the set of all points $z \in \mathsf{E}$ such that the nearest point mapping from X to z fails to have existence (resp. uniqueness, well posedness). If E is uniformly convex, we show that the set $A_{wp}(X)$ is σ-porous in E. Moreover, we prove that for most (in the sense of the Baire category) nonempty closed bounded subsets X of E the set $A_{wp}(X)$ is uncountable and dense in E, provided that E is separable, strictly convex, and of dimension ≥ 2. Finally, under the same assumptions on E, we prove that for most nonempty compact convex subsets X of E the ambiguous locus $A_u(X)$, relative to the farthest point mapping, is uncountable and dense in E.

1. Introduction

We denote by E a real Banach space with norm $\|\cdot\|$, and by 2^{E} the family of all nonempty subsets of E. If $X \subset \mathsf{E}$, by int X and diam X, $X \neq \phi$, we mean the interior of X and the diameter of X, respectively. For $z \in \mathsf{E}$ and $X \in 2^{\mathsf{E}}$, we denote by $P_X(z)$ the metric projection of z onto X, i.e.

$$P_X(z) = \{x \in X \mid \|x - z\| = d_X(z)\} ,$$

where $d_X(z) = \inf\{\|x - z\| \mid x \in X\}$. A point $x_0 \in X$ such that $\|x_0 - z\| = d_X(z)$ is called a *solution* of the nearest point problem

$$\min(z, X) . \tag{1.1}$$

A sequence $\{x_n\} \subset X$ such that $\lim_{n\to+\infty} \|x_n - z\| = d_X(z)$, is called a *minimizing sequence* of (1.1). A nearest point problem (1.1) is said *well posed* if it has a unique solution, say x_0, and every minimizing sequence of (1.1) converges to x_0.

S. P. Singh (ed.), Approximation Theory, Spline Functions and Applications, 341–349.

In a metric space Z, by $B(z,r)$ (resp. $\tilde{B}(z,r)$) we mean an open (resp. closed) ball with center $z \in Z$ and radius $r > 0$.

The following characterization of well posed problems is useful.

PROPOSITION 1.1 [10]. *Let $z \in \mathsf{E}$, and let X be a nonempty closed subset of E. Then, the nearest point problem* $\min(z,X)$ *is well posed if and only if*

$$\inf_{\sigma>0} diam\, L_{z,X}(\sigma) = 0 ,$$

where $L_{z,X}(\sigma) = \{x \in X \mid \|x - z\| \leq d_X(z) + \sigma\}$, $\sigma > 0$.

For $X \in 2^{\mathsf{E}}$, we set:

$$A_e(X) = \{z \in \mathsf{E} | P_X(z) = \phi\} ,$$
$$A_u(X) = \{z \in \mathsf{E} | P_X(z) \text{ contains at least 2 points}\} ,$$
$$A_{wp}(X) = \{z \in \mathsf{E} | \text{ the problem } \min(z,X) \text{ is not well posed}\} .$$

The set $A_e(X)$ (resp. $A_u(X)$, $A_{wp}(X)$) is called the *ambiguous locus* of *existence* (resp. *uniqueness, well posedness*) of X, for the nearest point mapping.

In this note we review some properties concerning ambiguous loci of sets $X \subset \mathsf{E}$. The paper consists of four Sections, with the Introduction. In Section 2 it is shown that, if E is uniformly convex, then the ambiguous locus $A_{wp}(X)$ of each nonempty bounded closed set $X \subset \mathsf{E}$ is σ-porous. In Section 3 we prove that, if E is a strictly convex separable Banach space of dimension ≥ 2 then,for most nonempty bounded closed sets $X \subset \mathsf{E}$, the ambiguous locus $A_{wp}(X)$ is uncountable and dense in E. Moreover, under the same assumptions on E, it is shown that for most nonempty compact sets $X \subset \mathsf{E}$ the ambiguous locus $A_u(X)$ is uncountable and dense in E. Finally, in Section 4, we prove that for most nonempty compact convex sets $X \subset \mathsf{E}$ (E as in Section 3) the ambiguous locus $A_u(X)$, relative to the furthest point mapping, is uncountable and dense in E.

2. Porosity of ambiguous loci

A subset X of E is called *porous* if there exist $\alpha > 0$ and $r_0 > 0$ such that for every $x \in \mathsf{E}$ and every $r \in]0, r_0]$ there is a $y \in \mathsf{E}$ such that $B(y, \alpha r) \subset B(x,r) \cap (\mathsf{E} \setminus X)$. A set X is called *σ-porous* if it is a countable union of porous subsets of E.

Clearly, a σ-porous set $X \subset \mathsf{E}$ is of the Baire first category, the converse being false, in general. Furthermore, if $\mathsf{E} = \mathbf{R}^n$, then each σ-porous set $X \subset \mathsf{E}$ is of Lebesgue measure zero.

Put

$$\mathcal{B}(\mathsf{E}) = \{X \in 2^{\mathsf{E}} | X \text{ is bounded closed}\} .$$

The space $\mathcal{B}(\mathsf{E})$ is equipped with the Hausdorff metric, under which it is complete.

LEMMA 2.1. *Let* E *be a uniformly convex Banach space with modulus of convexity* δ. *Let* $x \in \mathsf{E}$ *and* $r > 0$. *Let* $y \in B(x,r)$, $y \neq x$. *Then, for every* $0 < \sigma < 2\|y-x\|$, *we have*

$$\operatorname{diam} D(x,y;r,\sigma) \leq 2\sigma + 2(r - \|y-x\|)\delta^{-1}\left(\frac{\sigma}{2\|y-x\|}\right),$$

where $D(x,y;r,\sigma) = \tilde{B}(y, r - \|y-x\| + \sigma)\backslash B(x,r)$, *and* δ^{-1} *denotes the inverse function of* δ.

By using Lemma 2.1, one can prove:

LEMMA 2.2. *Let* E *be a uniformly convex Banach space. Let* $z \in \mathsf{E}$ *and* $X \in \mathcal{B}(\mathsf{E})$ *be such that the problem* $\min(z,X)$ *has a unique solution* x_z. *Let* $I_z = \{y \in \mathsf{E} | y = tx_z + (1-t)z, t \in]0,1/2]\}$. *Then, given* $\varepsilon > 0$ *and* $y \in I_z$, *there exists a* $\delta_{z,y}(\varepsilon) > 0$ *such that for every* $u \in B(y, \delta_{z,y}(\varepsilon))$ *we have*

$$\operatorname{diam} L_{u,X}(\delta_{z,y}(\varepsilon)) < \varepsilon .$$

THEOREM 2.1 [7]. *Let* E *be a uniformly convex Banach space. Then, for every* $X \in \mathcal{B}(\mathsf{E})$, *the set* $A_{wp}(X)$ *is* σ*-porous in* E.

Proof. Let $X \in \mathcal{B}(\mathsf{E})$. By virtue of a classical theorem of Stečkin [18], the set

$$E' = \{z \in \mathsf{E} | \min(z,X) \text{ has a unique solution}\}$$

is dense in E. Define

$$E^* = \bigcap_{k\in\mathbb{N}} \bigcup_{z\in E'} \bigcup_{y\in I_z} B(y, \delta_{z,y}(\varepsilon_k)),$$

where $\varepsilon_k = 1/2^k$, and $\delta_{z,y}(\varepsilon_k)$ is given by Lemma 2.2. By virtue of Proposition 1.1, we have

$$A_{wp}(X) \subset \mathsf{E}\backslash E^* .$$

Thus, to complete the proof, it suffices to show that $\mathsf{E}\backslash E^*$, is σ-porous in E.

For $h, k \in \mathbb{N}$, set:

$$E_k = E\backslash \bigcup_{z\in E'} \bigcup_{y\in I_z} B(y, \delta_{z,y}(\varepsilon_k)),$$

$$E_{kh} = \left\{z \in E_k | \frac{1}{h} < d_X(z) < h\right\} .$$

We have

$$\mathsf{E}\backslash E^{*} = \bigcup_{k\in\mathbb{N}} E_k = \bigcup_{k\in\mathbb{N}} \bigcup_{h\in\mathbb{N}} E_{kh} .$$

By using Lemma 2.2, one can show that each set E_{kh} is porous in E, with $\alpha = \min\{\sigma_0, \varepsilon_k/24\}$ and $r_0 = 1/(2h)$, where $\sigma_0 \in]0, 1/2[$ is such that $\delta^{-1}(12\sigma_0) \leq \varepsilon_k/(4h)$. This completes the proof.

Remark 2.1. The notion of a set of Baire first category is based on the notion of a nowhere dense set. The latter was strengthened by Dolzhenko [8] in 1967, (see also Zajíček [20]) who introduced the notion of a porous set, a notion essentially known to Denjoy. For a comprehensive study of nearest point problems $\min(z, X)$, we refer to Borwein and Fitzpatrick [3] and Dontchev and Zolezzi [9]. Variants of Theorem 2.1 have been obtained by Bartke and Berens [2] and Zajíček [20], by using a different approach.

3. Ambiguous loci of the nearest point mapping

A Banach space E is said to satisfy the *Kadec-Klee property* if for every sequence $\{x_n\} \subset \mathsf{E}$ converging weakly to an $x \in \mathsf{E}$, with $\lim_{n\to+\infty} \|x_n\| = \|x\|$, we have $\lim_{n\to+\infty} \|x_n - x\| = 0$.

The following theorem is essentially due to Stečkin, Lau, and Konjagin.

THEOREM 3.1 [18], [15], [17]. *Let E be a reflexive Banach space satisfying the Kadec-Klee property. Then, for every $X \in \mathcal{B}(\mathsf{E})$, the set $A_e(X)$ is of Baire first category.*

The hypothesis of Theorem 3.1 cannot be weakened. Indeed, we have:

THEOREM 3.2 [16], [3]. *Let E be a Banach space which is not both reflexive and satisfying the Kadec-Klee property. Then, there exists a set $X \in \mathcal{B}(\mathsf{E})$ such that $\mathrm{int}A_e(X) \neq \phi$.*

Now we are going to discuss some further results concerning ambiguous loci of sets $X \subset \mathsf{E}$. From now on, we suppose $\dim \mathsf{E} \geq 2$.

A subset X of E is called *everywhere uncountable* in E if, for every $x \in \mathsf{E}$ and $r > 0$, the set $A \cap B(x, r)$ is nonempty and uncountable.

LEMMA 3.1. *Let E be a strictly convex Banach space. Let $a \in \mathsf{E}$, $R > 0$, and $\lambda_0 \in]0, 1[$. Let $y_1, y_2 \in \mathsf{E}$, $y_1 \neq y_2$, be such that $\|y_1 - a\| = \|y_2 - a\| = R$. For $\lambda \in]0, 1[$, set*

$$z_\lambda(t) = a + (1-t)\lambda(y_1 - a) + t\lambda(y_2 - a) , \quad t \in [0, 1] .$$

Then, there exist $\theta > 0$ and $0 < \eta < \min\{\lambda_0, 1-\lambda_0\}$ such that, for every nonempty set $A_i \subset B(y_i, \theta)$, $i = 1,2$ and for every $\lambda \in [\lambda_0 - \eta, \lambda_0 + \eta]$, there exists a $t_\lambda \in]0,1[$ such that

$$d_{A_1}(z_\lambda(t_\lambda)) = d_{A_2}(z_\lambda(t_\lambda)) .$$

THEOREM 3.3 [4]. *Let* E *be a separable strictly convex Banach space. Then the set*

$$\mathcal{B}^0 = \{X \in \mathcal{B}(\mathsf{E}) | A_{wp}(X) \text{ is everywhere uncountable in } \mathsf{E}\}$$

is residual in $\mathcal{B}(\mathsf{E})$.

Proof. For $a \in \mathsf{E}$ and $r > 0$, set

$$\mathcal{N}_{a,r} = \{X \in \mathcal{B}(\mathsf{E}) | A_{wp}(X) \cap B(a,r) \text{ is empty or at most countable}\} .$$

By using Lemma 3.1 and Proposition 1.1, one can show that the set $\mathcal{N}_{a,r}$ is nowhere dense in $\mathcal{B}(\mathsf{E})$. Let $D \subset \mathsf{E}$ be a countable set dense in E. Let Q^+ be the set of all strictly positive rationals. Define

$$\mathcal{B}^* = \bigcap_{\substack{a \in D \\ r \in Q^+}} (\mathcal{B}(\mathsf{E}) \setminus \mathcal{N}_{a,r}) .$$

Clearly, by the Baire category theorem, the set $\mathcal{B}^*$ is dense in $\mathcal{B}(\mathsf{E})$. Let $X \in \mathcal{B}^*$. Let $x \in \mathsf{E}$ and $s > 0$ be arbitrary. Take $a \in D$ and $r \in Q^+$ such that $B(a,r) \subset B(x,s)$. Since $X \notin \mathcal{N}_{a,r}$, the set $A_{wp}(X)$ is nonempty and uncountable, thus $X \in \mathcal{B}^0$. Hence $\mathcal{B}^* \subset \mathcal{B}^0$, and $\mathcal{B}^0$ is residual in $\mathcal{B}(\mathsf{E})$, for $\mathcal{B}^*$ is so. This completes the proof.

Set

$$\mathcal{K}(\mathsf{E}) = \{X \in 2^{\mathsf{E}} | X \text{ is compact}\} .$$

The space $\mathcal{K}(\mathsf{E})$ is endowed with the Hausdorff distance, under which it is complete.

THEOREM 3.4 [4]. *Let* E *be as in Theorem 3.3. Then the set*

$$\mathcal{K}^0 = \{X \in \mathcal{K}(\mathsf{E}) | A_u(X) \text{ is everywhere uncountable in } \mathsf{E}\}$$

is residual in $\mathcal{K}(\mathsf{E})$.

Remark 3.1. The use of the Baire category theorem in problems of geometry seems to go back to Klee [14]. Further developments can be found, among others, in Gruber and Zamfirescu [12], Gruber and Kenderov [13], Zamfirescu [21]. Closed subsets X of E with ambiguous loci $A_e(X)$ dense in E have been discovered by Konjagin [16]. Zamfirescu [22] has shown that for most $X \in \mathcal{K}(\mathbf{R}^n)$, the ambiguous locus $A_u(X)$ is dense in R^n. The uncountability of $A_u(X)$, for most $X \in \mathcal{K}(\mathsf{R}^n)$, has been conjectured by Zamfirescu, in a private communication to the Authors. In [4] it is shown that there exist a residual set $\mathcal{K}^0 \subset \mathcal{K}(\mathsf{R}^n)$ and a set $\Omega_0 \subset \mathsf{R}^n$ of Lebesgue measure zero, such that for every $X \in \mathcal{K}^0$ the ambiguous locus $A_u(X)$ is contained in Ω_0. Some other results on ambiguous loci can be found in [5].

4. Ambiguous loci of the farthest point mapping

For $z \in \mathsf{E}$ and X a nonempty bounded subset of E, set

$$Q_X(z) = \{x \in X \mid \|x - z\| = e_X(z)\} ,$$

where $e_X(z) = \sup\{\|x - z\| \mid x \in X\}$. A point $x_0 \in X$ such that $\|x_0 - z\| = e_X(z)$ is called a *solution* of the farthest point problem

$$\max(z, X) . \tag{4.1}$$

For the problem (4.1) the notions of a maximizing sequence and of well posedness are given as for the nearest point problem (1.1).

The following characterization of well posed problems is useful.

PROPOSITION 4.1 [10]. *Let $z \in \mathsf{E}$, and let X be a nonempty bounded closed subset of E. Then, the farthest point problem* $\max(z, X)$ *is well posed if and only if*

$$\inf_{\sigma > 0} \mathrm{diam}\, M_{z,X}(\sigma) = 0 ,$$

where $M_{z,X}(\sigma) = \{x \in X \mid \|x - z\| \geq e_X(z) - \sigma\}$, $\sigma > 0$.

For X a nonempty bounded subset of E, we set:

$$A_e(X) = \{z \in \mathsf{E} \mid Q_X(z) = \phi\} ,$$
$$A_u(X) = \{z \in \mathsf{E} \mid Q_X(z) \text{ contains at least 2 points}\} ,$$
$$A_{wp}(X) = \{z \in \mathsf{E} \mid \text{ the problem } \max(z, X) \text{ is not well posed}\} .$$

The set $A_e(X)$ (resp. $A_u(X)$, $A_{wp}(X)$) is called the *ambiguous locus* of *existence* (resp. *uniqueness, well posedness*) of X, for the farthest point mapping.

By using the above characterization of well posedness, the results of the previous sections concerning ambiguous loci of sets $X \subset \mathsf{E}$, for nearest point problems, can be extended to the case of farthest point problems. Now, we discuss a case in which ambiguous loci do exist for farthest point problems, but not for nearest point problems.

From now on, we suppose $\dim \mathsf{E} \geq 2$. Set

$$\mathcal{C}(\mathsf{E}) = \{X \in 2^{\mathsf{E}} | X \text{ is compact convex}\} .$$

The space $\mathcal{C}(\mathsf{E})$ is equipped with the Hausdorff distance, under which it is complete.

LEMMA 4.1. *Let $a \in \mathsf{E}$ and $0 < r < R$ be arbitrary. Let $y_1, y_2 \in \mathsf{E}$, $y_1 \neq y_2$, be such that $\|y_1 - a\| = \|y_2 - a\| = R$. Let $X \subset \tilde{B}(a,r)$, $X \in \mathcal{C}(\mathsf{E})$, be any. Set $\Delta = [d/8, d/4]$, where $d = (R-r)/R$. Define:*

$$Z = \overline{co}(X \cup \{y_1, y_2\})$$

and

$$b_\theta(t) = (1-t)a_1(\theta) + ta_2(\theta) , \quad t \in [0,1] ,$$

where $a_i(\theta) = a + \theta(y_i - a)$, $i = 1, 2$, and $\theta \in \Delta$. Then, for every $\varepsilon > 0$, there exists a $\sigma > 0$ such that, for every $Y \in B(Z, \sigma)$ and every $\theta \in \Delta$, we have:

(i) $Q_Y(b_\theta(0)) \subset B(y_2, \varepsilon)$, $\quad Q_Y(b_\theta(1)) \subset B(y_1, \varepsilon)$,

(ii) $Q_Y(b_\theta(t)) \subset B(y_1, \varepsilon) \cup B(y_2, \varepsilon)$, *for every $t \in [0,1]$.*

THEOREM 4.1 [6]. *Let E be a strictly convex separable Banach space. Then the set*

$$\mathcal{C}^0 = \{X \in \mathcal{C}(\mathsf{E}) | A_u(X) \textit{ is everywhere uncountable in } \mathsf{E}\}$$

is residual in E.

Proof. For $a \in \mathsf{E}$ and $r > 0$, set

$$\mathcal{N}_{a,r} = \{X \in \mathcal{C}(\mathsf{E}) | A_u(X) \cap B(a,r) \text{ is empty or at most countable}\} .$$

By using Lemma 4.1, one can show that $\mathcal{N}_{a,r}$ is nowhere dense in $\mathcal{C}(\mathsf{E})$. Then, denoting by D a countable dense subset of E, and by Q^+ the set of the strictly positive rationals, define

$$\mathcal{C}^* = \bigcap_{\substack{a \in D \\ s \in Q^+}} (\mathcal{C}(\mathsf{E}) \setminus \mathcal{N}_{a,s}) .$$

As in the proof of Theorem 3.3, one can show that $\mathcal{C}^* \subset \mathcal{C}^0$, completing the proof.

Remark 4.1. If X is a nonempty bounded closed subset of a uniformly convex Banach space E, then by a result of Asplund [1] and Edelstein [11], the ambiguous locus $A_u(X)$ is of Baire first category in E. Moreover, it has been proved in [7] that $A_u(X)$ is actually a set σ-porous in E.

Remark 4.2. Theorem 4.1 has no analog, for the nearest point mapping. In fact, if E is as in Theorem 4.1, then for every $X \in \mathcal{C}(\mathsf{E})$ the ambiguous locus $A_u(X)$, for the nearest point mapping, is empty.

References

[1] Asplund, E. (1966), *Farthest points in reflexive locally uniformly rotund Banach spaces*, Israel J. Math. 4, pp. 213-216.

[2] Bartke, K. and Berens, H. (1986), *Eine Beschreibung der Nichteindentigkeitsmenge für die beste Approximation in der Euklidischen Ebene*, J. Approx. Theory 47, pp. 54-74.

[3] Borwein, J.M. and Fitzpatrick, S. (1989), *Existence of nearest points in Banach spaces*, Can. J. Math. 41, pp. 702-720.

[4] De Blasi, F.S. and Myjak, J. *Ambiguous loci of the nearest point mapping in Banach spaces*, (submitted).

[5] De Blasi, F.S. and Myjak, J. *Some typical properties of compact sets in Banach spaces*, (submitted).

[6] De Blasi, F.S. and Myjak, J. *Ambiguous loci of the farthest point mapping from compact convex sets*, (submitted).

[7] De Blasi, F.S., Myjak, J., Papini, P.L. *Porous sets in best approximation theory*, J. London Math. Soc. (to appear).

[8] Dolzhenko, E. (1967) *Boundary properties of arbitrary functions*, Izv. Akad. Nauk SSSR Ser. Mat., 31, pp. 3-14.

[9] Dontchev, A. and Zolezzi, T. *Well posed optimization problems*, (to appear).

[10] Furi, M. and Vignoli, A. (1970), *About well posed optimization problems for functionals in metric spaces*, J. Optim. Theory Appl., 5, pp. 225-229.

[11] Edelstein, M. (1966), *Farthest points of sets in uniformly convex Banach spaces*, Israel J. Math., 4, pp. 171-176.

[12] Gruber, P. and Zamfirescu, T. (1990), *Generic properties of compact starshaped*, sets, Proc. Amer. Soc., 108, pp. 207-214.

[13] Gruber, P. and Kenderov, P. (1982), *Approximation of convex bodies by polytopes*, Rend. Circ. Mat. Palermo, 31, pp. 195-225.

[14] Klee, V.L. (1959), *Some new results on smoothness and rotundity in normed linear paces*, Math. Ann., 139, pp. 51-63.

[15] Ka-Sing Lau (1978), *Almost Chebyshev subspaces in reflexive Banach spaces*, Indiana Univ. Math. J., 27, pp. 791-795.

[16] Konjagin, S.V. (1980), *On approximation properties of closed sets in Banach spaces and the characterization of strongly convex spaces*, Soviet Math. Dokl., 21, pp. 418-422.

[17] Konjagin, S.V. (1983), *Sets of points of nonemptyness and continuity of the metric projection*, Matemat. Zametki, 33, pp. 331-338.

[18] Stečkin, S. (1963), *Approximative properties of subsets of Banach spaces* (Russian), Rev. Roum. Math. Pures Appl., 8, pp. 5-8.

[19] Zajíček, T. (1985), *On the Fréchet differentiability of distance functions*, Rend. Circ. Mat. Palermo (2)5, Supplemento, pp. 161-165.

[20] Zajíček, L. (1987/88), *Porosity and σ-porosity*, Real Analysis Exchange, 13(2), pp. 314-350.

[21] Zamfirescu, T. (1985), *Using Baire category in geometry*, Rend. Sem. Mat. Univers. Politecn. Torino 43, 1, pp. 67-88.

[22] Zamfirescu, T. (1990), *The nearest point mapping is single valued nearly everywhere*, Arch. Math. 51, pp. 563-566.

A THEOREM ON BEST APPROXIMATIONS IN TOPOLOGICAL VECTOR SPACES

E. De Pascale - G. Trombetta
University of Calabria
Department of Mathematics
87036 Arcavacata di Rende(CS), Italy

ABSTRACT. We prove a best approximation theorem of Fan type in not necessarily locally convex topological vector spaces. The main tool used is a noncompactness measure, which is invariant with respect to taking the convex hull. We compare our result with a similar result demonstrated by Sehgal and Singh.

1. INTRODUCTION

In what follows E will denote a real Hausdorff topological vector space. In [5] A. Idzik gave the following definition:

Definition 1.1 (cf. Definition 2.2 of [5]) A set $B \subset E$ is convexly totally bounded (c. t. b. for short), if, for every neighbourhood V of $0 \in E$, there exist a finite subset $\{x_i, i \in I\} \subset B$ and a finite family of convex sets $\{C_i, i \in I\}$ such that $C_i \subset V$ for each $i \in I$ and $B \subset \cup \{x_i + C_i, i \in I\}$.

Using the Definition 1.1, he proved:

Theorem 1.2 (cf. Theorem 2.4 of [5]) Let B be a convex subset of E. Assume that $K \subset B$ is a compact set and $f: B \to K$ a continuous function. If $\overline{f(B)}$ is c. t. b., then f has a fixed point in K.

In [2] we introduced a measure of noncompactness, related with the Theorem 1.2. Unfortunately such measure is not invariant with respect to taking the convex hull. This property is of great importance in proving fixed point theorems. For this reason, the authors, in the paper [3] together with H. Weber, have been compelled to introduce the definition stated below.
From here on we will assume, for the sake of simplicity, that the space E is metrizable; d will denote a metric for the topology of E and $B(0, \varepsilon) = \{x \in E, d(0,x) \leq \varepsilon\}$ the closed ball of center 0 and radius ε.

Definition 1.3 For $B \subset E$, $\overline{\gamma_s}(B) = \inf \{\varepsilon > 0$, there exist a finite subset F of E and a convex subset C of $B(0, \varepsilon)$ such that $B \subset F + C\}$ (note that $\inf \emptyset = +\infty$). In particular we call B strongly convexly

S. P. Singh (ed.), Approximation Theory, Spline Functions and Applications, 351–355.

totally bounded (s.c.t.b. for short) if $\overline{\gamma}_S(B) = 0$.

$\overline{\gamma}_S$ is a noncompactness measure invariant with respect to taking the convex hull, even when E is not locally convex.
In [3] we proved a fixed point theorem of Darbo type [1] for functions which are $\overline{\gamma}_S$-contractive. In the present paper we observe that the fixed point theorem in [3] still holds in the context of convex valued multifunctions, which are $\overline{\gamma}_S$-condensing. As consequence we will obtain the main result of this paper: a Fan's best approximation theorem for $\overline{\gamma}_S$-condensing multifunctions.

2. FIXED POINTS

To begin few words about terminology, most of which is standard. We denote by $f : X \multimap Y$ a multifunction between two topological spaces X and Y, with nonempty values. f is uppersemicontinuous (u. s. c.) if, for every closed subset C of Y, $f^{-1}(C) = \{x \in X,\ f(x) \cap C \neq \emptyset\}$ is closed. f is lowersemicontinuous (l. s. c.) if, for every open subset C of Y, $f^{-1}(C)$ is open. f is continuous if it is both u. s. c. and l. s. c. If f is compact valued (i. e. for each $x \in X$, f(x) is compact) and Y is compact and regular, then f is u. s. c. iff the graph of f is closed in X x Y.

<u>Definition 2.1.</u> Let B be a subset of E. A multifunction $f: B \multimap E$ is $\overline{\gamma}_S$-condensing if for every bounded (in the sense of topological vector spaces) subset Z of B, with $\overline{\gamma}_S(Z) \leq \overline{\gamma}_S(f(Z))$, we have f(Z) s. c. t. b.

The following Lemma is, in a sense, classic (for a proof see for example Theorem 1 in [4])

<u>Lemma 2.2.</u> Let B a nonempty convex subset of E and $f : B \multimap B$ such that $\overline{co}\, f(B) \subset B$. Then for every $u \in f(B)$ there exists $L \subset B$ with $L = \overline{co}\, \{f(L) \cup \{u\}\}$.

The subsequent Theorem extends a Theorem proved in [3].

<u>Theorem 2.3</u> Let B a nonempty complete convex subset of E and $f : B \multimap B$ a convex valued γ_S-condensing multifunctions. If f has closed graph and $\overline{co}\, f(B)$ is a bounded subset of B, then f has a fixed point.

<u>Proof.</u> For a fixed $u \in f(B)$, let L be the set whose existence is guaranteed by Lemma 2.2. Since f is γ_S-condensing and $\overline{\gamma}_S(L) = \overline{\gamma}_S(f(L))$, we have that f(L) is s. c. t. b. Let $g : L \multimap L$ be the multifunction whose graph is defined by graph g = graph f $\cap$ (LxL). The graph of g is closed and g(L) is contained in the compact set f(L): so g is u. s. c. By Theorem 4.3 of [6] g has a fixed point x, which is fixed for f too, because g is a submultifunction of f (i. e. for each $x \in L$ $g(x) \subset f(x)$).□

The following fixed point Theorem, more general then Theorem 2.3, still holds.

<u>Theorem 2.4</u> Let $B \subset E$ be a complete convex set and W be a closed

neighbourhood of $u \in B$. Let $f : B \cap W \multimap B$ be a multifunction $\overline{\gamma}_S$-condensing. If f has closed graph, convex values, $\text{co}\, f(B \cap W)$ is bounded and the following boundary condition holds: "$x \in \partial W \cap B$ and $x \in tf(x)+(1-t)u \Rightarrow$ there exists $s > 1$ such that $x \in sf(x) = (1-s)u$", then f has a fixed point.

Proof. Let $X = \{x \in W \cap B$, "there exists $t \in [0,1]$ such that $x \in tf(x)+(1-t)u$"$\}$. The set X is closed. If $x \in X \cap (\partial W \cap B)$ there are $t \in [0,1]$ and $s > 1$ such that $x \in tf(x) + (1-t)u \cap sf(x) + (1-s)u$. Let $\alpha \in [0,1]$ be such that $\alpha s + (1-\alpha)t = 1$. Since $f(x)$ is convex we have

$$x \in \alpha[sf(x) + (1-s)u] + (1-\alpha)[tf(x) + (1-t)u] \subset f(x),$$

and so we are done.

So we can suppose $X \cap (\partial W \cap B) = \emptyset$. Since the space E is completely regular there exists a continuous function $\lambda : E \to [0,1]$ such that $\lambda(x)=0$ for every $x \in X$ and $\lambda(x) = 1$ for every $x \in \partial W \cap B$. Let

$$g(x) = \begin{cases} (1-\lambda(x))f(x) + \lambda(x)u & \text{if } x \in W \cap B \\ u & \text{if } x \in B \setminus W. \end{cases}$$

Then $g: B \multimap B$ has closed graph and $\text{co}\, g(B)$ is bounded. We claim that g is $\overline{\gamma}_S$-condensing. In fact, assume that $Z \subset B$ is bounded and $\overline{\gamma}_S(Z) \leq \overline{\gamma}_S(g(Z))$. We have

$\overline{\gamma}_S(Z \cap W) \leq \overline{\gamma}_S(Z) \leq \overline{\gamma}_S(g(Z)) = \overline{\gamma}_S(g(Z \cap W) \cup \{u\}) \leq \overline{\gamma}_S(\text{co}\{f(Z \cap W) \cup \{u\}\}) = \overline{\gamma}_S(f(Z \cap W))$.

Consequently $f(Z \cap W)$ is s. c. t. b. and the set $g(Z)$ is of the same type. By Theorem 2.3 there exists $x \in B$ such that $x \in g(x)$. If $x \in B \setminus W$ then $x = u \in W \cap B$, which is impossible. So $x \in W \cap B$ and consequently $x \in X$. So we have $\lambda(x) = 0$ and $x \in f(x)$.□

In concluding we will make some necessary remarks about the Definition 2.1 of γ_S-condensing functions. If in the implication of the Definitions 2.1

(1) Z bounded and $\overline{\gamma}_S(Z) \leq \overline{\gamma}_S(f(Z)) \Rightarrow f(Z)$ s. c. t. b.

we strengthen the hypothesis on Z (for example we can ask Z metrically bounded), we obtain obviously a larger class of functions.

Nevertheless to prove a fixed point theorem in this larger class of functions is more difficult, as we must expect. In fact some additional hypothesis are needed to obtain that the set L in the proof of Theorem 2.3 belongs to the class of sets Z considered in the definition of γ_S-condensing. A similar consideration holds if the hypothesis on Z in (1) are weakened.

Another problem in the Definition 2.1 is the substitution of (1) with

(2) Z bounded and $\overline{\gamma}_S(Z) \leq \overline{\gamma}_S(f(Z)) \Rightarrow Z$ s. c. t. b.

(1) and (2) are related by the conjecture:

(3) Z s. c. t. b. and f u. s. c. $\Rightarrow f(Z)$ s. c. t. b.

The authors have raised a similar question in [2]. Actually we can answer negatively to (3).

Counterexample
Let S[0,1] be the space of Lebesgue measurable functions equipped wit the topology of convergence in measure. Let A_n a sequence of measurable subsets of [0,1] such that $\mu(A_n) \to 0$, where μ denotes the Lesbegue mesure. Let $\{a_n\}$, $\{b_n\}$ sequences of real numbers choiced in such a way the set $A = \{x \varepsilon S[0,1], x_n = a_n \chi_{A_n}\}$ is s. c. t. b. and the set $B = \{x \varepsilon S[0,1], y_n = b_n \chi_{A_n}\}$ is not s. c. t. b.

We proved in [3] that such choice is possible.
The function $f: A \cup \{0\} \to B \cup \{0\}$ defined by $f(x_n) = y_n$ for every n and $f(0) = 0$ is continuous, but sends the set A s. c. t. b. in the set B not s. c. t. b.

3. BEST APPROXIMATIONS

In this paragraph we suppose that the space E is not only metrizable but also equipped with a continuous seminorm p.

Definition 3.1 A subset B of E is approximatively p-compact iff for each $y \varepsilon E$ and a net $\{x_\alpha\}$ in B satisfying $p(x_\alpha - y) \to d_p(y,B) = \inf\{p(y-z), z \varepsilon B\}$ there is a subnet $\{x_\beta\}$ and $x \varepsilon B$ such that $x_\beta \to x$.

For more details and informations about approximatively p-compactness see [7], [8] and references therein.

Proposition 3.2 If B is an approximatively p-compact subset of E, then for each $y \varepsilon E$, $P(y)=\{x \varepsilon B, p(y-x) = d_p(y,B)\}$ is nonempty and the multifunction $P:E \multimap B$ is u. s. c.

For a proof see Reich [7].

Theorem 3.3 Let B a nonempty approximatively p-compact, convex, bounded and complete subset of E. Suppose $f : B \multimap E$ be a continuous $\overline{\gamma}_s$-condensing multifunction with convex and compact values. If the metric projection $P : E \multimap B$ is $\overline{\gamma}_s$-nonexpansive on f(B) (i. e. for every $Z \subset f(B)$, $\overline{\gamma}_s(P(Z)) \leq \overline{\gamma}_s(Z)$), then there exists an $x \varepsilon B$ with $d_p(x, f(x)) = d_p(f(x), B)$.

Proof. Define a mapping $g : B \multimap B$ by $g(x) = \cup \{P(y), y \varepsilon f(x)$ and $d_p(f(x),B) = d_p(y,B)\}$. Since f(x) is compact, g(x) is a nonempty subset of P(f(x)). Further f(x) convex implies g(x) convex. In fact if u and v are in g(x), then there exist $y_1, y_2 \varepsilon f(x)$, $u \varepsilon P(y_1)$and $v \varepsilon P(y_2)$ and $p(y_1-u) = d_p(y_1,B) = d_p(f(x),B) = d_p(y_2,B) = p(y_2-v)$. For $t \varepsilon [0,1]$ we have $w(t) = ty_1 + (1-t)y_2 \varepsilon f(x) \cap B$ and $d_p(w(t),B) \leq d_p(w(t), tu + (1-t)v) \leq td_p(y_1,u) + (1-t)d_p(y_2, v) = d_p(f(x), B) \leq d_p(w(t),B)$. Let Z be a bounded subset of B such that $\overline{\gamma}_s(Z) \leq \overline{\gamma}_s(g(Z))$. We have $\overline{\gamma}_s(g(Z)) \leq \overline{\gamma}_s(P(f(Z))) \leq \overline{\gamma}_s(f(Z))$. Consequently f(Z) is s. c. t. b. and P(f(Z)) is of the same type together its subset g(Z). We show that g has closed graph. Let $\{x_\alpha\}$ be a net converging to x and $\{z_\alpha\}$ be a net converging to z such that $z_\alpha \varepsilon g(x_\alpha)$. By the definition of g, there exist $y_\alpha \varepsilon f(x_\alpha)$ such that $z_\alpha \varepsilon P(y_\alpha)$ and

$d_p(f(x_\alpha),B) = d_p(y_\alpha),B)$. Since f is u. s. c. and compact valued we can suppose $y_\alpha \to y \in f(x)$. Since P is u. s. c. and compact valued we can suppose $Z_\alpha \to z \in f(y)$. Further d_p and f are continuous and so $d_p(y_\alpha,B) = d_p(f(x_\alpha),B) \to d_p(f(x),B) = d_p(y,B)$. Consequently $z \in g(x)$. At last, since $\overline{co}\, g(B)$ is bounded we can apply to g Theorem 2.1. to obtain a fixed point $x \in g(x)$, i. e. there exists $x \in B$ such that $d_p(x,f(x)) = d_p(f(x),B)$.□

It is interesting to compare our Theorem 3.3 with the following one, demonstrated by Sehgal and Singh ([8]).

Theorem 3.4 Let B an approximatively p-compact, convex subset of a Hausdorff locally convex space E and $f : B \multimap E$ a continous multifunction with closed and convex values. If f(B) is relatively compact then there exists an $x \in B$ whith $d_p(x,f(x)) = d_p(f(x),B)$.

In our Theorem we have removed the compactness condition on the range of f and the local convexity of the space E.
Note that when E is locally convex, the metric projection P is automatically $\overline{\gamma}_s$-nonexpansive on every relatively compact subset X of E. On the other hand we have assumed E metrizable in all our paper. This restriction on E can be removed, working in the usual way, with a family of semimetrics, generating the topology of E, instead of only with one metric. We have not chosen to do so to avoid notational difficulties.

REFERENCES

1. Darbo G., "*Punti uniti in trasformazioni a codominio non compatto*", Rend. Sem. Mat. Univ. Padova 24 (1955), 84-92.

2. De Pascale E. and Trombetta G., "*Fixed points and best approximation for convexly condensing functions in topological vector spaces*", Rend. Mat. Appl.,Serie VII, 11 (1991), 175-186.

3. De Pascale E., Trombetta G and Weber H., to appear.

4. Hadzic O., "*Some properties of measures of noncompactness in paranormed spaces*", Proc. Amer. Math. Soc. 102 (1988), 843-849.

5. Idzik A., "*On γ-almost fixed theorems. The single valued case*", Bull. Polish Acad. Sci. Math. 35 (1987), 461-464.

6. Idzik A., "*Almost fixed point theorems*", Proc. Amer. Math. Soc. 104 (1988), 779-784.

7. Reich S., "*Approximate selections, best approximations, fixed points and invariant sets*", J. Math. Anal. Appl. 62 (1978), 104-113.

8. Sehgal V. M. and Singh S.P., "*A generalization to multifunctions of Fan's best approximation theorem*", Proc. Amer. Math. Soc. 102 (1988), 534-537.

ON THE CHARACTERIZATION OF TOTALLY POSITIVE MATRICES

M. GASCA AND J.M. PEÑA

Departamento de Matemática Aplicada
University of Zaragoza, Spain

Abstract. We present a survey of recent results on the characterization of totally positive and strictly totally positive matrices. Included are some new characterizations which we have obtained in recent papers by using Neville elimination.

1. Introduction.

In 1987 M. Gasca and G. Mühlbach [9, 15] used the idea of Aitken-Neville interpolation to develop aternative elimination strategies to Gaussian elimination for matrix factorization. The simplest of them was called Neville elimination. In recent years we have studied this elimination strategy [10-12] and proved that it is well suited for the study of totally positive matrices. In particular, we have obtained several characterizations of those matrices.

Totally positive matrices have interesting applications in many fields. For example, Gantmacher and Krein [7,8] showed their relation to vibrations of mechanical systems. Also applications appear in the theory of spline functions and computer aided geometric design (variation diminishing properties, intepolation matrices, etc.). For this reason, remarkable papers on total positivity due to specialists on these fields ([2, 5] among others) have appeared. Applications and theoretical results on total positivity can be found in the fundamental book by S. Karlin [13], influenced by the previous important papers by I.J. Schoenberg.

The survey by T. Ando [1] presents a very complete list of references on totally positive matrices before 1987. One of the main points in the study of this class of matrices has been that of characterizing them in practical terms. In this sense, in 1973 and 1976, C. Cryer [3, 4] obtained interesting characterizations by factorizations and by the nonnegativity of some minors (instead of that of all them, as claimed in the definition). Here we list different types of characterizations, with special emphasis in our own results.

S. P. Singh (ed.), Approximation Theory, Spline Functions and Applications, 357–364.

2. Definitions and notations.

We restrict our consideration to square matrices. The extension to rectangular matrices, when available, is straightforward.

Let A be a real square matrix of order n. When $1 \leq k \leq n$, $Q_{k,n}$ will denote the set of strictly increasing sequences of k natural numbers less than or equal to n:

$$(2.1) \qquad \alpha = (\alpha_i)_{1 \leq i \leq n} \in Q_{k,n} \quad \text{if} \quad (1 \leq)\alpha_1 < \alpha_2 < \ldots < \alpha_n(\leq n).$$

The dispersion $d(\alpha)$ of α is defined by

$$(2.2) \qquad d(\alpha) := \sum_{i=1}^{k-1}(\alpha_{i+1} - \alpha_i - 1) = \alpha_k - \alpha_1 - (k-1),$$

with $d(\alpha) = 0$ for $\alpha \in Q_{1,n}$. $d(\alpha) = 0$ means that α consists of k consecutive integers.

For $\alpha \in Q_{k,n}$, $\beta \in Q_{l,n}$, we denote by $A[\alpha|\beta]$ the $k \times l$ submatrix of A containing rows numbered by α and columns numbered by β.

We recall that A is said to be totally positive (resp. strictly totally positive) iff all its minors are nonnegative (positive). Totally positive (strictly totally positive) matrices will be referred to as TP(STP) matrices.

If A is a lower (resp. upper) triangular matrix, the minors $\det A[\alpha|\beta]$ with $\beta_k \leq \alpha_k$ ($\alpha_k \leq \beta_k$) for all k are called nontrivial minors of A, because all the other minors are equal to zero. We say that A is ΔSTP if A is triangular and its nontrivial minors are all positive.

The essence of Neville elimination is to produce zeros in a column of a matrix by adding to each row an appropriate multiple of the previous one (instead of using a fixed row with a fixed pivot as in Gaussian elimination). Reorderings of the rows are necessary when a nonzero element which is going to be transformed into zero has zero elements above it in its column. The process is described in detail in [10-12].

The complete Neville elimination consists in transforming A in two steps: first A is transformed into an upper triangular matrix U with upper echelon form (see [10]) and then the Neville elimination is applied to U^T. Equivalently we can perform the Neville elimination of U *by columns*. In the case of nonsingular matrices, which is our main interest, U is a nonsingular upper triangular matrix, and the Neville elimination of U^T leads to a nonsingular diagonal matrix.

When after appropriate reorderings of the rows we add to the $(i+1)$th row of A the ith row multiplied by α to get zero in the place $(i+1, k)$, we say that α is the (i,k) multiplier of the Neville elimination of A. The multiplier would be zero if the $(i+1,k)$ entry is already zero (see [12] for a precise definition).

3. Determinantal characterization of totally positive matrices.

Following the definition of TP (resp. STP) matrices, one should check the nonnegativity (positivity) of all minors, whose number is $\binom{2n}{n} - 1$ for a matrix of order n. However, a result due to Fekete [6] (see also [8, 13] and Theorem 2.1 of [3]) states:

– A square matrix of order n is STP iff, for all $1 \leq k \leq n$

$$\det A[\alpha|\beta] > 0 \quad \forall \alpha, \beta \in Q_{k,n} \quad \text{with} \quad d(\alpha) = d(\beta) = 0. \tag{3.1}$$

This means that the positivity of all minors with consecutive rows and columns, whose number is $\frac{n(n+1)(2n+1)}{6}$, is sufficient for A to be STP. For example for $n = 6$ this number is 91 instead of $\binom{12}{6} - 1 = 923$.

This result was improved by us in the Theorem 4.1 of [10]:

– A square matrix of order n is STP iff, for all $1 \leq k \leq n$,

$$\begin{cases} \det A[\alpha|1, 2, \ldots, k] > 0 & \forall \alpha \in Q_{k,n} \quad \text{with } d(\alpha) = 0, \\ \det A[1, 2, \ldots, k|\beta] > 0 & \forall \beta \in Q_{k,n} \quad \text{with } d(\beta) = 0. \end{cases} \tag{3.2}$$

The number of minors to be checked with this criterion is $n(n+1)$, that is 30 for $n = 6$.

This result can be interpreted in the sense that the minors appearing in (3.2), called by us in [10] column-initial and row-initial minors, play in total positivity a similar role to that of the leading principal minors $A[1, 2, \ldots, k|1, 2, \ldots, k]$ in positive definiteness of symmetric real matrices.

The first part (resp. the second part) of (3.2) is necessary and sufficient for a lower (upper) triangular matrix to be ΔSTP (see [3]).

The corresponding result for TP matrices was given in Theorem 1.3 of [4]:

– If A is an $n \times n$ matrix of rank m, then A is TP iff, for all $1 \leq k \leq n$,

$$\det A[\alpha|\beta] \geq 0 \qquad \forall \alpha, \beta \in Q_{k,n} \quad \text{with} \quad d(\beta) \leq n - m. \tag{3.3}$$

In the particular case of a nonsingular matrix A, (3.3) states that only minors with consecutive columns have to considered:

$$\det A[\alpha|\beta] \geq 0 \quad \forall \alpha, \beta \in Q_{k,n} \quad \text{with} \quad d(\beta) = 0. \tag{3.4}$$

In the case of a nonsingular lower (resp upper) triangular matrix A, according to Theorem 1.4 of [4], A is TP iff, for all $1 \leq k \leq n$,

$$\det A[\alpha|1, 2, \ldots, k] \geq 0 \qquad \forall \alpha \in Q_{k,n}. \tag{3.5}$$

For tridiagonal matrices it is known from [8] (see also [4], proof of Theorem 4.2) that they are TP iff they are nonnegative and, for all k,

$$\det A[1,2,\dots,k|1,2,\dots,k] \geq 0. \tag{3.6}$$

Consequently, a triangular tridiagonal matrix (that is bidiagonal) is TP iff all its elements are nonnegative.

The characterization of nonsingular TP matrices given by (3.4) was improved in Theorem 3.2 of [10], but again, it has been very recently improved in [11], where the asymmetry of (3.4) and that of Theorem 3.2 of [10] have been replaced by the following result, reminiscent of (3.2):

– A nonsingular matrix A of order n is TP iff it satisfies, for each $1 \leq k \leq n$,

$$\begin{cases} \det A[\alpha|1,2,\dots,k] \geq 0 & \forall \alpha \in Q_{k,n}, \\ \det A[1,2,\dots,k|\beta] \geq 0 & \forall \beta \in Q_{k,n}, \\ \det A[1,2,\dots,k|1,2,\dots,k] > 0. \end{cases} \tag{3.7}$$

If we have a TP matrix A, Theorem 4.3 of [10] gives a very simple condition for A to be STP:

– Let A be a TP matrix of order n. Then A is STP iff it satisfies, for each $1 \leq k \leq n$,

$$\begin{cases} \det A[1,2,\dots,k|n-k+1,n-k+2,\dots,n] > 0 \\ \det A[n-k+1,n-k+2,\dots,n|1,2,\dots,k] > 0. \end{cases} \tag{3.8}$$

4. Algorithmic characterization of totally positive matrices.

Neville elimination, as described briefly in Section 2 and more in detail in [10, 11, 12], has been the way of getting most of our results on total positivity. However it can be directly used to give equivalent characterizations and to compute the necessary minors with low computational cost. So, it was already proved in [15] that Gaussian elimination uses $O(n^5)$ operations to check the strict total positivity of an $n \times n$ matrix, while Neville elimination uses $O(n^4)$. However as pointed out in Remark 4.2 of [10], by (3.2) it can be checked by Neville elimination with only $O(n^3)$ operations (see also [12]).

– A square matrix A of order n is STP iff the complete Neville elimination of A can be carried out without row or column exchanges, all multipliers being positive and obtaining a diagonal matrix with positive diagonal entries.

For total positivity there is a similar characterization (Theorem 5.4 of [10]) replacing the positivity of the multipliers and of the diagonal entries by nonnegativity, and allowing row or column exchanges, under condition that the row or column which has had to be moved to the bottom is zero (see [10] for more details).

The particular case of nonsingular TP matrices (Corollary 5.5 of [10], see also [12]) is completely similar to the STP characterization, replacing only the positivity of the multipliers by nonnegativity.

In the following section we will find matricial descriptions of these characterizations. An algorithm of the type of the above mentioned one for TP matrices, given in terms of factorizations, was given in [4].

5. Factorization of totally positive matrices.

Some of the most well-known characterizations of TP and STP matrices are related to their *LU* factorization. C. Cryer, in [4], extended to TP matrices what was known for STP matrices, thus obtaining the following result:

– A square matrix A is TP (resp. STP) iff it has an *LU* factorization such that L and U are TP (ΔSTP).

Here, as usual, L (resp. U) denotes a lower (upper) triangular matrix.

In the same paper, C. Cryer pointed out (Remark 4.1 of [4]), as we have said at the end of the previous section, that the matrix A is STP iff it can be written in the form

$$A = \prod_{r=1}^{N} L_r \prod_{S=1}^{M} U_s \tag{5.1}$$

where each L_r (resp. U_s) is a TP lower (upper) tridiagonal matrix.

Observe that this result does not mention the relation of N or M with the order n of the matrix A.

The matricial description of Neville elimination [12] allows us to give the following result:

– Let A be a nonsingular matrix of order n. Then A is STP iff it can be expressed in the form:

$$A = H_1 \dots H_{n-1} D K_{n-1} K_{n-2} \dots K_1, \tag{5.2}$$

where for each $i = 1, 2, \ldots, n-1$

$$
(5.3) \quad H_i = \begin{bmatrix} 1 & & & & & & \\ 0 & 1 & & & & & \\ & \ddots & \ddots & & & & \\ & & 0 & 1 & & & \\ & & & h_{i+1}^{(i)} & 1 & & \\ & & & & \ddots & \ddots & \\ & & & & & h_n^{(i)} & 1 \end{bmatrix}, \quad K_i = \begin{bmatrix} 1 & 0 & & & & & \\ & \ddots & \ddots & & & & \\ & & 1 & 0 & & & \\ & & & 1 & k_i^{(i)} & & \\ & & & & \ddots & \ddots & \\ & & & & & 1 & k_{n-1}^{(i)} \\ & & & & & & 1 \end{bmatrix},
$$

with $h_r^{(i)} > 0$ for $r > i$, $k_r^{(i)} > 0$ for $r > i-1$ and D is a diagonal matrix with positive diagonal entries.

There is a similar characterization for a nonsingular matrix A to be TP replacing the positivity of the off-diagonal entries of (5.3) by nonnegativity with an additional condition:

$$
(5.4) \qquad \begin{cases} h_j^{(i)} = 0 \Longrightarrow h_r^{(i)} = 0 & \forall r > j, \\ k_l^{(i)} = 0 \Longrightarrow k_r^{(i)} = 0 & \forall r > l. \end{cases}
$$

However, it seems that there were no results in literature on QR factorization (that is the factorization of a matrix as a product of an orthogonal matrix and an upper triangular matrix) in spite of the interest of this factorization in Numerical Analysis. In [12] we obtained a new characterization of TP matrices in terms of it. But first we have to define a special class of matrices ([12]). A nonsingular matrix A is said to be a γ-matrix (strictly γ-matrix) if it admits an LDU factorization with LD and U^{-1} TP (resp. ΔSTP). Then we have:

– A nonsingular matrix A is TP (resp. STP) iff there exist two orthogonal γ-matrices (strictly γ-matrices) Q_1, Q_2 and two nonsingular upper triangular TP (ΔSTP) matrices R_1, R_2 such that

$$
(5.5) \qquad A = Q_1 R_1, \qquad A^T = Q_2 R_2.
$$

6. Characterization of totally positive matrices by their inverses.

Another characterization of a TP matrix based again upon the Neville elimination process has been obtained in [10] and [12]. In [10], Proposition 5.6, we give the following result:

– Let A be a nonsingular matrix of order n. Then A is TP iff A^{-1} is a product of $2n-1$ bidiagonal matrices with positive elements on (and only on) the main diagonal.

This has been given in a much more precise way in [12], where we have proved that a nonsingular matrix A is STP iff A^{-1} can be expressed in the form

$$A^{-1} = H_1' \dots H_{n-1}' D' K_{n-1}' \dots K_1', \tag{6.1}$$

where H_i', K_i' have the form (5.3) with elements $h_r'^{(i)}$, $k_r'^{(i)}$ negative and D' a diagonal matrix with positive diagonal entries. In fact, the set $\{h_r'^{(i)} | 1 \le i \le n-1\}$, $i+1 \le r \le n$ consists of the opposite numbers to those of the set $\{h_r^{(i)} | 1 \le i \le n-1,\ i+1 \le r \le n\}$ of (5.3), and the same happens with $\{k_r'^{(i)} | 1 \le i \le n-1,\ i \le r \le n-1\}$ and $\{k_r^{(i)} | 1 \le i \le n-1,\ i \le r \le n-1\}$, but this does not means that $h_r'^{(i)} = -h_r^{(i)}$ or $k_r'^{(i)} = -k_r^{(i)}$. D' is the inverse of the diagonal matrix D of (5.2).

A nonsingular matrix A is TP iff A^{-1} can be expressed in the form (6.1) with elements $h_r'^{(i)}$, $k_r'^{(i)}$ nonpositive satisfying the condition (5.4) and with the diagonal matrix D' with positive diagonal entries.

These characterizations mean that the complete Neville elimiantion of A^{-1} can be performed without row or colums exchanges, with nonpositive multipliers in the case of TP matrices (negative for STP matrices) and positive final diagonal elements.

References

[1] T. Ando, Totally positive matrices. *Linear Algebra Appl.* 90:165-219 (1987).

[2] A.S. Cavaretta; W. Dahmen; C.A. Micchelli and P.W. Smith, A factorization theorem for banded matrices. *Linear Algebra Appl.* 39:229-245 (1981).

[3] C. Cryer, *LU*-factorization of totally positive matrices. *Linear Algebra Appl.*7:83-92 (1973).

[4] C. Cryer, Some properties of totally positive matrices. *Linear Algebra Appl.* 15:1-25 (1976).

[5] C. de Boor and A. Pinkus, The approximation of a totally positive band matrix by a strictly totally positive one. *Linear Algebra Appl.*43: 81-98 (1982).

[6] M. Fekete, Über ein Problem von Laguerre,*Rend. Conti Palermo* 34: 89-100 (1913).

[7] F.R. Gantmacher and M.G. Krein, Sur les matrices completement non-negatives et oscillatoires,*Compositio Math.*445-276 (1937).

[8] F.R. Gantmacher and M.G. Krein,*Oszillationsmatrizen, Oszillationskerne und kleine Schwingungen mechanischer Systeme.*Akademie-Verlag, Berlin (1960)

[9] M. Gasca and G. Mühlbach, Generalized Schur complements and a test for total positivity, *Appl. Numer. Math.*3: 215-232 (1987).

[10] M. Gasca and J.M. Peña, Total positivity and Neville elimination, to appear in *Linear Algebra Appl.*(1991).

[11] M. Gasca and J.M. Peña, Total positivity, QR factorization and Neville elimination. Preprint Univ. Zaragoza (1991).

[12] M. Gasca and J.M. Peña, A matricial description of Neville elimination with applications to total positivity. Preprint Univ. Zaragoza (1991).

[13] S. Karlin, *Total Positivity.* Stanford U.P., Stanford, California (1968).

[14] K. Metelmann, *Ein Kriterium für den Nachweis der Totalnichtnegativität von Bandmatrizen,* Linear Algebra Appl. 7: 163-171 (1973).

[15] G. Mühlbach and M. Gasca, A test for strict total positivity via Neville elimination, in *Current Trends in Matrix Theory* (F. Uhlig and R. Groue, Eds.), Elsevier Science, 1987 pp. 225-232.

ITERATIVE METHODS FOR THE GENERAL ORDER COMPLEMENTARITY PROBLEM

G. ISAC
Department of Mathematics
Royal Military College of Saint-Jean
Saint-Jean-sur-Richelieu, Québec
Canada, JOJ 1RO

ABSTRACT. We study some iterative methods for the General Order Complementarity Problem associating some heterotonic operators.

1. INTRODUCTION

The study of complementarity problems is now an interesting domain of Applied Mathematics. With respect to their applications we have three important types of complementarity problems: the **Explicit Complementarity Problem,** the **Implicit Complementarity Problem** and the **Order Complementarity Problem.**

The Explicit Complementarity Problem is very much studied and it has important applications in: Optimization, Game theory, Economics, Engineering, Elasticity theory etc. [4], [10], [11], [14].

The Implicit Complementarity Problem was defined in 1973 by Bensoussan and Lions [2] as the mathematical model of some stochastic optimal control problems.

It was studied in [3], [11], [12], [23] etc. A new class of complementarity problem was recently defined, this is the class of Order Complementarity Problems.

The Linear Order Complementarity Problem was studied in [5] and the nonlinear case in [11], [15]. Now, in this paper we consider the **General Order Complementarity Problem.** This problem is very important since in many practical problems we need to use the complementarity condition simultaneously with respect to several operators.

We study the localization and the appoximation of solutions of a General Order Complementarity Problem using the fixed points and the coupled fixed points of some associated **heterotonic operators.**

2. PRELIMINARIES

We denote by E a Banach space or a locally convex space. Given a closed pointed convex cone $K \subset E$ we denote by "$\leq$" the ordering defined by K, that is, "$x \leq y$" if and only

S. P. Singh (ed.), Approximation Theory, Spline Functions and Applications, 365–380.

if "$y - x \in K$". We have $K = \{x \in E | 0 \leq x\}$.

We suppose that with respect to the ordering "$\leq$" E is a **vector lattice,** that is, for every $x, y \in E$ there exist $x \wedge y$ and $x \vee y$, where "$\wedge$" (resp. "$\vee$") is the "inf" (resp. the "sup"). Supposing E to be a locally convex space we say that K is **normal**, if for every two arbitrary nets $\{x_i\}_{i\in I}$, $\{y_i\}_{i\in I}$ satisfying, $0 \leq x_i \leq y_i$, for every $i \in I$ and $\lim_{i\in I} y_i = 0$, we have that $\{x_i\}_{i\in I}$ is convergent and $\lim_{i\in I} x_i = 0$.

We say that K is **regular** if every order bounded and increasing net $\{x_i\}_{i\in I}$ in K is convergent. When E is a Banach space the nets are replaced by sequences.

The reader find the principal properties of vector lattices and others characterizations of normal or regular cones in [24]. When E is a vector lattice we can define for every $x \in E, x^+ = 0 \vee x, x^- = 0 \vee (-x)$ and $|x| = x\vee(-x)$. We have $x = x^+ - x^-$ and $|x| = x^+ + x^-$. For every x, $y \in E$ we have also,
$|x^+ - y^+| \leq |x - y|$ and $x \vee y = -[(-x) \wedge (-y)]$.

Definition 1

If $(E, \| \|, \leq)$ is an ordered Banach space which is a vector lattice, we say that the norm $\| \|$ is a Riesz norm if;

i) $\| |x| \| = \|x\|$ **, for all** $x \in E$ **(the norm is absolute)**

ii) $0 \leq x \leq y$ **implies** $\|x\| \leq \|y\|$ **, for all** $x, y \in E$ **(the norm is increasing).**

Remark

It is easy to show that a norm $\| \|$ on E is Riesz if and only if $|x| \leq |y|$ implies $\|x\|, \leq \|y\|$ for all $x, y \in E$.

For the inequality $||x| - |y|| \leq |x - y|$ we have that if the norm $\| \|$ is a Riesz norm then the absolute value function is nonexpansive, that is, the absolute value is continuous and we can show that the lattice operations are continuous too.

For others notions of ordered topological vector spaces we recommend [24] and for some notions of nonlinear analysis [26].

3. THE GENERAL ORDER COMPLEMENTARITY PROBLEM

Given n operators $T_1, T_2, \ldots, T_n$:E $\rightarrow$ E and a nonempty subset $D \subset E$, the **General Order Complementarity Problem** associated to $\{T_i\}_1^n$ and D is:

$$GOCP(\{T_i\}_1^m, K, D): \quad \begin{array}{l} \textit{find } x_* \in D \textit{ such that} \\ \wedge (T_1(x_*), T_2(x_*), \ldots, T_n(x_*)) = 0 \end{array}$$

The feasible set of the problem $GOCP(\{T_i\}_1^n, K, D)$ is $F = \{x \in D | T_i(x) \in K,$

for all i = 1, 2, ..., n}.

In many practical problem D = K and in this case we denote our problem by $GOCP(\{T_i\}_1^n, K)$. This problem is very important since it is the mathematical model for many interesting practical problem. We give some examples.

A. The Generalized Complementarity Problem (Cottle-Dantzig) (1970).

Suppose given a function $f:R^n \to R^m$ where $m = \sum_{j=1}^{n} m_j;\ m \geq n;\quad f = (f^j)_{j=1}^n$ and $f^j:R^n \to R^{m_j}$.

The Generalized Complementarity Problem is to find $z \in R^n$ such that for j = 1, 2, ..., n we have

$$\left.\begin{array}{l} z_j \geq 0 \\ f_i^j(z) \geq 0,\ i = 1, 2, \ldots, m_j \\ z_j \cdot \prod_{i=1}^{m_j} f_i^j(z) = 0 \end{array}\right\} \tag{1}$$

Problem (1) was considered in [7] with $f^j = q^j + M^j z$; j = 1, 2, ..., n, where M^j is a $m_j \times n$-*matrix*, $q^j \in R^{m_j}$ and $\sum_{j=1}^{n} m_j = m$.

If we denote $m_* = \max_j\{m_j\}$ and we define the mappings $T_1, T_2, \ldots, T_{m_*}:R^n \to R^n$ as the columns of the following matrix

$$A = \begin{bmatrix} f_1^1(z) & f_2^1(z) & \cdots & f_{m_1}^1(z) & \alpha & \cdots & \alpha \\ f_1^2(z) & f_2^2(z) & \cdots & \alpha & \alpha & \cdots & \alpha \\ \vdots & \vdots & & & & & \vdots \\ f_1^{j_*}(z) & f_2^{j_*}(z) & \cdots & \cdots & \cdots & \cdots & f_{m_*}^{j_*}(z) \\ \vdots & \vdots & & & & & \vdots \\ f_1^n(z) & f_2^n(z) & \cdots & \cdots & f_{m_n}^n(z) & \cdots & \alpha \end{bmatrix}$$

where α is an arbitrary real number such that $\alpha > 0$, then problem (1) is exactly the problem $GOCP(I, T_1, T_2, \ldots, T_{m_x}, R_+^n)$. Problem (1) was studied in [7], [8], [19] and [25].

B. The Mixed Lubrication

Consider the case of mixed lubrication in the context of a journal bearing with elastic support. In this case we have two operators,

$$T_1(X) = H_O + a + L(X) \text{ and}$$
$$T_2(X) = -R(X) + U\left(\frac{\partial H}{\partial x}\right) + V\left(\frac{\partial H}{\partial t}\right)$$

where $R(X) = \nabla . e^{GX} H^3(x,y,t) \nabla X$.

The variables have the following significations: X is the contact pressure, L is a linear integral operator, a is a constant (depending of problem), H is the film thickness, U is the entrainment velocity, G is a piezo-viscous coefficient, x is the spatial variable in the direction of rotation, y is the transverse spatial variable and V is a positive constant depending of problem.

The experiments show that the physical phenomenon is described by three regions:

$$X \geq O,\ T_1(X) = 0,\ T_2(X) \geq 0;\ (\textit{solid-to-solid-contact}) \quad (2)$$
$$X = 0,\ T_1(X) \geq 0,\ T_2(X) \geq 0;\ (\textit{cavity point}) \quad (3)$$
$$X \geq 0,\ T_1(X) \geq 0,\ T_2(X) = 0;\ (\textit{lubrication point}). \quad (4)$$

Considering $E = H^1(\Omega)$ (defined over $L^2(\Omega)$) and $K = \{u \in H^1(\Omega) | u \geq 0$ *a.e. on* $\Omega\}$, the problem to compute the pressure is to solve the problem GOCP (I, T_1, T_2, K). This problem was considered in [16], [20], [21] and recently in [15].

C. The Dynamic Complementarity Problem

This problem was defined by Harrison and Reiman in 1981 [see the references of [18]] and it is very important since this model is a unifying framework for fluid and diffusion approximation of stochastic flow network.

Consider $E = R^m$ with the Euclidean structure and the ordering defined by $K = R^m_+$.

Suppose given $x = \{x(0), x(1), ..., x(n), ...\}$ a sequence in R^m with $x(0) \geq 0$ and R a m × m - matrix.

The **Discrete Dynamic Complementarity Problem** is,

$$\left.\begin{array}{l} \textit{find the sequence } y = \{y(0), y(1), ..., y(n), ...\} \textit{ such that,} \\ (i)\ \ z(n) = x(n) + Ry(n) \geq 0\ ;\ \forall n = 0, 1. ... \\ (ii)\ \ y(0) = 0,\ \Delta y(n) - y(n-1) \geq 0 \textit{ and} \\ (iii)\ \ \langle z(n), \Delta y(n) \rangle = 0;\ \forall n = 0, 1, ... \end{array}\right\} \quad (5)$$

(we consider by convention y(-1) = 0. From (ii) we have that each coordinate $y_j = \{y_j(n) | n = 0, 1, ...\}$ of y is nondecreasing and from (iii) we have that $y_j(n-1)$ strictly increases to $y_j(n)$ only when $z_j(n) = 0$ for j = 1, 2, ..., m.

Consider the vector space $S = \{x | x{:}N \rightarrow R^m\}$, ordered by $K = \{x \in S | x(n) \geq 0;$ $\forall n \in N\}$, where N = 0, 1, 2, ..., n, ...

S is a Fréchet space with respect to the topology defined by the family of seminorms $\{p_r\}_{r\in N}$, where $p_r(x) = \sup_{0\le n\le r} \|x(n)\|$. Also, S is a vector lattice and K is normal and regular.

Now, we define the following operators:

$$T_1(y)(n) = x(n) + Ry(n);\ for\ all\ n\in N,$$
$$T_2(y) = \{0, y(1)-y(0), ..., y(n)-y(n-1), ...\}.$$

The Discrete Dynamic Complementarity Problem is exactly the problem GOCP(T_1, T_2, K, D) where $D = \{y \in S | y(0) = 0\}$. This problem was considered in [18] and [13].

D. The Global Reproduction of an Economical System with several Technologies.

Consider an economy consisting of n production sectors. Every sector is constrainted to use the production of the others. The demand of the sector j(j = 1, 2, ..., n), for the technology k(k = 1, 2, ..., m) for the product of the sector i is given by the function $f_{ij}^k(x_j)$, where x_j is the level of the gross activity performed in the sector j.

We suppose: 1) $f_{ij}^k(x_j)$ are continuous, 2) $f_{ij}^k(0) = 0$ and

3) $0 \le u_j \le v_j$ implies $f_{ij}^k(u_j) \le f_{ij}^k(v_j)$, for every k = 1, 2, ..., m.

The balances between total activities and final demands by the technology k are given by:

$$\left.\begin{array}{l} x_i = \sum_{j=1}^{n} f_{ij}^k(x_j) + y_i \\ i = 1, 2, ..., n \end{array}\right\} \qquad (6)$$

where y_i is the final demand for the sector i.

We define $f_j^k(x_j) = \left[f_{ij}^k(x_j)\right]_{i=1}^n$ for every j = 1, ..., n and $F^k(x) = x - \sum_{j=1}^{n} f_j^k(x_j)$ for every k = 1, ..., m where $x = (x_1, ... x_n)^t$.

We say that a final demand $y^o = (y_1^o,...,y_n^o)^t$ is **attainable** if the set

$$S_{y^o} = \{x \in R_+^n | F^1(x) - y^o \ge 0, ..., F_m(x) - y^o \ge 0\}$$

is nonempty.

For this model, the problem is to show that given $y^o > 0$ the problem $GOCP(T_1, ..., T_m, R_+^n)$ has a solution $x^o > 0$ which is the least element of S_{y^o}, where $T_1(x) = F^1(x) - y^o, ..., T_m(x) = F^m(x) - y^o$. In this case the production x^o is realizing y^o with minimum social cost.

4. THE GENERAL ORDER COMPLEMENTARITY PROBLEM AND HETEROTONIC OPERATORS.

Given m operators $T_1, T_2, \ldots, T_m: E \to E$ and a nonempty set $D \subset E$ we consider the problem $GOCP\ (\{T_i\}_1^m, K, D)$ and we define the operators

$$H(x) = \vee(x - T_1(x), x - T_2(x), \ldots, x - T_m(x)); \text{ for all } x \in E,$$
$$G(x) = \wedge\,(x + T_1(x), x + T_2(x), \ldots, x + T_m(x)); \text{ for all } x \in E$$

Proposition 1

The element $x_o \in D$ is a solution of the problem $GOCP(T_i\}_1^m, K, D)$ if and only if, x_o is a fixed point of H or, if and only if, x_o is a fixed point of G.

Proof. The proof is an elementary calculus. ■

From **Proposition 1** we obtain that it is important to study the operators H and G.

In this sense we consider the following general cases. Let $F_1, F_2, \ldots, F_m$ be m operators from E into E. We denote,

$$F_\wedge(x) = \wedge(F_1(x), F_2(x), \ldots, F_m(x)); \text{ for all } x \in E,$$
$$F_\vee(x) = \vee(F_1(x), F_2(x), \ldots, F_m(x)); \text{ for all } x \in E$$

Definition 2 [Opoitsev] [22]

We say that $T: E \to E$ is heterotonic on a set $D \subset E$ if and only if there exists an operator $\hat{T}: ExE \to E$ such that,

i) $\hat{T}(x, x) = T(x)$; *for all* $x \in D$,

ii) $\hat{T}(x, \cdot)$ **is monotone increasing on D,**

iii) $\hat{T}(\cdot\,, y)$ **is monotone decreasing on D.**

When we say that T is heterotonic we consider that $\hat{T}$ is well defined.

Remarks

1) A monotone increasing (resp. decreasing) operator is heterotonic.
2) If T is heterotonic the choice of $\hat{T}$ is not unique.
3) The sum and the composition of two heterotonic operators is a heterotonic operator.
4) If T is a heterotonic operator and x_* is a fixed point of T then we have
$x_* = T(x_*) = \hat{T}(x_*, x_*)$.

Given a heterotonic operator $T: E \to E$ we say that (x_*, y_*) is a coupled fixed point of T if $\hat{T}(x_*, y_*) = x_*$ and $\hat{T}(y_*, x_*) = y_*$.

This concept was introduced by **Lakshmikantham** and studied in [6], [9] etc.

Remarks

1) Every fixed point is a coupled fixed point.
2) The set of coupled fixed point localizes the set of fixed point.

Definition 3

We say that a coupled fixed point (x_*, y_*) of a heterotonic operator T is minimal and maximal on D if for every coupled fixed point $(\bar{x}, \bar{y})$ of T on D we have $x_* \le \bar{x} \le y_*$ and $x_* \le \bar{y} \le y_*$.

Definition 4

We say that the conical segment $< u_o, v_o > = \{x \in E | u_o \le x \le v_o\}$ is strongly invariant for the heterotonic operator T if $u_o \le \hat{T}(u_o, v_o)$ and $\hat{T}(v_o, u_o) \le v_o$.

Theorem 2

If $F_i = R_i + S_i$; i = 1, 2, ..., m, where R_i is increasing and S_i is decreasing then $F_\wedge$ and $F_\vee$ are heterotonic operators.

Proof. We define,

$$\hat{F}_\wedge(x, y) = \wedge (R_1(x) + S_1(y), \ldots, R_m(x) + S_m(y)); \textit{ and}$$
$$\hat{F}_\vee(x, y) = \vee(R_1(x) + S_1(y), \ldots, R_m(x) + S_m(y)); \textit{ for all } x, y \in E. \quad \blacksquare$$

More general we have also the following result.

Theorem 3

If F_i, for every i = 1, 2, ..., m is heterotonic then $F_\wedge$ and $F_\vee$ are heterotonic. ■

Consider now that $(E, \| \, \|$ is a Banach space.

Definition 5

Let $\sigma = \sigma(t)$ be an increasing real continuous function defined on R_+ and such that $\sigma(0) = 0$. Let $D \subset E$ be a nonempty set. We say that T:D → E is *σ-Hölder* continuous if $\|T(x) - T(y)\| \le \sigma(\|x - y\|)$, for all $x, y \in D$.

Remark

We say that T is of Hölder type if there is a σ such that T is *σ-Hölder* continuous.

Theorem 4

If $(E, \| \|, K)$ is an ordered Banach space which is a vector lattice and the norm $\| \|$ is Riesz, then for every $F_1, F_2, ..., F_m$ operators of Hölder type we have that $F_\wedge$ and $F_\vee$ are operators of Hölder type too.

Proof. It is sufficient to prove the theorem for m = 2. So, suppose that F_1 (resp. F_2) is σ_1 *(resp.* σ_2*)-Hölder* continuous. We have
$F_1(x) \vee F_2(x) = F_1(x) + [F_2(x) - F_1(x)]^+$ *and* $F_1(y) \vee F_2(y) = F_1(y) + [F_2(y) - F_1(y)]^+$ which imply $|F_1(x) \vee F_2(x) - F_1(y) \vee F_2(y)| = |(F_1(x) - F_1(x)) + [F_2(x) - F_1(x)]^+ - [F_2(y) - F_1(y)]^+ \leq 2|F_1(x) - F_1(y)| + |F_2(x) - F_2(y)|$.

Since the norm of E is Riesz we deduce, $\|F_1(x) \vee F_2(x) - F_1(y) \vee F_2(y)\| \leq \leq 2\|F_1(x) - F_1(y)\| + \|F_2(x) - F_2(y)\| \leq 2\sigma_1(\|x - y\|) + \sigma_2(\|x - y\|) = \sigma(\|x - y\|)$, where $\sigma(t) = 2\sigma_1(t) + \sigma_2(t)$, $t \in [0, +\infty)$ is continuous, increasing and $\sigma(0) = 0$. We obtain the same conclusion for $F_\wedge$ since
$F_\wedge(x) = F_1(x) \wedge F_2(x) = -[(-F_1(x)) \vee (-F_2(x))]$ *for all* $x \in E$. ■

By a similar calculus as in the proof of Theorem 4 we obtain also the following result.

Theorem 5

If $(E, \| \|, K)$ is a Banach space which is a vector lattice, the norm $\| \|$ is Riesz and for every i = 1, 2, ..., m, $F_i = R_i + S_i$, where R_i and S_i are of Hölder type then

$$\hat{F}_\vee(x, y) = \vee(R_1(x) + S_1(y), ..., R_m(x) + S_m(y)) \text{ and}$$
$$\hat{F}_\wedge(x, y) = \wedge(R_1(x) + S_1(y), ..., R_m(x) + S_m(y))$$

are of Holder type if we consider on E x E the norm $\|(x, y)\| = \|x\| + \|y\|$. ■

The next result gives a localization and the existence of a solution of the problem $GOCP(\{T_i\}_1^m, K, D)$. Suppose that E is a vector lattice and the norm $\| \|$ is a Riesz norm.

Theorem 6

Let $(E, \| \|, K)$ be a uniformly convex Banach space and suppose $T_1, T_2, ..., T_m$ to be heterotonic operators. The operator H or G associated to the problem $GOCP(\{T_i\}_1^m, K, D)$ is heterotonic and we denote it by T.

If the following assuptions are satisfied:

1) $\hat{T}$ is continuous

2) there exist $x_o, y_o \in D$ such that $< x_o, y_o >$ is strongly invariant for T and $<x_o, y_o> \subseteq D$,

3) T is non-expansive, or condensin or continuous and $dim\ E < +\infty$,

then there exists a coupled fixed point (x_*, y_*) of T minimal and maximal in $<x_o, y_{o*}>$ and a solution u_* of the problem $GOCP(\{T_i\}_1^m, K, D)$ such that $x_* \leq u_* \leq y_*$.

Proof. Using the points x_o, y_o we define the sequences $\{x_n\}_{n\in N}$ and $\{y_n\}_{n\in N}$ by:

$$\left.\begin{array}{l} x_{n+1} = \hat{T}(x_n, y_n);\ \forall n \in N \\ y_{n+1} = \hat{T}(y_n, x_n);\ \forall n \in N \end{array}\right\}.$$

We have, $x_o \le x_1 = \hat{T}(x_o,y_o)$; $y = \hat{T}(y_o,x_o) \le y_o$ and $x_1 = \hat{T}(x_o,y_o) \le y_1 = \hat{T}(y_o,x_o)$ that is we have $x_o \le x_1 \le y_1 \le y_o$. By induction we show that $x_n \le x_{n+1} \le y_{n+1} \le y_n$, for every $n \in N$.

Using the fact that T is heterotonic we can show that $T(<x_n, y_n>) \subset <x_n, y_n>$ for every $n \in N$. Since the norm is Riesz K is normal and regular (since E is reflexive).

Hence because $\{x_n\}$ is increasing, $\{y_n\}$ decreasing and they are order bounded there exist $x_* = \lim_{n\to\infty} x_n$ and $y_* = \lim_{n\to\infty} y_n$.

We have $x_* \le y_*$ and from the definition of $\{x_n\}$ and $\{y_n\}$ and the continuity of $\hat{T}$ we obtain, $x_* = \hat{T}(x_*, y_*)$ and $y_* = \hat{T}(y_*, x_*)$, that is (x_*, y_*) is a coupled fixed point of T.

Let $(\bar{x}, \bar{y})$ be another coupled fixed point of T in $<x_o, y_o>$. We can show that $x_1 \le \bar{x} \le y_1$, $x_1 \le \bar{y} \le y_1$ and by induction $x_n \le \bar{x} \le y_n$, $x_n \le \bar{y} \le y_n$. Computing the limit we deduce $x_* \le \bar{x}, \bar{y} \le y_*$, that is (x, y_*) is minimal and maximal on $<x_o, y_o>$.

Since, we can show that $T(<x_*, y_*>) \subset <x_*. y_*>$, using Browder's or Sadovski's or Schauder's fixed point theorem [26] we obtain that T has a fixed point $u_* \in <x_*, y_*>$ which is a solution of the problem $GOCP(\{T_i\}_1^m, K, D)$ and the theorem is proved. ■

Definition 6

We say that a heterotonic operator $T:E \to E$ is $\alpha-$ (concave, convex) operator $(0 < \alpha < 1)$ if for every $0 < \lambda < 1$ and $x, y \in D$ we have,

1) $\lambda^\alpha \hat{T}(x, y) \le \hat{T}(\lambda x, y)$,
2) $\hat{T}(x, \lambda y) \le \lambda^\alpha \hat{T}(x, y)$.

The concepts of α*-concave* and α*-convex* operator depending of one variable were studied by Krasnoselskii in [17].

Remark

We can show that condition 1 (resp. 2) of **Definition 6** is equivalent to the following 1') (resp. 2'):

1') $\hat{T}(\mu x, y) \le \mu^\alpha \hat{T}(x, y)$; *for all* $\mu > 1$ *and* $x, y \in D$,
2') $\hat{T}(x, \mu y) \ge \mu^{-\alpha} \hat{T}(x, y)$; *for all* $\mu > 1$ *and* $x, y \in D$.

When $(E(\tau, K)$ is an ordered locally convex space with the topolgy τ defined by a family of seminorms $\{p_\alpha\}_{\alpha\in A}$ when K is **normal** we can suppose that every seminorm p_α has the following property:

$$0 \leq x \leq y \Rightarrow p_\alpha(x) \leq p_\alpha(y), \textit{ for every } x, y \in K.$$

We suppose that the family $\{p_\alpha\}_{\alpha \in A}$ is **sufficient.**

Theorem 7

Let $(E(\tau), \{p_\alpha\}_{\alpha \in A})$ be a locally convex space ordered by a regular normal, pointed closed convex cone K. Suppose given m hetertonic operators $T_1, T_2, ..., T_m: E \to E$ and consider the problem $GOCP(\{T_i\}_1^m, K)$. Denote by T the heterotonic operator H or G associated to this problem.

Suppose that T is α- (concave, convex) and $\hat{T}$ is continuous.

If there exist $\mu_0 > 1$ and $u_o > 0$ such that computing $x_o = \mu_o^{-1} u_0, y_o = \mu_o u_o, x_1 = \hat{T}(x_o, y_o)$ and $y_1 = \hat{T}(y_o, x_o)$ we have that $x_o \leq x_1 \leq y_1 \leq y_o$ then there exists x_* such that $T(x_*) = \hat{T}(x_*, x_*) = x_*$ (that is x_* is a solution of the problem $GOCP(\{T_i\}_1^m, K)$). Moreover,

$$x_* = \lim_{n \to \infty} x_n = \lim_{n \to \infty} y_n, \text{ where}$$

$$\left.\begin{aligned} x_n &= \hat{T}(x_{n-1}, y_{n-1}); \forall n \in N \\ y_n &= \hat{T}(y_{n-1}, x_{n-1}); \forall n \in N \end{aligned}\right\}$$

and for every $\alpha \in A$ we have,

$$p_\alpha(x_* - x_n) \leq \mu_o \left(1 - \frac{1}{\mu_o^{2\alpha_n}}\right) p_\alpha(u_o); \forall n \in N.$$

Proof. Consider the sequences $\{x_n\}_{n \in N}$ and $\{y_n\}_{n \in N}$ defined above. We have

$$x_{n-1} \leq x_n \leq y_n \leq y_{n-1}; \textit{ for all } n \in N \tag{7}$$

Indeed, for n = 1 (7) is true by assumption.

Supppose (7) true for n and we prove that it is true for n + 1. We have,

$$\begin{aligned} x_n &= \hat{T}(x_{n-1}, y_{n-1}) \leq \hat{T}(x_n, y_{n-1}) \leq \hat{T}(x_n, y_n) = x_{n+1} \\ x_{n+1} &= \hat{T}(x_n, y_n) \leq \hat{T}(y_n, x_n) = y_{n+1} \textit{ and} \\ y_{n+1} &= \hat{T}(y_n, x_n) \leq \hat{T}(y_{n-1}, x_{n-1}) = y_n. \end{aligned}$$

We prove now that, $\mu_o^{-2\alpha^n} y_n \leq x_n; \textit{ for all } n \in N.$ (8)

Indeed, we have, $\mu_o^{-2\alpha^n} y_o = \mu_o^{-2}(\mu_o u_o) = \mu_o^{-1} \mu_o = x_o.$

Suppose that (8) is true for n and we prove that it is true for n + 1. We have,

$\mu_o^{-2\alpha^{n+1}} y_{n+1} = ; \mu_o^{-2\alpha^{n+1}} \hat{T}(y_n, x_n) \leq \hat{T}(\mu_o^{-2\alpha^n} y_n, x_n) \leq \hat{T}(x_n, x_n) \leq \hat{T}(x_n, \mu_o^{-2\alpha^n} y_n) \leq$
$\mu_o^{-2\alpha^{n+1}} \hat{T}(x_n, y_n) = \mu_o^{-2\alpha^{n+1}} \cdot x_{n+1} \leq x_{n+1}$. *Hence* (8) *is true.*

Since we have, $x_o \leq x_1 \leq \ldots \leq x_n \leq \ldots \leq y_n \leq \ldots \leq y_1 \leq y_o$ we obtain (using the fact that K is regular) that there exist $x_* = \lim_{n\to\infty} x_n$ and $y_* = \lim_{n\to\infty} y_n$ and

$x_n \leq x_* \leq y_* \leq y_n$, for every $n \in N$, which implies $0 \leq y_* - x_* \leq y_n - x_n \leq (1 - \mu_o^{-2\alpha^n}) \leq (1 - \mu_o^{-2\alpha^n})\, y_o \leq \mu_o(1 - \mu_o^{-2\alpha^n}) u_o$.

Since K is normal we obtain, $p_\alpha(y_* - x_*) \leq \mu_o(1 - \mu_o^{-2\alpha^n}) p_\alpha(u_o)$, for all $n \in N$ and all $\alpha \in A$ and hence $x_* = y_*$.

Now since, $0 \leq x_* - x_n \leq y_n - x_n$, for all $n \in N$, we get,

$p_\alpha(x_* - x_n) \leq \mu_o(1 - \mu_o^{-2\alpha^n}) p_\alpha(u_o)$, for all $n \in N$ and all $\alpha \in A$.

Finally, using the continuity of $\hat{T}$ we have $x_* = \hat{T}(x_*, x_*) = T(x_*)$ and the proof is finished. ■

Remark

In the proof of **Theorem 7** the assumption

$$x_o \leq x_1 \leq y_1 \leq y_o \tag{9}$$

is important.

We give now some examples when (9) is satisfied.

a) Suppose $\mu_o > 1$ and $u_o > 0$. Compute $x_o = \mu_o^{-1} u_o$, $y_o = \mu_o u_o$, $x_1 = \hat{T}(x_o, y_o)$ and $y_1 = \hat{T}(y_o, x_o)$.

If $0 < \alpha < \dfrac{1}{2}$ and $\mu_o^{2\alpha-1} u_o \leq \hat{T}(u_o, u_o) \leq \mu_o^{1-2\alpha} u_o$, then we can show that $x_o \leq x_1 \leq y_1 \leq y_o$.

b) Very interesting is the case when E is a Banach space, K regular and *Int K* $\neq \phi$.

In this case K is normal and **if T is heterotonic** α-***(concave, convex)***, with $0 < \alpha < \dfrac{1}{2}$ and $\hat{T}$:*(Int K)*$\times$*(Int K)* $\to$ *Int K*, then for every $z_o \in$ *Int K* there is $0 < t_o < 1$ such that $t_o^{\frac{1}{2}-\alpha} z_o \leq \hat{T}(z_o, z_o) \leq t_o^{-\frac{1}{2}+\alpha} \cdot z_o$.

If we put, in this case, $u_o = z_o$ and $\mu_o = t_o^{-\frac{1}{2}}$, then we have $\mu_o^{2\alpha-1} \cdot u_o \leq \hat{T}(u_o, u_o) \leq \mu_o^{1-2\alpha} \cdot u_o$.

Also, in this case we can show that T has exactly one fixed point in *Int* K.

Consider now, another interesting case.

Suppose that $\hat{T}(x,y) = \Lambda[F_1(x) + F_2(y), G_1(x) + G_2(y)]$

If for every $0 < \lambda < 1$ we have,

$$(\lambda x) + F_2\left(\frac{1}{\lambda}y\right) \geq \lambda^{r_1}\ [F_1(x) + F_2(y)]\ and$$

$$G_1(\lambda x) + G_2\left(\frac{1}{\lambda}\right) \geq \lambda^{r_2}[G_1(x) + G_1(x) + G_2(y)]$$

then we have $\hat{T}\left(\lambda x, \frac{1}{\lambda}y\right) \geq \lambda^{r_o}\hat{T}(x, y)$, where $r_o = \max(r_1, r_2)$. This result is important since **Theorem 7** is true if the assumption that "T is α-(concave, convex)" is replace by "there exists 0 < a < 1 such that $\hat{T}(tx, t^{-1}y) \geq t^a\hat{T}(x, y)$, for all 0 < t < 1 and $x, y \in K$.

5. THE GENERAL ORDER COMPLEMENTARITY PROBLEM AND THE MANN-TOEPLITZ ITERATIONS

We study now the approximation of solutions of the problem $GOCP(\{T_i\}_1^m, K, D)$ when E is a Banach space, by the Mann-Toeplitz iterations. We denote again by T the operator H or G associated to the problem $GOCP(\{T_i\}_1^m, K, D)$.

Theorem 8

Let $D \subset E$ be a nonempty closed convex set and T:D $\rightarrow$ D a mapping such that I-T is σ-Hölder continuous. Let $\{x_n\}_{n\in N}$ be the Mann-Toeplitz itertions associated to T, that is,

$$\left.\begin{aligned} &x_o \in D \text{ and for every } n \in N,\\ &x_{n+1} = (1 - \alpha_n)x_n + \alpha_n T(x_n) \end{aligned}\right\}$$

where $\{\alpha_n\}$ satisfies the following properties:

1) $\alpha_o = 1;$

2) $0 \leq \alpha_n \leq 1$, **for all** $n \in N$ **and**

3) $\sum_{n=0}^{\infty} \alpha_n = +\infty.$

If there exist $n_o \in N$ and $r_o > 0$ such that $r_o \leq \alpha_n$ for every $n \geq n_o$ and $\{x_n\}_{n\in N}$ is convergent to x_*, then x_* is a solution of the problem $GOCP(\{T_i\}_1^m, K, D$.

Proof. We will show that x_* is a fixed point of T. Since D is iclosed we have that $x_* \in D$. For every $n \geq n_o$ we have,

$$\|T(x_*) - T(x_n)\| = \|T(x_*) - x_* + x_* - x_n + x_n - T(x_n)\| \leq$$
$$\leq \|(x_n - T(x_n)) - (x_* - T(x_*))\| + \|x_* - x_n\| \leq$$
$$\leq \|x_* - x_n\| + \sigma(\|x_* - x_n\|).$$

But, since for every $n \geq n_o$ we have $\frac{1}{\alpha_n} \leq \frac{1}{r_o}$ we deduce,

$$\|x_* - T(x_*)\| = \|(x_* - x_n) + (x_n - T(x_n)) + (T(x_n) - T(x_*)\| \leq$$

$$\leq \|x_* - x_n\| + \|x_n - T(x_n)\| + \|T(x_n) - T(x_*)\| \leq$$

$$\leq \|x_* - x_n\| + \frac{1}{\alpha_n}\|x_n - x_{n+1}\| + \|x_* - x_n\| + \sigma(\|x_* - x_n\|) =$$

$$= 2\|x_* - x_n\| + \sigma(\|x_* - x_n\|) + \frac{1}{\alpha_n}\|x_n - x_{n+1}\| \leq$$

$\leq 2\|x_* - x_n\| + \sigma(\|x_* - x_n\|) + \frac{1}{r_o}\|x_n - x_{n+1}\|$. Now, computing the limit we obtain that $T(x_*) = x_*$ and the proof is finished. ■

Remark

Theorem 8 has an interesting consequence for the problem $GOCP(\{T_i\}_1^m, K)$, when K is regular. In this case the definition of $\{x_n\}$ implies that,

$$x_{n+1} - x_n = \alpha_n [T(x_n) - x_n]; \textit{ for all } n \in N.$$

Since $\alpha_n > 0$, when $T(x_n) - x_n \leq 0$, for all $n \in N$ we obtain that $\{x_n\}$ is decreasing and hence it is convergent. The condition $T(x_n) - x_n \leq 0$ when T(x) = H(x) is equivalent to $T_i(x_n) \geq 0$ for all i = 1, 2, ..., m, that is x_n is **feasible**. So, we obtain the following result.

Corollary

If the all assumptions of Theorem 8 are satisfied, K is regular and for every $n \in H$ x_n is feasible then $\{x_n\}$ is convergent to a solution of the problem $GOCP(\{T_i\}_1^m, K)$.

6. THE GENERAL ORDER COMPLEMENTARITY PROBLEM AND THE PROJECTIVE METRIC

The first projective metric was defined by Hilbert in 1895 [see the references of [14]].

We use in this section an extension to locally convex spaces of another projective metric defined by Thompson [14].

Let $(E(\tau), K)$ be an ordered locally convex space where K is **normal** and **sequentially complete.**

We say that $x, y \in K$ (not both zero) are **linked** if there exist finite numbers $\lambda, \mu > 0$ such that $x \leq \lambda y$ and $y \leq \mu x$. This is an equivalence relation. Given $z \in K$ we denote by $\xi(z)$ the equivalence class of z. Generally, $\xi(z)$ is not closed and $0 \notin \xi(z)$ for every $z \neq 0, z \in K$.

On every $\xi(z)$ $(z \in K\backslash\{0\}$ we define the function $\alpha(x, y) = \log \{\max (\alpha, \beta)\}$, for every $x, y \in \xi(z)$, where $\alpha = inf \{\lambda \mid x \leq \lambda y\}$ and $\beta = inf\{\mu \mid y \leq \mu x\}$. We can show that if $x, y \in \xi(z)$ and $x \neq y$ then one of the numbers α, β is larger than unity. The function α is a distance on $\xi(z)$ and we can show that $(\xi(z), d)$ with $z \in K\backslash\{0\}$ is a **complete metric space.**

Consider now the problem $GOCP(\{T_i\}_1^m, K)$ where $T_1, T_2, ..., T_m:E \rightarrow E$ are **heterotonic operators.** Consider the space E x E ordered by $\tilde{K} = \{(x, y) \in E \times E | x \geq 0 \; and \; y \leq 0\}$. In our case $\tilde{K}$ is normal and sequentially complete.

Using the **Banach's contraction principles** we obtain the following result.

We denote by T the heterotonic operator H or G and we define $T_o(x, y) = (\hat{T}(x, y), \hat{T}(y, x))$ for all $x, y \in E$. The operator $\hat{T}$ is from E x E into E x E and it is increasing with repect to the ordering defined by $\tilde{K}$.

Theorem 9

Consider a locally convex space $E(\tau)$, orderd by a normal, closed and sequentially complete cone K and the problem $GOCP(\{T_i\}_1^m, K)$ where T_1, T_2 $T_m:E \rightarrow E$ are heterotonic. If the following assumptions are satisfied:

1) there exist $p_1, p_2 \in (0, 1)$ such that $T_o(\lambda(x, y)) \leq \lambda^p T_o(x, y)$, for every $(x, y) \in \tilde{K}$ and $\lambda \in R_+$, where $p = p_1$ if $\lambda < 1$, and $p = p_2$ if $\lambda > 1$

2) there exist $(x_o, y_o) \in \tilde{K}\backslash(0, 0)$ and $\lambda, \mu > 0$ such that $\mu(x_o, y_o) \leq T_o(x_o, y_o) \leq \lambda(x_o, y_o)$.

Then T has a coupled fixed point (x_*, y_*) in $\xi((x_o, y_o))$ which is unique in this component. We have $y_* \leq 0 \leq x_*$ and if T is compact then the problem $GOCP(\{T_i\}_1^m, K, <y_*, x_*>)$ has a solution. Moreover, $(x_*, y_*) = \lim_{n\to\infty}(x_n, y_n)$ (the limit is with respect to the topology τ), where $\{(x_n, y_n)\}$ is defined by, $(x_n, y_n) = T_o(x_{n-1}, y_{n-1})$.

Also, $d((x_n, y_n), (x_*, y_*)) \leq \dfrac{p^n}{1-p} d((x_o, y_o), (x_1, y_1))$ where $p = \max(p_1, p_2)$. ■

7. CONCLUSION

The heterotnic operators are used for the first time, in this paper, in the study of the problem $GOCP(\{T_i\}_1^m, K, D)$ and our results show that this idea must be continued.

REFERENCES

[1] Banas, J. and Goebel, K. (1980) "Measure of Noncompactness in Banach Spaces, Lecture Notes in Pure and Appl. Math." Vol. 60, Marcel Dekker Inc., New York and Basel.

[2] Bensoussan, A. et Lions, J.L. (1973) "Nouvelle formulation de problèmes de contrôle impulsionnel et applications", C.R. Acad. Sci. Paris, Série A-B 276, 1189-1192.

[3] Bensoussan, A., Gourset M. et Lions, J.L. (1973) "Contrôle impulsionnel et inéquations quasi-variationnelles stationnaires", C.R. Acad. Sci. Paris, Série A-B 276, 1279-1284.

[4] Bershchanskii, Y.M. and Meerov, M.V. (1983) "The Complementarity Problem: Theory and Method of Solution", Automat Remote Control 44 Nr 6, Part 1, 687-710.

[5] Borwein, J.M. and Dempster, M.A.H. (1989) "The Linear Order Complementarity Problem" Math. Operations Research 14 Nr 3, 534-558.

[6] Chen, Y.Z. (1991) "Existence Theorems of Coupled Fixed Points", J. Math. Anal. Appl., 154, 142-150.

[7] Cottle, R.W. and Dantzig, G.B. (1970) "A Generalization of the Linear Complementarity Problem", J. Combinatorial Theory, 8, 79-90.

[8] Ebiefung, A.A. and Kostreva, M.M. (1990) "Global Solvability of Generalized Linear Complementarity Problems and a Related Class of Polynomial Complementarity Problems", Technical Report Nr 595, Clemson University.

[9] Guo, D. and Lakshmikantham, V. (1987) "Coupled Fixed Points of Nonlinear Operators with Applications", Nonlinear Anal. Theory, Meth. Appl. 11 Nr 5, 623-632.

[10] Isac, G. (1985) "Problèmes de complémentarité (en dimension infinité)", Publ. Dép. Math. Univ. Limoges, Limoges, France.

[11] Isac, G. (1986)"Complementarity Problem and Coincidence Equations on Convex Cones", Boll. Un. Mat. Ital. 6 Nr 5-B, 925-943.

[12] Isac, G. (1990) "A Special Variational Inequality and the Implicit Complementarity Problem", J. Fac. Science Univ. Tokyo, Sec. 1A 37 Nr 1, 109-127.

[13] Isac, G. and Goeleven, D. (1991) "The General Order Complementarity Problem: Models and Iterative Methods", Preprint.

[14] Isac, G. and Németh, A.B. (1990) "Projection Methods, Isotone Projection Cones and the Complementarity Problem", J. Math. Anal. Appl., 153, 258-275.

[15] Isac, G. and Kostreva M.M. (1990) "The Generalized Order Complementarity Problem", To Appear, J. Opt. Theory Appl.

[16] Kostreva, M.M. (1984) "Elasto-Hydrodynamic Lubrication: a Nonlinear Complementarity Problem", International J. Numer. Fluids, 4, 377-397.

[17] Krasnoselski, M.A. (1964) "Positive Solutions of Operators Equations", Groningen, Noordhoff.

[18] Mandelbaum, A. (1990) "The Dynamic Complementarity Problem", Preprint, Graduate School of Business, Stanford University.

[19] Mangasarian, O.L. (1978) "Generalized Linear Complementarity Problems as Linear Programs", Technical report Nr. 339, University of Wisconsin-Madison.

[20] Oh, K.P. (1984) "The Numerical Solution of Dynamically Loaded Elasto-hydrodynamic Contact as a Nonlinear Complementarity Problem", J. Tribology 106, 88-95.

[21] Oh, K.P. (1986) "The Formulation of the Mixed Lubrication Problem as a Generalized Nonlinear Complementarity Problem", J. Tribology 108, 598-604.

[22] Opoitsev, V.I. (1979) "A Generalization of the Theory of Monotone and Concave Operators", Trans. Moscow, Math. Soc. Nr. 2, 243-279.

[23] Pang, J.S. (1982) "On the Convergence of a Basic Iterative Method for the Implicit Complementarity Problem", J. Optim. Theory Appl. 37 Nr 2, 149-162.

[24] Peressini, A.L. (1967) "Ordered Topological Vector Spaces", Harper & Row, New York, Evanston and London.

[25] Szanc, B.P. (1989) "The Generalized Complementarity Problem", Ph.D. Thesis, Rensselaer Polytechnic Institute Troy, New York.

[26] Zeidler, E. (1985) "Nonlinear Functional Analysis and its Applications", Vol.I, Springer-Verlag, New York, Berlin, Heidelberg, Tokyo.

WAVELETS, SPLINES AND DIVERGENCE-FREE VECTOR FUNCTIONS.

Pierre-Gilles LEMARIE-RIEUSSET

The aim of this lecture is to give a quick review on wavelets and spline theory, a very quick one since Professor Chui already gave you an extended lecture on spline wavelets [7]. To avoid too much redundancy with Chui's talk, I will speak as few as possible about "classical wavelet theory" - namely the orthonormal wavelet bases provided by the multi-resolution analysis scheme - and a little more about heretical wavelet theories, such as bi-orthogonal wavelets or the pre-wavelets of G. Battle. A very nice example of how to apply such heretical wavelets will be given in the study of divergence-free vector wavelets.

I. Wavelet theory in 1D.

I.1. Orthonormal wavelet bases.

The orthonormal wavelet bases were introduced in 1985 by Y. Meyer and LR [22], but the theory was entirely modified by the introduction of S. Mallat's multi-resolution analysis scheme in 1986 [23]. This notion of multi-resolution analysis, together with the bases of compactly supported wavelets of I. Daubechies [11], led to efficient algorithms (the fast wavelet transform) with a wide-spreading range of applications. The key reference to wavelet theory is the book by Y. Meyer [24].

Roughly speaking, there exist by now two ways of constructing wavelet bases : the multi-resolution analysis scheme of S. Mallat, which is the most popular way in the world of wavelet users (because of its connexion to fast transforms), and the direct construction of pre-wavelets through the solution of a variational problem (the approach of G. Battle).

The multi-resolution analysis scheme of S. Mallat is based on the notion of a pre-scaling function. A *pre-scaling function* is a real-valued function $g(x)$ such that :

(1.1) $g(x)$ is rapidly decreasing ($\forall k \in \mathbb{N}$, $x^k g(x) \in L^2(\mathbb{R})$)

(1.2) For some positive constant A, we have : $\forall \xi \in \mathbb{R}$, $\sum_{k\in\mathbb{Z}} \mid \hat{g}(\xi + 2k\pi) \mid^2 \geq A$

(1.3) For some $(\alpha_k) \in \ell^2(\mathbb{Z})$, $g(\frac{x}{2}) = \sum_{k\in\mathbb{Z}} \alpha_k g(x-k)$.

Then it is easily proved (see [20] by instance) that, if m_0 is defined as

(2.1) $m_0(\xi) = \sum_{k\in\mathbb{Z}} \frac{1}{2}\alpha_k e^{-ik\xi}$

S. P. Singh (ed.), Approximation Theory, Spline Functions and Applications, 381–390.

then

(2.2) $\quad \hat{g}(0) \neq 0$ and $\hat{g}(\xi) = \prod_{j=1}^{\infty} m_0(\frac{\xi}{2^j}).\hat{g}(0)$

(where the Fourier transform $\hat{g}$ is defined as $\hat{g}(\xi) = \int g(x)e^{-ix\xi}dx)$ and that the set of the $g(2^jx-k)$, $j \in \mathbb{Z}$, $k \in \mathbb{Z}$, is total in $L^2(\mathbb{R})$ (i.e. that the closed linear span of those functions is $L^2(\mathbb{R})$). We then define V_j as the closed linear span of the $g(2^jx-k)$, $k \in \mathbb{Z}$. The sequence of $(V_j)_{j\in\mathbb{Z}}$ is a *multi-resolution analysis*, in the sense that :

(3.1) $\quad V_j \subset V_{j+1}$, $\bigcap_{j\in\mathbb{Z}} V_j = \{0\}$, $\bigcup_{j\in\mathbb{Z}} V_j$ is dense in $L^2(\mathbb{R})$.

(3.2) $\quad f \in V_j \Leftrightarrow f(2x) \in V_{j+1}$

(3.3) $\quad V_0$ has a Riesz basis $g(x-k)$, $k \in \mathbb{Z}$.

To the multi-resolution analysis (V_j) is associated an approximation process $f \to (P_jf)_{j\in\mathbb{Z}}$, where P_j is the orthogonal projection of L^2 onto V_j. We then have the obvious compatibility relationship $P_j(P_{j+1}f) = P_{j+1}(P_jf) = P_jf$ and the convergence properties (for $f \in L^2$) $\| P_jf \|_2 \underset{-\infty}{\longrightarrow} 0$ and $\| P_jf - f \|_2 \underset{+\infty}{\longrightarrow} 0$. This orthogonal projection can be described in terms of orthonormal bases. More precisely, a *scaling function* is a function $\varphi(x)$ in V_0 such that the $\varphi(x-k)$, $k \in \mathbb{Z}$, are an orthonormal basis of V_0. (Such functions exist ; we can take by instance : $\hat{\varphi}(\xi) = \frac{\hat{g}(\xi)}{(\sum_{k\in\mathbb{Z}}|\hat{g}(\xi+2k\pi)|^2)^{1/2}}$). Of course, we have $\hat{\varphi}(\xi) = \hat{\varphi}(0)\prod_{j=1}^{\infty} M_0(\frac{\xi}{2^j})$ for some 2π-periodical function.

A *wavelet* is then a function ψ such that the $\psi(x-k)$, $k \in \mathbb{Z}$, are an orthonormal basis of W_0, the orthogonal complement of V_0 in V_1. (Such function exists ; we can take by instance $\hat{\psi}(2\xi) = e^{-i\xi}\bar{M}_0(\xi+\pi)\hat{\varphi}(\xi)$).

From (3.1) and (3.2), we then have that the $2^{j/2}\psi(2^jx-k)$, $j \in \mathbb{Z}$, $k \in \mathbb{Z}$, are an orthonormal basis of $L^2(\mathbb{R})$.

The approach of G. Battle is completely different [3]. He introduces the notion of a *pre-wavelet*, namely a function Λ such that :

(4.1) $\quad$ the set $\Lambda(2^jx-k)$, $j \in \mathbb{Z}$, $k \in \mathbb{Z}$, is total in $L^2(\mathbb{R})$

(4.2) $\quad < \Lambda(2^jx-k) \mid \Lambda(2^{j'}x-k') >= 0$ for $j \neq j'$ (inter-scale orthogonality)

(4.3) $\quad$ the $\Lambda(x-k)$ are a Riesz basis of some closed subspace W_0 of $L^2(\mathbb{R})$.

As we will see later, he constructs such functions through the solution of a variational problem. Then a *wavelet* will be a function ψ such that the $\psi(x-k)$, $k \in \mathbb{Z}$, are an orthonormal basis of W_0 (take by instance $\hat{\psi}(\xi) = \frac{\hat{\Lambda}(\xi)}{(\sum_{k\in\mathbb{Z}}|\hat{\Lambda}(\xi+2k\pi)|^2)^{1/2}}$). The $2^{j/2}\psi(2^jx-k)$, $j \in \mathbb{Z}$, $k \in \mathbb{Z}$, are then an orthonormal wavelet basis of $L^2(\mathbb{R})$.

I.2. Bi-orthogonal wavelets.

The formalism of bi-orthogonal wavelets was introduced in the thesis of J. C. Feauveau [13]. It is based on the notion of conjugate pre-scaling functions. A couple of *conjugate pre-scaling functions* is a couple of real-valued function (g, g^*) such that :

(5.1) g and g^* are compactly supported and of class C^ϵ for some $\epsilon > 0$.

(5.2) $< g(x-k) \mid g^*(x-\ell) >= \delta_{k,\ell}$

(5.3) $g(\frac{x}{2})$ is a linear combination of the $g(x-k)$, $k \in \mathbb{Z}$.

(5.4) $g^*(\frac{x}{2})$ is a linear combination of the $g^*(x-k)$, $k \in \mathbb{Z}$.

(5.5) $\hat{g}(0) = \hat{g}^*(0) = 1$.

Then g generates a multi-resolution analysis (V_j) and g^* a multi-resolution analysis (V_j^*). We may then consider the approximation process $f \to (P_j f)_{j\in\mathbb{Z}}$ where P_j is the projection of $L^2(\mathbb{R})$ onto V_j in the direction of $(V_j^*)^\perp$. The *oblique projection* P_j is then given by the formula :

$$P_j f(x) = \sum_{k\in\mathbb{Z}} 2^j < f(y) \mid g^*(2^j y - k) > g(2^j x - k). \tag{6}$$

Again, we have the compatibility relationships $P_{j+1} \circ P_j = P_j \circ P_{j+1} = P_j$ and the convergence properties $\| P_j f \|_2 \underset{-\infty}{\longrightarrow} 0$, $\| P_j f - f \|_2 \underset{+\infty}{\longrightarrow} 0$. We have also *oblique wavelets* γ, γ^* such that :

$$P_{j+1} f - P_j f = \sum_{k\in\mathbb{Z}} 2^j < f(y) \mid \gamma^*(2^j y - k) > \gamma(2^j x - k). \tag{7}$$

The whole pattern is very close from the multi-resolution analysis scheme of S. Mallat. There exist trigonometrical polynomials $\alpha(\xi)$ and $\alpha^*(\xi)$ such that :

$$\hat{g}(\xi) = \prod_{j=1}^{\infty} \alpha\left(\frac{\xi}{2^j}\right) \quad \text{and} \quad \hat{g}^*(\xi) = \prod_{j=1}^{\infty} \alpha^*\left(\frac{\xi}{2^j}\right) \; ; \tag{8.1}$$

the conjugacy relationship is then expressed by :

$$\alpha(\xi)\bar{\alpha}^*(\xi) + \alpha(\xi+\pi)\bar{\alpha}^*(\xi+\pi) = 1 \; ; \tag{8.2}$$

the oblique wavelets γ and γ^* can be taken as :

$$\hat{\gamma}(2\xi) = e^{-i\xi}\bar{\alpha}^*(\xi+\pi)\hat{g}(\xi) \quad \text{and} \quad \hat{\gamma}^*(2\xi) = e^{-i\xi}\bar{\alpha}(\xi+\pi)\hat{g}^*(\xi) \tag{8.3}$$

and the $2^{j/2}\gamma(2^j x - k)$, $j \in \mathbb{Z}$, $k \in \mathbb{Z}$, and the $2^{j/2}\gamma^*(2^j x - k)$, $j \in \mathbb{Z}$, $k \in \mathbb{Z}$, are then bi-orthogonal unconditional bases of $L^2(\mathbb{R})$. Moreover we have also fast (oblique) wavelet transforms with finite-impulse-responsed filters (for both decomposition and reconstruction).

A very impressive property of bi-orthogonal wavelets is that the approximation process is compatible with differentiation, as proved in [20].

More precisely, if (g_1, g_1^*) is a couple of conjugate pre-scaling functions such that g_1 is $C^{1+\epsilon}$ for some $\epsilon > 0$, then there exists a couple (g_0, g_0^*) of conjugate pre-scaling functions such that :

$$\frac{d}{dx} g_1 = g_0(x) - g_0(x-1) \quad \text{and} \quad \frac{d}{dx} g_0^* = g_1^*(x+1) - g_1^*(x) \tag{9.1}$$

$$\frac{d}{dx} \circ P_j^{(1)} = P_j^{(0)} \circ \frac{d}{dx}. \tag{9.2}$$

The first formula expresses that in V_0 differentiation is equivalent to a finite difference operator. The second formula expresses that, in a sense, approximation by oblique projections and differentiation commute.

These formulae can also be applied to the oblique wavelets defined by (8.3) and to the oblique projector $Q_j = P_{j+1} - P_j$:

$$\frac{d}{dx} \gamma_1 = 4\gamma_0 \quad \text{and} \quad \frac{d}{dx} \gamma_0^* = -4\gamma_1^* \tag{10.1}$$

$$\frac{d}{dx} \circ Q_j^{(1)} = Q_j^{(0)} \circ \frac{d}{dx}. \tag{10.2}$$

I.3. Spline wavelets.

A very natural example of multi-resolution analysis is the sequence of the spaces V_j of square-integrable spline functions of degree N with nodes in $\frac{1}{2^j}\mathbb{Z}$. The pre-scaling function g generating V_0 can be taken as *the normalized B-spline* defined by $\hat{g}(\xi) = \left(\frac{1-e^{-i\xi}}{i\xi}\right)^{N+1}$.

From now on, we may define many different pre-scaling, scaling, pre-wavelets or wavelet functions for this multi-resolution analysis (and also oblique wavelets, thereafter).

Of course, we might just take

$$\hat{\varphi}(\xi) = \frac{\hat{g}(\xi)}{(\sum \mid \hat{g}(\xi + 2k\pi) \mid^2)^{1/2}} \quad \text{and} \quad \hat{\psi}(2\xi) = e^{-i\xi} \bar{M}_0(\xi + \pi) \hat{\varphi}(\xi)$$

as suggested in Section I.1. We then obtain the scaling function and wavelet of *LR* [16]. The main interest of this choice is that φ is then *even* (up to a translation), so that the filters associated to the wavelet transforms will have linear phase ; the main inconvenient is that the filters are neither F.I.R. nor recursive I.I.R., so that we have computational burden.

Now, we have, if N is odd, a very natural pre-scaling function, the *Lagrangian spline* L of I. J. Schoenberg [25] defined by $L(0) = 1$, $L(k) = 0$ for $k \in \mathbb{Z}^*$. Let us now be strange mathematicians for a while and consider the well-known Dirac function δ which is 0 outside 0 and such that $\int_{-\infty}^{+\infty} \delta(x)dx = 1$; let us jump from V_j^N, the multi-resolution analysis of splines of degree N, to V_j^{2N+1} and consider the conjugate pre-scaling "functions" L_{2N+1} (Lagrangian spline) and δ (Dirac function). We have a nice oblique wavelet Λ_{2N+1} defined by (8.3) as $\Lambda_{2N+1}(x) = L_{2N+1}(2x-1)$. Let us now differentiate $(N+1)$-times the spline L_{2N+1} and integrate $(N+1)$-times the "function" δ with the help of formula (9.1). Differentiating $(N+1)$-times a spline of degree $2N+1$ or integrating $(N+1)$-times a Dirac function gives a spline of degree N ; we thus obtain a description of the orthonormal multi-resolution analysis V_j^N. So we are on a safe ground again. What did we obtain ? Applying the integration formula (9.1) $N+1$-times to δ gives just a normalized B-spline (up to a translation), so that applying the derivation formula (9.1) $N+1$-times to L_{2N+1} gives just the *dual function* of the normalized B-spline. (This function has been described by C. K. Chui and J. Z. Wang in [8] and by M. Unser et al. in [1]). Moreover, applying the derivation formula (10.1) $N+1$-times to the oblique wavelet Λ_{2N+1} gives a very interesting pre-wavelet in W_0^N : this is just *the pre-wavelet of G. Battle* [3] (that $L_{2N+1}(2x-1)$ is solution of a variational problem in a very well-known fact in spline theory, this is namely Holladay's theorem), also recently described by C. K. Chui and J. Z. Wang [9].

There are still others interesting basis functions, such as the *Compactly supported spline pre-wavelet* (described by LR [15], P. Auscher [2], C. K. Chui and J. Z. Wang [8] and M. Unser et al. [1]) or the *Causal spline scaling function* of J. O. Strömberg [26] (for which the associated filter is a *recursive stable IIR filter)*, and the *Causal spline wavelet.*

Thus, there are many available bases for the orthonormal setting. Moreover, we can associate to V_j a dual multi-resolution analysis V_j^* through the formalism of bi-orthogonal wavelets. If we require the conjugate pre-scaling functions to be compactly supported (in order to deal with FIR filters and thus to have fast transforms), g has to be the normalized B-spline (so that $\alpha(\xi) = \left(\frac{1+e^{-i\xi}}{2}\right)^{N+1}$), and g^* can never be a spline. g^* is a compactly supported function given by formula (8.1) where α^* is bound to satisfy (8.2) ; the existence and regularity of such a function is discussed in [10] and [12]. There are solutions of arbitrarily great regularity ; the associated oblique spline wavelet is then very much oscillating.

II. Wavelet theory in multi-D.

II.1. Tensor products.

The most easy way of defining a multi-resolution analysis in $L^2(\mathbb{R}^d)$ is to take the closure of the tensorial product of one-dimensional multi-resolution analyses. I will just describe the $2D$-setting to avoid notational burden.

Let us then consider two one-dimensional multi-resolution analyses $V_j^{(0)}$ (associated to a couple of conjugate pre-scaling functions (g_0, g_0^*)) and $V_j^{(1)}$ (associated to (g_1, g_1^*)). Then define $V_j \subset L^2(\mathbb{R}^2)$ as the closure of $V_j^{(0)} \otimes V_j^{(1)}$. This is a multi-resolution analysis of $L^2(\mathbb{R}^2)$ in the sense that it satisfies (3.1) to (3.3) (with $\mathbb{R}$ replaced by $\mathbb{R}^2$ and $k \in \mathbb{Z}$ by $k \in \mathbb{Z}^2$).

We then have an oblique projection P_j of $L^2(\mathbb{R}^2)$ onto V_j defined by

(11.1) $$g = g_0 \otimes g_1, \quad g^* = g_0^* \otimes g_1^*$$

(11.2) $$P_j f(x) = \sum_{k \in \mathbb{Z}^2} 2^{2j} < f(y) \mid g^*(2^j y - k) > g(2^j x - k)$$

(or, equivalently, $P_j = P_j^{(0)} \otimes P_j^{(1)}$). Writing $P_{j+1}^{(i)} = P_j^{(i)} \oplus Q_j^{(i)}$, we have : $P_{j+1} = P_j \oplus P_j^{(0)} \otimes Q_j^{(1)} \oplus Q_j^{(0)} \otimes P_j^{(1)} \oplus Q_j^{(0)} \otimes Q_j^{(1)}$, so that we have three oblique wavelets : $g_0 \otimes \gamma_1$, $\gamma_0 \otimes g_1$, $\gamma_0 \otimes \gamma_1$ corresponding to three oblique projectors. We then obtain an unconditional basis of $L^2(\mathbb{R})$ from those three wavelets through dyadic dilations and translations.

Of course, we still have fast wavelet transforms with separable FIR filters.

II.2. Generalized splines.

A more intrinsic way of defining a multi-resolution analysis on $L^2(\mathbb{R}^d)$ was introduced independently by G. Battle [4] and LR as described in the book of Y. Meyer, chapter III.5 [24]).

Let us define $V_j^{(N)}$ as the space of square-integrable generalized splines with nodes in $\frac{1}{2^j}\mathbb{Z}^d$, id est : $f \in V_j^{(N)}$ iff it satisfies :

(12.1) $f \in H^N(\mathbb{R}^d)$ (Sobolev space) with $N > \frac{d}{2}$

(12.2) $(-\Delta)^N f = \sum_{k \in \frac{1}{2^j}\mathbb{Z}^d} a_k \delta(x-k)$ for some $(a_k) \in \ell^2(\frac{1}{2^j}\mathbb{Z}^d)$ where Δ is the Laplacian operator and $\delta(x)$ the Dirac mass at $x = 0$. Then the $V_j^{(N)}$, $j \in \mathbb{Z}$, are a multi-resolution analysis of $L^2(\mathbb{R}^d)$, so that we have a very efficient tool to generate wavelets.

The approach adopted by LR fits in the setting of multi-resolution analysis. In this setting, we have to produce a Riesz basis of $V_0^{(N)}$. But it is very easy to see that there exists a Lagrangian generalized spline, i.e. a spline $L_N \in V_0^{(N)}$ such that $L_N(0) = 1$ and $L_N(k) = 0$ for $k \in \mathbb{Z}^d \backslash \{0\}$. Moreover L_N has exponential decay. Then the general theory of multi-resolution analysis in $\mathbb{R}^d$ [24] or specific computations for the case of a Lagrangian interpolating pre-scaling function [19] allow us to exhibit generalized spline wavelets.

Most of the proofs can be done by multi-variate Fourier series, but we have also a more geometrical proof, adapted to the case of a more general set of nodes (generalized splines with nodes in a set X such that for two positive constants A, B, the balls $B(x, A)$, $x \in X$, are disjoint and the balls $B(x, B)$, $x \in X$, are a covering of $\mathbb{R}^d$) [17]. One can extend this approach to other geometrical settings, as some Lie groups [18] (where one should use translations and dilations adapted to the Lie structure) or compact manifolds [14] (where one should replace the invariance through translation and dilation of the basis by asymptotic estimates).

The approach adopted by G. Battle is one more time to exhibit directly pre-wavelets by mean of a variational problem. Indeed, it is very easy to see that if L_{2N} is the Lagrangian generalized spline in $V_0^{(2N)}$, then $W_0^{(N)}$ has a Riesz basis composed with the functions $(-\Delta)^N L_{2N}(2x-k)$, $k \in \mathbb{Z}^d \backslash 2\mathbb{Z}^d$, and that this Riesz basis is derived through $\mathbb{Z}^d$-translations from a finite set of $2^d - 1$ functions. Similarly, if we consider the Lagrangian generalized spline $L_{2N+1} \in V_0^{(2N+1)}$, then the $(-\Delta)^N L_{2N+1}(2x-k)$, $k \in \mathbb{Z}^d \backslash 2\mathbb{Z}^d$, are a pre-wavelet basis for the scalar product $< -\Delta f \mid f >$. Those pre-wavelets have exponential decay and can be very easily orthonormalized (for $< -\Delta f \mid f >$).

III. Divergence-free vector wavelets.

III.1. Two-D divergence free orthonormal vector wavelets.

G. Battle and P. Federbush have recently announced the construction of divergence-free orthonormal vector wavelet bases in two dimensions [5] and three dimensions [6]. The two-D construction is straightforward and is based on the following lemma : if $\psi_{\epsilon,j,k}(x) = \psi_\epsilon(2^j x - k)$ $(1 \leq \epsilon \leq 3,\ j \in \mathbb{Z},\ k \in \mathbb{Z}^2)$ is an orthonormal basis for the scalar product $< -\Delta f \mid f >$ in $\mathbb{R}^2$, then the functions

$$\vec{\psi}_{\epsilon,j,k} = \left(-2^j \frac{\partial}{\partial x_2} \psi_\epsilon(2^j x - k), 2^j \frac{\partial}{\partial x_1} \psi_\epsilon(2^j x - k)\right)$$

are an orthonormal basis for the space $\vec{H}$ of divergence-free square-integrable functions : $\vec{H} = \{(f_1, f_2) \in L^2(\mathbb{R}^2)^2 / \frac{\partial}{\partial x_1} f_1 + \frac{\partial}{\partial x_2} f_2 = 0\}$.

We may then exhibit C^∞ vector wavelets with rapid decay at infinity just by taking the usual C^∞ wavelets of Y. Meyer and LR [22] and defining $\vec{\psi}_{\epsilon,j,k}$ as $(-R_{x_2}\psi_{\epsilon,j,k}, R_{x_1}\psi_{\epsilon,j,k})$ where the R_{x_i} are the Riesz transforms ; we may also exhibit C^N vector wavelets with exponential decay by considering the generalized spline wavelet basis of G. Battle for $< -\Delta f \mid f >$.

III.2. Unconditional bases of divergence free vector wavelets in dD.

Let us see now a very different approach of the problem developped in [21]. The aim is to define (oblique) multi-resolution analyses for $L^2(\mathbb{R}^d)$ (with projectors P_j) and $L^2(\mathbb{R}^d)^d$ (with projectors $\vec{P}_j$) such that we have a commutation formula :

$$\vec{\nabla} \cdot (\vec{P}_j \vec{f}) = P_j(\vec{\nabla} \cdot \vec{f}). \tag{13}$$

Then the approximation process $(\vec{P}_j)$, available for any (locally) square-integrable vector function, will be an internal process in the space $\vec{H}$ of divergence-free vector functions.

The construction of such multi-resolution analyses will be very easy. Let me give some details, for $d = 2$ to avoid notational burden. We choose a couple (g_1, g_1^*) of conjugate pre-scaling functions such that we may differentiate g_1 as $\frac{dg_1}{dx} = g_0(x) - g_0(x-1)$. We note $V_j^{(i)}$, $P_j^{(i)}$, $Q_j^{(i)}$, γ_i, γ_i^* the spaces, projectors and oblique wavelets associated to (g_i, g_i^*). Then we just define V_j as $V_j = V_j^{(0)} \hat{\otimes} V_j^{(0)}$ (with projector $P_j = P_j^{(0)} \otimes P_j^{(0)}$) and $\vec{V}_j$ as $\vec{V}_j = (V_j^{(1)} \hat{\otimes} V_j^{(0)}, V_j^{(0)} \hat{\otimes} V_j^{(1)})$ (with projectors $\vec{P}_j = (P_j^{(1)} \otimes P_j^{(0)}, P_j^{(0)} \otimes P_j^{(1)})$, where the symbol $\hat{\otimes}$ stands for the L^2-closure of the tensor product. Now, (13) is just a straightforward application of (9.2).

Now, we have got a very nice $\vec{V}_j$ space to analyze $\vec{H}$ (with the property that $\vec{P}_j(\vec{H}) = \vec{V}_j \cap \vec{H}$). But the details spaces are also very well suited to the description of $\vec{H}$, since the details spaces are derivatives spaces. Indeed, we have $\vec{V}_1 = \vec{V}_0 \oplus \vec{W}_0$ where

$$\begin{aligned}\vec{W}_0 &= (V_0^{(1)} \hat{\otimes} W_0^{(0)}, V_0^{(0)} \hat{\otimes} W_0^{(1)}) \oplus (W_0^{(1)} \hat{\otimes} V_0^{(0)}, W_0^{(0)} \hat{\otimes} V_0^{(1)}) \otimes (W_0^{(1)} \hat{\otimes} W_0^{(0)}, W_0^{(0)} \hat{\otimes} W_0^{(1)}) \\ &= \left(\frac{\partial}{\partial x_2}(V_0^{(1)} \hat{\otimes} W_0^{(1)}), V_0^{(0)} \hat{\otimes} W_0^{(1)}\right) \oplus \left(W_0^{(1)} \hat{\otimes} V_0^{(0)}, \frac{\partial}{\partial x_1}(W_0^{(1)} \hat{\otimes} V_0^{(1)})\right) \\ &\qquad \oplus \left(\frac{\partial}{\partial x_2}(W_0^{(1)} \hat{\otimes} W_0^{(1)}), \frac{\partial}{\partial x_1}(W_0^{(1)} \hat{\otimes} W_0^{(1)})\right),\end{aligned}$$

hence :

$$\vec{W}_0 \cap \vec{H} = \left(-\frac{\partial}{\partial x_2}, \frac{\partial}{\partial x_1}\right) \left(V_0^{(1)} \hat{\otimes} W_0^{(1)} \oplus W_0^{(1)} \hat{\otimes} V_0^{(1)} \oplus W_0^{(1)} \hat{\otimes} W_0^{(1)}\right).$$

Hence, we can obtain a basis for $\vec{W}_0 \cap \vec{H}$ from the oblique wavelets of the multi-resolution analysis $V_j^{(1)} \hat{\otimes} V_j^{(1)}$.

Similarly, we obtain the following result in d dimensions :

THEOREM. *Let $\vec{H}$ be $\vec{H} = \{\vec{f} \in L^2(\mathbb{R}^d)^d / \vec{\nabla} \cdot \vec{f} = 0\}$. Then there exist :*

* $(2^d - 1)(d - 1)$ *functions $\vec{\psi}_\epsilon$ such that :*
- *i) $\vec{\psi}_\epsilon \in \vec{H}$*
- *ii) $\vec{\psi}_\epsilon$ is compactly supported*
- *iii) $\vec{\psi}_\epsilon$ is of class C^N (N fixed, arbitrarily great).*

* $(2^d - 1)(d - 1)$ *functions $\vec{\theta}_\epsilon$ in $L^2(\mathbb{R}^d)^d$ with compact support such that :*

$$\forall \vec{f} \in \vec{H} \quad \vec{f} = \sum_\epsilon \sum_{j \in \mathbb{Z}} \sum_{k \in \mathbb{Z}^d} 2^{jd} < \vec{f}(y) \mid \vec{\theta}_\epsilon(2^j y - k) > \vec{\psi}_\epsilon(2^j x - k)$$

and that, for $p \in \mathbb{N}$, for $p \in \mathbb{N}$, $0 \le p \le N-1$, for $\vec{f} \in \vec{H} \cap H^p(\mathbb{R}^d)^d$

$$\| \vec{f} \|_{H^p(\mathbb{R}^d)^d} \simeq \left\{ \sum_\epsilon \sum_j \sum_k 2^{jd}(1 + 4^{jp}) \,|< \vec{f} \mid \vec{\theta}_\epsilon(2^j y - k) >|^2 \right\}.$$

References

[1] A. ALDROUBI, M. EDEN & M. UNSER A family of polynomial spline wavelet transform. Preprint, 1990.

[2] P. AUSCHER Ondelettes fractales et applications. Thèse, Paris IX, 1989.

[3] G. BATTLE A block spin construction of ondelettes. Part I : Lemarié functions. *Comm. Math. Phys.* 110 (1987), 601-615.

[4] G. BATTLE A block spin construction of ondelettes. Part II : The QFT connection. *Comm. Math. Phys.* 114 (1988), 93-102.

[5] G. BATTLE & P. FEDERBUSH A note on divergence-free vector wavelets. Preprint, *T.A.M.U.*, 1991.

[6] G. BATTLE & P. FEDERBUSH Divergence-free vector wavelets. Preprint, *T.A.M.U.*, 1991.

[7] C. K. CHUI An introduction to spline wavelets. Lecture at the NATO-ASI on "Approximation theory, splines and applications", Maratea, 1991.

[8] C. K. CHUI & J. Z. WANG On compactly supported spline wavelets and a duality principle. To appear in *Trans. Amer. Math. Soc.*

[9] C. K. CHUI & J. Z. WANG A cardinal spline approach to wavelets. To appear in *Proc. Amer. Math. Soc.*

[10] A. COHEN, I. DAUBECHIES & J. C. FEAUVEAU Bi-orthogonal bases of compactly supported wavelets. Preprint, *ATT & Bell Laboratories,* 1990.

[11] I. DAUBECHIES Orthonormal bases of compactly supported wavelets. *Comm. Pure Appl. Math.* 46 (1988), 909-996.

[12] I. DAUBECHIES & J. LAGARIAS Two-scale difference equations. Preprint, *ATT & Bell Laboratories*, 1989.

[13] J. C. FEAUVEAU Analyse multi-résolution par ondelettes non orthogonales et banc de filtres numériques. Thèse, Paris XI, 1990.

[14] S. JAFFARD Construction et propriétés des bases d'ondelettes. Thèse, Ecole Polytechnique, 1989.

[15] P. G. LEMARIE Construction d'ondelettes splines. Unpublished, 1987.

[16] P. G. LEMARIE Ondelettes à localisation exponentielle. *J. Math. Pures & Appl.* 67 (1988), 227-236.

[17] P. G. LEMARIE Théorie L^2 des surfaces splines. Unpublished 1987.

[18] P. G. LEMARIE Bases d'ondelettes sur les groupes de Lie stratifiés. *Bull. Soc. Math. France* 117 (1989), 211-232.

[19] P. G. LEMARIE Some remarks on wavelets and interpolation theory. Preprint, Paris XI, 1990.

[20] P. G. LEMARIE Fonctions à support compact dans les analyses multi-résolutions. To appear in *Revista Matematica Ibero-americana.*

[21] P. G. LEMARIE-RIEUSSET Analyses multi-résolutions non orthogonales et ondelettes vecteurs à divergence nulle. Preprint, Paris XI, 1991.

[22] P. G. LEMARIE & Y. MEYER Ondelettes et bases hilbertiennes. *Rev. Mat. Ibero-americana* 2 (1986), 1-18.

[23] S. MALLAT A theory for multi-resolution signal decomposition : the wavelet representation. *IEEE PAMI* 11 (1989), 674-693.

[24] Y. MEYER *Ondelettes et opérateurs,* tome 1. Paris, Hermann, 1990.

[25] I. J. SCHOENBERG Cardinal spline interpolation, *CBMS-NSF. Series in Applied Math.* $\neq$ 12, SIAM Publ., Philadelphia, 1973.

[26] J. O. STROMBERG A modified Franklin system and higher-order systems of $\mathbb{R}^n$ as unconditional bases for Hardy spaces, in Conf. on Harmonic Anal. in honor of A. Zygmund, vol. 2, Waldsworth, 1983, 475-494.

CNRS UA D 0757
Université de Paris-Sud
Mathématiques - Bâtiment 425
91405 ORSAY CEDEX
(France)

AN APPROACH TO MEROMORPHIC APPROXIMATION IN A STEIN MANIFOLD

Clement H. Lutterodt
Department of Mathematics
Howard University
Washington, DC 20059

ABSTRACT: We employ our lifting lemma for sections of meromorphic sheaves, to transfer by means of a sheaf epimorphism, some rational approximation results in a polydisc in $\mathbb{C}^N$ to a meromorphic approximation in an analytic polyhedral neighborhood in a Stein manifold. The main results are associated with convergence in measure.

Key words and phrases: Embedding and lifting, Stein manifold, convergence in measures.

AMS Classification: 41A20, 32A20, 32E30.

§1. Introduction:

The problem of meromorphic approximation in $\mathbb{C}^N$ remains complicated and rudimentary. However, in 1974, a significant break–through was worked out in [2], where power series techniques were engaged to obtain results over a Stein manifold. Recently, by means of a lifting lemma developed in our paper [5], we have been able to obtain a meromorphic extension of Oka–Weil approximation. It is the same lifting lemma that partly provides us with the machinery for the weaker approximation results in a Stein manifold, discussed in this paper.

The main results of this paper are theorems 4.4 and 4.5; theorem 4.5 being the analog of theorem 4.4 in an analytic polyhedral neighborhood in a Stein manifold. Theorem 4.3 is another convergence in measure result that has featured in an earlier paper [6]. However, its proof has been restructured in view of the analysis given in the appendix. The main body of the paper has been divided into three parts; section 2, section 3 and section 4. Section 2 puts all preliminary ideas including definitions together; section 3 dwells on the embedding of a Stein manifolds into $\mathbb{C}^N$ and its ramifications in terms of lifting; the main lemmas are housed here; section 4 deals exclusively with convergence for the $\mathfrak{M}^{II}$ – submodules: the appendix provides an accurate proof of a lemma that is central to all the convergence results.

S. P. Singh (ed.), Approximation Theory, Spline Functions and Applications, 391–403.

§2. Definitions, Notations and other Preliminaries

We begin this section with a definition of a Stein manifold as an important complex manifold on which a lot of interesting analysis is carried out.

Definition 2.1: An n–dimensional complex manifold $(X, {}_X\mathfrak{O})$ is called a Stein manifold of dimension n if

(i) X has a countable topology;
(ii) X is holomorphically convex; that is, $K \subset X$ is compact $\Rightarrow \hat{K} := \{z \in X: |f(z)| \leq \|f\|_K;\ \forall f \in {}_X\mathfrak{O}(X)\} \subset X$ is compact;
(iii) X is holomorphically separable; that is, for any points $z_1, z_2 \in X$ with $z_1 \neq z_2$, $\exists\, f \in {}_X\mathfrak{O}(X)$ such that $f(z_1) \neq f(z_2)$;
(iv) X has local coordinates defined by holomorphic functions.

Here ${}_X\mathfrak{O}$ represents the structure sheaf of germs of local holomorphic sections on X and ${}_X\mathfrak{O}(X)$ is the module of global holomorphic sections on X. We let ${}_X\mathfrak{M}$ be the sheaf of germs of local meromorphic sections on X and ${}_X\mathfrak{M}(X)$ be its corresponding module of meromorphic sections over X. We similarly let $\mathfrak{O}$ be the structure sheaf of $\mathbb{C}^N$ as a complex manifold and let $\mathfrak{M}$ be the sheaf of germs of local meromorphic functions on $\mathbb{C}^N$. Their corresponding modules of sections are $\mathfrak{O}(\mathbb{C}^N)$ and $\mathfrak{M}(\mathbb{C}^N)$ respectively. In respect of (ii) of definition 2.1, we point out that a compact subset K of X is holomorphically convex if $K = \hat{K}$.

Next we let $\mathfrak{O}(\Delta_\rho^N)$ denote the restriction of $\mathfrak{O}(\mathbb{C}^N)$ to $\Delta_\rho^N := \{\zeta \in \mathbb{C}^N: |\zeta_j| < \rho,\ j = 1,\ldots, N\}$ is a polydisc centered at $0 \in \mathbb{C}^N$. We shall introduce two submodules $\mathfrak{M}^1(\Delta_\rho^N)$ and $\mathfrak{M}^{II}(\Delta_\rho^N)$ of $\mathfrak{M}(\Delta_\rho^N)$; the submodule $\mathfrak{M}^{II}(\Delta_\rho^N)$ is, in fact, a submodule of $\mathfrak{M}^1(\Delta_\rho^N)$. This will be made clear shortly. We characterize elements in $\mathfrak{M}^1(\Delta_\rho^N)$ as follows:

(i) Each $F \in \mathfrak{M}^1(\Delta_\rho^N)$ is holomorphic at $0 \in \Delta_\rho^N$.
(ii) Each $F \in \mathfrak{M}^1(\Delta_\rho^N)$ is such that $\exists$ a non–homogeneous normalized polynomial q of minimal multiple degree for which

$$Z(F^{-1}) \cap \Delta_\rho^N = Z(q) \cap \Delta_\rho^N \tag{2.1}$$

where Z(g) is the zero set of the holomorphic function g in $\mathbb{C}^N$. Here the equality (2.1) means that the polar set of F is determined on Δ_ρ^N by the zero set of the polynomial q. Thus Fq is holomorphic in Δ_ρ^N and $Z(Fq) \neq Z(q)$ in Δ_ρ^N. There is an analog of $\mathfrak{M}^1(\Delta_\rho^N)$ over an analytic polyhedral neighborhood A_ρ in X, a Stein manifold; and it is denoted by ${}_X\mathfrak{M}^1(A_\rho)$. The existence of ${}_X\mathfrak{M}^1(A_\rho)$ is guaranteed by our lifting lemma 3.3, which is an indirect consequence of embedding X in $\mathbb{C}^N$ with $N \geq 2n + 1$. The $\mathfrak{M}^{II}(\Delta_\rho^N)$ consists of functions in $\mathfrak{M}^1(\Delta_\rho^N)$ whose polar sets in Δ_ρ^N are determined by normalized, non–homogeneous polynomials with minimal multiple "uniform" degree at most $\underline{d}$; uniform in the sense that $\underline{d} = (d,..., d)$, where $d \in I := \{0,1,2,...\}$

Next we let $I^N = I \times ... \times I$, N copies of I and introduce a partial ordering on I^N as follows: for each $\alpha,\beta \in I^N$ $\alpha \preceq \beta \Leftrightarrow \alpha_j \leq \beta_j$, $j = 1,2,..., N$. For each $\mu \in I^N$, we let

$$E_\mu := \{\gamma \in I^N: 0 \preceq \gamma \preceq \mu\}$$

This is a finite subset of I^N. Also for each μ and $\nu \in I^N$ we define another finite subset of I^N as follows:

Definition 2.2. A subset $E^{\mu\nu}$ of I^N is called a maximal index set if

(i) $E^{\mu\nu} = E_\mu \cup A_\nu$, $E_\mu \cap A_\nu = \phi$

(ii) For each $\lambda \in E^{\mu\nu}$, $E_\lambda \subset E^{\mu\nu}$

(iii) The cardinality of $E^{\mu\nu}$ satisfies the inequality

$$|E^{\mu\nu}| \geq \prod_{j=1}^{N} (\mu_j + 1) + \prod_{j=1}^{N} (\nu_j + 1) - 1$$

(v) Each axis of $E^{\mu\nu}$ has a Padé index set.

We remark that since the cardinality of E_μ is given by $|E_\mu| = \prod_{j=1}^{N} (\mu_j + 1)$, from (iii) $|A_\nu| \geq \prod_{j=1}^{N} (\nu_j + 1) - 1$. Again for each fixed $\mu,\nu \in I^N$, we let $\mathfrak{R}_{\mu\nu}$ be the class rational functions in $\mathbb{C}^N$ defined by quotients of two non– homogeneous polynomials $P_\mu(\zeta)$ and $Q_\nu(\zeta)$ in $\mathbb{C}^N$ of multiple degree at most μ in the numerator and ν in the denominator. Furthermore each element $P_\mu(\zeta)/Q_\nu(\zeta) \in \mathfrak{R}_{\mu\nu}$ is such that $Q_\nu(0) \neq 0$ and $P_\mu(\zeta)$, $Q_\nu(\zeta)$ are relatively prime except on set of codimension at least two in $\mathbb{C}^N$ of their common zeros.

Definition 2.3. Let $F \in \mathcal{O}(U)$, $0 \in U \subset\subset \Delta^N_\rho$, $(\mathfrak{M}^1|U = \mathcal{O}|U)$.
$P_\mu(\zeta)/Q_\nu(\zeta) \in \mathfrak{R}_{\mu\nu}$ is called a (μ,ν)–rational approximant to F at 0 if

$$\text{(i)}\quad \partial^{|\lambda|}(Q_\nu(\zeta)F(\zeta) - P_\mu(\zeta))|_{\zeta=0} = 0 \quad \forall\lambda \in E_\mu \tag{2.2}$$

$$\text{(ii)}\quad \partial^{|\lambda|}_{A_\nu}(Q_\nu(\zeta)F(\zeta))|_{\zeta=0} = \sum_{\substack{\gamma\leq\lambda \\ \gamma,\lambda\in A_\nu}} \Pi\binom{\lambda}{\gamma}\, \partial^{|\lambda|}(Q_\nu(\zeta))$$

$$\partial^{|\lambda-\gamma|}(F(\zeta))|_{\zeta=0} = 0 \tag{2.3}$$

Here $\partial^{|\lambda|} \equiv \partial^{|\lambda|}/(\partial\zeta_1^{\lambda_1} \ldots \partial_n^{\lambda_n})$ and $\Pi\binom{\lambda}{\gamma} = \binom{\lambda_1}{\gamma_1} \times\ldots\times \binom{\lambda_N}{\gamma_N}$.

Each maximal index set $E^{\mu\nu}$ (there are clearly many per each pair (μ,ν)) is associated with a unique (μ,ν) rational approximant. We refer to those (μ,ν)–rational approximants associated with maximal sets $E^{\mu\nu}$, that have normalized denominator as unisolvent rational approximants (URA). These are typically written as $\pi_{\mu\nu}(\zeta) = P_{\mu\nu}(\zeta)/Q_{\mu\nu}(\zeta)$. In the discussion of the main result of this paper we use the following form of $(\mu,\underline{\nu})$ URA's: $\pi_{\mu\underline{\nu}}(\zeta) = P_{\mu\underline{\nu}}(\zeta)/Q_{\mu\underline{\nu}}(\zeta)$ where $\underline{\nu} = (\sigma,\ldots,\sigma)$, $\sigma \in I$.

§3. Embedding and Lifting

Here we briefly recapitulate the foundations of our lifting lemma in [5]. It was in fact the culmination of embedding X in $\mathbb{C}^N$, $(N \geq 2n + 1$, $n = \dim X)$.

Let $\Lambda: X \to \mathbb{C}^N$ be a holomorphic proper embedding of X into $\mathbb{C}^N$, with $\Lambda(z) = (g_1(z),\ldots, g_N(z)) \in \mathbb{C}^N$, where $g_i \in {}_X\mathcal{O}(X)$, z's are local coordinates in X and $\operatorname{rank}(\frac{\partial g_i}{\partial z_j}) = n$ everywhere in X. Λ is proper in the sense that any χ compact in $\mathbb{C}^N$ has $\Lambda^{-1}(\chi)$ compact in X.

Now with respect to a polydisc $\Delta^N_\rho := \{\zeta \in \mathbb{C}^N: |\zeta_j| < \rho,\ j = 1,\ldots, N\}$, $\rho \in \mathbb{R}_+$, the embedding produces a biholomorphism between

$A_\rho := \{z \in X: |g_j(z)| < \rho, j = 1,..., N\} \subset\subset X$ and the closed subset $B_\rho := \{\zeta \in \Delta_\rho^N: \zeta_j = g_j(z), j = 1,..., N\}$ of Δ_ρ^N. That is $\Lambda_{|A_\rho}: A_\rho \to B_\rho$ is biholomorphic and this, in turn induces the isomorphism $\Lambda^*: {}_{B_\rho}\mathfrak{O}(B_\rho) \to {}_X\mathfrak{O}(\Lambda^{-1}(B_\rho)) = {}_X\mathfrak{O}(A_\rho)$. Between Δ_ρ^N and B_ρ, $\exists$ a natural projection r: $\Delta_\rho^N \to B_\rho$. A direct consequence of Cartan's theorem B, then furnishes us with the following restriction epimorphism $r^*: \mathfrak{O}(\Delta_\rho^N) \to {}_{B_\rho}\mathfrak{O}(B_\rho)$, where $\mathfrak{O}$, ${}_X\mathfrak{O}$, ${}_{B_\rho}\mathfrak{O}$ are in fact coherent sheaves. Combining Λ^* and r^* suitably leads to the following:

Lemma 3.1 Given Λ^* and r^* as above, with Λ^* an isomorphism and r^*, a restriction epimorphism, $\exists$ an epimorphism Λ_1^* such that the diagram

$$\begin{array}{ccc} & \mathfrak{O}(\Delta_\rho^N) & \\ \Lambda_1^* \swarrow & & \searrow r^* \\ {}_X\mathfrak{O}(A_\rho) & \xleftarrow[\Lambda^*]{} & \mathfrak{O}(B_\rho) \end{array} \tag{3.1}$$

commutes.

Our 'lifting lemma' is a meromorphic analog of the above lemma. The extension from the holomorphic to the meromorphic was facilitated by the following modified version of a lifting theorem of Grauert (& Remmert [4])

Theorem 3.2 (Grauert). Given $\Lambda_{|A_\rho}: A_\rho \to B_\rho$ is biholomorphic and every nowhere dense subset in B_ρ lifts to a nowhere dense subset of A_ρ. Then the isomorphism

$$\Lambda^*: {}_{B_\rho}\mathfrak{O}(B_\rho) \longrightarrow {}_X\mathfrak{O}(\Lambda^{-1}(B_\rho)) = {}_X\mathfrak{O}(A_\rho)$$

lifts to a unique extension isomorphism

$$\hat{\Lambda}^*: {}_{B_\rho}\mathfrak{M}(B_\rho) \longrightarrow {}_X\mathfrak{M}(\Lambda^{-1}(B_\rho)) = {}_X\mathfrak{M}(A_\rho)$$

There is also an extension of the restriction epimorphism r^* to $\hat{r}^*: \mathfrak{M}(\Delta_\rho^N) \to {}_{B_\rho}\mathfrak{M}(B_\rho)$ based on the simple fact that Δ_ρ^N is itself Stein. With these preliminaries settled, we now state our lemma as follows:

Lemma 3.3 (Lutterodt). Given the extensions of the isomorphism Λ^* and the restriction epimorphism r^*, as $\hat{\Lambda}^*$ and $\hat{r}^*$ respectively, $\exists$ an epimorphism $\hat{\Lambda}_1^*$ such that the diagram

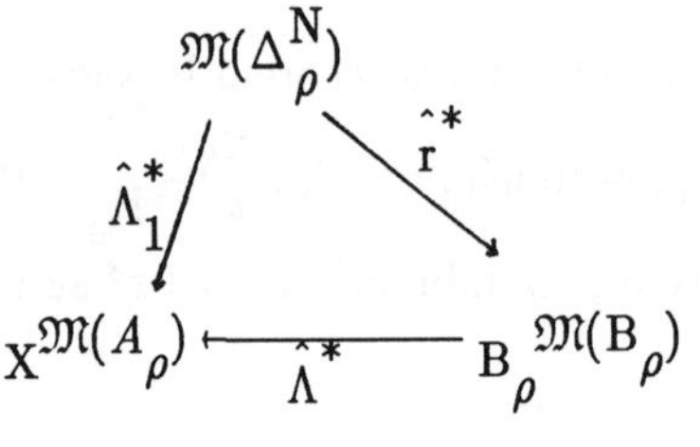

$$\begin{array}{ccc} & \mathfrak{M}(\Delta_\rho^N) & \\ \hat{\Lambda}_1^* \swarrow & & \searrow \hat{r}^* \\ {}_X\mathfrak{M}(A_\rho) & \xleftarrow[\hat{\Lambda}^*]{} & {}_{B_\rho}\mathfrak{M}(B_\rho) \end{array} \tag{3.2}$$

commutes. (see [5] for full details).

There is an induced similar relation between $\mathfrak{M}^1(\Delta_\rho^N)$, ${}_X\mathfrak{M}^1(A_\rho)$ and ${}_{B_\rho}\mathfrak{M}^1(B_\rho)$ which in turn leads to further induced similar relation between $\mathfrak{M}^{II}(\Delta_\rho^N)$, ${}_X\mathfrak{M}^{II}(A_\rho)$ and ${}_{B_\rho}\mathfrak{M}^{II}(B_\rho)$.

Next we introduce a lemma due to Bishop [1] paraphrased according to Narasimham [7]. It plays a central role in relation to our main theorem.

Lemma 3.4 (Bishop): Let χ be a compact subset of $\mathbb{C}^N$. Let $P_{\underline{d}}(\zeta)$ be a normalized polynomial of multiple degree $\underline{d} = (d,\dots, d)$ in $\mathbb{C}^N$. Let $0 < \eta < 1$ be given. Then $\exists$ $c = c(\chi,N)$ such that $Z_\eta := \{\zeta \in \chi\colon |P_{\underline{d}}(\zeta)| < \eta^d\}$ has a $\mathbb{C}^N$ – Lebesgue measure satisfying

$$m_N(Z_\eta) \le c\eta^{2/N} \tag{3.3}$$

Note that the power of η in the above inequality is independent of the degree of the polynomial $P_{\underline{d}}$. Its dependence is on the dimension of the underlying space.

Although Lemma 3.4 is stated in relation to $\mathbb{C}^N$, our focus in this paper is on the polydisc Δ_ρ^N. We want to examine the effect of the factored map $\Lambda^{-1} \circ r$ on $Z_\eta \cap \Delta_\rho^N$; recall that r is a natural projection from Δ_ρ^N to $B_\rho = \Lambda|_{A_\rho}(A_\rho)$ and $\Lambda|_{A_\rho}$ is biholomorphic. To simplify notation $Z_\eta \cap \Delta_\rho^N$, we modify the definition of Z_η in relation to a compact set χ in Δ^N_ρ instead of $\mathbb{C}^N$. Thus take $\chi \subset \Delta^N_{\rho'} \subset\subset \Delta^N_\rho$ $(\rho' < \rho)$ compact, we let $Z_\eta := \{\zeta \in \chi: |P_{\underline{d}}(\zeta)| < \eta^d\}$.

We shall assume $K = \hat{K}$ in X (i.e. compact and holomorphically convex) and let A_ρ be any analytical polyhedral neighborhood of K in X with $A_\rho \subset\subset X$. We let $\tilde{\chi} = \Lambda_{|A_\rho}(K)$. Then $\tilde{\chi}$ is compact in B_ρ; furthermore, we let $\tilde{\chi}$ be the compact image of χ compact in Δ^N_ρ under the natural projection r. Then under $\Lambda^{-1} \circ r$ map

$$Z_\eta \xrightarrow{\ r\ } \tilde{Z}_\eta \xrightarrow{\ \Lambda^{-1}\ } \check{Z}_\eta \tag{3.4}$$

where $r(Z_\eta) = \tilde{Z}_\eta := \{\zeta \in \tilde{\chi}: |P_{\underline{d}}(\zeta)| < \eta^d\}$ of reduced dimension from N to n. $\check{Z}_\eta := \{z \in K: |P_{\underline{d}} \circ \Lambda(z)| < \eta^d\}$.

<u>Lemma 3.5</u>. Let $\check{Z}_\eta$ be the lifted image of Z_η under the map $\Lambda^{-1} \circ r$. Then $\exists\, c_1 = c_1(K,n)$ such that with respect to an induced measure $m^\#_n$ on A_ρ associated with m_n on B_ρ

$$m^\#_n(\check{Z}_\eta) \leq c_1 \eta^{2/n}. \tag{3.5}$$

<u>Proof</u>: Since B_ρ is a submanifold of dimension n in Δ^N_ρ there is an intrinsic $\mathbb{C}^n$–Lebesgue measure m_n on B_ρ, $m_{N|B_\rho} = m_n$, such that with respect to the set $\tilde{Z}_\eta$ in B_ρ, where $\tilde{Z}_\eta := \{\zeta \in \tilde{\chi}: |P_{\underline{d}}(\zeta)| < \eta^d\}$ we have

$$m_n(\tilde{Z}_\eta) \leq c_0 \eta^{2/n} \tag{3.6}$$

for some $c_0 = c_0(\tilde{\chi},n) > 0$ constant in accordance with lemma 3.4 in the reduced submanifold B_ρ. Now with respect to the measure m_n on B_ρ, since the map $\Lambda^{-1}_{|B_\rho} : B_\rho \to A_\rho$ is biholomorphic, there is an induced map $\Lambda_\#$ such that

$$\Lambda_{\#}(m_n)A = m_n(\Lambda(A)) := m_n^{\#}(A) \tag{3.7}$$

$A, \Lambda(A)$ measurable in the sense given in Federer [3] p. 54. From (3.7) and (3.6) the desired result follows where $c_0 = c_0(\tilde{\chi}, n) = c_0(\Lambda(K), n) = c_1(K, n)$. ◁

§4. Convergence results

The results in this section are exclusively for the $\mathfrak{M}^{II}$ – submodules among the $\mathfrak{M}^1$'s. This means that results established for the $\mathfrak{M}^1$'s will apply to the $\mathfrak{M}^{II}$'s and not necessarily conversely. The first two theorems in this section are uniform convergence results analogous to those established for the $\mathfrak{M}^1$'s (see [5]).

Theorem 4.1. Let $\underline{\nu} = (\sigma, \ldots, \sigma) \in I^N$ be fixed. Suppose $F \in \mathfrak{M}^{II}(\Delta_\rho^N)$ i.e. $\exists$ a normalized non–homogeneous polynomial $q_{\underline{\nu}}$ of minimal degree $\underline{\nu} = (\sigma, \ldots, \sigma)$ such that $Z(F^{-1}) \cap \Delta_\rho^N = Z(q_{\underline{\nu}}) \cap \Delta_\rho^N$. Suppose $\pi_{\mu\underline{\nu}}(\zeta)$ is a $(\mu, \underline{\nu})$ – URA to F at 0, and let χ be any compact subset of Δ_ρ^N. Then as $\mu' = \min_{1 \leq j \leq N} (\mu_j) \to \infty$

(i) $Z(\pi_{\mu\underline{\nu}}^{-1}) \cap \Delta_\rho^N \to Z(F^{-1}) \cap \Delta_\rho^N$

(ii) $\pi_{\mu\underline{\nu}}(\zeta) \to F(\zeta)$ compactly on $\chi \backslash Z(F^{-1})$

One of the key lemmas that makes the above result possible is given in the appendix with a corrective term to fill in a gap that had existed in previous proofs.

The next result is an $\mathfrak{M}^{II}$ – analog of our extended Oka–Weil theorem.

Theorem 4.2 (Lutterodt): Let $K = \hat{K}$ in X, a Stein manifold, and let $A_\rho \subset\subset X$ be an analytic polyhedral neighborhood of K. Then $\forall f \in {}_X\mathfrak{M}^{II}(K)$ with a unique extension in ${}_X\mathfrak{M}^{II}(A_\rho)$, $\exists\, f_m \in \mathfrak{M}^{II}(A_\rho)$ such that as $m \to \infty$

(i) $Z(f_m^{-1}) \cap A_\rho \to Z(f^{-1}) \cap A_\rho$

(ii) $f_m \to f$ compactly on $K \backslash Z(f^{-1})$.

The proof of theorem 4.2 is wholly analogous to its $\mathfrak{M}^1$–counterpart in [5] and therefore will not be repeated here. We remark that unlike theorem 4.1, theorem 4.2 is a truly uniform meromorphic approximation on X, since both f and f_m are following subset $\Omega_\eta^{\mu\underline{\nu}}$ of χ:

$$\Omega_\eta^{\mu\underline{\nu}} := \{\zeta \in \chi : |F(\zeta) - \pi_{\mu\underline{\nu}}(\zeta)|^{1/\mu'} \geq \frac{1}{\eta}\} \tag{4.4}$$

From (4.3) and (4.4) we get on setting $L_{\underline{d}}(\zeta) = Q_{\mu\underline{\nu}}(\zeta) q_{\underline{\omega}}(\zeta)$ where $\underline{d} = \underline{\nu} + \underline{\omega} = (\sigma + \omega, \ldots, \sigma + \omega)$, that

$$|L_{\underline{d}}(\zeta)|^{1/d} < (\frac{B\rho''}{\rho'})^{1/d} (\frac{\rho''}{\rho'} \eta)^{\mu'/d} \tag{4.5}$$

Since $\mu' \to \infty \Rightarrow \sigma \to \infty$ but $\frac{\sigma}{\mu'} \to 0$, recall $\sigma = o(\mu')$ and ω being fixed, $\exists\, \mu_0' \in I$ such that $\mu' > \mu_0'$ we obtain ($\mu' >> d$)

$$|L_{\underline{d}}(\zeta)|^{1/d} < (\frac{\rho''}{\rho'} \eta)^{\mu'/d} < \eta \tag{4.6}$$

Now $\Omega_\eta^{\mu\underline{\nu}} \subset \{\zeta \in \chi : |L_{\underline{d}}(\zeta)| < \eta^d\}$ and by lemma 3.4, $\exists\, c_0 = c_0(\chi, N)$ such that the $\mathbb{C}^N$ – Lebesgue measure satisfies

$$m_N(\Omega_\eta^{\mu\underline{\nu}}) \leq c_0 \eta^{2/N}.$$

This conclude the proof.◂

The next result is a special case dealing with "diagonal approximants"

Theorem 4.4: Let the initial hypothesis be the same as in Theorem 4.3. Suppose $\pi_{\underline{\mu\mu}}$ with $\underline{\mu} = (\sigma, \ldots, \sigma)$, is a 'diagonal' $(\underline{\mu}, \underline{\mu})$ URA to F at the origin in $\Delta_{\rho'}^N$. Then on any compact subset χ of $\Delta_{\rho'}^N$, $\exists\, \sigma_0 \in I$ and $c_1 = c_1(\chi, N)$ such that $\sigma > \sigma_0$

$$m_N\{\zeta \in \chi : |F(\zeta) - \pi_{\underline{\mu\mu}}(\zeta)|^{1/\sigma} \geq \frac{1}{\eta}\} \leq c_1 \eta^{2/N} \tag{4.7}$$

Proof. Here we follow the approach employed in the proof of the preceding theorem, so the analog of inequality (4.2) becomes "pull–backs" of $F \in \mathfrak{M}^{II}(\Delta_\rho^N)$ and its $(\mu,\underline{\nu})$–URA $\pi_{\mu\underline{\nu}}$ respectively making both of them meromorphic on X and in particular on $K(\subset A_\rho \subset\subset X)$ compact.

Next we pass to a set of weak convergence results that rely on Bishop's lemma.

Theorem 4.3: Let $\eta \in (0,1)$ be given. Let $\omega \in I$ be fixed. Suppose $F \in \mathfrak{M}^{II}(\Delta_\rho^N)$, i.e. $\exists\, q_{\underline{\omega}}$ a normalized non–homogeneous polynomial of minimal degree $\underline{\omega} = (\omega,\ldots, \omega)$ such that $Z(F^{-1}) \cap \Delta_\rho^N = Z(q_{\underline{\omega}}) \cap \Delta_\rho^N$. Suppose $\pi_{\mu\underline{\nu}}$ is a $(\mu,\underline{\nu})$ URA to F at the origin, with $\underline{\nu} = (\sigma,\ldots, \sigma)$, $\sigma \in I$ and $0 < \omega < \sigma$, $\sigma \to \infty$ as $\mu' = \min_{1 \le j \le N} (\mu_j) \to \infty$ where $\sigma = o(\mu')$. Then on any compact subset χ of Δ_ρ^N, $\exists$ $\mu_0' \in I$, $c_0 = c_0(\chi,N)$ such that $\mu' > \mu_0'$

$$m_N\{\zeta \in \chi\colon |F(\zeta) - \pi_{\mu\underline{\nu}}(\zeta)|^{1/\mu'} \ge \tfrac{1}{\eta}\} \le c_0 \eta^{2/N} \tag{4.1}$$

Proof: We start with $H_{\mu\nu\underline{\omega}}(\zeta) = Q_{\mu\underline{\nu}}(\zeta)F(\zeta)q_{\underline{\omega}}(\zeta) - P_{\mu\nu\underline{\omega}}(\zeta)q_{\underline{\omega}}(\zeta)$ as given in the appendix. Since $H_{\mu\nu\underline{\omega}}(\zeta) \in \mathfrak{O}(\Delta_\rho^N)$ and therefore $H_{\mu\nu\underline{\omega}}(\zeta) \in \mathfrak{O}(\Delta_{\rho'}^N) \cap C(\overline{\Delta}_{\rho'}^N)$ where $\Delta_{\rho'}^N \subset\subset \Delta_\rho^N$ $(\rho' < \rho)$, we can compute Cauchy's estimates of the coefficients of the power series of $H_{\mu\nu\underline{\omega}}$. From appendix (ap 11) we obtain the inequality with $\zeta \in \Delta_{\rho''}^N \subset\subset \Delta_{\rho'}^N$,

$$|H_{\mu\nu\underline{\omega}}(\zeta)| \le \frac{2MN}{(1 - \frac{\rho''}{\rho'})^N} \left(\frac{\rho''}{\rho'}\right)^{\mu'+1} \tag{4.2}$$

If we let $B = B(M,N,\rho'',\rho') = \dfrac{2MN}{(1 - \frac{\rho''}{\rho'})^N}$, then we can modify (4.2) into the following form:

$$|F(\zeta) - \pi_{\mu\underline{\nu}}(\zeta)| \le \frac{B}{|Q_{\mu\underline{\nu}}(\zeta)\, q_{\underline{\omega}}(\zeta)|} \left(\frac{\rho''}{\rho'}\right)^{\mu'+1} \tag{4.3}$$

Now for any compact subset χ of $\Delta_{\rho'}^N$, (for instance $\chi = \overline{\Delta}_{\rho''}^N$) we define the

$$H_{\underline{\mu}\underline{\mu}\underline{\omega}}(\zeta) \le \frac{2MN\ )}{(1\ -\ \frac{\rho''}{\rho'})^N} (\frac{\rho''}{\rho'})^{\sigma+1} \tag{4.8}$$

where $\underline{\mu} = (\sigma,\ldots,\sigma)$ and $\zeta \in \Delta^N_{\rho''} \subset\subset \Delta^N_{\rho'}$, ($\rho'' < \rho'$). Defining B as before, the inequality (4.8) is put into the following form

$$|F(\zeta) - \pi_{\underline{\mu}\underline{\mu}}(\zeta)| \le \frac{B}{|Q_{\underline{\mu}\underline{\mu}}(\zeta)\, q_{\underline{\omega}}(\zeta)|} (\frac{\rho''}{\rho'})^{\sigma+1} \tag{4.9}$$

Again take any compact subset χ of $\Delta^N_{\rho'}$, (for instance $\chi = \overline{\Delta}^N_{\rho''}$) and define $\Omega^{\underline{\mu}}_{\eta}$ of χ by

$$\Omega^{\underline{\mu}}_{\eta} := \{\zeta \in \chi \colon |F(\zeta) - \pi_{\underline{\mu}\underline{\mu}}(\zeta)|^{1/\sigma} \ge \frac{1}{\eta}\} \tag{4.10}$$

where $\eta \in (0,1)$ is given. Again from (4.9), we get

$$|Q_{\underline{\mu}\underline{\mu}}(\zeta) q_{\underline{\omega}}(\zeta)| \le (\frac{B\rho''}{\rho'}) (\frac{\rho''}{\rho'}\eta)^{\sigma} \tag{4.11}$$

Recalling that ω is fixed in I, we find

$$\overline{\lim_{\sigma\to\infty}} |Q_{\underline{\mu}\underline{\mu}}(\zeta)|^{1/\sigma} \le \frac{\rho''}{\rho'}\eta < \eta \tag{4.12}$$

and so given $\epsilon > 0$, $\exists\ \sigma_0 \in I$ such that $\sigma > \sigma_0$

$$|Q_{\underline{\mu}\underline{\mu}}(\zeta)|^{1/\sigma} < \eta + \epsilon \tag{4.13}$$

This leads to $\Omega^{\underline{\mu}}_{\eta} \subset \{\zeta \in \chi \colon |Q_{\underline{\mu}\underline{\mu}}(\zeta)| < (\eta+\epsilon)^{\sigma}\}$. Hence by Lemma 3.4, $\exists\ c_1 = c_1(\chi,N)$ such that

$$m_N \Omega^{\underline{\mu}}_{\eta} \le c_1(\eta+\epsilon)^{2/N}.$$

The desired result follows on letting $\epsilon \to 0$. ◅

Next we consider the analog of theorem 4.4 on a Stein manifold X.

Theorem 4.5. Let $K = \hat{K}$ in a Stein manifold. Let $A_\rho \subset\subset X$, be a analytic polyhedral neighborhood of K in X. Then for each $\eta \in (0,1)$ and $f \in \mathfrak{M}^{II}(K)$ with a unique extension in $\mathfrak{M}^{II}(A_\rho)$, $\exists\, f_\sigma \in \mathfrak{M}^{II}(A_\rho)$, $\sigma_0 > 0$ and a constant $c_1 = c_1(K,n) > 0$ such that $\sigma > \sigma_0$

$$m_n^{\#}\{z \in K: |f - f_\sigma|^{1/\sigma} \geq \frac{1}{\eta}\} < c_1 \eta^{2/n} \tag{4.15}$$

where $m_n^{\#}$ is the measure induced on A_ρ by m_n on B_ρ under the biholomorphic map $\Lambda^{-1}: B_\rho \to A_\rho$.

Remark: The real difference between the above theorem and theorem 4.2, lies in the significant growth that occurs in polar set of f_σ. That, in turn, tends to stifle any kind of almost uniform convergence on any compact holomorphically convex subsets.

Proof of Theorem 4.5. Given $f \in {}_X\mathfrak{M}^{II}(K)$, with a unique extension in ${}_X\mathfrak{M}^{II}(A_\rho)$ where $K = \hat{K}$ and A_ρ is an analytic polyhedral neighborhood of K, we choose $A_{\rho'} \subset\subset A_\rho$ $(\rho' < \rho)$ so that $K \subset A_{\rho'}$. Then $f \in {}_X\mathfrak{M}^{II}(A_{\rho'})$. Let $\Lambda(K) = \tilde{\chi} \subset B_{\rho'} \subset\subset B_\rho$. Then $\exists\, \chi \subset \Delta^N_{\rho'} \subset\subset \Delta^N_\rho$ compact such that $r(\chi) = \tilde{\chi}$. Recall that $r: \Delta^N_{\rho'} \to B_{\rho'}$ is a natural projection and $\Lambda | A_{\rho'}: A_{\rho'} \to B_{\rho'}$ is biholomorphic. Now $f \circ \Lambda^{-1} \in \mathfrak{M}^{II}(B_{\rho'})$ and since $\hat{\Lambda}_1^*$ factors according to $\hat{\Lambda}_1^* = \hat{\Lambda}^* \circ \hat{r}^*$, Lemma 3.3 guarantees under $\hat{r}^*: \mathfrak{M}^{II}(\Delta^N_{\rho'}) \to \mathfrak{M}^{II}(B_{\rho'})$, an $F \in \mathfrak{M}^{II}(\Delta^N_{\rho'})$ such that $F = f \circ \Lambda^{-1} \in \mathfrak{M}^{II}(B_\rho)$. For such an $F \in \mathfrak{M}^{II}(\Delta^N_{\rho'})$ by theorem 4.4, given $\eta \in (0,1)$ and $\chi \subset \Delta^N_{\rho'} \subset\subset \Delta^N_\rho$ compact, $\exists$ a $(\underline{\mu},\underline{\mu})$–URA $\pi_{\underline{\mu}\underline{\mu}}$ to F at $0 \in \Delta^N_{\rho'}$, $c_0 = c_0(N,\chi)$ and $\sigma_0 > 0$ such that $\Omega^{\underline{\mu}}_\eta := \{\zeta \in \chi: |F - \pi_{\underline{\mu}\underline{\mu}}|^{1/\sigma} \geq \eta^{-1}\}$ satisfies for $\sigma > \sigma_0$

$$m_N \Omega^{\underline{\mu}}_\eta \leq c_0 \eta^{2/N} \tag{4.16}$$

Under the composite map $\Lambda^{-1} \circ r$, $\Omega_{\eta}^{\underline{\mu}}$ is lifted on to $\check{\Omega}_{\eta}^{\underline{\mu}} := \{z \in K:$ $|F \circ \Lambda - \pi_{\underline{\mu\mu}} \circ \Lambda|^{1/\sigma} \geq \eta^{-1}\}$. Now by lemma 3.5, $\exists\, c_1 = c_1(n,K) > 0$, and an induced $\mathbb{C}^n$ – Lebesgue measure $m_n^{\#}$ on A_ρ such that

$$m_n^{\#}(\check{\Omega}_{\eta}^{\underline{\mu}}) \leq c_1 \eta^{2/n} \tag{4.17}$$

The desired result is immediate if one takes $f = F \circ \Lambda$ and $f_\sigma = \pi_{\underline{\mu\mu}} \circ \Lambda$ where $\underline{\mu} = (\sigma, \ldots, \sigma)$. ◅

References:

[1] Bishop, E. (1963) 'Holomorphic Completions, Analytic Continuations and Interpolation of semi–norms', Ann. Math. 78 #3, 468–500.

[2] Chirka, M. (1974) 'Expansions in series and the degree of rational approximation for holomorphic functions with analytic singularities,' Math USSR Sbornik 22, 323–332.

[3] Federer, H. (1969), Geometric Measure Theory, Springer–Verlag, NY.

[4] Grauert, H. and Remmert, R. (1984). Coherent Analytic Sheaves, Springer–Verlag, NY.

[5] Lutterodt, C.H. (1991) 'A Meromorphic Extension of Oka–Weil Approximation in a Stein Manifold', Complex Variables 16, pp. 153–162.

[6] Lutterodt, C.H. (1985), 'On Convergence of (μ,ν)–sequences of Unisolvent Rational Approximants to Meromorphic functions in $\mathbb{C}^n$', Internat. J. Math. & Math. Sci. 8, #4, 641–652.

[7] Narasimhan, R. (1971), Several Complex Variables, Univ. of Chicago Press.

Acknowledgement: This research was partially supported by NSF grants RM #8912667 and INT–#8914788.

APPROXIMATING FIXED POINTS FOR NONEXPANSIVE MAPS IN HILBERT SPACES

Giuseppe Marino
Università della Calabria
Dipartimento di Matematica
87036 Arcavacata di Rende (CS), ITALY

ABSTRACT. Let C be a closed convex subset of a Hilbert space H, $f:C\longrightarrow H$ a nonexpansive map, $f_{t,c}(x):=tf(x)+(1-t)c$ ($t\in[0,1]$, $c\in C$) and $P_C:H\longrightarrow C$ the proximity map on C. We show that if f has fixed points, then the set of fixed points of f coincides with the set of fixed points of P_Cf. From this we deduce that if f has fixed points, then the nets of fixed points of $(P_Cf)_{t,c}$ and $P_Cf_{t,c}$ converge strongly (as $t\rightarrow 1^-$) to the fixed point of f closest to c, for each $c\in C$ fixed.

1. INTRODUCTION

In the following we will label:
- H a Hilbert space.
- C a closed convex subset of H.
- $P_C:H\rightarrow C$ the proximity map on C defined by $||P_C(x)-x||=d(x,C)=\min_{y\in C}||x-y||$.
- $f:C\rightarrow H$ a nonexpansive map (i.e. $||f(x)-f(y)||\leq||x-y||$ for each $x,y\in C$).
- $g:C\rightarrow H$ a map.
- $F(g):=\{x\in C:\ x=g(x)\}$.
- $g_{t,c}:C\rightarrow H$ the map defined by $g_{t,c}(x):=tg(x)+(1-t)c$, $t\in[0,1)$, $c\in C$.
- $x_{t,c}$ the unique fixed point of the strict contraction $(P_Cf)_{t,c}$, i.e. $x_{t,c}=tP_C(f(x_{t,c}))+(1-t)c$.
- $y_{t,c}$ the unique fixed point of the strict contraction $P_Cf_{t,c}$, i.e. $y_{t,c}=P_C(tf(y_{t,c})+(1-t)c)$.

In this note we examine the question: when the nets $(x_{t,c})$ and $(y_{t,c})$ do converge strongly (as $t\rightarrow 1^-$) to a fixed point of f?

S. P. Singh (ed.), *Approximation Theory, Spline Functions and Applications*, 405–409.

The answer is given in Theorem 5 below and it is unexpectedly simple: if and only if $F(f)$ is nonempty.

Of course if f is a self-mapping on C, then the maps $(P_C f)_{t,c}$ and $P_C f_{t,c}$ coincide with $f_{t,c}$ and so $x_{t,c}=y_{t,c}$. In this case Browder [1] and Halpern [2] have shown that if C is bounded, then $x_{t,c}$ converges strongly (as $t \to 1^-$) to the fixed point of f closest to c. A similar result was obtained by Reich [3] for a nonexpansive self-mapping defined on a closed convex subset of a uniformly smooth Banach space.
Returning to Hilbert spaces, if f is not a self-mapping, then Browder's theorem does not work, but if one assumes that $f(\partial C) \subset C$, then it is easy to see that $y_{t,c}=f_{t,c}(y_{t,c})$. From this observation, Singh and Watson [4] have obtained the following result, which extends the Browder's theorem to the case of not self-mappings: If $f(C)$ is bounded and $f(\partial C) \subset C$, then $y_{t,c}$ converges strongly (as $t \to 1^-$) to the fixed point of f closest to c.
The theorem 5 given below obviously includes the above results and seems to be the first that uses the asymptotic behavior of $(P_C f)_{t,c}$ and $P_C f_{t,c}$ to obtain explicit iterative methods (see Corollary 6 below) to pick out fixed points of f (at least for the cases in which one explicitly knows the proximity map).

The proofs of our results are very simple and use the following known facts:
(A) P_C is nonexpansive ([5]).
(B) $F(f)$ is closed and convex ([6]).
(C) $I-f$ is demiclosed, i.e. if $v_n \rightharpoonup v$ and $v_n - f(v_n) \to w$ then $w=v-f(v)$ [1]
(D) If $v,w,c \in H$, $t \in [0,1)$ and $\|[w-(1-t)c]/t - v\| \leq \|w-v\|$ then
$\|w-c\| \leq \|v-c\|$
(see proof of result of [4], in which this is shown for $c=0$. The proof for $c \neq 0$ is analogous).

2. RESULTS

Lemma 1 (geometric) Let $x \in H$. Then $x \in C$ if and only if there exists $y \in C$ such that $\|x-y\| \leq \|P_C(x)-y\|$.
Proof. Sufficiency.
Consider the ball $B(y, \|y-P_C(x)\|)$ of center y and radius $\|y-P_C(x)\|$ and the segment $S=[y,P_C(x)]$. Of course

$$x \in B(y, \|y-P_C(x)\|) \text{ and } d(x,C)=d(x,S). \tag{1}$$

Let now $v \in B(y, \|y-P_C(x)\|)$, $v \neq P_C(x)$. Then the inner product

$$(v-P_C(x), y-P_C(x))$$

is strictly positive.
So, if $w_t:=ty+(1-t)P_C(x)$ is a generic point of S, once defined the function $h:[0,1]\to\mathbf{R}$ by $h(t):=||v-w_t||^2$, we have that $h'(0)<0$, in such a way that $d(v,S)<||v-P_C(x)||$. From (1) follows thus $x=P_C(x)$.

Lemma 2. Let $F(P_Cf)$ be nonempty. Then the following are equivalent:

(i) There exists $y^*\in C$ such that $||f(x)-y^*||\leq||x-y^*||$ for each $x\in C$.
(ii) For any $x\in C$ there exists a $y\in C$ such that $||f(x)-y||\leq||x-y||$.
(iii) For any $x\in\partial C$ there exists a $y\in C$ such that $||f(x)-y||\leq||x-y||$.
(iv) $F(P_Cf)=F(f)$.
(v) $F(f)$ is nonempty.

Proof. (iii) implies (iv). If $u=P_C(f(u))$ belongs to ∂C, apply Lemma 1 to $x=f(u)$. The other implications are trivial.

Theorem 3. Let $F(f)$ be nonempty. Then $F(P_Cf)=F(f)$.

Proof. It is an immediate consequence of Lemma 2.

Lemma 4. Let $F(f)$ be nonempty, $c\in C$ and $(z_{t,c})$ indifferently the net $(x_{t,c})$ or $(y_{t,c})$. Then $||z_{t,c}-c||\leq||v-c||$ for each $v\in F(f)$.

Proof. Case 1: $z_{t,c}=y_{t,c}$. Then $||[y_{t,c}-(1-t)c]/t-v||=$
$||[y_{t,c}-(tv+(1-t)c)]/t||=||[P_C(f_{t,c}(y_{t,c}))-P_C(tv+(1-t)c)]/t||\leq$
$||[f_{t,c}(y_{t,c})-(tv+(1-t)c)]/t||=||f(y_{t,c})-v||\leq||y_{t,c}-v||$, so the claim follows by (D) of introduction.

Case 2: $z_{t,c}=x_{t,c}$. Then $||[x_{t,c}-(1-t)c]/t-v||=||P_C(f(x_{t,c}))-P_C(f(v))||\leq$ $||x_{t,c}-v||$ and so also in this case the claim follows by (D).

Theorem 5. Let $F(f)$ be nonempty, $c\in C$ and $(z_{t,c})$ indifferently the net $(x_{t,c})$ or $(y_{t,c})$. Then $z_{t,c}$ converges strongly (as $t\to 1^-$) to the fixed point of f closest to c.

Proof. Case 1: $z_{t,c}=y_{t,c}$. For absurd the claim is not true. Let $\hat{z}$ the point of $F(f)$ closest to c. There exists a sequence $(t_n)\subset[0,1]$ such that

$t_n \to 1$ for $n \to \infty$ and

$$||y_{t_n,c} - \hat{z}|| \geq r \tag{2}$$

for a certain $r > 0$.
From Lemma 4, the sequence $(y_{t_n,c})$ is bounded. Let $(y^*_{t_n,c})$ be a subsequence of $(y_{t_n,c})$ weakly convergent to $w \in C$. From the fact that $||(I-P_C f)(y^*_{t_n,c})|| \longrightarrow 0$ and from the demiclosure of $I-P_C f$, it follows $w \in F(P_C f)$ and so, from Theorem 3, $w \in F(f)$. Now, since $y^*_{t_n,c} \rightharpoonup w$, we have $||w-c|| \leq \liminf ||y^*_{t_n,c} - c|| \leq$ (Lemma 4) $\leq ||v-c||$ for each $v \in F(f)$.
In particular, $||w-c|| \leq ||\hat{z}-c||$, yielding $w = \hat{z}$ since $\hat{z}$ is the unique point of $F(f)$ closest to c. Summarizing, we have proved that $y^*_{t_n,c} \rightharpoonup \hat{z}$.
Prove now that $y^*_{t_n,c} \to \hat{z}$. Indeed,

$$||c-\hat{z}||^2 \geq \text{(Lemma 4)} \geq ||y^*_{t_n,c} - c||^2 = ||y^*_{t_n,c} - \hat{z}||^2 + ||c-\hat{z}||^2 + 2(y^*_{t_n,c} - \hat{z}, \hat{z} - c)$$

i.e. $0 \leq ||y^*_{t_n,c} - \hat{z}||^2 \leq 2(y^*_{t_n,c} - \hat{z}, c - \hat{z}) \to 0$ for $n \to \infty$. This contradict (2).
Case 2: $z_{t,c} = x_{t,c}$. Analogous to case 1.

From our Theorem 5 and from the proof of Theorem 3 of [2], we obtain the following corollary:

Corollary 6. Let $t_n = 1 - n^{-a}$, where $a \in (0,1)$ and $n \in \mathbf{N}$. Let $c \in C$ fixed. Then the sequence (x_n) defined recoursively by

$$x_{n+1} = t_n P_C(f(x_n)) + (1-t_n)c, \quad n \in \mathbf{N}$$

converges strongly, for $n \to \infty$, to the fixed point of f closest to c.

Remark 1. One can suspect that if $F(f)$ is nonempty, then $F(f_{t,c})$ is nonempty too at least for t sufficiently close to 1. This is true if $F(f) \setminus \partial C$ is nonempty, but it is not true if $F(f) \subset \partial C$ (easy counterexamples can be constructed in two-dimensional spaces too).

Remark 2. The Lemma 2 yields a proof very short and simple of a result of Schöneberg (Theorem 6 of [7]).

REFERENCES

1. Browder F.E, (1967) "Convergence of approximants to fixed points of nonexpansive nonlinear mappings in Banach spaces", Arch. Rational Mech. Anal. 24, 82-90.
2. Halpern B., (1967) "Fixed points of nonexpanding maps", Bull. Amer. Math. Soc. 73, 957-961.
3. Reich S., (1980) "Strong convergence theorems for resolvents of accretive operators in Banach spaces", J. Math. Anal. Appl. 75, 287-292.
4. Singh S.P. and Watson B., (1986) "On approximating fixed points", Proc. Symp. Pure Math. 45 part 2, 393-395.
5. Cheney E.W. and Goldstein A.A., (1959) "Proximity maps for convex sets" Proc. Amer.Math. Soc. 10, 448-450.
6. Browder F.E., (1965) "Fixed point theorem for noncompact mappings in Hilbert spaces" Proc. Nat. Acad. Sci. U.S.A. 53, 1272-1276.
7. Singh S.P. and Watson B., (1983) "Proximity maps and fixed points" J. Approx. Theory 39, 72-76.

On Approximation and Interpolation of Convex Functions*

Marian Neamtu†
Department of Applied Mathematics
University of Twente
P. O. Box 217, 7500 AE Enschede
The Netherlands

Abstract

Some negative results concerning the convexity preserving approximation and interpolation of multivariate functions are presented. We prove that the approximation based on both interpolation and local operators cannot be convexity preserving, provided the approximation space is (locally) finite dimensional. In both cases we can dispense with the asssumption of the linearity of the approximation operator and the assumption that the approximation space is a space of piecewise polynomials. Some consequences for the construction of shape preserving approximations are discussed.

1 Introduction

In this paper we present some negative results concerning the problem of approximation and interpolation of multivariate convex functions. We consider continuous convex functions $f \in C(\Omega)$, defined on a closed convex domain $\Omega \subset \mathbf{R}^s, s \geq 1$. By *convexity* of f it is understood that for arbitrary points $x^i, i = 0, \ldots, s$ in Ω it holds

$$f(\sum_{i=0}^{s} \lambda_i x^i) \leq \sum_{i=0}^{s} \lambda_i f(x^i),$$

whenever $\lambda_i \geq 0, \sum_{i=0}^{s} \lambda_i = 1$. We investigate here the problem of *convexity preservation* of the approximation, which means that the approximating function is convex if the approximated function is. Some considerations presented here were inspired by a result of Dahlberg and Johansson [4].

The paper is organized as follows. In Section 2 we consider the interpolation problem. We show that an interpolation operator preserves convexity only under some very restrictive conditions on the data sites and on the space of interpolating functions. In Section 3, we are concerned with the following. Let F, S be two function–spaces and $A : F \to S$ an operator. In [4] it has been shown that if S is a space of piecewise polynomials and if A is both a linear and local operator, then A is not convexity preserving. Here we prove a stronger

*This paper is in final form and will not appear elsewhere.
†Supported by the Netherlands Foundation of Mathematics (SMC) and the Netherlands Organization for the Advancement of Research (NWO).

S. P. Singh (ed.), Approximation Theory, Spline Functions and Applications, 411–418.

result. We show, that if S is a space which is *locally finite dimensional* then locality of A alone implies that A is not convexity preserving. Moreover, our definition of locality is more general than the one in [4]. Finally, in Section 4 we discuss some consequences of the negative facts presented in Sections 2 and 3 for the construction of convexity preserving approximations.

2 Can an interpolation operator be convexity preserving?

Throughout the paper we will employ the following notation. We will assume that Ω is a compact convex subset of $\mathbf{R}^s$. $C^k(\Omega), 0 \leq k \leq \infty$ will denote the space of all functions continuously differentiable in Ω up to order k.

In this section we consider the following problem. Let $X := \{x^i\}_{i=1}^n, n \in \mathbf{N}$ be a finite set of distinct points in $\Omega \subset \mathbf{R}^s, s > 1$, also called the *data sites.* A set of data $\{f_i\}_{i=1}^n$ associated with the data sites is called *convex* if there exists a convex continuous function f in Ω such that $f(x^i) = f_i, i = 1, \ldots, n$. The objective is to find, for a given set of convex data, a convex interpolating function from a given space S of functions. We show that if S is a finite dimensional subspace of $C(\Omega)$, then an interpolation operator can be convexity preserving only under some very restrictive conditions on the data sites and on S.

In the following we consider a convexity preserving interpolation operator I from $C^k(\Omega)$, $k \in \mathbf{Z}_+$ to S such that $If(x^i) = f(x^i), i = 1, \ldots, n$. In general, if S is a space of smooth functions, it has only sense to take $k \geq 1$, since otherwise the interpolation operator is not convexity preserving. This follows from the theorems below. We first consider the univariate case.

Theorem 2.1 *Let $s = 1$. An interpolation operator $I : C^k(\Omega) \to S, k \in \mathbf{Z}_+$ is convexity preserving only if $n \leq 4$ or if S contains a non-smooth function.*

Proof. Without loss of generality we assume $x^1 < \cdots < x^n$. If $n = 4$ we can take e.g., $S = \operatorname{span}\{1, x, (x^2 - x)_+^2, (x - x^3)_+^2\}$, which is a space of smooth functions. Here, we employed the notation $f_+ := \max\{0, f\}$. The Lagrange interpolation based on the nodes $x^1, \ldots, x^4$ by functions from S is obviously convexity preserving (in fact, linear). A similar construction can be given for $n < 4$.

Next, consider the case $n > 4$. We show that I is a convexity preserving interpolation operator only if S contains a certain piecewise linear (*i.e.*, non–smooth) function. Let f be given as $f(x) = (x - x^3)_+, x \in \Omega$ and let $\{f^i\}_{i=1}^\infty$ be a sequence of convex C^k–functions converging, in $C(\Omega)$, to f. Due to the special choice of the function f and the fact that $\{If^i\}_{i=1}^\infty$ is a sequence of convex functions interpolating $\{f^i\}_{i=1}^\infty$, it is evident that f is the limit of $\{If^i\}_{i=1}^\infty$ in $C(\Omega)$. Since S has finite dimension, it is closed in $C(\Omega)$, and therefore $f \in S$. ∎

In the following we call an operator I a C^1 operator if $If \in C^1(\Omega)$ whenever $f \in C^1(\Omega)$.

Corollary 2.1 *Let $n > 4$ and let S be a finite dimensional subspace of $C(\Omega)$. Then there does not exist a C^1 interpolation operator $I : C(\Omega) \to S$, which is convexity preserving.*

Proof. If a sequence of C^1 functions from a finite dimensional space S converges (in $C(\Omega)$), then the limit is a C^1 function contained in S. Therefore, the sequence $\{If^i\}_{i=1}^{\infty}$ of C^1 functions from the proof of Theorem 2.1 (with $k = 1$) converges to a C^1 function, which is a contradiction with the fact that f is not smooth. ■

Remark 2.1 Observe, that we did not require from I to be continuous. The only property we have used is the convexity preservation and the fact that I is a C^1 interpolation operator.

Remark 2.2 In the case when S a space of piecewise polynomials, this result was proved in [9]. There, however, a different argument has been employed which made a particular use of the fact that S is a space of piecewise polynomials. Here we only assume that S is finite dimensional.

The following result generalizes Corollary 2.1 to the multivariate case $s > 1$. Let X_0 be the set of all elements from X contained in the interior of co(X), the convex hull of X. $|X_0|$ will denote the cardinality of X_0.

Theorem 2.2 *Let $|X_0| > s+1, \mathrm{vol}_s(\mathrm{co}(X_0)) \neq 0$ and let S be a finite dimensional subspace of $C(\Omega)$. Then there does not exist a C^1 interpolation operator $I : C(\Omega) \to S$, which is convexity preserving.*

Proof. The proof is based on the following fact. Given arbitrary $s + 2$ (distinct) points in $\mathbf{R}^s$ not all lying on an $(s-1)$–dimensional hyperplane, there exist s points among them, spanning a hyperplane which (strictly) separates the remaining two points. Therefore, since $|X_0| > s + 1$ and $\mathrm{vol}_s(\mathrm{co}(X_0)) \neq 0$, we can find an $(s-1)$–dimensional hyperplane spanned by some vectors from X_0, and two vectors $x^-, x^+ \in X_0$ which are strictly separated by this hyperplane. Let this hyperplane be given by the equation $l(x) = \langle e, x\rangle + e_0 = 0, x \in \mathbf{R}^s, e \in \mathbf{R}^s, e_0 \in \mathbf{R}$. Without loss of generality we may assume $l(x^-) < 0$ and $l(x^+) > 0$. Consider the function $f(x) = l_+(x) := \max\{0, l(x)\}, x \in \Omega$. Obviously, any interpolant of f based on the nodes X is equal to f in the regions $\mathrm{co}(X^-)$ and $\mathrm{co}(X^+)$, where $X^- := \{x \in X \mid l(x) \leq 0\}, X^+ := \{x \in X \mid l(x) \geq 0\}$. This is because x^-, x^+ are located strictly inside $\mathrm{co}(X^-), \mathrm{co}(X^+)$, respectively. The proof now follows along the same lines as the proofs of Theorem 2.1 and Corollary 2.1. We consider a sequence $\{f^i\}_{i=1}^{\infty}$ of convex C^1 functions converging to f. Since the C^1 interpolation operator I is convexity preserving, $\{If^i\}_{i=1}^{\infty}$ is a sequence of convex C^1 functions from S. However, since S is finite dimensional, this is a contradiction with the fact that this sequence converges to f on $\mathrm{co}(X^-) \cup \mathrm{co}(X^+)$. ■

Remark 2.3 It is possible to strengthen the restrictions in Theorem 2.2 on the location of the data sites. However, already the given requirement on the existence of an interpolation operator, namely $|X_0| \leq s + 1$, is very restrictive and it rules out a prevalent number of typical data locations, including data on regular grids.

3 Local convexity preserving approximation

In this section we address the problem of local convexity preserving approximation. We wish to present a stronger version of a result of Dahlberg and Johansson [4] and Dahlberg [3] by considering a more general definition of locality of an operator.

Let A be an operator from $C(\mathbf{R}^s)$ to $C(\mathbf{R}^s)$. We call A *local*, if $Af - Ag \in C_0(\mathbf{R}^s)$ whenever $f - g \in C_0(\mathbf{R}^s), f, g \in C(\mathbf{R}^s)$, where $C_0(\mathbf{R}^s)$ is the space of compactly supported continuous functions. We call A *uniformly local*, if for every $\delta > 0$ there is an $\varepsilon > 0$ such that for every $f, g \in C(\mathbf{R}^s)$ and every $x \in \mathbf{R}^s$, $(f - g)(y) = 0, y \in \mathbf{R}^s \backslash B_\delta(x)$ implies $(Af - Ag)(y) = 0, y \in \mathbf{R}^s \backslash B_\varepsilon(x)$. Here, $B_\alpha(x) := \{y \in \mathbf{R}^s, \|y - x\|_2 < \alpha\}$ denotes the open ball centered in x with radius α. We say that A *preserves constants*, if $Af = f$ whenever f is a constant function. Finally, we call a subspace $S \subset C(\mathbf{R}^s)$ *locally finite dimensional*, if $\dim(S|_\Omega) < \infty$ for every bounded set $\Omega \subset \mathbf{R}^s$.

Theorem 3.1 *Let $A : C(\mathbf{R}^s) \to S$ be an operator such that*

(*i*) *A is uniformly local,*

(*ii*) *A preserves constants,*

(*iii*) *S is locally finite dimensional,*

(*iv*) *A is not trivial* i.e., *$Af \equiv 0$ only if $f \equiv 0$.*

Then A is not convexity preserving.

Proof. Let $e \in \mathbf{R}^s$ and $f_e(x) := (\langle e, x \rangle)_+, x \in \mathbf{R}^s$. Observe, that f_e is convex and equals zero for all x in the half-space $\{x \in \mathbf{R}^s \mid \langle e, x \rangle \leq 0\}$. Let $g_e := Af_e$. Consider the set $G_e := \{x \in \mathbf{R}^s; g_e(x) = 0\}$. Since A preserves constants and is uniformly local, there exists a real number $\bar{c}_e$ such that

$$\{y \in \mathbf{R}^s; \langle e, y \rangle + \bar{c}_e \leq 0\} \subset G_e.$$

Moreover, G_e is convex and, since A is not a trivial operator, $G_e \not\equiv \mathbf{R}^s$. But this means that there exists a constant $c_e \in \mathbf{R}$ such that

$$G_e = \{y \in \mathbf{R}^s, \langle e, y \rangle + c_e \leq 0\}.$$

Hence, g_e is a function of the form

$$g_e(x) = 0, \quad x \in \{y \in \mathbf{R}^s \mid \langle e, y \rangle + c_e \leq 0\} \tag{3.1}$$
$$g_e(x) > 0, \quad x \in \{y \in \mathbf{R}^s \mid \langle e, y \rangle + c_e > 0\}. \tag{3.2}$$

Therefore, S contains for all $e \in \mathbf{R}^s$ at least one continuous function g_e which is identically zero in one half-space determined by the hyperplane $\langle e, x \rangle + c_e = 0$ and positive in the complementary (open) half-space. The assertion of the theorem is now a direct consequence of Lemma 3.1 below. ∎

Lemma 3.1 *If the space S contains, for all $e \in \mathbf{R}^s$, a function g_e satisfying (3.1) and (3.2), then S is not locally finite dimensional.*

Proof. Let $G := \{g_e, e \in E\}$ be a set of functions satisfying (3.1) and (3.2), where $E := \{e = (e_1, \ldots, e_s) \in \mathbf{R}^s, e_1 + \cdots + e_s = 1\}$. In order to prove the lemma it will be sufficient to show that for every $\delta > 0$ there exists a point $\xi \in \mathbf{R}^s$ such that $\operatorname{span}(G|_{B_\delta(\xi)})$ is infinite

dimensional. Since E is not countable, it is a matter of an elementary argument to show that there exists a line $t \subset \mathbf{R}^s$ such that the set $t_G := \{x^e, e \in E^t\}$ is not countable, where x^e denotes the intersection point of t with $h_e := \{x \in \mathbf{R}^s, \langle e, x \rangle + c_e = 0\}, e \in E^t$ and E^t is the set of all $e \in E$ such that h_e is not parallel with t and such that $e_1 \neq e_2$ implies $x^{e_1} \neq x^{e_2}$, for all $e_1, e_2 \in E^t$. Therefore, for every $\delta > 0$ there exists a point $\xi \in t$ such that $B_\delta(\xi) \cap t_G$ is not countable. We show that the linear span of the univariate functions $g_e^t := g_e|_t, e \in E^t$ is infinite dimensional in $B_\delta(\xi) \cap t$. Without loss of generality we may assume that t is the line given by $x_2 = 0, \ldots, x_s = 0$, where $x_1, \ldots, x_s$ denote Cartesian coordinates of a vector $x \in \mathbf{R}^s$. Consider the set S^- of all functions from $\{g_e^t, e \in E^t\}$ such that whenever $g_e^t \in S^-$, then $g_e^t(x_1) = 0$, for all $x_1 \leq x_1^e$. We may assume that S^- is an infinite set. Otherwise, we would consider instead S^- the set S^+ of all functions $\{g_e^t, e \in E^t\}$ such that whenever $g_e^t \in S^+$, then $g_e^t(x_1) = 0$, for all $x_1 \geq x_1^e$. Consider an infinite sequence $g_{e_1}^t, g_{e_2}^t, \ldots$ of functions from S^- such that $e_1 < e_2 < \ldots$ and such that $x_1^{e_1}, x_1^{e_2}, \ldots \in B_\delta(\xi) \cap t_G$. Obviously any finite subsequence of this sequence forms a set of linearly independent functions in $B_\delta(\xi) \cap t$. Therefore, the dimension of the linear span of all functions from S^- is infinite in $B_\delta(\xi) \cap t$ and thus, so is the dimension of $\operatorname{span}(G|_{B_\delta(\xi)})$.
■

Remark 3.1 Note that the above assumptions on A are weaker than those imposed in [4]. From the proof of Theorem 3.1 it can be seen that the conditions (i) and (ii) may still be weakened and replaced by the requirement that A preserves locally at least one constant function e.g., the zero function.

In the remainder of this section we consider the problem of local interpolation. Let $X := \{x^i\}_{i=1}^\infty \subset \mathbf{R}^s$ be an infinite set of *quasi-uniformly* distributed data sites i.e., there exist $D_1, D_2 > 0$ such that $X \cap B_{D_2}(t) \neq \emptyset$, for every $t \in \mathbf{R}^s$ and $X \cap B_{D_1}(x) = x$, for every $x \in X$. Hence X contains no accumulation points and $\cup_{n \in \mathbf{N}} \operatorname{co}(\{x^1, \ldots, x^n\}) = \mathbf{R}^s$. In the following, let $\mathbf{R}^\infty := \{F = \{f_i\}_{i=1}^\infty, f_i \in \mathbf{R}\}$.

Theorem 3.2 *Let $I : \mathbf{R}^\infty \to C(\mathbf{R}^s)$ be a convexity preserving interpolation operator i.e., if $F = \{f_i\}_{i=1}^\infty$ is a set of convex data corresponding to the data sites from X then IF is a convex function and $(IF)(x^i) = f_i, i \in \mathbf{N}$. Then I is not local.*

We need first the following

Lemma 3.2 *Let X be a set of quasi-uniformly distributed data sites and let $D > 0$ and $t \in \mathbf{R}^s$. There exists a sequence of s-tuples $\{x^{1,i}, \ldots, x^{s,i}\}_{i=1}^\infty \subset X$, such that $h := \lim_{i \to \infty} \operatorname{co}(\{x^{1,i}, \ldots, x^{s,i}\})$ is a hyperplane in $\mathbf{R}^s$ not passing through any $x \in X \cap B_D(t)$ and such that $t \in h$.*

Proof. Consider a hyperplane h passing through t and having empty intersection with the (finite) set $X \cap B_D(t)$, and a sequence $\{t^{1,i}, \ldots, t^{s,i}\}_{i=1}^\infty$ of points from h such that $t \in \operatorname{co}(\{t^{1,i}, \ldots, t^{s,i}\})$ and such that $\lim_{i \to \infty} \operatorname{co}(\{t^{1,i}, \ldots, t^{s,i}\})$ is a hyperplane in $\mathbf{R}^s$. Let $x^{j,i}$ be a data site contained in $X \cap B_{D_2}(t^{j,i}), j = 1, \ldots, s, i \in \mathbf{N}$. Then obviously $\{x^{1,i}, \ldots, x^{s,i}\}_{i=n}^\infty$, for n sufficiently large, is the desired sequence. ■

Proof of Theorem 3.2: We show that for every $D \geq D_2 > 0$ and every $t \in \mathbf{R}^s \backslash X$, one can find convex data $F \in \mathbf{R}^\infty$, such that if $(IF)(t)$ depends only on the data in $X \cap B_D(t)$, then IF is not convex.

Consider any hyperplane h passing through t, having the properties as in Lemma 3.2, which is given by the equation $\langle e, x - t\rangle = 0, x \in \mathbf{R}^s$, for some $e \in \mathbf{R}^s, \|e\|_2 = 1$. Let $\varepsilon := d(X \cap B_D(x), h)$, $d(S_1, S_2)$ the Euclidean distance of two sets S_1, S_2. By construction, $\varepsilon > 0$. Let us define the following two functions f^l, f^u.

$$f^l(x) :=| \langle e, x - t\rangle | - \varepsilon, \quad f^u(x) := (f^l(x))_+, \quad x \in \mathbf{R}^s.$$

The essence of this construction is that both these functions are convex and $f^l\,|_{X \cap B_D(t)} = f^u\,|_{X \cap B_D(t)}$. This means that on basis of the data in $B_D(t)$, one cannot determine whether the underlying function is f^l or f^u. We next show, that the convexity preservation of I enforces $(IF^l)(t) = -\varepsilon \neq 0 = (IF^u)(t)$, where $F^l := f^l|_X$ and $F^u := f^u|_X$. This implies, that $(IF^l)(t), (IF^u)(t)$ do not depend only on data from $B_D(t)$. Consider first the function f^l. Let $\{x^{1,i}, \ldots, x^{s,i}\}_{i=1}^\infty \subset X$ be a sequence of s-tuples from Lemma 3.2 such that $h = \lim_{i\to\infty} \mathrm{co}(\{x^{1,i}, \ldots, x^{s,i}\})$. Then, $\lim_{i\to\infty} f^l(x^{j,i}) = -\varepsilon, j = 1, \ldots, s$. Since IF^l must be convex and $t \in \lim_{i\to\infty} \mathrm{co}(\{x^{1,i}, \ldots, x^{s,i}\})$, it follows that $(IF^l)(t) = -\varepsilon$. By an analogous argument it can be shown that $(IF^u)(t) = 0$, which finishes the proof of Theorem 3.2. ∎

4 Concluding remarks

In the previous two sections we have presented a number of negative results concerning the approximation and interpolation of convex functions. We wish to conclude this paper by recording some consequences for the construction of multivariate convexity preserving approximations.

In general, the presented results imply that in the *multivariate* case standard approaches do not lead to approximations which reflect the qualitative property of convexity of the underlying functions and discrete data. The reasons for this can be briefly summarized as follows.

(*i*) C^1 interpolation of discrete data is, in general, *not* convexity preserving if the interpolation space is (locally) finite dimensional.

(*ii*) Convexity preserving interpolation is *not* local.

(*iii*) Convexity preserving approximation which preserves constants is *not* local if the approximation space is locally finite dimensional.

The conclusion is therefore that a convexity preserving approximation method must be either *global* or the approximation/interpolation space must be *(locally) infinite dimensional.*

The above facts explain the difficulties associated with the construction of algorithms for convexity preserving approximation. Recently, there were proposed two different methods, described in [1, 2] and [6]. They are both designed to solve the problem of convexity preserving bivariate C^1 interpolation. The so called variational approach represents yet another potential technique to attack the shape preserving problems [11]. We point out, that all three approaches are global, which is in accordance with our results.

In the context of convexity preserving interpolation of discrete convex data it is natural to consider the so called *convex triangulation* of the data sites [10, 5, 7]. This refers to a *simplicial partition* of the parameter domain in question, which is such that the piecewise linear interpolant of the data based on it is convex. It is known, that the construction of the convex triangulation is a global process and thus, so is the construction of the convex piecewise linear interpolant. However, ones the convex triangulation is given, it is natural to ask whether a convexity preserving C^1 interpolation method can be *quasi-local* in the sense that it is local with respect to the given convex triangulation. It turns out, that such a method exists. In fact, the method suggested in [6] is a constructive proof for the existence of a quasi-local convexity preserving interpolation method. This method does not use piecewise polynomials, however. We point out, that it is possible to show that there actually exists a quasi-local convexity preserving method for C^1 interpolation of scattered data, which is based on piecewise polynomials. We intend to present details on this elsewhere.

Finally, we point out that similar questions were studied from a more general point of view in [8].

Acknowledgments. I wish to thank Dr. Bernd Mulansky for his valuable comments and suggestions, which improved the final version of the paper.

References

[1] R. Andersson, E. Andersson, and M. Boman. The automatic generation of convex surfaces. In R. R. Martin, editor, *Mathematics of Surfaces II*, pages 427–445. Oxford University Press, 1987.

[2] R. Andersson, E. Andersson, M. Boman, T. Elmroth, B. Dahlberg, and B. Johansson. Automatic construction of surfaces with prescribed shape. *Computer Aided Design*, 20(6):317-324, 1988.

[3] B. E. J. Dahlberg. Construction of surfaces with prescribed shape. In C. K. Chui, L. L. Schumaker, and J. D. Ward, editors, *Approximation Theory VI vol. 1*, pages 157–159, New York, 1989. Academic Press.

[4] B. E. J. Dahlberg and B. Johansson. Shape preserving approximations. In R. R. Martin, editor, *Mathematics of Surfaces II*, pages 419–426. Oxford University Press, 1987.

[5] W. Dahmen and C. A. Micchelli. Convexity of multivariate Bernstein polynomials and box spline surfaces. *Studia Scientiarum Mathematicarum Hungarica*, 23:265–287, 1988.

[6] N. Dyn, D. Levin, and D. Liu. Interpolatory convexity preserving subdivision schemes for curves and surfaces. preprint, 1990.

[7] B. Mulansky. Interpolation of scattered data by a bivariate convex function. I: Piecewise linear C^0–interpolation. Memorandum no. 858, University of Twente, 1990.

[8] B. Mulansky and M. Neamtu. On existence of shape preserving interpolation operators. preprint, 1991.

[9] E. Passow and J. A. Roulier. Monotone and convex spline interpolation. *SIAM Journal Numerical Analysis*, 14:904–909, 1977.

[10] D. S. Scott. The complexity of interpolating given data in three–space with a convex function in two variables. *Journal of Approximation Theory*, 42:52–63, 1984.

[11] F. I. Utreras. Constrained surface construction. In L. L. Schumaker C. K. Chui and F. I. Utreras, editors, *Topics in Multivariate Approximation*, pages 233–254, New York, 1987. Academic Press.

CONVERGENCE OF APPROXIMATING FIXED POINT SETS FOR MULTIVALUED NONEXPANSIVE MAPPINGS

Paolamaria Pietramala
Università della Calabria
Dipartimento di Matematica
87036 Arcavacata di Rende (CS), ITALY

ABSTRACT. Let K be a closed convex subset of a Hilbert space H and $T:K \multimap K$ a nonexpansive multivalued map with a unique fixed point z such that $\{z\} = T(z)$. It is shown that we can construct a sequence of approximating fixed point sets converging in the sense of Mosco to z.

Let H be a Hilbert space, K a closed convex subset of H, T a multivalued nonexpansive map from K in the family of nonempty compact subsets of K. It is our object in this paper to show that in a specific case it is possible to construct a net of approximating sets converging in the sense of Mosco to a fixed point of T.

Our investigation is prompted by the papers of Browder [1], Reich [2], Singh and Watson [3], in which analogous problems are treated for single-valued mappings. In particular, in [1] it is shown that: If K is a closed convex bounded subset of a Hilbert space and $T:K \to K$ is a nonexpansive map, then for any $\hat{x} \in K$, the net $(x_t)_{t\in[0,1)}$ of the fixed points of the contraction maps $T_{t,\hat{x}}$ defined by $T_{t,\hat{x}}(x)=tT(x)+(1-t)\hat{x}$, converges strongly in K, as t approaches 1, to the fixed point of T in K closest to $\hat{x}$. The paper [3] extends this result to the case of not self-mappings (but $T(\partial K)\subset K$) and K not necessarily bounded (but T(K) bounded).

The following example of multivalued self-map defined on a closed convex bounded subset of a finite-dimensional Hilbert space shows that the recalled results cannot be extended to genuine multivalued case.

Let $H=\mathbf{R}^2$, $K=[0,1]\times[0,1]$ and T the nonexpansive map defined by $T((a,b))=$triangle whose vertices are (0,0), (a,0), (0,b), for $(a,b)\in K$. Thus, for $(\hat{x},\hat{y})\in K$, the point $((1-t)\hat{x},(1-t)\hat{y})$ is a fixed point of the map $T_{t,(\hat{x},\hat{y})}$ for all $t\in[0,1)$ and we have $((1-t)\hat{x},(1-t)\hat{y}) \to (0,0)$ as t approaches 1. If $\hat{x}>\hat{y}$ $(\hat{x}<\hat{y})$ then the fixed point of T closest to $(\hat{x},\hat{y})$ is $(\hat{x},0)$ $((0,\hat{y}))$, but the net of the fixed points sets of $T_{t,(\hat{x},\hat{y})}$ does

S. P. Singh (ed.), Approximation Theory, Spline Functions and Applications, 419–422.

not converge to $(\hat{x},0)$ $((0,\hat{y}))$ even in the weaker convergence of sets, that is the Kuratowski convergence.

In the setting of Hilbert spaces, our result is formulated for nonexpansive maps T that have a unique fixed point z and this point satisfies $\{z\} = T(z)$. The precise generality of the class of functions satisfying this condition is not known but it has been studied, for example, in [4], [5], [6]. More recently the interest in optimization theory for such type of maps has prompted a corresponding interest in fixed point theory, since in [7] it has been shown that the maximization of a multivalued map T with respect to a cone, which subsumes ordinary and Pareto optimization, is equivalent to a fixed point problem of determining y such that $\{y\} = T(y)$.

Now we introduce some necessary notations and definitions. Let K be a closed convex subset of a Hilbert space H. We denote by **CB**(H) the family of nonempty closed bounded subsets of H and by **K**(K) the family of nonempty compact subsets of K. For $A \in \mathbf{CB}(H)$ we define $d(x,A)=\inf_{y\in A} ||x-y||$.

For any $A,B \in \mathbf{CB}(H)$ we note with D(A,B) the Hausdorff distance induced by the norm of H, i.e. $D(A,B)=\max[\sup_{a\in A} d(a,B), \sup_{b\in B} d(b,A)]$.

Remark 1. Let $A,B \in \mathbf{K}(K)$. It is well known that for any $a \in A$ there exists $b \in B$ such that $||a-b|| \leq D(A,B)$. Hence, if $B=\{b\}$, we have that for all a in A, $||a-b|| \leq D(A,B)$.

We denote by $\rightarrow$ and $\rightharpoonup$ strong and weak convergence respectively. Let (A_n) be a sequence of closed subset of H. We define the inner limit of $(A_n)^n$ by

$$\liminf A_n := \left\{ x \in H: \exists \text{ a sequence } (x_n),\ x_n \in A_n \text{ and } x_n \rightarrow x \right\}$$

and the weak-outer limit of (A_n) by

$$\text{w-limsup } A_n := \left\{ x \in H: \exists \text{ a subsequence } (A_{n'}) \text{ of } (A_n) \text{ and a sequence } (x_{n'}),\ x_{n'} \in A_{n'} \text{ such that } x_{n'} \rightharpoonup x \right\}.$$

We will say that (A_n) converges to A in the sense of Mosco (written $A_n \rightarrow A$) if $\liminf A_n = \text{w-limsup } A_n = A$.

A net $(A_t)_{t \in [0,1)}$ of closed subsets of H converges to A in the sense defined before if every sequence (A_{t_n}), $t_n \rightarrow 1$ as $n \rightarrow \infty$, converges in such sense to A.

A multivalued map $T:K \rightarrow \mathbf{K}(K)$ is said to be lipschitzian if there exists $L \geq 0$ such that $D(T(x),T(y)) \leq L||x-y||$ for every x,y in K. T is said to be a contraction if $L<1$ and nonexpansive if $L=1$. A map $T:K \rightarrow \mathbf{K}(K)$ is said to be demiclosed if $x_n \rightharpoonup x$, $y_n \rightarrow y$ and $y_n \in T(x_n)$ imply $y \in T(x)$.

Let $T:K \rightarrow \mathbf{K}(K)$ be a nonexpansive map. For $\hat{x} \in K$ and $t \in [0,1)$ we denote by $T_{t,\hat{x}}$ the contraction map defined by $T_{t,\hat{x}}(x)=tT(x)+(1-t)\hat{x}$ for any x in K.

Finally, we denote by

$$F(T)=\{x\in K:\ x\in T(x)\}$$

and

$$F(T_{t,\hat{x}})=\{x\in K:\ x\in T_{t,\hat{x}}(x)\}$$

the sets of fixed points of T and $T_{t,\hat{x}}$ respectively.

Theorem. Let K be a closed convex subset of a Hilbert space H. Let $T:K\to \mathbf{K}(K)$ be a nonexpansive map such that $F(T)=\{z\}$ and this point z satisfies $T(z)=\{z\}$. Then for every $\hat{x}\in K$ we have that $F(T_{t,\hat{x}})\to F(T)$ as $t\to 1$.

Proof. We have to prove that $F(T_{t_n,\hat{x}})\to\{z\}$ as $n\to\infty$ for every sequence $t_n\to 1$. Since we have always liminf $\subset$ w-limsup, it remains to prove that w-limsup $F(T_{t_n,\hat{x}})\subset\{z\}$ and $\{z\}\subset$ liminf $F(T_{t_n,\hat{x}})$.

Step 1. w-limsup $F(T_{t_n,\hat{x}})\subset\{z\}$.

Let $x\in$ w-limsup $F(T_{t_n,\hat{x}})$, then there exist a subsequence $(t_{n'})$ of (t_n) and a sequence $(y_{n'})$, $y_{n'}\in F(T_{t_{n'},\hat{x}})$ such that $y_{n'}\rightharpoonup x$. Since $y_{n'}$ is in $F(T_{t_{n'},\hat{x}})$, there exists $w_{n'}\in T(y_{n'})$ such that $y_{n'}=t_{n'}w_{n'}+(1-t_{n'})\hat{x}$. Thus $||y_{n'}-w_{n'}||=(1-t_{n'})||w_{n'}-\hat{x}||\to 0$ as $t_{n'}\to 1$ because the sequence $(w_{n'})$ is bounded. Since $I-T$ is demiclosed [8], it follows that $0\in(I-T)(x)$, hence $x=z$.

Step 2. $\{z\}\subset$ liminf $F(T_{t_n,\hat{x}})$.

Let $x_n\in F(T_{t_n,\hat{x}})$. We prove that $x_n\to z$. For absurd, suppose that there exist $\varepsilon>0$ and a subsequence $(t_{n'})$ of (t_n) such that

$$||x_{n'}-z||\geqslant\varepsilon. \tag{1}$$

From $x_{n'}=t_{n'}w_{n'}+(1-t_{n'})\hat{x}$, $w_{n'}\in T(x_{n'})$ it follows that

$$||[x_{n'}-(1-t_{n'})\hat{x}]/t_{n'}-z||=||w_{n'}-z||.$$

Furthermore, from Remark 1, we have $||w_{n'}-z||\le D(T(x_{n'}),T(z))$ and the non expansivity of T yields $||w_{n'}-z||\le||x_{n'}-z||$. Hence

$$||(x_{n'}-\hat{x})/t_{n'}-(z-\hat{x})||^2\le||(x_{n'}-\hat{x})+(\hat{x}-z)||^2$$

which implies $||x_{n'}-\hat{x}||^2\le[2t_{n'}/(1+t_{n'})](x_{n'}-\hat{x},z-\hat{x})\le(x_{n'}-\hat{x},z-\hat{x})$, so that

$$||x_{n'}-\hat{x}||^2 \leq ||x_{n'}-\hat{x}|| \; ||z-\hat{x}||. \tag{2}$$

If it would be $x_{n'}=\hat{x}$ for a certain n', we should have $\hat{x}=t_{n'}w_{n'}+\hat{x}-t_{n'}\hat{x} = w_{n'} \in T(x_{n'})=T(\hat{x})$. Then $\hat{x}=z$, contradicting (1). Thus, from (2) it follows

$$||x_{n'}-\hat{x}|| \leq ||z-\hat{x}|| \tag{3}$$

which implies that the subsequence $(x_{n'})$ is bounded. Hence there exists a subsequence $(x_{n''})$ of $(x_{n'})$ such that $x_{n''} \rightharpoonup x$. Proceeding as in the proof of Step 1, we obtain x=z. At this point, the well known relation $||z-\hat{x}|| \leq \liminf ||x_{n''}-\hat{x}||$ and (see (3)) $\limsup ||x_{n''}-\hat{x}|| \leq ||z-\hat{x}||$ imply $||x_{n''}-\hat{x}|| \to ||z-\hat{x}||$. Hence, we have $x_{n''} \to z$, contradicting (1).

Remark 2. The same proof of the Theorem above also yields the convergence of $F(T_{t,\hat{x}})$ to F(T) in the sense of Fisher (for Fisher convergence definition see [9]). Moreover, from Propositions 1 and 8 in [9], one can deduce also the convergence in the sense of Hausdorff, Wijsman and Kuratowski.

Remark 3. In the following example our theorem works.
Let $H=\mathbf{R}$, $K=[0,\infty)$, $T:K \to \mathbf{K}(K)$ be the nonexpansive map defined by

$$T(x)=[0,x/2].$$

REFERENCES

1. Browder F.E.,(1967) "Convergence of approximants to fixed points of nonexpansive nonlinear mappings in Banach spaces", Arch.Rational Mech.Anal. 24, 82-90.
2. Reich S.,(1980) "Strong convergence theorems for resolvents of accretive operators in Banach spaces", J.Math.Anal.Appl. 75, 287-292.
3. Singh S.P. and Watson B.,(1986) "On approximating fixed points", Proc.Symp.Pure Math. 45 part 2, 393-395.
4. Reich S.,(1972) "Fixed points of contractive functions" Boll.UMI 5, 26-42.
5. Ciric L.B.,(1972) "Fixed points for generalized multivalued contractions" Mat.Vesnik, N.Ser. 9 (24), 265-272.
6. Iséki K.,(1974) "Multivalued contraction mappings in complete metric spaces" Math.Sem.Notes, 2, 45-49.
7. Corley H.W.,(1986) "Some hybrid fixed point theorems related to optimization" J.Math.Anal.Appl. 120, 528-532.
8. Lami Dozo E.,(1973) "Multivalued nonexpansive mappings and Opial's condition" Proc.Amer.Math.Soc. 38, 286-292.
9. Baronti M. and Papini P.L.,(1986) "Convergence of sequences of sets", Functional Analysis and Approximation Theory, Proc.Internat.Conf.Bombay, 1985, International Series of Numerical Analysis, Basel.

A SUBDIVISION ALGORITHM FOR NON–UNIFORM B–SPLINES

Ruibin Qu and John A. Gregory
Department of Mathematics and Statistics
Brunel University, Uxbridge
Middlesex, UB8 3PH
Britain

ABSTRACT. *In this paper, a (recursive) subdivision algorithm for non–uniform B–spline curves with simple knots is formulated so that B–spline curves with non–uniform knot partitions can also be coped with in a "free–form manner". Since it provides enough parameters to adjust the shape of the resulting curves, the algorithm is quite useful in interactive CAD/CAGD systems. The formulation is based on the "adapted parametrization" technique which was first used in our analysis of a non-uniform corner cutting algorithm in order to seek conditions under which the algorithm produces smooth curves. Another method to derive the algorithm, which uses the Böhm's knot insertion technique, is also presented. The subdivision algorithm could also be derived from the Oslo algorithm.*

§1. Introduction

Since B–splines play a very important role in curve and surface design and computation, a variety of algorithms and techniques are introduced to meet the needs for solving various kinds of problems. It is generally accepted that the de Boor-Cox algorithm [3] is the most popular choice for the pointwise evaluation of both uniform and non–uniform B–splines and, the uniform subdivision technique [8] (*e.g.*, Chaikin's algorithm [6]), or the Lane-Riesenfeld algorithm [4,11], provides a very suitable tool for shape design and manipulation. Unfortunately, the latter method can only treat uniform B–splines, that is, B–splines with uniform knot partitions.

In [10,13], we studied a simple non-uniform corner cutting scheme for the generation of smooth curves. The scheme is defined by the following refinement equations:

S. P. Singh (ed.), Approximation Theory, Spline Functions and Applications, 423–436.

$$
\begin{cases}
\mathbf{P}_{2i}^{k+1} & = (1-\alpha_i^k)\mathbf{P}_i^k + \alpha_i^k \mathbf{P}_{i+1}^k, \\
\mathbf{P}_{2i+1}^{k+1} & = \beta_i^k \mathbf{P}_i^k + (1-\beta_i^k)\mathbf{P}_{i+1}^k,
\end{cases}
\tag{1.1}
$$

where, $\{\mathbf{P}_i^k\}$ is the *control polygon* in $\mathbf{R}^N$ $(N \geq 2)$ at level k and the coefficients α_i^k and β_i^k are assumed to satisfy

$$
\alpha_i^k, \beta_i^k > 0, \quad \alpha_i^k + \beta_i^k < 1, \quad \forall i \in \mathbf{Z}, \quad k \geq 0. \tag{1.2}
$$

Figure 1 shows the subdividing process of the corner cutting scheme.

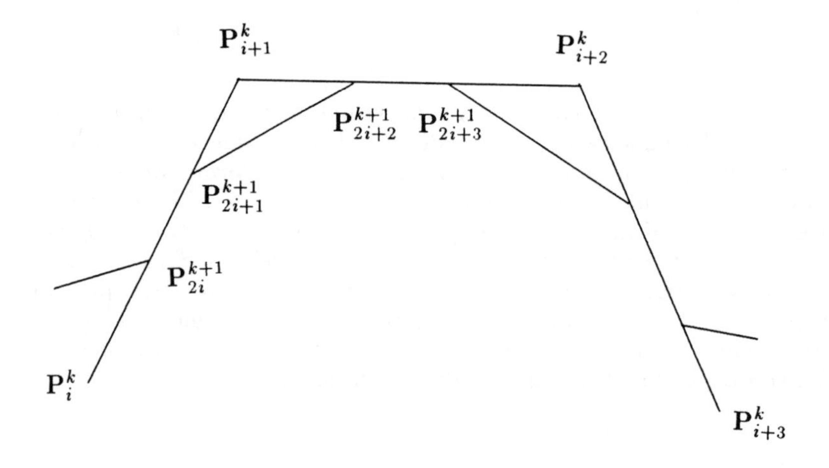

Figure 1. The geometric construction of the corner cutting scheme.

It was found that any smooth parabolic B-spline curve could be produced by this scheme if the proportions $\{\alpha_i^k, \beta_i^k\}$ of the corner cutting process were properly chosen [13]. For example, the scheme will produce a B-spline curve with control points (de Boor points) $\{\mathbf{P}_i^0\}$ and strictly increasing knots $\{x_i^0 : x_i^0 < x_{i+1}^0, \forall i \in \mathbf{Z}\}$ if the parameters are chosen such that

$$
\begin{cases}
\alpha_i^k & = \frac{(x_i^k - x_{i-1}^k)}{(x_{i+1}^k - x_{i-1}^k)} \theta_{i-1}^k, \\
\beta_i^k & = \frac{(x_{i+1}^k - x_i^k)}{(x_{i+1}^k - x_{i-1}^k)} \left(1 - \theta_i^k\right),
\end{cases}
\tag{1.3}
$$

where, the knots $\{x_i^k\}$ are given by the following refinement formulae:

$$\begin{cases} x_{2i}^{k+1} & = x_i^k, \\ x_{2i+1}^{k+1} & = (1-\theta_i^k)x_i^k + \theta_i^k x_{i+1}^k, \end{cases} \tag{1.4}$$

for all $\{\theta_i^k\}$ satisfying $0 < \theta_0 \leq \theta_i^k \leq \theta_1 < 1$ for some constants θ_0 and θ_1. Hence, the monotonicity of the knot sequence $\{x_i^k : x_i^k < x_{i+1}^k, \forall i \in \mathbf{Z}, k \geq 0\}$ will be preserved. This strictly monotonicity of knot sequence will be assumed throughout this paper.

From this quadratic B-spline algorithm, a question that arises immediately is that could a cubic (or even any) B-spline curve be generated in a similar way?

In the next section we will discuss a generalisation of scheme (1.1) to find a subdivision algorithm for non-uniform B-spline curves. It will be shown that the parametrization of the control polygons plays a central role in our analysis.

Remark 1. *If $x_i^0 = i, \theta_i^k = \frac{1}{2}$ for all i and k, then from (1.3) we have $\alpha_i^k = \beta_i^k = \frac{1}{4}, \forall i \in \mathbf{Z}, k \geq 0$ and thus scheme (1.1) becomes the Chaikin's algorithm [6], which generates uniform quadratic B-splines.*

Remark 2. *A similar scheme to (1.1) is also studied in [12].*

§2. Motivation and technique

Although Chaikin's algorithm (*cf.* [6]), Catmull-Clark's algorithm (for uniform cubic B-spline curves) and the Lane-Riesenfeld algorithm (*cf.* [4]) have been used for a long time, it seems that no similar subdivision algorithm for non-uniform B-spline curves has yet been developed, especially for cubic and quartic B-spline curves which are commonly used. From the above non-uniform corner cutting algorithm (1.1) and the structures of parabolic or piecewise parabolic curves, we formulated a subdivision algorithm for smooth quadratic B-spline curves [13]. The most important tool for the construction of the algorithm is the parametrization of the curves.

To construct the Subdivision Algorithm (SA for short) for higher order B-splines, we hope to *integrate* scheme (1.1) just as in the uniform subdivision case (*cf.* [5,8]). It is hoped that this should work in the non-uniform case. Hence, the problem becomes how to integrate a non-uniform subdivision scheme. The main difficulty is to establish the relations between the parametrization of the scheme and the parametrizations of its related schemes such as its *integrated* and *divided difference* schemes.

The difficulty can be overcome when it is considered from another point of view. Our success of the construction of the algorithm is based on the *Adapted Parametrisation technique* and the well known result–the Greville's identity for B-splines. A brief description of the formulation of the algorithm is as follows. More details about the construction and the *adapted parametrization technique* can be found in [10,13].

First, we state a result obtained in [10,13]. Suppose that for $k \geq 0$, the control polygon $\{\mathbf{P}_i^k\}$ is parametrized at the parameter values $\{t_i^k : t_i^k < t_{i+1}^k, i \in \mathbf{Z}\}$. Then the adapted parametrization of the control polygon sequence $\{\mathbf{P}_i^k\}$ means that the parametric values $\{t_i^k\}$ are determined by the subdivision scheme itself, *i.e.*, $\{t_i^k\}$ satisfy

$$\begin{cases} t_{2i}^{k+1} &= (1-\alpha_i^k)t_i^k + \alpha_i^k t_{i+1}^k, \\ t_{2i+1}^{k+1} &= \beta_i^k t_i^k + (1-\beta_i^k)t_{i+1}^k, \end{cases} \tag{2.1}$$

with the same coefficients α_i^k and β_i^k as in scheme (1.1). Conditions (1.2) guarantee that $\{t_i^k : t_i^k < t_{i+1}^k, i \in \mathbf{Z}\}$ for all $k \geq 1$. Hence, the *divided differences* can be defined:

$$\mathbf{D}_i^k := \frac{\mathbf{P}_{i+1}^k - \mathbf{P}_i^k}{t_{i+1}^k - t_i^k}. \tag{2.2}$$

and its control polygons, which are defined by the divided difference data $\{\mathbf{D}_i^k : i \in \mathbf{Z}, k \geq 0\}$, can be formed. A study of the control polygon sequence $\{\mathbf{D}_i^k\}$ which are also parametrized at $\{t_i^k\}$, gives the following result (*cf.* [10,13]):

Proposition 3. *If the continuous curve sequence* $\{\mathbf{D}^k(t)\}$ *converges uniformly to a continuous curve* $\mathbf{D}(t)$*, then* $\{\mathbf{P}^k(t)\}$ *converges uniformly to a differentiable curve* $\mathbf{P}(t)$*. Furthermore, we have*

$$\mathbf{P}'(t) = \mathbf{D}(t). \quad \blacksquare \tag{2.3}$$

Remark 4. *For* $k \geq 0$*, the control polygon curve* $\mathbf{P}^k(t)$ *is defined as the* C^0 *piecewise linear interpolant satisfying*

$$\mathbf{P}^k(t_i^k) = \mathbf{P}_i^k, \qquad \forall i \in \mathbf{Z} \tag{2.4}$$

and $\mathbf{D}^k(t)$ *is defined similarly. It is proved that this type of parametrization is regular (cf. [10]).*

This result suggests that smoother curves, C^2 say, could be produced by a subdivision scheme similar to (1.1) provided that its divided difference scheme can be written in form (1.1), and with the same adapted parametrization, produces C^1 curves. In next section, we will construct a subdivision algorithm for higher order B-splines.

§3. Subdivision scheme for cubic and quartic B-splines

Now we *integrate* the quadratic B-spline subdivision algorithm (1.1) to obtain the cubic B-spline algorithm. From the Lane-Riesenfeld algorithm, we assume that the scheme is in a form (*cf.* Figure 2):

$$\begin{cases} \mathbf{P}_{2i}^{k+1} &= \gamma_i^k \mathbf{P}_i^k + (1-\gamma_i^k)\mathbf{P}_{i+1}^k, \\ \mathbf{P}_{2i+1}^{k+1} &= \kappa_i^k \mathbf{P}_i^k + (1-\kappa_i^k-\rho_i^k)\mathbf{P}_{i+1}^k + \rho_i^k \mathbf{P}_{i+2}^k, \end{cases} \tag{3.1}$$

with its adapted parametrization $\{\xi_i^k\}$ given by

$$\begin{cases} \xi_{2i}^{k+1} &= \gamma_i^k \xi_i^k + (1-\gamma_i^k)\xi_{i+1}^k, \\ \xi_{2i+1}^{k+1} &= \kappa_i^k \xi_i^k + (1-\kappa_i^k-\rho_i^k)\xi_{i+1}^k + \rho_i^k \xi_{i+2}^k, \end{cases} \tag{3.2}$$

where, the initial parametric values $\{\xi_i^0\}$ are assumed to be strictly monotonic, *i.e.*, $\{\xi_i^0 : \xi_i^0 < \xi_{i+1}^0, \forall i \in \mathbf{Z}\}$.

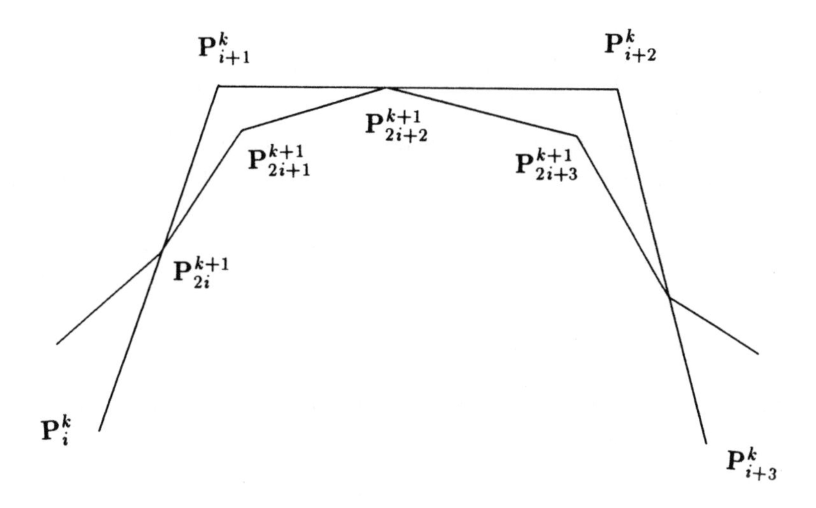

Figure 2. The construction of the subdivision scheme for cubic splines.

The divided difference scheme of (3.1), with its adapted parametrization (3.2), can then be written in form (1.1) with parameters α_i^k and β_i^k given by

$$\begin{cases} \alpha_i^k &= \dfrac{\gamma_i^k \Delta\xi_{i+1}^k}{[(\gamma_i^k - \kappa_i^k)\Delta\xi_i^k + \rho_i^k \Delta\xi_{i+1}^k]}, \\ \beta_i^k &= \dfrac{\kappa_i^k \Delta\xi_i^k}{[\gamma_i^k \Delta\xi_i^k + (1-\kappa_{i+1}^k - \rho_i^k)\Delta\xi_{i+1}^k]}, \end{cases} \tag{3.3}$$

where, Δ is the forward difference operator: $\Delta\xi_i^k := \xi_{i+1}^k - \xi_i^k$ etc.

Since the scheme is a continuous refinement of the control polygons, just like the Chaikin's algorithm, Greville's identity for B-splines suggests that the knots $\{x_i^k\}$ of the spline curve and its parametric points $\{t_i^k\}$ and $\{\xi_i^k\}$ should be interrelated by

$$\begin{cases} \xi_i^k &= \frac{1}{3}(x_{i-2}^k + x_{i-1}^k + x_i^k), \\ t_i^k &= \frac{1}{2}(x_{i-1}^k + x_i^k), \end{cases} \tag{3.4}$$

for all $i \in \mathbf{Z}$ and $k \geq 0$. These constraints require that the coefficients γ_i^k, κ_i^k and ρ_i^k in (3.1) should satisfy

$$\begin{cases} \gamma_i^k &= \dfrac{x_{i+1}^k - x_{2i-1}^{k+1}}{x_{i+1}^k - x_{i-2}^k}, \\ \kappa_i^k &= \dfrac{x_{i+1}^k - x_{2i+1}^{k+1}}{x_{i+1}^k - x_{i-1}^k}\gamma_i^k, \\ \rho_i^k &= \dfrac{x_{2i-1}^{k+1} - x_{i-1}^k}{x_{i+1}^k - x_{i-1}^k}\left(1 - \gamma_{i+1}^k\right), \end{cases} \tag{3.5}$$

where, the knots $\{x_i^k\}$ are given by the refinement equations (1.4).

Similarly, we can construct the subdivision algorithm for quartic B-splines. The scheme is given by

$$\begin{cases} \mathbf{P}_{2i}^{k+1} &= a_i^k \mathbf{P}_i^k + (1 - a_i^k - b_i^k)\mathbf{P}_{i+1}^k + b_i^k \mathbf{P}_{i+2}^k, \\ \mathbf{P}_{2i+1}^{k+1} &= c_i^k \mathbf{P}_i^k + (1 - c_i^k - d_i^k)\mathbf{P}_{i+1}^k + d_i^k \mathbf{P}_{i+2}^k, \end{cases} \tag{3.6}$$

where, $\{\mathbf{P}_i^k\}$ is the *control polygon* at level k and the coefficients in (3.6) are given by

$$\begin{cases} a_i^k &= \dfrac{(x_{i+2}^k - x_{2i-1}^{k+1})(x_{i+2}^k - x_{2i+1}^{k+1})}{(x_{i+2}^k - x_{i-2}^k)(x_{i+2}^k - x_{i-1}^k)}, \\ b_i^k &= \dfrac{(x_{2i-1}^{k+1} - x_{i-1}^k)(x_{2i+1}^{k+1} - x_{i-1}^k)}{(x_{i+2}^k - x_{i-1}^k)(x_{i+3}^k - x_{i-1}^k)}, \\ c_i^k &= \dfrac{(x_{2i+3}^{k+1} - x_{2i+1}^{k+1})(x_{i+2}^k - x_{2i+1}^{k+1})}{(x_{i+2}^k - x_{i-2}^k)(x_{i+3}^k - x_{i-1}^k)}, \\ d_i^k &= \dfrac{(x_{2i+1}^{k+1} - x_{i-1}^k)(x_{2i+3}^{k+1} - x_{i-1}^k)}{(x_{i+2}^k - x_{i-1}^k)(x_{i+3}^k - x_{i-1}^k)}. \end{cases} \tag{3.7}$$

From the above discussion, we now conclude the following results. A proof of them will be given in the next section.

Theorem 5. *The non-uniform scheme (3.2) produces a cubic B-spline curve with knots $\{x_i^0\}$ and control points $\{\mathbf{P}_i^0\}$ provided that the coefficients $\{\gamma_i^k, \kappa_i^k, \rho_i^k\}$ are given by (3.5) and the refined knots $\{x_i^k\}$ satisfy (1.4).* ∎

Remark 6. *If the initial knots are equally spaced and the subdivision parameter* $\theta_i^k = \frac{1}{2}, \forall i \in \mathbf{Z}, k \geq 0$, *then* $\gamma_i^k = \frac{1}{2}, \kappa_i^k = \rho_i^k = \frac{1}{8}, \forall i \in \mathbf{Z}, k \geq 0$, *and thus scheme (3.1) becomes the Catmull-Clark's algorithm for uniform cubic B-spline curves (cf.* [11,13]).

Theorem 7. *Scheme (3.6) produces a quartic B-spline curve with knots* $\{x_i^0\}$ *and control points* $\{\mathbf{P}_i^0\}$ *provided that the coefficients* $\{a_i^k, b_i^k, c_i^k, d_i^k\}$ *are given by (3.7) and the refined knots* $\{x_i^k\}$ *satisfy (1.4).* ■

Remark 8. *If the initial knots are equally spaced and the subdivision parameter* $\theta_i^k = \frac{1}{2}, \forall i \in \mathbf{Z}, k \geq 0$, *then* $a_i^k = d_i^k = \frac{5}{16}$ *and* $b_i^k = c_i^k = \frac{1}{16}, \forall i \in \mathbf{Z}, k \geq 0$, *and thus scheme (3.6) becomes the subdivision algorithm for uniform quartic B-splines (cf.* [4,11,13]).

For higher order B-spline curves, a similar subdivision scheme can also be derived which will be discussed in the next section. The idea is the same but the calculations of the corresponding coefficients are more complicated. Hence, we will use the knot insertion technique to construct the subdivision algorithm for higher B-spline curves with simple knots.

§4. Subdivision scheme for B-splines of order n

In this section, we formulate systematically a general subdivision scheme for B-spline curves of order n with simple knots. The recurrence relations of B-spline basis functions play an important rule in the construction.

Let $\{B_{i,n}^k(x)\}$ denote the normalized B-spline basis of order n with simple knots $\{x_i^k\}$ and $\{B_{i,n}^{k+1}(x)\}$ be the normalized B-spline basis of order n with knots $\{x_i^{k+1}\}$, where $\{x_i^k\}$ and $\{x_i^{k+1}\}$ satisfy the strictly refinement relation (1.4). Then, from the definition of the B-spline basis (*cf.* [3]), we have the following linear relation (*cf.* Figure 3, where $n = 3$):

$$B_{i,n}^k(x) = \sum_{j=0}^{n} b_{i,j,n}^k B_{2i+j,n}^{k+1}(x), \qquad i \in \mathbf{Z}, \quad k = 0,1,2,..., \quad \forall n > 1, \tag{4.1}$$

where, $\{b_{i,j,n}^k\}$ are also called *discrete B-splines* (*cf.* [7]). It can be shown that $b_{i,j,n}^k$ is determined by knots $\{x_m^{k+1}, m = 2i, ..., 2i + 2n\}$ only. The explicit expression for $b_{i,j,n}^k$ will be given recursively later in this section.

To formulate the subdivision algorithm, we suppose a B-spline curve $\mathbf{P}(x)$ of order n is given in the following form:

$$\mathbf{P}(x) = \sum_i \mathbf{P}_i^k B_{i,n}^k(x), \qquad \forall k \geq 0. \tag{4.2}$$

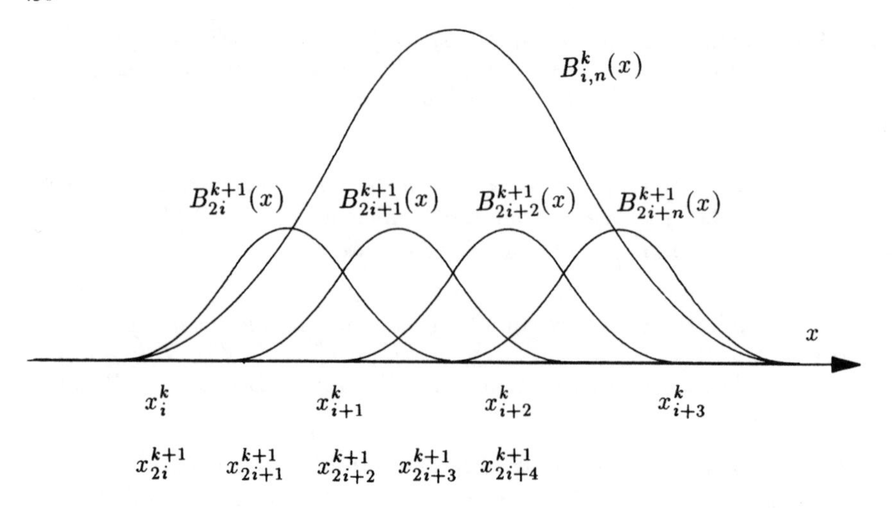

Figure 3. The relations between the normalised B-spline bases.

Then, substituting $B_{i,n}^{k}(x)$ by (4.1), we obtain, for $k \geq 0$:

$$\begin{aligned} \mathbf{P}(x) &= \sum_{i} \mathbf{P}_{i}^{k} B_{i,n}^{k}(x) \\ &= \sum_{i} \mathbf{P}_{i}^{k} \left(\sum_{j=0}^{n} b_{i,j,n}^{k} B_{2i+j,n}^{k+1}(x) \right) \\ &= \sum_{i} \mathbf{P}_{i}^{k+1} B_{i,n}^{k+1}(x), \end{aligned} \tag{4.3}$$

where, refined control polygons $\{\mathbf{P}_{i}^{k+1}\}$ are obtained by the following 2–step recursive formulae:

$$\begin{cases} \mathbf{P}_{2i}^{k+1} = \displaystyle\sum_{j=0}^{[n/2]} b_{i-j,2j,n}^{k} \mathbf{P}_{i-j}^{k}, \\ \mathbf{P}_{2i+1}^{k+1} = \displaystyle\sum_{j=0}^{[(n-1)/2]} b_{i-j,2j+1,n}^{k} \mathbf{P}_{i-j}^{k}. \end{cases} \tag{4.4}$$

Here, $\left[\frac{n}{2}\right]$ is the maximum integer less than or equal to $\frac{n}{2}$ and $\left[\frac{n-1}{2}\right]$ is defined similarly.

It is now obvious that equations in (4.4) define a non-uniform subdivision algorithm for B-spline curves with simple knots. Moreover, the B-spline relation coefficients, *i.e.*, the discrete B-splines $\{b_{i,j,n}^{k}\}$ are just the weights of the subdivision scheme.

For completeness, we derive the recurrence relations of $\{b^k_{i,j,n}\}$ on n, the order of the splines. For simplicity, we assume that

$$b^k_{i,j,n} = 0, \qquad \text{for} \quad j < 0 \quad \text{or} \quad j > n, \quad i \in \mathbf{Z}. \tag{4.5}$$

For $n = 1$ and 2, it can be shown easily that for $i \in \mathbf{Z}, k \geq 0$:

$$b^k_{i,0,1} = b^k_{i,1,1} = 1, \tag{4.6}$$

and

$$\begin{cases} b^k_{i,0,2} &= \dfrac{x^{k+1}_{2i+1} - x^{k+1}_{2i}}{x^k_{i+1} - x^k_i}, \\ b^k_{i,1,2} &= 1, \\ b^k_{i,2,2} &= \dfrac{x^{k+1}_{2i+4} - x^{k+1}_{2i+3}}{x^k_{i+2} - x^k_{i+1}}. \end{cases} \tag{4.7}$$

For $n \geq 3$, we have, for $i \in \mathbf{Z}$, $k \geq 0$, by differentiating both sides of equation (4.1):

$$\frac{dB^k_{i,n}(x)}{dx} = \sum_{j=0}^{n} b^k_{i,j,n} \frac{dB^{k+1}_{2i+j,n}(x)}{dx}. \tag{4.8}$$

Substituting the above derivative terms by the following formula for the derivative of B-splines (*cf.* [3]):

$$\frac{dB^k_{i,n}(x)}{dx} = (n-1)\left\{ \frac{B^k_{i,n-1}(x)}{x^k_{i+n-1} - x^k_i} - \frac{B^k_{i+1,n-1}(x)}{x^k_{i+n} - x^k_{i+1}} \right\}, \tag{4.9}$$

then replacing the terms $B^k_{i,n-1}(x)$ and $B^k_{i+1,n-1}(x)$ by (4.1) for $n-1$, and rearranging the terms, we obtain:

$$\frac{\sum_{j=0}^{n-1} b^k_{i,j,n-1} B^{k+1}_{2i+j,n-1}(x)}{x^k_{i+n-1} - x^k_i} - \frac{\sum_{j=0}^{n-1} b^k_{i+1,j,n-1} B^{k+1}_{2i+j+2,n-1}(x)}{x^k_{i+n} - x^k_{i+1}}$$

$$= \sum_{j=0}^{n} b^k_{i,j,n} \left\{ \frac{B^{k+1}_{2i+j,n-1}(x)}{x^{k+1}_{2i+j+n-1} - x^{k+1}_{2i+j}} - \frac{B^{k+1}_{2i+j+1,n-1}(x)}{x^{k+1}_{2i+j+n} - x^{k+1}_{2i+j+1}} \right\}. \tag{4.10}$$

Because the B-spline basis $\{B_{i,n-1}^{k+1}(x)\}_{-\infty}^{+\infty}$ are linearly independent for all $x \in \mathbf{R}$, equation (4.10) can only be true if and only if the coefficients of $B_{i,n-1}^{k+1}(x), i \in \mathbf{Z}$, are the same in both sides. This gives the following recurrence relations for $b_{i,j,n}^{k}$ $(n \geq 3)$:

$$\begin{aligned} b_{i,j,n}^{k} = b_{i,j-1,n}^{k} &+ \frac{x_{2i+j+n-1}^{k+1} - x_{2i+j}^{k+1}}{x_{2i+2n-2}^{k+1} - x_{2i}^{k+1}} b_{i,j,n-1}^{k} \\ &- \frac{x_{2i+j+n-1}^{k+1} - x_{2i+j}^{k+1}}{x_{2i+2n}^{k+1} - x_{2i+2}^{k+1}} b_{i+1,j-2,n-1}^{k}, \\ &j = 0, 1, ..., n. \end{aligned} \tag{4.11}$$

From (4.11), (4.6), (4.7) and the assumption (4.5), all the weights, or the *discrete B-splines*, $\{b_{i,j,n}^{k}\}$ can be obtained recursively. A special case of (4.11) is the uniform case whereby all the knots are equally spaced. In this case, (4.11) becomes quite simple:

$$\begin{aligned} b_{i,j,n}^{k} = b_{i,j,n-1}^{k} + \tfrac{1}{2}\left(b_{i,j,n-1}^{k} - b_{i+1,j-2,n-1}^{k}\right), \\ j = 0, 1, ..., n, \quad n \geq 3. \end{aligned} \tag{4.12}$$

This linear difference equation, together with the initial conditions (4.6), (4.7) and convention (4.5), has a unique solution:

$$b_{i,j,n}^{k} = \frac{(n-1)!}{2^{n-1} j!(n-j-1)!}, \quad 0 \leq j \leq n, \quad i \in \mathbf{Z}, \quad k \geq 0, \quad n \geq 1, \tag{4.13}$$

which are just the binomial coefficient weights.

This result is the same as the line averaging algorithm for uniform B-splines described by Dyn, Gregory [8] and Levin and Lane & Riesenfeld in [11].

Remark 9. *Another relation for $\{b_{i,j,n}^{k}\}$ can also be formulated by applying the recurrence relation for the B-spline basis $B_{i,n}^{k}(x)$ and $B_{i,n}^{k+1}(x)$ to both sides of equation (4.1). Thus (4.11) can be written in a very nice form:*

$$\begin{aligned} b_{i,j,n}^{k} = \frac{x_{2i+j+n-1}^{k+1} - x_{i}^{k}}{x_{i+n-1}^{k} - x_{i}^{k}} b_{i,j,n-1}^{k} + \frac{x_{i+n}^{k} - x_{2i+j+n-1}^{k+1}}{x_{i+n}^{k} - x_{i+1}^{k}} b_{i+1,j-2,n-1}^{k}, \\ j = 0, 1, \ldots, n, n = 2, 3, ..., k = 1, 2, 3, ... \end{aligned} \tag{4.14}$$

From the above formulations, we conclude:

Theorem 10. *The subdivision algorithm (4.4), with* $\{b^k_{i,j,n}\}$ *given by (4.5–4.7, 4.11), generates a B–spline curve of order n with control points* $\{\mathbf{P}^0_i\}$ *and knots* $\{x^0_i\}$. ■

§5. Remarks

Remark 11. *The quadratic B-spline scheme can be proved by many methods. A simple geometric proof comes directly from the properties of parabolic curves.*

Remark 12. *The subdivision scheme for B-spline curves is just a refinement scheme. They can be regarded as a special case of the Böhm's knot insertion algorithm (simultaneous knot insertion).*

Remark 13. *The subdivision scheme can be generalized to surfaces. Hence, any tensor-product B-spline surface (with simple knots) can be computed by a corresponding subdivision algorithm (either uniform or non-uniform algorithm).*

Remark 14. *An important application of the algorithm is that, due to its flexibility, it is very useful in interactive design. For instance, from the same control polygon, different curves or surfaces can be produced if different knots are chosen. Furthermore, by adjusting some appropriate control points, the required curves or surfaces can be controlled easily.*

Remark 15. *The non-uniform subdivision algorithm is a special case of the Oslo algorithm (cf.* [7]) for B-splines. Hence, if the initial knots are equally spaced and the new knots are spread uniformly, then the scheme degenerates to the Lane-Riesenfeld's line average algorithm.

Remark 16. *(4.11) and (4.14) are still true for splines with multiple knots.*

Remark 17. *From (4.14), it is obvious that the subdivision weightings* $\{b^k_{i,j,n}\}$ *are positive and hence many properties of B-spline curves can be proved by this subdivision process.*

§6. Examples

Six graphic examples are given in this section. These figures are drawn by the Nicolet Drum Plotter at Brunel University in the United Kingdom. For simplicity, only the initial control polygons and the limit curves (in fact, the control polygons after the sixth subdivision) are displayed and the knot parameters $\theta^k_i \equiv \frac{1}{2}\ \forall i, k$. Most of the initial control polygons are the same: just the unit square, *i.e.*, there are only four initial data points, the corner points of the unit square. The control points are marked by small triangles centred at these points.

In Figure 4, two different quadratic spline curves are produced by choosing different knots (shape controls). Both uniform and non-uniform cubic and quartic spline curves generated by scheme (3.1) and (3.6) are shown in Figure 5 and Figure 6. In these figures, the non-uniform splines have the same periodic knot sequence: $1.0, 2.0, 2.5, 3.0, 4.0, 5.0, 5.5, \ldots$ It is obvious that these *shape controls* provide plenty parameters for free–form curves.

§7. Conclusion

Based on the analysis of the non-uniform corner cutting scheme (1.1) and the adapted parametrization technique, a non-uniform subdivision algorithm for B-spline curves of any order with simple knots are derived. The key to the success of this analysis is the Greville's identity for B-splines which relates the non-uniform parametrization of the scheme and the parametrization of its divided difference scheme. Other relatively simple proofs of the results can be obtained by using either the curve refinement techniques (*cf.* [7]) or the Böhm's knot insertion idea.

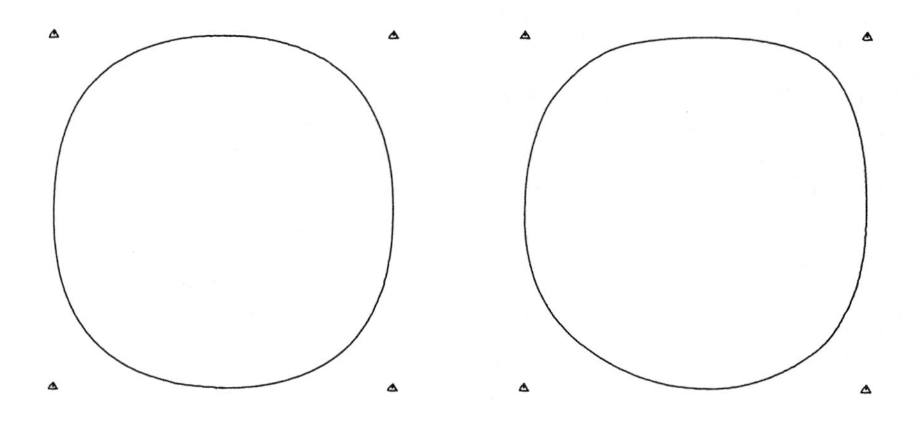

Uniform case　　　　Non-uniform case

Figure 4. Quadratic spline curves produced by the corner cutting scheme.

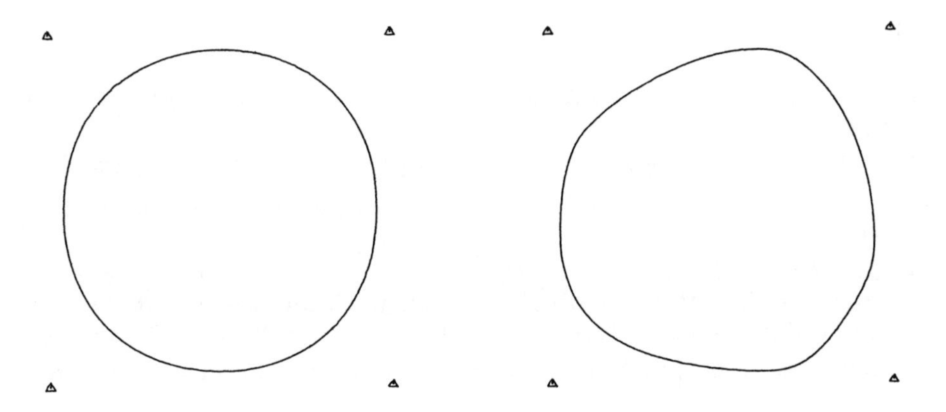

Figure 5. Curves produced by the subdivision scheme for cubic splines.

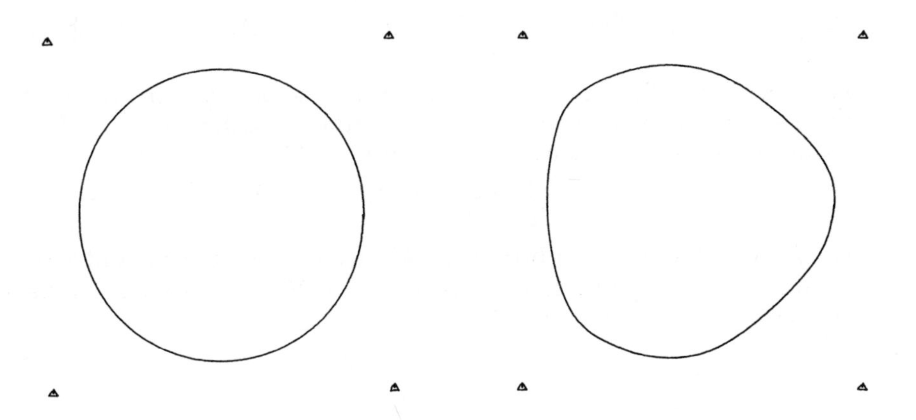

Figure 6. Curves produced by the subdivision scheme for quartic splines.

References

1. Böhm, W., "Inserting new knots into B-spline curves", CAD 12, 1980, 199–216.
2. De Boor, C., "Corner cutting always works", CAGD 4, 1987, 269–278.
3. De Boor, C., "On calculating with B-splines", J. Appx. Theory 6, 1972, 50–62.
4. Cavaretta, A. C. and Micchelli, C. M., "The design of curves and surfaces by subdivision", in **Mathematical Methods in Computer Aided Geometric Design**, Lyche, T. & Schumaker, L. L. eds, New York, 1989, 115–154.
5. Cavaretta, A. C., Micchelli, C. M. and Dahmen, W., "Regular Subdivision", preprint.
6. Chaikin, G. M., "An algorithm for high speed curve generation", Computer Graphics and Image Processing 3, 1974, 346–349.
7. Cohen, E., Lyche, T. and Riesenfeld, R., "Discrete B-splines and subdivision techniques in computer aided geometric design and computer graphics", Computer Graphics and Image Processing 14, 1980, 87-111.
8. Dyn, N., Gregory, J. A. and Levin, D., "Analysys of uniform binary subdivision scheme for curve design", to appear in Constructive Approximation.
9. Goodman,T. N. T. and Micchelli, C. A., "Corner cutting algorithms for the Bézier representation of free form curves", Rep. No. 54611, 1986, IBM Research centre, Yorktown Heights, New York.
10. Gregory, J. A. and Qu, R. B., "Non-uniform corner cutting", to appear in CAGD.
11. Lane, J. M. and Riesenfeld, R. F., "A theoretical development for the computer generation of piecewise polynomial surfaces", IEEE Trans. on Pattern Analysis and Machine Intellegence 2, 1980, 35–46.
12. Lu Wei, Jin Tonggun and Liang Youdong, "A new method for curve and surface modelling", to appear in CAGD.
13. Qu, R. B., **"Recursive Subdivision Algorithms for Curve and Surface Design"**, Ph.D Thesis, 1990, Department of Mathematics and Statistics, Brunel University, Uxbridge, Middlesex, Britain.

SOME APPLICATIONS OF AN APPROXIMATION THEOREM FOR FIXED POINTS OF MULTI-VALUED CONTRACTIONS

BIAGIO RICCERI
Department of Mathematics
University of Messina
98166 Sant'Agata-Messina
Italy

ABSTRACT. In this lecture we report some applications (concerning nonlinear functional analysis, differential inclusions and control theory) of an approximation theorem by O.Naselli Ricceri ([4], Theorem 3.4) for fixed points of multi-valued contractions.

Let (X,d) be a metric space. For $x_0 \in X$ and non-empty $A, C \subseteq X$, put:

$$d(x_0, A) = \inf_{y \in A} d(x_0, y)$$

$$d_H(A,C) = \max \{\sup_{x \in A} d(x,C), \sup_{x \in C} d(x,A)\} .$$

A non-empty-valued multifunction $F: X \to 2^X$ is said to be *Lipschitzian* provided there is some constant $L \geq 0$ (called Lipschitz constant) such that

$$d_H(F(x), F(y)) \leq L d(x,y)$$

for all $x, y \in X$. When $L < 1$, F is said to be a *multi-valued contraction*. We denote by Fix(F) the set of all fixed points of F. That is to say:

$$\text{Fix}(F) = \{x \in X: x \in F(x)\} .$$

It is well known that if X is complete and if F is a multi-valued

S. P. Singh (ed.), Approximation Theory, Spline Functions and Applications, 437–444.

contraction with closed values, then the set Fix(F) is non-empty and closed. Given a sequence $\{A_n\}$ of non-empty subsets of X, we put:

$$\operatorname{Li}_{n\to\infty} A_n = \{x\in X: \lim_{n\to\infty} d(x,A_n)=0\} .$$

So, a point $x\in X$ belongs to $\operatorname{Li}_{n\to\infty} A_n$ if and only if there exists a sequence $\{x_n\}$ in X such that $\lim_{n\to\infty} d(x,x_n)=0$ and $x_n\in A_n$ for all $n\in\mathbb{N}$.

In [4], as a corollary of a more general result, O.Naselli Ricceri obtained the following approximation theorem:

THEOREM 1 ([4], Theorem 3.4). - *Let* (X,d) *be a complete metric space and let* $F, F_1, F_2, \ldots$ *be a sequence of multi-valued contractions, from* X *into itself, with closed values, having a same Lipschitz constant* $L<1$. *Further, assume that*

$$F(x)\subseteq \operatorname{Li}_{n\to\infty} F_n(x)$$

for all $x\in X$.
Then, one has

$$\operatorname{Fix}(F)\subseteq \operatorname{Li}_{n\to\infty} \operatorname{Fix}(F_n) .$$

Owing to its generality, Theorem 1 can be usefully applied in many different settings. The aim of this lecture is just to report three samples of application of Theorem 1.

The first application concerns nonlinear functional analysis. So, let $(X,\|\cdot\|_X)$, $(Y,\|\cdot\|_Y)$ be two real Banach spaces. We denote by $\mathcal{L}(X,Y)$ the space of all continuous linear operators from X into Y, endowed with the usual norm. For any surjective $\Phi\in\mathcal{L}(X,Y)$, put:

$$\alpha_\Phi = \sup\ \{d(0,\Phi^{-1}(y)): y\in Y,\ \|y\|_Y\le 1\} ,$$

d being, of course, the metric induced by $\|\cdot\|_X$.
By the open mapping theorem, we have $\alpha_\Phi<+\infty$.

THEOREM 2 ([10], Théorème 2). - *Let* $(X,\|\cdot\|_X)$, $(Y,\|\cdot\|_Y)$ *be two real*

Banach spaces; $\{\Phi_n\}$ *a sequence of continuous linear operators from* X *onto* Y *converging, in* $\mathcal{L}(X,Y)$, *to a surjective operator* Φ; $\{\Psi_n\}$ *a sequence of Lipschitzian operators from* X *into* Y, *each of which with Lipschitz constant* L_n, *converging pointwise to an operator* Ψ. *Assume that*

$$\sup_{n\in\mathbb{N}} \alpha_{\Phi_n} L_n<1 \ .$$

Under such assumptions, for each $y\in Y$ *and each sequence* $\{y_n\}$ *in* Y *converging to* y, *one has*

$$\emptyset\neq(\Phi+\Psi)^{-1}(y)\subseteq \operatorname*{Li}_{n\to\infty} (\Phi_n+\Psi_n)^{-1}(y_n) \ .$$

SKETCH OF PROOF. For each $n\in\mathbb{N}$, $x\in X$, $y\in Y$, put:

$$F_n(x)=\Phi_n^{-1}(y_n-\Psi_n(x))$$

$$F(x)=\Phi^{-1}(y-\Psi(x)) \ .$$

Of course, one has

$$(\Phi_n+\Psi_n)^{-1}(y_n)=\mathrm{Fix}(F_n)$$

$$(\Phi+\Psi)^{-1}(y)=\mathrm{Fix}(F) \ .$$

It is seen that each multifunction F_n $(n\in\mathbb{N})$ is Lipschitzian, with Lipschitz constant $\alpha_{\Phi_n} L_n$. On the other hand, one has $\lim_{n\to\infty} \alpha_{\Phi_n}=\alpha_{\Phi}$. This implies that also F is a multi-valued contraction. Further, it is possible to prove that

$$F(x)\subseteq \operatorname*{Li}_{n\to\infty} F_n(x)$$

for all $x\in X$ (this is the most delicate part of the proof). Now, our conclusion follows directly from Theorem 1. ■

The next application of Theorem 1 concerns differential inclusions. So, let $(E,\|\cdot\|)$ be a separable Banach space and [a,b] be a compact real interval. We denote by $AC([a,b],E)$ the space of all strongly absolutely continuous functions $\varphi\colon [a,b]\longrightarrow E$ for which the strong derivative φ' does exist a.e. in $[a,b]$. We consider $AC([a,b],E)$ equipped with the norm

$$\|\varphi\|_{AC}= \max_{t\in[a,b]} \|\varphi(t)\|+\int_a^b \|\varphi'(t)\|dt .$$

For each $t_0\in[a,b]$, $x_0\in E$ and each multifunction $F\colon [a,b]\times E\longrightarrow 2^E$, we put:

$$\Gamma(t_0,x_0,F)=\{\varphi\in AC([a,b],E)\colon \varphi'(t)\in F(t,\varphi(t)) \text{ a.e. in } [a,b],$$

$$\varphi(t_0)=x_0\} .$$

THEOREM 3 ([9], Theorem 3.1). - *Let* $F,F_1,F_2,\dots$ *be a sequence of multifunctions from* $[a,b]\times E$ *into* E, *with non-empty closed values, satisfying the following conditions:*

(i) *for every* $x\in E$, *each multifunction* $F(\cdot,x)$, $F_n(\cdot,x)$ $(n\in\mathbb{N})$ *is Lebesgue measurable in* $[a,b]$;

(ii) *there is some* $g\in L^1([a,b])$ *such that*

$$\max\{d_H(F(t,x),F(t,y)),\ d_H(F_n(t,x),F_n(t,y))\}\le g(t)\|x-y\|$$

for a.e. $t\in[a,b]$ *and for all* $x,y\in E$, $n\in\mathbb{N}$ (*d being the metric induced by* $\|\cdot\|$);

(iii) *one has*

$$F(t,x)\subseteq \operatorname*{Li}_{n\to\infty} F_n(t,x)$$

for a.e. $t\in[a,b]$ *and for all* $x\in E$.

Then, the following assertions are equivalent:

(a) *The set functions* $A\longrightarrow\int_A d(0,F_n(t,0))dt$ $(n\in\mathbb{N})$ *are equi-absolutely continuous.*

(b) *For every* $t_0\in[a,b]$, $x_0\in[a,b]$, *the sets* $\Gamma(t_0,x_0,F)$, $\Gamma(t_0,x_0,F_n)$ $(n\in\mathbb{N})$ *are non-empty and closed, and one has*

$$\Gamma(t_0,x_0,F) \subseteq \operatorname*{Li}_{n\to\infty} \Gamma(t_0,x_0,F_n) .$$

SKETCH OF PROOF. We sketch only the proof of the implication (a)⇒(b). So, let (a) hold. Given $t_0\in[a,b], x_0\in E$, fix $M>1$ and consider $L^1([a,b],E)$ endowed with the norm

$$\|\varphi\|_1 = \int_a^b \exp\left(-2M\left|\int_{t_0}^t g(\tau)d\tau\right|\right)\|\varphi(t)\|dt .$$

Of course, this norm is equivalent to the usual one. Now, for each $\psi\in L^1([a,b],E)$, $n\in\mathbb{N}$, put:

$$\Phi_n(\psi) = \left\{\varphi\in L^1([a,b],E):\ \varphi(t)\in F_n\left(t, x_0+\int_{t_0}^t \psi(\tau)d\tau\right) \text{ a.e. in } [a,b]\right\}$$

$$\Phi(\psi) = \left\{\varphi\in L^1([a,b],E):\ \varphi(t)\in F\left(t, x_0+\int_{t_0}^t \psi(\tau)d\tau\right) \text{ a.e. in } [a,b]\right\} ,$$

the integral $\int_{t_0}^t \psi(\tau)d\tau$ being taken in the sense of Bochner. One proves that the multifunctions Φ, Φ_n ($n\in\mathbb{N}$) have non-empty closed values and that they are multi-valued contractions (with respect to the metric induced by $\|\cdot\|_1$) with Lipschitz constant $\frac{1}{M}$. Moreover, making essential use of (a), one proves that

$$\Phi(\psi) \subseteq \operatorname*{Li}_{n\to\infty} \Phi_n(\psi)$$

for all $\psi\in L^1([a,b],E)$. Consequently, by Theorem 1, one has

$$\operatorname{Fix}(\Phi) \subseteq \operatorname*{Li}_{n\to\infty} \operatorname{Fix}(\Phi_n) .$$

Now, consider the operator $T: L^1([a,b],E) \longrightarrow AC([a,b],E)$ defined by putting

$$T(\psi)(t) = x_0 + \int_{t_0}^t \psi(\tau)d\tau$$

for all $\psi\in L^1([a,b],E)$, $t\in[a,b]$.

Of course, T is an one-to-one affine continuous operator which maps closed sets onto closed sets. So, our conclusion follows observing that $\Gamma(t_0,x_0,F)=T(\mathrm{Fix}(\Phi))$, $\Gamma(t_0,x_0,F_n)=T(\mathrm{Fix}(\Phi_n))$ $(n\in\mathbb{N})$. ∎

The last application of Theorem 1 here reported concerns control theory and is obtained via Theorem 3.

Let $n,m\in\mathbb{N}$. We denote by $\mathbb{R}^{n,m}$ the space of all real $n\times m$-matrices. We will consider the spaces $\mathbb{R}^n$, $\mathbb{R}^m$ endowed with their Euclidean norms. The norm of a matrix $C\in\mathbb{R}^{n,m}$ is denoted by $\|C\|$, that is $\|C\|=\sup\{\|Cu\|_{\mathbb{R}^n}:u\in\mathbb{R}^m, \|u\|_{\mathbb{R}^m}\le 1\}$. The symbol $\mathfrak{M}([a,b],\mathbb{R}^{n,m})$ denotes the space of all measurable functions $B:[a,b]\longrightarrow\mathbb{R}^{n,m}$.

On a given compact interval $[a,b]\subseteq\mathbb{R}$, consider the linear control system

$$x'=A(t)x+B(t)u \tag{1}$$

where A is a function from $[a,b]$ into $\mathbb{R}^{n,n}$ and B is a function from $[a,b]$ into $\mathbb{R}^{n,m}$.

We say that the system (1) is completely controllable if for every $x_0,x_1\in\mathbb{R}^n$, there exist $u\in L^\infty([a,b],\mathbb{R}^m)$ and $\varphi\in AC([a,b],\mathbb{R}^n)$ such that

$$\begin{cases} \varphi'(t)=A(t)\varphi(t)+B(t)u(t) & \text{a.e. in } [a,b] \\ \varphi(a)=x_0 \\ \varphi(b)=x_1. \end{cases}$$

THEOREM 4 ([5], Theorem 1). - *Let* $A\in L^1([a,b],\mathbb{R}^{n,n})$, $B\in\mathfrak{M}([a,b],\mathbb{R}^{n,m})$ *and let the system* (1) *be completely controllable. Then, there exists* $\rho>0$ *such that, for every* $\tilde{A}\in L^1([a,b],\mathbb{R}^{n,n})$, $\tilde{B}\in\mathfrak{M}([a,b],\mathbb{R}^{n,m})$ *satisfying*

$$\int_a^b\|\tilde{A}(t)-A(t)\|dt+\int_a^b\frac{\|\tilde{B}(t)-B(t)\|}{1+\|\tilde{B}(t)-B(t)\|}dt<\rho\ ,$$

the system

$$x'=\tilde{A}(t)x+\tilde{B}(t)u$$

is completely controllable.

SKETCH OF PROOF. Arguing by contradiction, assume that there are a sequence $\{A_k\}$ in $L^1([a,b],\mathbb{R}^{n,n})$ and a sequence $\{B_k\}$ in $\mathfrak{M}([a,b],\mathbb{R}^{n,m})$ such that, for each $k\in\mathbb{N}$, one has

$$\int_a^b \|A_k(t)-A(t)\|dt+\int_a^b \frac{\|B_k(t)-B(t)\|}{1+\|B_k(t)-B(t)\|}\,dt<\frac{1}{k}$$

and the system

$$x'=A_k(t)x+B_k(t)u$$

is not completely controllable.
So, in particular, there are an increasing sequence $\{k_r\}$ in $\mathbb{N}$ and a function $g\in L^1([a,b])$ such that

$$\lim_{r\to\infty} \|A_{k_r}(t)-A(t)\|+\|B_{k_r}(t)-B(t)\|=0 \text{ and } \|A_{k_r}(t)\|\le g(t)$$

a.e. in $[a,b]$, for all $r\in\mathbb{N}$.
Now, for each $t\in[a,b]$, $x\in\mathbb{R}^n$, $r\in\mathbb{N}$, put

$$F_r(t,x)=A_{k_r}(t)x+B_{k_r}(t)S$$

$$F(t,x)=A(t)x+B(t)S$$

where S is the closed unit ball of $\mathbb{R}^m$.
Taking into account the preceding remark, one sees that the multifunctions F,F_r $(r\in\mathbb{N})$ satisfy conditions (i), (ii), (iii) and (a) of Theorem 3. Consequently, keeping the same notation as in Theorem 3, if we put:

$$V_r=\{\varphi(b):\ \varphi\in\Gamma(a,0,F_r)\}$$

$$V=\{\varphi(b):\ \varphi\in\Gamma(a,0,F)\}\ ,$$

we have

$$V\subseteq \operatorname*{Li}_{r\to\infty} V_r\ .$$

Since the system (1) is completely controllable, we have $\mathrm{int}(V)\neq\emptyset$. From this and from the preceding inclusion, it follows that $\mathrm{int}(V_r)\neq\emptyset$ for r large enough, against the fact the each system $x'=A_k(t)x+B_k(t)u$ is not completely controllable. ∎

REFERENCES

[1] G.BONANNO and S.A.MARANO, *Random differential inclusions depending on a parameter*, J.Math. Anal. Appl., to appear.

[2] J.P.DAUER, *Perturbations of linear control systems*, SIAM J. Control Optim., **9** (1971), 393-400.

[3] S.A.MARANO, *Classical solutions of partial differential inclusions in Banach spaces*, Appl. Anal., to appear.

[4] O.NASELLI RICCERI, *A-fixed points of multi-valued contractions*, J. Math. Anal. Appl., **135** (1988), 406-418.

[5] O.NASELLI RICCERI, *A theorem on the controllability of perturbed linear control systems*, Rend. Accad. Naz. Lincei, **83** (1989), 89-91.

[6] O.NASELLI RICCERI, *On the controllability of a class of differential inclusions depending on a parameter*, J. Optim. Th. Appl., **65** (1990), 281-288.

[7] O.NASELLI RICCERI, *On some classes of implicit partial differential equations in Banach spaces*, Rend. Accad. Naz. Sci. XL, Mem. Mat., 14 (1990), 67-85.

[8] O.NASELLI RICCERI, *Classical solutions of the problem* $x'\in F(t,x,x')$, $x(t_0)=x_0$, $x'(t_0)=y_0$ *in Banach spaces*, Funkc. Ekv., to appear.

[9] O.NASELLI RICCERI and B.RICCERI, *Differential inclusions depending on a parameter*, Bull. Pol. Acad. Sci. Math., **37** (1989), 665-671.

[10] B.RICCERI, *Structure, approximation et dépendance continue des solutions de certaines équations non linéaires*, C.R.Acad. Sci. Paris, Série I, **305** (1987), 45-47.

[11] B.RICCERI, *Sur les solutions classiques du problème de Darboux pour certaines équations aux dérivées partielles sous forme implicite dans les espaces de Banach*, C.R.Acad, Sci. Paris, Série I, **307** (1988), 325-328.

GEOMETRICAL DIFFERENTIATION AND HIGH–ACCURACY CURVE INTERPOLATION

ROBERT SCHABACK
Institut für Numerische und Angewandte Mathematik
Universität Göttingen
Lotzestraße 16-18
W-3400 Göttingen
Germany

ABSTRACT: Let f be a smooth curve in $\mathbb{R}^d$, parametrized by arclength. If a large sample of data points $p_i = f(t_i)$ at unknown parameter values $t_i < t_{i+1}$ is given, one can use local n–th degree polynomial interpolation at parameters $s_i = \|P_i - P_\ell\| \text{sgn}(i - \ell)$ of data points P_i around a fixed point P_ℓ to calculate approximations to the derivatives $f^{(j)}(t_\ell)$ with accuracy $\mathcal{O}(h^{n+1-j})$, where $h := \max(t_i - t_{i-1})$ and $0 \leq j \leq k-1 \leq n$. Using these as data for properly parametrized Hermite interpolation problems for polynomials of degree $\leq 2k-1 \leq n$ between successive data points, one can construct GC^{k-1} interpolants of f with accuracy $\mathcal{O}(h^{2k})$.

1. Introduction

The classical problem of numerical differentiation consists in finding an approximation of the j–th derivative $f^{(j)}(t^*)$ of some smooth real–valued function f on $[a, b] \subset \mathbb{R}$ in a given point $t^* \in [a, b]$, if $n + 1$ nodes

$$a \leq t_0 < t_1 < \ldots < t_n \leq b$$

and $n + 1$ real function values

$$f(t_0), f(t_1), \ldots, f(t_n)$$

are given. The standard approach simply takes the j–th derivative of the n–th degree polynomial p interpolating these data, and the error is easily evaluated from the representation

$$f(t) - p(t) = \left(\prod_{i=0}^{n} (t - t_i) \right) \Delta^{n+1}(t_0, t_1, \ldots, t_n, t) f, \tag{1.1}$$

where $\Delta^i(t_0, \ldots, t_i) f$ is the i–th divided difference of f with respect to the nodes $t_0, t_1, \ldots, t_i$. The j–th derivative of (1.1) at $t^* \in [t_0, t_n]$ can then be bounded by

$$|f^{(j)}(t^*) - p^{(j)}(t^*)| \leq c \cdot h_t^{n+1-j} \cdot \max_{0 \leq i \leq j} \|f^{(n+1+i)}\|_{\infty, [a,b]} \tag{1.2}$$

S. P. Singh (ed.), Approximation Theory, Spline Functions and Applications, 445–462.

with

$$h_t := \max_{1 \le i \le n} (t_i - t_{i-1}) \tag{1.3}$$

and a constant c which does not depend on f and the node distribution. Other approaches, like Sard's optimal approximation of linear functionals [4][5], and Micchelli's optimal recovery schemes [3], try to find a formula of a certain type, e.g.:

$$f'(t^*) \approx \sum_{i=0}^{n} \alpha_i f(t_i)$$

where the weights α_i are chosen to minimize the error in some well–defined sense.

In Computer–Aided–Design applications the situation is different. The given data only consist of an ordered set of points $P_0, P_1, \ldots, P_n$ in $\mathbb{R}^d$, which can be considered as a sample from the range $R := f([a,b])$ of a smooth and regular curve $f : [a,b] \to \mathbb{R}^d$. In particular, the points P_i may be written as $P_i = f(s_i)$ for some parameter values s_i which are not available and depend on the parametrization of f. Of course, the s_i might be chosen arbitrarily, but this will introduce some additional and hypothetical information. By **geometrical differentiation** we denote methods that construct data like tangent directions, curvature or torsion values at the P_i by exclusive use of the point sequence $P_0, \ldots, P_n$ and the geometry of the range R of the curve.

Given the range R of f, a canonical parametrization of f by arclength t can theoretically be constructed, and this parametrization depends only on R. Thus $P_i = f(t_i)$ can be assumed for the unknown arclength parametrization in order to derive error estimates.

The "**mesh width**" of the sample can be described by either

$$h_s := \max_{1 \le i \le n} \|P_i - P_{i-1}\|_2 = \max_{1 \le i \le n} \|f(t_i) - f(t_{i-1})\|_2 \tag{1.4}$$

or (1.3) as the maximum of chordlengths or arclengths between successive points.

Clearly, chordlength is numerically accessible while arclength is not. However, once arclength is small enough, the two are equivalent in the sense used for the notion of equivalence of norms:

Lemma 1.1 *If $f : [0,L] \to \mathbb{R}^d$ is a C^1 curve, parametrized by arclength, then there is a constant $h_0(f) \in (0,L)$ such that for any two arguments t and $t+h$ with*

$$0 \le t < t+h \le L, \;\; 0 < h \le h_0(f)$$

the inequalities

$$\frac{1}{\sqrt{d+1}} h \le \|f(t+h) - f(t)\|_2 \le \sqrt{d} h \tag{1.5}$$

hold.

Proof. If we consider just one coordinate x of f, we have

$$|x(t+h) - x(t)| = |x'(\tau)|h, \quad t < \tau < t+h, \tag{1.6}$$

and since arclength parametrization implies $|x'(t)|^2 \leq \|f'\|_2^2 = 1$, we get the right–hand side of (1.5) by summing squares of components. Since $\|f'\|_2^2 = 1$ holds everywhere, there is a constant $h_0(f)$ such that on every subinterval I of $[0, L]$ of length $h \leq h_0(f)$ there is some component of f' whose absolute value is at least $(d+1)^{-1/2}$ on I. If x is such a component for given points t and $t+h$ with $h \leq h_0(t)$, then (1.6) implies

$$\|f(t+h) - f(t)\|^2 \geq |x(t+h) - x(t)|^2 \geq \frac{1}{d+1} h^2.$$

□

The main consequence of Lemma 1.1 is the equivalence of $\mathcal{O}(h_s^k)$ and $\mathcal{O}(h_t^k)$ error estimates for $h_s \to 0$ or $h_t \to 0$: both arclength and chordlength can be used to handle the asymptotics of error bounds. Thus we will generally use h as a symbol to mean h_t or h_s, if a fixed multiplicative constant does not matter.

Now let F be a real–valued functional, not necessarily linear, on a set S of smooth regular curves, parametrized on some interval $[0, L]$, e.g.: curvature

$$F(f) = \kappa_f(\tau) = \frac{\|f'(\tau) \times f''(\tau)\|}{\|f'(\tau)\|_2^3}, \tag{1.7}$$

and let G_h be an approximation of F, based on data with density h. The quality of G_h can be measured by comparison to F on smooth regular curves f, parametrized by arclength, in the sense of

Definition 1.2 *A functional G_h is an m–th order approximation of F with respect to S and $h \to 0$, if there is a constant c such that for all $f \in S$ there are positive constants $h_0(f)$ and $K(f)$ with*

$$|F(f) - G_h(f)| \leq c \cdot h^m \cdot K(f)$$

for all $h \in (0, h_0(f)]$. □

The goal of this paper is to develop a general method for constructing high–oder approximations for geometric data of smooth and regular curves, e.g.: tangent directions, curvature or torsion values, or derivatives thereof with respect to arclength. Interpolation of curves will be a major application, because there are good high–order methods [1] requiring tangent or curvature data which must be constructed from positional data, if they are not available from other sources.

2. Local Polynomial Interpolants

If arclength values t_i of the data $P_i = f(t_i)$ were known and if the functional F had the form $F(f) = f^{(j)}(t^*)$, then vector-valued polynomial interpolation would be a very convenient tool to construct approximations of F. As a variation of this idea, one can consider polynomial interpolation at approximations s_i of the actual arclengths t_i. This requires a straightforward variation of the standard error estimate (1.2) for polynomial interpolation:

Lemma 2.1 *Let f be a C^{n+k} curve on $[a,b] \subset \mathbb{R}$, $1 \leq k \leq n$, parametrized by arclength. Consider n–th degree polynomial interpolation to given data $P_i = f(t_i)$, $0 \leq i \leq n$, at perturbed parameter values s_i, $0 \leq i \leq n$, satisfying*

$$\varphi(s_i) = t_i, \qquad 0 \leq i \leq n, \qquad c \leq s_0 < s_1 < \ldots < s_n \leq d \tag{2.1}$$

for a strictly monotonic reparametrization function

$$\varphi : [c,d] \to [a,b], \quad \varphi \in C^{n+k}[c,d].$$

Then the interpolant p to $P_i = f(t_i)$ at s_i has the classical error representation (1.1) *in the form*

$$(f \circ \varphi)(s) - p(s) = \left(\prod_{i=0}^{n}(s - s_i)\right) \Delta^{n+1}(s_0, \ldots, s_n, s)(f \circ \varphi) \tag{2.2}$$

and the derivatives satisfy

$$\|(f \circ \varphi)^{(j)}(s) - p^{(j)}(s)\| \leq h_s^{n+1-j} \cdot K(n,k,f \circ \varphi)$$

for all $s \in [s_0, s_n]$, all $j \in \{0, \ldots, k-1\}$, where

$$h_s := \max_{1 \leq i \leq n} (s_i - s_{i-1}).$$

□

Here, K is dependent on the data distribution, because it contains derivatives of $f \circ \varphi$ up to order $n+k$. Furthermore, (2.2) is still dependent on the reparametrization function φ, and the next two sections will address this drawback.

3. Smoothly refinable parametrizations

We now consider strategies for determining "good" parameter values s_i. If convergence orders of Lemma 2.1 are to be kept as large as possible, the reparametrization functions φ of (2.1) should have derivatives of order up to $n+k$, which can be bounded independently of the density or position of the s_i and t_i, if the mesh width in the sense of (1.4) and (1.3) tends to zero. We then call such a strategy **smoothly refinable** of order $n+k$.

Example 3.1 The most obvious parametrization strategy uses successive chordlengths

$$s_i - s_{i-1} = \|P_i - P_{i-1}\|_2 = \|f(t_i) - f(t_{i-1})\|,$$

and sets $s_0 = t_0 = 0$ without loss of generality. For uniformly distributed data on a circular arc one can easily show that parametrization by successive chordlengths is smoothly refinable of arbitrary order. In general, successive chordlengths can be smoothly refinable only up to order three, as may be shown by taking divided differences of non–uniform samples of circular data. □

This eliminates successive chordlength parametrization as a tool for higher order geometric differentiation.

To overcome the difficulties with successive chordlength parametrization, one can take chordlengths with respect to a **fixed** point P_ℓ, $0 \leq \ell \leq n$, and define

$$s_i = s_\ell + \|P_i - P_\ell\| \cdot \operatorname{sgn}(i - \ell), \qquad 0 \leq i \leq n.$$

We call this a **locally centered chordlength parametrization** and get

Lemma 3.2 *If $f \in C^m[a, b]$, $m > 2$, is parametrized by arclength, then any locally centered chordlength parametrization is smoothly refinable of order m.*

Proof: Let $P_\ell = f(0)$ be the center of a given chordlength parametrization, and define the real–valued function

$$s(t) := \|f(t) - f(0)\|_2 \operatorname{sgn}(t), \qquad t \in [a, b] \ni 0.$$

Then $s(t_i) = s_i$ holds for $0 \leq i \leq n$, and $s^{-1} = \varphi$ is our candidate for a reparametrization function. Clearly, a simple Taylor expansion implies that

$$s^2(t) = \|f(t) - f(0)\|_2^2 = t^2 + t^4 \cdot q(t)$$

is a smooth function with $q \in C^{m-2}[a, b]$, and $s(t)$ has the form

$$s(t) = t\sqrt{1 + t^2 q(t)},$$

the square root taken to be positive. Around $t = 0$ the function $s(t)$ is in C^m and strictly monotonic. Thus $\varphi = s^{-1}$ exists and shares these properties. □

Applications of locally centered chordlength parametrization should make sure that the mesh width h_s of data P_i is small enough to make the s_i monotonic with respect to i.

Combining Lemmas 2.1 and 3.1 we get

Theorem 3.3 *For sufficiently dense samples of data $P_i = f(t_i)$ from a regular C^{n+k} curve, the j–th derivatives of local n–th degree polynomial interpolants at centered chordlengths are approximations of order h^{n+1-j}, $0 \leq j \leq k - 1 \leq n$, to derivatives $(f \circ \varphi)^{(j)}$, where varphi is the reparametrization function of centered chordlength parametrization.* □

The details of the numerical realization are summarized in steps A1–A3 of the algorithm in section 7.

4. Elimination of parametrization

The results of the previous section yield high–order approximations of $(f \circ \varphi)^{(j)}$, but not of f itself. To eliminate the (unknown) reparametrization function φ, we define

$$g(s) = (f \circ \varphi)(s) \tag{4.1}$$

and use $g'(s) = (f' \circ \varphi)(s) \cdot \varphi'(s)$ and $\|f' \circ \varphi\|_2 = 1$ to get derivatives

$$\begin{aligned} \varphi'(s) &= \|g'(s)\|_2 \\ \varphi^{(j)}(s) &= \frac{d^{j-1}}{ds^{j-1}} \|g'(s)\|_2, \quad j = 2, 3, \ldots \end{aligned} \tag{4.2}$$

of φ from the derivatives of g. A local polynomial interpolant p at centered chordlength abscissae will be a good approximation to $g = f \circ \varphi$, and by application of (4.2) to p instead of g we can construct approximations of the derivatives of φ at centered chordlength abscissae.

To get derivatives of f instead of $f \circ \varphi$, we simply take derivatives of (4.1) and use the information on φ' that are deduced from (4.2). Clearly,

$$(f' \circ \varphi)(s) = g'(s)/\|g'(s)\|$$

directly yields f' at $t = \varphi(s)$, and higher derivatives require the solution of equations

$$\begin{aligned} g''(s) &= (f' \circ \varphi)(s)(\varphi'(s))^2 + (f' \circ \varphi)(s)\varphi''(s) \\ g^{(j)}(s) &= (f^{(j)} \circ \varphi)(s)(\varphi'(s))^j + (f' \circ \varphi)(s)\varphi^{(j)}(s) + \text{lower derivatives} \end{aligned} \tag{4.3}$$

for $f^{(j)}$ at $t = \varphi(s)$. This will produce approximations for $f^{(j)} \circ \varphi$ from $(f \circ \varphi)^{(j)}$, using the approximations for $\varphi^{(j)}$ as obtained from (4.2). The data $f^{(j)} \circ \varphi$ are used as geometric information for further use in Hermite interpolation processes of the following sections.

If the data are dense enough, and if locally centered chordlength parametrization is used, there will be no problems with (4.2) because $\varphi' \approx 1$ for small h. The sample of points P_i can be rejected as being not dense enough, if the numerical test $0 < \alpha \le \|\varphi'\| = \|g'\| \le \alpha^{-1}$ for some $\alpha \in (0,1)$ is not satisfied. Details are given in steps A4 and A5 of the algorithm in section 7.

If j–th derivatives of g contain an error of order h^{n+1-j}, some elementary calculations prove that j–th derivatives of φ and f, the latter taken at unknown arclength values $t = \varphi(s)$, also have errors of order $\mathcal{O}(h^{n+1-j})$, provided that the data are dense enough. The resulting derivatives of f at arclength parameters are now (asymptotically) independent of the parametrization chosen to supply the intermediate derivative of $f \circ \varphi$. Of course, this strategy for this elimination of parametrization effects will work in general, not just for the centered chordlength parametrization of the previous section.

Another approach to eliminate parametrization effects is to calculate curvature, torsion (or derivatives thereof with respect to arclength) directly from the derivatives of the curve

$g = f \circ \varphi$. Since these results do not depend on parametrization, the contribution of φ is asymptotically eliminated.

In both cases there may be numerical problems due to cancellation effects and roundoff, if h is still large and high-order derivatives are calculated.

5. Application to curve interpolation

The previous section provided $\mathcal{O}(h^{n+1-j})$ approximations to j-th derivatives of curves $f \in C^{n+k}$, $1 \leq j \leq k-1$, parametrized by arclength. These can be put into existing Hermite interpolation schemes to generate piecewise interpolating curves. This is put on a rigorous basis by

Theorem 5.1 *Let $f \in C^{n+k}[0, h_0]$ with $h_0 > 0$ and $1 \leq k \leq n$ be given, and let p_h be the two-point Hermite interpolation polynomial of degree $\leq 2k-1$ to data*

$$\begin{aligned} y_{0,j} &= f^{(j)}(0) + \eta_{0,j}, \quad |\eta_{0,j}| \leq ch^{n+1-j},\ 0 \leq j \leq k-1 \\ y_{h,j} &= f^{(j)}(h) + \eta_{h,j}, \quad |\eta_{n,j}| \leq ch^{n+1-j},\ 0 \leq j \leq k-1 \end{aligned} \tag{5.1}$$

on $[0, h]$ for $h \in (0, h_0]$. Then there exists a constant C, independent of h, such that the error bound

$$|f(t) - p_h(t)| \leq C \cdot h^{\min(2k, n+1)}$$

holds for all $t \in [0, h]$.

Proof: Let q_h be the Hermite interpolation polynomial to exact data of f. Then the error has the classical representation

$$f(t) - q_h(t) = t^k (h-t)^k \Delta^{2k}(k\#0, k\#h, t) f$$

involving the $(2k)$-th generalized divided difference of f with repeated arguments, i.e.: the notation $k\#x$ means k repetitions of x as an argument. Since f is in C^{2k}, we get

$$|f(t) - q_h(t)| \leq C_1 h^{2k}, \quad 0 \leq t \leq h,$$

where $C_1 = \dfrac{1}{(2k)!} \|f^{(2k)}\|_{\infty,[0,h_0]}$.

Both p_h and q_h can be represented via divided differences, and thus

$$\begin{aligned} & p_h(t) - q_h(t) = \\ & \sum_{j=0}^{2k-1} t^{m(j)} (h-t)^{j-m(j)} \Delta^j (m(j+1)\#0,\ (j+1-m(j+1))\#h)(p_h - q_h) \end{aligned}$$

where $m(j) = \min(j, k)$. Therefore it suffices to prove

$$|\Delta^j (m(j+1)\#0, (j+1-m(j+1))\#h)(p_h - q_h)| \leq C_2 h^{n+1-j} \tag{5.2}$$

inductively for $j = 0, 1, \ldots, 2k-1$. But these divided differences can be evaluated on the η values of (5.1), and all j–th divided differences with fully coalescing arguments are of order $\mathcal{O}(h^{n+1-j})$ by definition, where $0 \leq j \leq k-1$. If divided differences are formed via the usual recursive relation, the order drops by one. This proves (5.2). □

Theorem 5.1 has quite a number of applications. First consider the de–Boor–Höllig–Sabin method [1] for piecewise GC^2 interpolation of planar data by cubics. According to the first part of this paper, local interpolation by quintic polynomials at centered chordlengths will suffice to generate $\mathcal{O}(h^{6-j})$ estimates of j–th derivatives for $j = 1, 2$ to produce full order $\mathcal{O}(h^6)$ of the interpolation process. This follows from the proof technique in [1] and from Theorem 5.1 for $k = 3$, $n = 5$. Note that we require $f \in C^8$ for this method.

Piecewise GC^1 Hermite interpolation of planar data by quadratics was studied in [8], together with a direct method of determining tangent directions with accuracy $\mathcal{O}(h^3)$. The approach of the previous sections can also be used to supply such derivative estimates, using local cubic interpolants on centered chordlength parameters (case $k = 2$, $n = 3$ of Theorem 5.1).

For data in $\mathbb{R}^d$ with arbitrary d one can use the "$h/3$–rule" to determine a GC^1 piecewise cubic interpolant of accuracy $\mathcal{O}(h^4)$ from two positions P_0, P_1 and normalized tangent directions r_0, r_1 in P_0, P_1 by constructing Bernstein–Bézier control points

$$P_0, P_0 + \frac{h}{3} r_0, P_1 - \frac{h}{3} r_1, P_1$$

where $h = \|P_0 - P_1\|$ is the local chordlength. This method can be applied to purely positional data without loss of accuracy, if approximations of first derivatives of order $\mathcal{O}(h^3)$ are provided via first derivatives of local third–degree interpolants at centered chordlength parameters. This application will also be covered by Theorem 5.1 for $k = 2$, $n = 3$ together with the basic proof technique of [1]. Note that all of the simpler strategies for tangent estimation will not produce full fourth–order accuracy. For instance, the method of McConalogue [2] uses three–point estimates of tangents via quadratic interpolation, giving $\mathcal{O}(h^2)$ accuracy of tangent directions and an overall $\mathcal{O}(h^3)$ error of interpolation, as Theorem 5.1 shows for $k = n = 2$.

6. Two–point Hermite interpolation

In case of prescribed parameter values t_i one can always generate a piecewise polynomial C^{k-1} interpolant from Lagrange data by the following purely local algorithm:

1. performing local interpolation on sets of $2k$ consecutive data points by polynomials of degree at most $2k-1$;
2. taking derivatives of order $j = 0, 1, \ldots, k-1$ of these local interpolants,
3. solving a symmetric two–point Hermite interpolation problem for polynomials of degree $\leq 2k-1$ on each pair of consecutive data points, using the derivatives of the previous step.

If the data $P_i = f(t_i)$ are sampled from a C^{3k-1} function f, this process will have accuracy $\mathcal{O}(h^{2k})$, as follows from Lemma 2.1 and Theorem 5.1.

We now proceed to generalize this method to the parametric case. The data now consist of a large sample of points $P_i = f(t_i)$, $0 \leq i \leq N$ of a smooth and regular curve f with values in $\mathbb{R}^d$, parametrized by arclength, but we do not know the arclength values t_i. We can assume that the methods of the previous sections have been applied to yield approximative derivatives of f with respect to arclength up to order $k-1$ at the P_i. For this we make the implicit assumption that N is large enough with respect to k.

We now pick a pair $P_\ell, P_{\ell+1}$ of consecutive data points and want to apply polynomial Hermite interpolation of degree $\leq 2k-1$ between $P_\ell = f(t_\ell)$ and $P_{\ell+1} = f(t_{\ell+1})$. This makes it necessary to introduce some parametrization again, because the exact arclength $\tau_\ell = t_{\ell+1} - t_\ell$ is not known. Furthermore, the interpolation should produce an overall GC^{k-1} curve when several patches are joined together. This will require C^{k-1} continuity after a suitable reparametrization.

We simply use chordlength $\sigma_\ell := \|P_{\ell+1} - P_\ell\|$ to parametrize the interpolant locally over $[0, \sigma_\ell]$, using the reparametrization function φ_ℓ of locally centered chordlength at P_ℓ, i.e.

$$s_i = s_\ell + \|P_i - P_\ell\| \cdot \operatorname{sgn}(i - \ell),\; t_i = \varphi_\ell(s_i).$$

This will satisfy

$$\varphi_\ell(0) = t_\ell, \qquad \varphi_\ell(\sigma_\ell) = t_{\ell+1}$$

if we set $s_\ell = 0$ without loss of generality. Thus, we apply nonparametric two-point Hermite interpolation to $f \circ \varphi_\ell$.

Data at the left endpoint $s_\ell = 0$ should then be

$$\begin{aligned} f_\ell^{(0)} &:= P_\ell \\ (f \circ \varphi_\ell)^{(j)}&(s_\ell),\;\; 1 \leq j \leq k-1, \end{aligned} \tag{6.1}$$

while at the right endpoint $s_{\ell+1} = \sigma_\ell$ we should interpolate the values

$$\begin{aligned} f_{\ell+1}^{(0)} &:= P_{\ell+1}, \qquad 1 \leq j \leq k-1 \\ (f \circ \varphi_\ell)^{(j)}&(s_{\ell+1}) \;\; 1 \leq j \leq k-1 \end{aligned} \tag{6.2}$$

If f and φ_ℓ were known, including all derivatives of order $\leq k-1$, this approach would work easily. In fact, each interpolant would be a piecewise Hermite interpolant to k derivatives of f with respect to arclength after elimination of the local parametrizations, which are uniformly bounded together with their derivatives. The overall accuracy would still be $\mathcal{O}(h^{2k})$, as follows from the proof technique of de Boor, Höllig, and Sabin [1].

To make the method feasible in practice, we have to take a closer look at the numerical process which replaces $(f \circ \varphi_\ell)^{(j)}$ by accessible values. Let p_ℓ be the polynomial used for local Lagrange interpolation of degree $\leq 2k-1$ with centered chordlengths around P_ℓ, including $P_{\ell+1}$. This will interpolate $f \circ \varphi_\ell$ at centered chordlength abscissae such that

$$p_\ell^{(j)}(s) = (f \circ \varphi_\ell)^{(j)}(s)$$

holds exactly for $j = 0$ and certain chordlength values s_i including s_ℓ and $s_{\ell+1}$, while an error of order $2k-j$ occurs for $j > 0$ or arbitrary arguments near s_ℓ. The method of section 4 is then applied to get approximate values

$$\begin{aligned} f_\ell^{(j)} &:= f^{(j)}(t_\ell) + \mathcal{O}(h^{2k-j}), \quad 1 \leq j \leq k-1 \\ \varphi_{\ell,i}^{(j)} &:= \varphi_\ell^{(j)}(s_i) + \mathcal{O}(h^{2k-j}) \quad 1 \leq j \leq k-1 \end{aligned} \tag{6.3}$$

for $i = \ell, \ell+1$ as results of the numerical process behind (4.2) and (4.3). Details are in steps A4 and A5 of section 7.

Our goal is to use the data $f_\ell^{(j)}$ at P_ℓ for each value of ℓ, if the parametrization is removed. Because the removal of φ_ℓ from $p_\ell^{(j)}(s_\ell)$ will produce $f_\ell^{(j)}$, we can use the data $p_\ell^{(j)}(s_\ell)$ in (6.1) directly, but then in (6.2) we run into a problem caused by

$$(f \circ \varphi_\ell)^{(j)}(s_{\ell+1}) \neq p_{\ell+1}^{(j)}(s_{\ell+1}),$$

as used in the next segment. Note that the discrepancy in the above formula is only of order $\mathcal{O}(h^{2k-j})$, but we cannot ignore it without losing GC^{k-1} continuity.

We avoid this difficulty by modifying the actual interpolation data at $s_{\ell+1}$ in a suitable way: if we eliminate φ_ℓ from the actual data at $s_{\ell+1}$, we should arrive at $f_{\ell+1}^{(j)}$, the approximate chordlength derivative of f at $s_{\ell+1}$ in the next segment. This is accomplished by using (4.3) backwards, starting from the derivatives

$$\begin{aligned} f_{\ell+1}^{(j)} &= f^{(j)}(t_{\ell+1}) + \mathcal{O}(h^{2k-j}), \quad 1 \leq j \leq k-1 \\ \varphi_{\ell,\ell+1}^{(j)} &= \varphi_\ell^{(j)}(s_{\ell+1}) + \mathcal{O}(h^{2k-j}), \quad 1 \leq j \leq k-1 \end{aligned}$$

and synthesizing an $\mathcal{O}(h^{2k-j})$ approximation $\hat{f}_{\ell+1}^{(j)}$ of $(f \circ \varphi_\ell)^{(j)}(s_{\ell+1})$ by evaluating the chain rule for (4.3). Details are in steps B1 and B2 of the algorithm in section 7.

We still have to prove that this process maintains overall GC^{k-1} continuity and $\mathcal{O}(h^{2k})$ accuracy. The latter fact is clear because our modifications do not spoil the accuracy, being of order $\mathcal{O}(h^{2k-j})$ when referring to j–th derivatives. To prove CC^{k-1} continuity we introduce the reparametrization function ψ_ℓ which is a Hermite interpolant of the numerically obtained derivative values $\varphi_{\ell,i}^{(j)}$ of (6.3) together with $\varphi_{\ell,i}^{(0)} := \varphi_\ell(s_i)$ for $i = \ell, \ell+1$. Then $\varphi_\ell - \psi_\ell = \mathcal{O}(h^{2k})$ holds between s_ℓ and $s_{\ell+1}$, because the derivatives of φ_ℓ are uniformly bounded. If h is small enough, ψ_ℓ will be strictly monotonic. If q_ℓ is the Hermite interpolant to the data $p_\ell^{(j)}(s_\ell)$ and $\hat{f}_{\ell+1}^{(j)}$, $0 \leq j \leq k-1$ at $s = s_\ell$ and $s = s_{\ell+1}$, respectively, our construction guarantees that $q_\ell \circ \psi_\ell^{-1}$ interpolates the data $f_\ell^{(j)}$ and $f_{\ell+1}^{(j)}$, $0 \leq j \leq k-1$ at $t = t_\ell$ and $t = t_{\ell+1}$, respectively, because we used the exact chain rule for derivative values of q_ℓ and ψ_ℓ. But this means that the functions q_ℓ form a GC^{k-1} curve, since after reparametrization by ψ_ℓ^{-1} they coincide of order $k-1$ at the breakpoints. We summarize:

Theorem 6.1 *Let f be a regular C^{3k-1} curve, parametrized by arclength. The the above process, given in algorithmic form in the next section, provides a piecewise polynomial GC^{k-1} interpolant of accuracy $\mathcal{O}(h^{2k})$.* □

Of course, the above approach is biased towards the left endpoint P_ℓ, because we used the local chordlength parametrization φ_ℓ centered at P_ℓ. A similar interpolation can be done on $[-\sigma_\ell, 0]$, using local chordlengths centered at $P_{\ell+1}$, and we found it practically useful to take means of the two solutions in order to maintain symmetry and to avoid instabilities due to the calculations based on (4.2) and (4.3). Furthermore, the local estimation of derivatives via interpolation at centered chordlengths should be of order $\le 2k$ instead of $2k-1$ to get symmetry-preserving formulae based on an odd number of points. Both modifications do not affect our theoretical results, and they are incorporated into the algorithmic formulation of the method in the next section.

7. Algorithm

For quick reference and easier programming, we summarize our method in algorithmic form.

Data: $N \in \mathbb{N}$, $d \in \mathbb{N}$, $P_0, \ldots, P_N \in \mathbb{R}^d$ with $P_i \neq P_{i-1}$ for $1 \le i \le N$. It is implicitly assumed that the data are a large sample of points $P_i = f(t_i)$ from the range of a smooth and regular curve with values in $\mathbb{R}^d$, parametrized by arclength, for arclength values satisfying $t_{i-1} < t_i$, $1 \le i \le N$.

Parameters: Choose numbers $k \in \mathbb{N}_{\geq 0}$ with $2k \le N$ and $\alpha \in (0,1)$. The final order of accuracy is $\mathcal{O}(h^{2k})$ if $f \in C^{3k}$, and overall GC^{k-1} continuity will be achieved. Large values of α increase the safety of the method, but will at the same time restrict the range of admissible applications.

Step A: Local Interpolation at centered chordlengths.

For all ℓ with $0 \le \ell \le N$ do :

A1: Calculate centered chordlength values around P_ℓ:

$$\begin{aligned} \ell^* &:= \max(k, \min(N-k, \ell)) = \left\{ \begin{array}{cc} k & 0 \le \ell < k \\ \ell & k \le \ell \le N-k \\ N-k & N-k < \ell \le N \end{array} \right\} \\ s_i &:= \|P_i - P_\ell\| \operatorname{sgn}(i-\ell), \quad \ell^* - k \le i \le \ell^* + k. \end{aligned}$$

A2: If the s_i are not strictly monotonic, give up. The data do not form a sufficiently dense sample from a smooth and regular curve. Use a more robust, but less accurate method.

A3: Calculate a local interpolant p_ℓ around P_ℓ:

$$p_\ell := \text{polynomial interpolant of data } (s_i, P_i),\ \ell^* - k \le i \le \ell^* + k, \text{ of degree } \le 2k$$

$$N_\ell := \left\{ \begin{array}{cc} \{\ell, \ell+1\} & \ell = 0 \\ \{\ell-1, \ell, \ell+1\} & 1 \le \ell < N \\ \{\ell-1, \ell\} & \ell = N \end{array} \right\}$$

Evaluate $p_\ell^{(j)}(s_i)$ for all $0 \le j < k$, all neighbors s_i for $i \in N_\ell$, and store these values.

A4: Use (4.2) to get approximate derivatives of the reparametrization function φ_ℓ via

$$\begin{array}{lll} \varphi_{\ell,i}^{(1)} & := \|p_\ell'(s_i)\|_2 & \text{for } k \ge 2 \\ \varphi_{\ell,i}^{(2)} & := p_\ell'(s_i)^T p_\ell''(s_i)/\varphi_{\ell,i}^{(1)} & \text{for } k \ge 3 \\ \varphi_{\ell,i}^{(3)} & := (p_\ell''(s_i)^T p_\ell''(s_i) + p_\ell'(s_i)^T p_\ell'''(s_i) - (\varphi_{\ell,i}^{(2)})^2)/\varphi_{\ell,i}^{(1)} & \text{for } k \ge 4 \\ \varphi_{\ell,i}^{(4)} & := (3p_\ell''(s_i)^T p_\ell'''(s_i) + p_\ell'(s_i)^T p_\ell^{(4)}(s_i) - 3\varphi_{\ell,i}^{(2)}\varphi_{\ell,i}^{(3)})/\varphi_{\ell,i}^{(1)} & \text{for } k \ge 5 \end{array}$$

etc.

for $i \in N_\ell$ and store these values. After evaluating $\varphi_{\ell,i}^{(1)}$, test for

$$0 < \alpha \le \varphi_{\ell,i}^{(1)} \le 1/\alpha$$

and give up, if the test fails. In this case the data are no sufficiently dense sample from a smooth and regular curve. Use a more robust, but less accurate method.

A5: Use (4.3) to eliminate the parametrization effect from derivatives of p_ℓ

$$\begin{array}{lll} f_\ell^{(0)} & := P_\ell & \text{for } k \ge 1 \\ f_\ell^{(1)} & := p_\ell'(s_\ell)/\varphi_{\ell,\ell}^{(1)} & \text{for } k \ge 2 \\ f_\ell^{(2)} & := (p_\ell''(s_\ell) - \varphi_{\ell,\ell}^{(2)} f_\ell^{(1)})/(\varphi_{\ell,\ell}^{(1)})^2 & \text{for } k \ge 3 \\ f_\ell^{(3)} & := (p_\ell'''(s_\ell) - \varphi_{\ell,\ell}^{(3)} f_\ell^{(1)} - 3\varphi_{\ell,\ell}^{(1)}\varphi_{\ell,\ell}^{(2)} f_\ell^{(2)})/(\varphi_{\ell,\ell}^{(1)})^3 & \text{for } k \ge 4 \\ f_\ell^{(4)} & := (p_\ell^{(4)}(s_\ell) - \varphi_{\ell,\ell}^{(4)} f_\ell^{(1)} - 4\varphi_{\ell,\ell}^{(1)}\varphi_{\ell,\ell}^{(3)} f_\ell^{(2)} & \\ & \quad -3(\varphi_{\ell,\ell}^{(2)})^2 f_\ell^{(2)} - 6(\varphi_{\ell,\ell}^{(1)})^2 \varphi_{\ell,\ell}^{(2)} f_\ell^{(3)})/(\varphi_{\ell,\ell}^{(1)})^4 & \text{for } k \ge 5 \end{array}$$

etc.

and store these values, which will be the "geometric" Hermite data for the next step.

Step B: Local two-point Hermite interpolation.

For all ℓ with $0 \le \ell \le N - 1$ do :

B1: Prepare data for interpolation between P_ℓ and $P_{\ell+1}$, using (4.3) and the reparametrization function φ_ℓ:

$$\begin{aligned}
\sigma_\ell &:= \|P_\ell - P_{\ell+1}\|_2 \\
\widehat{f}^{(0)}_{\ell+1} &:= f^{(0)}_{\ell+1} \\
\widehat{f}^{(1)}_{\ell+1} &:= \varphi^{(1)}_{\ell,\ell+1} f^{(1)}_{\ell+1} \\
\widehat{f}^{(2)}_{\ell+1} &:= \varphi^{(2)}_{\ell,\ell+1} f^{(1)}_{\ell+1} + (\varphi^{(1)}_{\ell,\ell+1})^2 f^{(2)}_{\ell+1} \\
\widehat{f}^{(3)}_{\ell+1} &:= \varphi^{(3)}_{\ell,\ell+1} f^{(1)}_{\ell+1} + 3\varphi^{(1)}_{\ell,\ell+1}\varphi^{(2)}_{\ell,\ell+1}(\sigma_\ell) f^{(2)}_{\ell+1} + (\varphi^{(1)}_{\ell,\ell+1})^3 f^{(3)}_{\ell+1} \\
\widehat{f}^{(4)}_{\ell+1} &:= \varphi^{(4)}_{\ell,\ell+1} f^{(1)}_{\ell+1} + 3(\varphi^{(2)}_{\ell,\ell+1})^2(\sigma_\ell) f^{(2)}_{\ell+1} + 4\varphi^{(1)}_{\ell,\ell+1}\varphi^{(3)}_{\ell,\ell+1} f^{(2)}_{\ell+1} \\
&\quad +5(\varphi^{(1)}_{\ell,\ell+1})^2\varphi^{(2)}_{\ell,\ell+1} f^{(3)}_{\ell+1} + (\varphi^{(1)}_{\ell,\ell+1})^4 f^{(4)}_{\ell+1}
\end{aligned}$$

etc.

for the appropriate values of k.

B2: Let $\widehat{q}_\ell$ be the nonparametric Hermite interpolant of degree $\leq 2k-1$ of data $p^{(j)}_\ell(s_\ell)$ at 0 and of $\widehat{f}^{(j)}_{\ell+1}$ at σ_ℓ, where $0 \leq j < k$. Store $\widehat{q}_\ell$ in some form or other.

B3: Prepare data for interpolation between P_ℓ and $P_{\ell+1}$, now using the reparametrization function $\varphi_{\ell+1}$:

$$\begin{aligned}
\widetilde{f}^{(0)}_\ell &:= f^{(0)}_\ell \\
\widetilde{f}^{(1)}_\ell &:= \varphi^{(1)}_{\ell+1,\ell} f^{(1)}_\ell \\
\widetilde{f}^{(2)}_\ell &:= \varphi^{(2)}_{\ell+1,\ell} f^{(1)}_\ell + (\varphi^{(1)}_{\ell+1,\ell})^2 f^{(2)}_\ell \\
\widetilde{f}^{(3)}_\ell &:= \varphi^{(3)}_{\ell+1,\ell} f^{(1)}_\ell + 3\varphi^{(1)}_{\ell+1,\ell}\varphi^{(2)}_{\ell+1,\ell} f^{(2)}_\ell + (\varphi^{(1)}_{\ell+1,\ell})^3 f^{(3)}_\ell \\
\widetilde{f}^{(4)}_\ell &:= \varphi^{(4)}_{\ell+1,\ell} f^{(1)}_\ell + 3(\varphi^{(2)}_{\ell+1,\ell})^2 f^{(2)}_\ell + 4\varphi^{(1)}_{\ell+1,\ell}\varphi^{(3)}_{\ell+1,\ell} f^{(2)}_\ell \\
&\quad +5(\varphi^{(1)}_{\ell+1,\ell})^2\varphi^{(2)}_{\ell+1,\ell} f^{(3)}_\ell + (\varphi^{(1)}_{\ell+1,\ell})^4 f^{(4)}_\ell
\end{aligned}$$

etc.

for the appropriate values of k.

B4: Let $\widetilde{q}_\ell$ be the nonparametric Hermite interpolant of degree $\leq 2k-1$ of data $\widetilde{f}^{(j)}_\ell$ at $-\sigma_\ell$ and of $p^{(j)}_{\ell+1}(s_{\ell+1})$ at 0, where $0 \leq j < k$. Store $\widetilde{q}_\ell$ in some form or other.

B5: For evaluation of the solution between P_ℓ and $P_{\ell+1}$, use the polynomial

$$q_\ell(s) := \frac{1}{2}\left(\widehat{q}_\ell(s) + \widetilde{q}_\ell(s-\sigma_\ell)\right), \; s \in [0, \sigma_\ell],$$

or a similar weighted mean between $\widehat{q}_\ell$ and $\widetilde{q}_\ell$.

Remarks

There can be m additional data points between the endpoints P_ℓ, $P_{\ell+1+m}$ of a local Hermite interpolation. The degree must then be $2k-1+m$ and the approximation order will be $\mathcal{O}(h^{2k+m})$, provided that geometric differentiation of data from $f \in C^{3k+m}$ is done with local polynomials of degree at least $2k-1+m$.

At the end of the range of points, e.g. on $[P_0, P_1]$ or $[P_{N-1}, P_N]$, one does not need derivative values, because no GC^{k-1} continuity must be guaranteed. Then a non–symmetric Hermite interpolation problem with chordlength parameters locally centered at P_1 or P_{N-1} is sufficient, giving the same order of accuracy. This improvement was not incorporated into the examples of the last section.

An alternative approach would avoid the reparametrization of the $f_\ell^{(j)}$ values by direct interpolation on $[0, t^*]$ for a high–accuracy estimate t^* of the actual arclength $t_{\ell+1} - t_\ell$. However, it then seems to be difficult to reach $\mathcal{O}(h^{2k})$ accuracy at comparable costs.

We note that our method is not convexity preserving in general, and it may produce bad results for coarse or noisy data sets. It is designed for sufficiently dense samples of exact data from smooth curves, and its major feature is its convergence order, which may be arbitrarily high for sufficiently smooth curves. So far, the convexity preserving method of highest approximation order is the rational GC^2 interpolation of [6][7], being of order four.

8. Examples

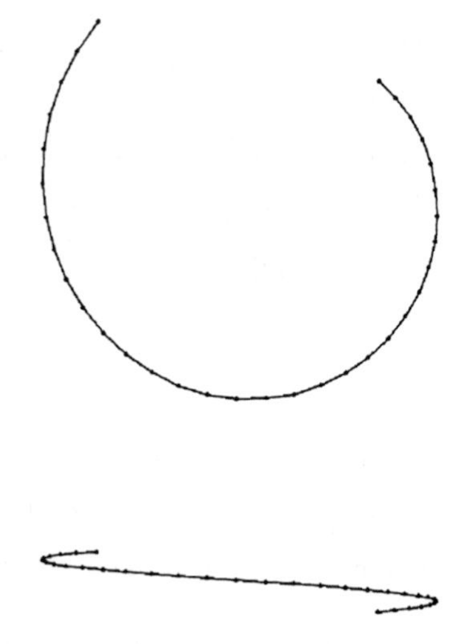

Figure 1: 33 points on spiral in $I\!R^3$, $k = 1$, polygonal interpolant

Figure 1 shows polygonal interpolation of 33 points sampled from part of a logarithmic spiral in $I\!R^3$, seen from above and from the side. The spiral takes a 270 degree turn and has monotonic and nonzero curvature and torsion. Reproduction of the curve shape was perfect within plot precision in all cases. Thus we plot some of the instances where problems like discontinuities or peaks for higher derivatives of curvature and torsion occur.

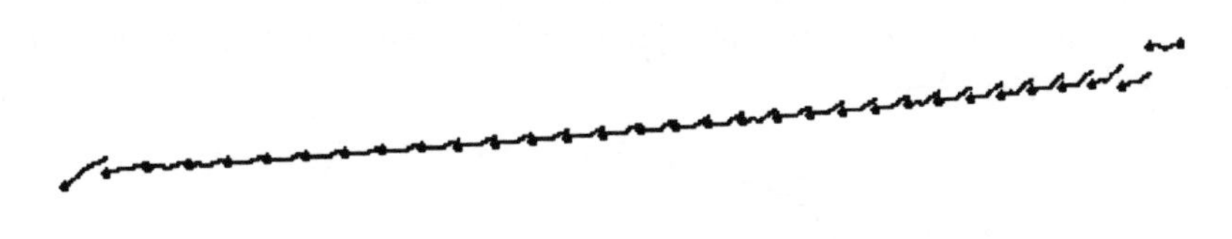

Figure 2: Same data, $k = 2$, first derivative of curvature

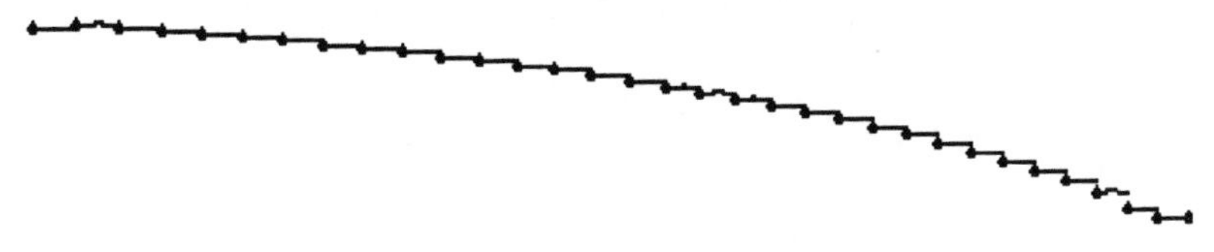

Figure 3: Same data, $k = 2$, torsion

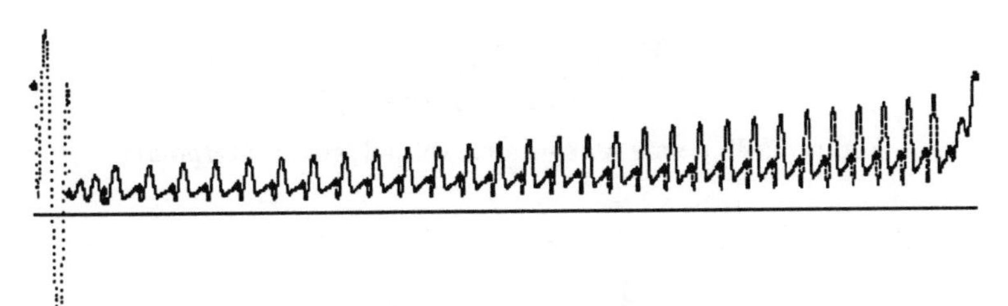

Figure 4: Same data, $k = 5$, second derivative of curvature

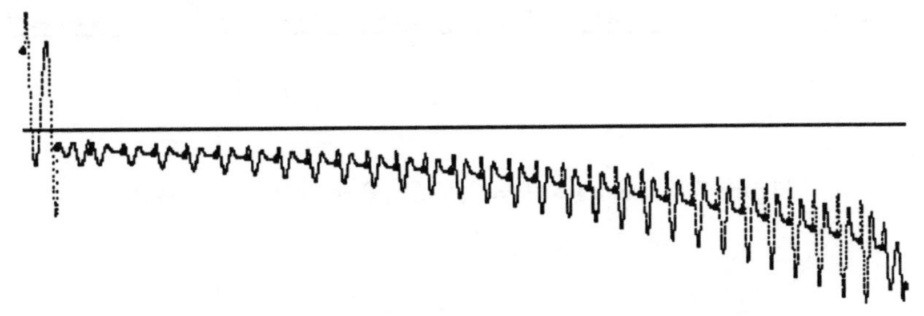

Figure 5: Same data, $k = 5$, first derivative of torsion

One can clearly see that the degree of the local interpolants is too high; there are "unnecessary" wiggles in the curves of Figures 4 and 5, which still have to be continuous by construction.

Figure 6 shows polygonal interpolation of 71 points sampled from part of a Lissajous type figure in $\mathbb{R}^3$, seen from above and from the side. There are two high peaks of curvature and torsion which should be reproduced by the interpolant. Again, we found that reproduction of the curve shape was perfect within plot precision in all cases, and thus we plot only some curvature and torsion data.

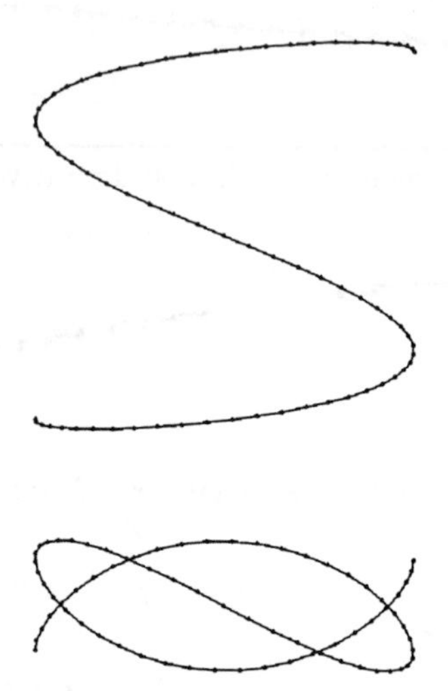

Figure 6: 71 points on space curve, polygonal interpolant

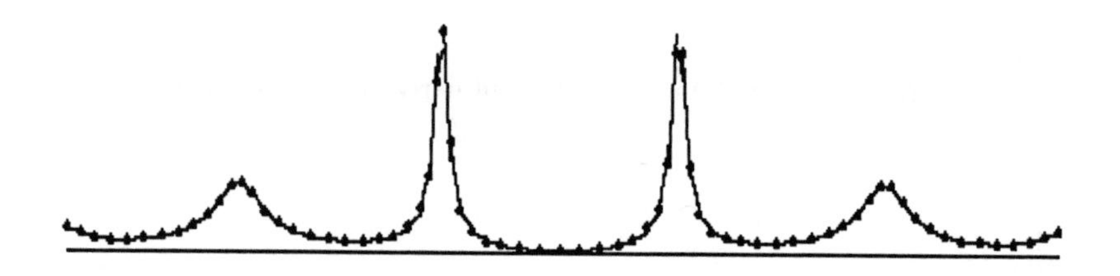

Figure 7: Same data, $k = 2$, GC^1, discontinuous curvature

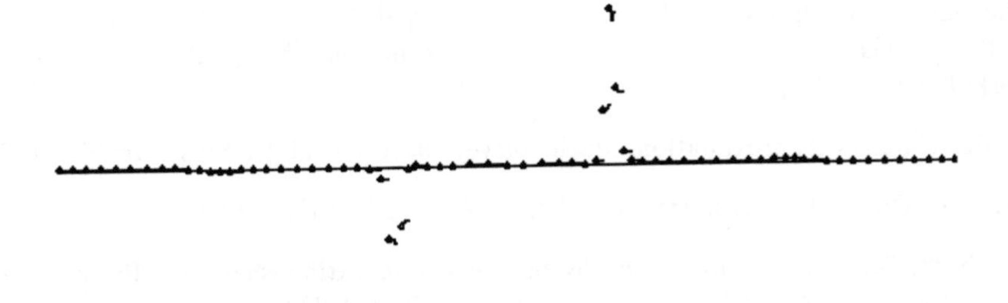

Figure 8: Same data, $k = 2$, GC^1, discontinuous torsion

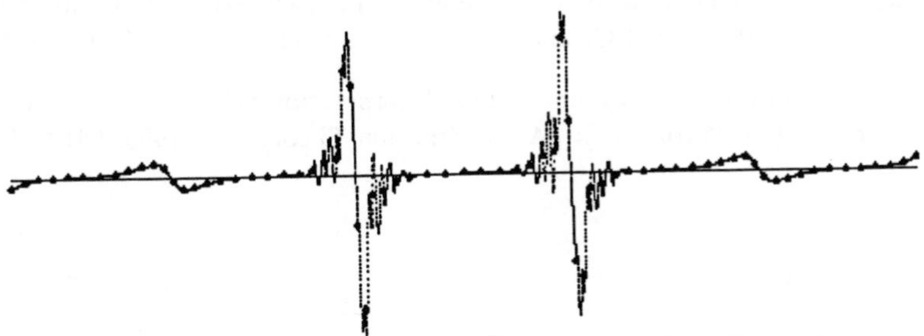

Figure 9: Same data, $k = 4$, GC^3, first derivative of curvature

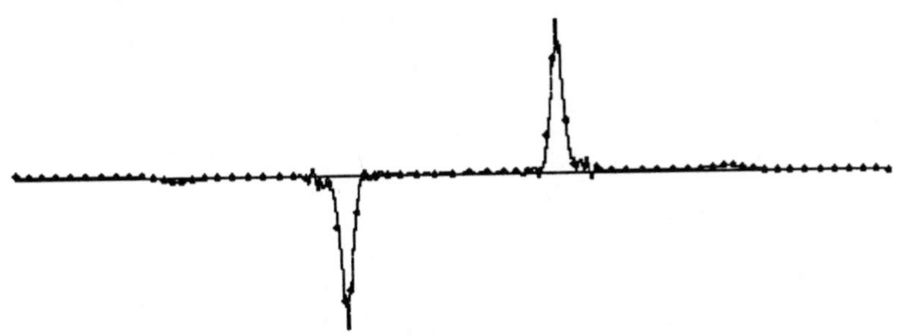

Figure 10: Same data, $k = 4$, GC^3, torsion

Further research should try to achieve high orders of accuracy with low polynomial degrees, e.g.: along the lines of the paper [1] by deBoor, Höllig, and Sabin, or by rational interpolation.

References

[1] deBoor, C., Höllig, K., and Sabin, M., High Accuracy Geometric Hermite Interpolation, Computer Aided Geometric Design 4, 269–278, 1987

[2] McConalogue, D.J., A quasi–intrinsic scheme for passing a smooth curve through a discrete set of points, The Computer Journal,13 (1970) 392–396

[3] Micchelli, C. A., Rivlin, Th. J., A Survey of Optimal Recovery, in C. A. Micchelli, Th. J. Rivlin (eds.): Optimal Estimation in Approximation Theory, Plenum Press, New York–London 1977

[4] Sard, A., Linear Approximation, Math. Surveys 9, Amer. Math. Soc., Providence 1963

[5] Sard, A., Optimal approximation, J. Funct. Anal. 1 (1967), 222-244

[6] Schaback, R., Interpolation in $\mathbb{R}^2$ by piecewise quadratic visually C^2 Bezier polynomials, Computer Aided Geometric Design 6 (1989) 219–233

[7] Schaback, R., On Global GC^2 Convexity Preserving Interpolation of Planar Curves by Piecewise Bezier Polynomials, in T. Lyche and L.L. Schumaker (eds.): "Mathematical Methods in Computer Aided Geometric Design", Academic Press 1989, 539–547

[8] Schaback, R., Convergence of Planar Curve Interpolation Schemes, in C.K. Chui, L.L. Schumaker and J.D. Ward (eds): "Approximation Theory VI", 1989, 581-584

ON BEST SIMULTANEOUS APPROXIMATION IN NORMED LINEAR SPACES

V. M. SEHGAL
University of Wyoming
Laramie, WY 82071, USA

and

S. P. SINGH
Memorial University of Newfoundland
St. John's, NF, Canada, A1C 5S7

ABSTRACT. A brief survey of the best simultaneous approximation in normed linear spaces is given. We give results which extend and unify some of the earlier work. Several results are obtained as corollaries.

1 Introduction

Diaz and McLaughlin [6] and Dunham [7] considered the problem of approximating two continuous functions simultaneously by the same elements of a nonempty family of real valued continuous functions on $[a,b]$. These results in a general setting were given by Holland, Sahney and Tzimbalario [11], Bosznay [5], Mach [14], and Milman [15]. Some further results were given by Dunham [8]. The problem of nonlinear best simultaneous approximation was considered by Blätt [4].

Recently a few mathematicians have studied relative Chebyshev centers, a concept closely related to the best simultaneous approximation, see e.g. Amir and Ziegler [2, 3], Franchetti and Cheney [9], Lambert and Milman [13], Smith and Ward [24, 25], Ward [26], and others.

Several interesting related results could not be included in this survey and therefore it is far from complete.

Definition 1.

Let C be a subset of a normed linear space X. Given a bounded subset F of X, define

$$d(F,C) = \inf_{c\in C}(\sup_{f\in F} \|f - c\|).$$

The set C is said to admit the best simultaneous approximation to F iff there exists a $c \in C$ such that $d(F,C) = \sup_{f\in F} \|f - c\|$.

The following results are given in [11].

S. P. Singh (ed.), Approximation Theory, Spline Functions and Applications, 463–470.

Lemma 1.1 *Let $K \subset X$, and F be a subset of a normed linear space X. Then*

$$\phi(k) = \sup_{f \in F} \|f - k\|$$

is a continuous functional on X.

Lemma 1.2 *If K is a finite dimensional subspace of a normed linear space X, then there exists a best simultaneous approximation $k^* \in K$ to any given subset $F \subset X$.*

Proof.
Since F is compact, there is a finite constant M such that $\|f\| \leq M$ for every $f \in F$.
We define the subset S of K by

$$S = S(0, 2M),$$

then

$$\inf_{k \in S} \sup_{f \in F} \|f - k\| = \inf_{k \in K} \sup_{f \in F} \|f - k\| \leq M.$$

Since S is compact, the continuous functional $\phi(k)$ attains its minimum over S for some k^* which will be the best simultaneous approximation to F.

Lemma 1.3 *Let K be a convex subset of X and $F \subset X$. If k_1 and k_2 in K are best simultaneous approximations to F by elements of K, then*

$$k^* = \lambda k_1 + (1 - \lambda) k_2 \; 0 \leq \lambda \leq 1$$

is also a best simultaneous approximation to F.

Holland, Sahney and Tzimbalario [11] proved the following two theorems. We include the proof for the sake of completeness.

Theorem 1.4 *Let K be a finite dimensional subspace of a strictly convex normed linear space X. Then there exists a unique best simultaneous approximation from the elements of K to any given compact subset $F \subset X$.*

Proof.
The existence of a best simultaneous approximation is guaranteed by Lemma 1.2. Suppose k_1 and $k_2 (k_1 \neq k_2)$ are two best simultaneous approximations to F, i.e.

$$\begin{aligned} & \inf_{k \in S} \sup_{f \in F} \|f - k\| \\ = \; & \sup_{f \in F} \|f - k_1\| \\ = \; & \sup_{f \in F} \|f - k_2\| \\ = \; & d, \text{ say,} \end{aligned}$$

By Lemma 3, $\frac{k_1 + k_2}{2}$ is also a best simultaneous approximation, i.e.

$$\sup \|f - \frac{k_1 + k_2}{2}\| = d.$$

Since F is compact, there exists an f_0 such that

$$\|f_0 - \frac{k_1 + k_2}{2}\| = d.$$

Now, since X is strictly convex and

$$\|f_0 - k_1\| = d \text{ and } \|f_0 - k_2\| = d,$$

we get

$$\|f_0 - \frac{k_1 + k_2}{2}\| < d,$$

which is a contradiction. So $k_1 = k_2$, a unique best simultaneous approximation.

Theorem 1.5 *Let K be a closed and convex subset of a uniformly convex Banach space X. For any compact subset $F \subset X$, there exists a unique best simultaneous approximation to F from the elements of K.*

Proof.
The outline of the proof follows as given below. Let

$$d = \inf_{k \in K} \sup_{f \in F} \|f - k\|$$

and $\{k_n\}$ any sequence of elements in K such that

$$\lim_{n \to \infty} \sup_{f \in F} \|f - k_n\| = d.$$

By using uniform convexity of X and the compactness of F it is shown that $\{k_n\}$ is a Cauchy sequence; hence, it converges to some k in X. Since K is closed, therefore, $k \in K$. The element k is a unique best simultaneous approximation follows by using strict convexity.

We give the following where uniform convexity condition is relaxed [19].

Theorem 1.6 *Let X be a strictly convex Banach space and C a weakly compact, convex subset of X. Then there exists a unique best simultaneous approximation from the elements of C to any given compact subset F of X.*

Definition 2. A bounded subset F of a normed linear space X is said to be remotal with respect to a subset $K \subset X$, if for each $k \in K$ there exists a point $f_0 \in F$ such that

$$\|k - f_0\| = \sup_{f \in F} \|k - f\|.$$

Definition 3. A subset F of a normed linear space X is said to be proximinal with respect to a subset $K \subset X$, if for each $f \in F$ there exists a point $k \in K$ such that

$$\|f - k\| = \inf_{x \in K} \|f - x\|.$$

Theorem 1 of [20] just gives existence and for uniqueness X has to be strictly convex. However, the following holds [21].

If X is a strictly convex normed linear space and C is a compact, convex subset of $X, F \subset X$ is a remotal set with respect to C, then there exists a unique best simultaneous approximation to F from the elements of C. (See also Narang [16]).

Theorem 1.4 has been given for a finite dimensional space. A natural question arises as follows: Is the hypothesis of finite dimensional really necessary?

We prove in [20] the following where we do not have uniqueness:

Theorem 1.7 *Let X be a normed linear space and C a reflexive subspace of X. Then for any nonempty bounded subset F of X, there exists a best simultaneous approximation in C.*

Bosznay [5] gave the following in a more general setting:

Theorem 1.8 *Let X be a strictly convex normed linear space and let M be a reflexive subspace of X. Then for every nonempty compact set $F \subset X$, there exists a unique simultaneous best approximation point in M.*

Sastry and Naidu [22] have proved the following:

Theorem 1.9 *If K is boundedly weakly sequentially compact subset of X, then there exists a best simultaneous approximation to a bounded subset $F \subset X$.*

Note 1. If X is strictly convex, K is convex, and F is a farthest point set with respect to K, then there exists at most one best simultaneous approximation to F from K.

Note 2. If X is strictly convex, K is boundedly weakly sequentially compact and convex, and F is a farthest point set with respect to K, then there exists a unique best simultaneous approximation to F from K.

Theorem 1.10 *If X is strictly convex and reflexive, K is closed and convex, and F is compact, then there exists a unique best simultaneous approximation to F from K.*

Proof.
Since a closed convex set in a reflexive Banach space is weakly compact then the proof follows from Note 1 and Note 2.

The following theorem suggested by Cheney is given in much more general setting. First we state the following

Lemma 1.11 *If F is a bounded subset of a normed space X, then the function $\phi : X \to R$ defined by*

$$\phi(x) = \sup_{y \in F} \|x - y\|$$

is finite valued and lower semicontinuous.

Proof.
The supremum of a family of continuous (or even loser semicontinuous) functions is lower semicontinuous (Rudin [18]).
Note. ϕ is even weakly lower semicontinuous because the map $x \to \|x-y\|$ is weakly lower semicontinuous. The latter follows from the fact that

$$\|x - y\| = \sup_{\substack{f \in x^* \\ \|f\| = 1}} f(x - y),$$

and each map $x \to f(x-y)$ is weakly lower semicontinuous. Thus, ϕ will assume its infimum on any weakly compact set. Hence the following.

Theorem 1.12 *If K is a weakly compact subset of a normed linear space X and if F is a bounded subset of X, then there exists a best simultaneous approximant to F in the set K ([21]).*

Ahuja and Narang [1] proved the following.

Theorem 1.13 *Let K be any subset of a normed linear space X. Then given any bounded subset F of X there exists a best simultaneous approximation to the set F by elements of K, if F is remotal with respect to K and K is proximinal with respect to F.*

If, in addition, K in the above theorem is convex and the space is strictly convex, then the best simultaneous approximation is unique.

Uniqueness follows as in Theorem 1.

For the following terms see Franchetti and Cheney [9]. Let X be a Banach space and A a subset of X. For each $x \in X$, the distance from x to A is defined by

$$d(x, A) = \inf_{a \in A} \|x - a\|.$$

If a set of elements B is given in X, one might like to approximate all of the elements of B simultaneously by a single element in A. This type of problem arises when a function being approximated is not known precisely, but is known to belong to a set.

The Chebyshev radius of a set B with respect to A is defined by

$$r_A(B) = \inf_{a \in A} \sup_{b \in B} \|a - b\|.$$

In approximating simultaneously all elements of B by a single element of A, one identifies elements of the set

$$E_A(B) = \{a \in A | \sup_{b \in B} \|a - b\| = r_A(B)\}.$$

The elements of $E_A(B)$ are solutions to the simultaneous approximation problem.

The set $E_A(B)$ is called the restricted center of B relative to A. The concept was first introduced by Garkavi [10]. In case $A = X$ a normed linear space, $E_X(B)$ is called the Chebyshev center.

The elements of $E_A(B)$ are also called simultaneous approximants of B or global approximants of B or restricted centers of B. If $E_A(B) \neq \phi$ for every bounded set B in X, it is said that A is proximinal for the global approximation of bounded sets.

Mach [15] has shown that in certain subspaces A of a given Banach space X a best simultaneous approximation exists for every bounded set $F \subset X$.

Amir and Ziegler [2, 3], Lambert and Milman [13], Smith and Ward [24, 25], Ward [26], and others have also given interesting results in this case.

Bosznay [5] has proved the continuity of the best simultaneous operator, whereas Sastry and Naidu [23] have discussed a weaker form of continuity in a more general setting.

We give the following and derive several known results as corollaries.

Theorem 1.14 *Let C be a weakly closed subset of a reflexive space X and F a bounded subset of X. Then C admits a best simultaneous approximation to F.*

Proof.
Define a map $\phi : C \to [0, \infty)$ by

$$\phi(c) = \sup_{f \in F} \|f - c\|.$$

Since for any real $r, \{c : \phi(c) \leq r\}$ is a weakly closed subset of C, it follows that ϕ is weakly lower semicontinuous on C. Let

$$d = \inf_{c \in C} \phi(c).$$

Since F is bounded, $\phi(c) \to \infty$ as $\|c\| \to \infty$. Consequently there is a real $r > 0$ such that $\phi(c) > d + 1$ for $\|c\| > r$. Let

$$K = \{x \in X : \|x\| \leq r\} \cap C.$$

Then K is a bounded subset of C and there exists a sequence $\{c_n\}$ in K with $\phi(c_n) \downarrow d$. Since C is weakly closed and X is reflexive, there is a $c_0 \in C$ and a subsequence $\{c_{n_i}\}$ with $c_{n_i} \to c_0$ weakly. Let, for each n,

$$A_n = \left\{c \in C : \phi(c) \leq d + \frac{1}{n}\right\}.$$

Since A_n is weakly closed and $\phi(c_{n_i}) \downarrow d$, it follows that $c_0 \in A_n$ for each n. Thus $\phi(c_0) = d$; that is, $d(F, C) = \sup_{f \in F} \|f - c_0\|$.

Corollary 1.15 *Let C be a closed bounded convex subset of a uniformly convex Banach space X, then for any bounded subset F of X, C admits a best simultaneous approximation to F ([11]).*

Corollary 1.16 *Let C be a reflexive subspace of a normed linear space X, then for any bounded subset F of X, C admits a best simultaneous approximation to F ([19]).*

Proof.
Since C is a reflexive subspace, C is closed and convex and hence C is weakly closed. The result follows from Theorem 1.14.

2 References

1. G. C. Ahuja and T. D., Narang, 'On best simultaneous approximation,' *Neiuw Arch. voor Wiskunde* (3) XXVII (1979), 255-261.

2. D. Amir and Z. Ziegler, 'Relative Chebyshev centers in normed linear spaces,' Part II, 38 (1983) 293-311.

3. D. Amir and Z. Ziegler, 'Characterization of relative Chebyshev centers,' *Approximation Theory III*, Ed. Cheney, E. W., Academic Press (1980) 157-162.

4. H.P. Blätt, 'Nicht-linear Gleichmassige simultan-approximation,' Dissertation, Univ. des Saarlandes, Saarbrücken (1970).

5. A. P. Bosznay, 'A remark on simultaneous approximation,' *J. Approx. Theory*, 23 (1978) 296-298.

6. J.B. Diaz and H.W. McLaughlin, 'On simultaneous Chebyshev approximation of a set of bounded complex valued functions,' *J. Approx. Theory*, 2 (1969) 419-432.

7. C. B. Dunham, 'Simultaneous Chebyshev approximation of functions on an interval,' *Proc. Amer. Math. Soc.*, 18 (1967) 472-477.

8. C. B. Dunham, 'Approximation with respect to Chebyshev norms,' *Aequationes Math.* 8 (1972) 267-270.

9. C. Franchetti and E. W. Cheney,'Simultaneous approximation and restricted Chebyshev centers in function spaces, preprint.

10. A. L. Garkavi, 'The Chebyshev center and the convex hull of a set,' *Usp. Mat. Nauk.* 18 (1964) 139-145.

11. A.S.B. Holland, J. Tzimbalario, and B. N. Sahney, 'On best simultaneous approximation,' *J. Indian Math. Soc.*, 40 (1976) 69-73.

12. A.S.B. Holland, B. N. Sahney, and J. Tzimbalario, 'On best simultaneous approximation,' *J. Approx Theory*, 17 (1976) 187-188.

13. J. M. Lambert and P.D. Milman, 'Restricted Chebyshev centers of bounded subsets in arbitrary Banach spaces,' *J. Approx. Theory*, 26 (1979) 71-78.

14. J. Mach, 'On the existence of best simultaneous approximation,' *J. Approx. Theory*, 25 (1979) 258-265.

15. P.D. Millman, 'On best simultaneous approximation in normed linear spaces,' *J. Approx. Theory*, 20 (1977) 223-238.

16. T.D. Narang, 'On best simultaneous approximation,' *J. Approx. Theory*, 39 (1983) 93-96.

17. G.M. Phillips, J.H. McCabe, and E. W. Cheney, 'On simultaneous Chebyshev approximation,' *J. Approx. Theory*, 27 (1979) 93-98.

18. W. Rudin, 'Real and complex analysis,' McGraw-Hill, New York (1966).

19. B. N. Sahney and S. P. Singh, 'On best simultaneous approximation in Banach spaces,' *J. Approx. Theory*, 35 (1982) 222-224.

20. B. N. Sahney and S. P. Singh, 'On best simultaneous approximation,' *Approximation Theory III*, Ed. Cheney, E. W., Academic Press (1980) 782-789.

21 B.N. Sahney and S. P. Singh, 'On best simultaneous Chebyshev approximation with additive weight functions and related results.' *Nonlinear Analysis and Applications*, (Ed. S.P. Singh & J.H. Burry) Marcel Dekker, New York (1982) 443-463.

22. K.P.R. Sastry and S.V. Naidu, 'On best simultaneous approximation in normed linear spaces,' *Proc. Nat. Acad. Sci. India*, 48 (1978) 249-250.

23. K.P.R. Sastry, and S.V. Naidu, 'Upper semi-continuity of the best simultaneous approximation operator,' *Pure and Applied Math. Sciences*, X (1979) 7-8.

24. P.W. Smith and J.D. Ward, 'Restricted centers in $C(\Omega)$,' *Proc. Amer. Math. Soc.*, 48 (1975) 165-172.

25. P.W. Smith and J.D. Ward, 'Restricted centers in suibalgebras of $C(X)$,' *J. Approx. Theory*, 15 (1975) 54-59.

26. J.D. Ward, 'Chebyshev centers in spaces of continuous functions,' *Pacif. J. Math.*, 52 (1974), 283-287.

Some Examples Concerning Projection Constants

Boris Shekhtman
Department of Mathematics
University of South Florida
Tampa, Florida 33620

ABSTRACT. We provide some simple examples of subspaces of L_∞ for which the projection constants do not behave nicely. In particular we show that the relative projection constant does not interpolate.

1 Introduction

Recently, I was asked several questions concerning projections of small norms onto subspaces of Banach spaces. These questions were motivated by "natural subspaces" arising in approximation theory such as polynomials and spline functions.

The purpose of this article is to give a few simple examples of "unnatural subspaces" of Banach spaces that do not have the properties in question. I will mostly deal with finite dimensional subspaces E of the spaces L_p, $1 \leq p \leq \infty$.

Sometimes, I will consider E as an algebraic space in which case E_p will stand for the collection of functions E considered as a subspaces of L_p. I will use the usual notations: If E is a subspace of X, then

$$\lambda(E,X) := \inf\{\|P\| \ : \ P \text{ is a projection from } X \text{ onto } E.\}$$
$$\lambda(E) = \sup\{\lambda(E,X) \ : \ X \subset E\}$$

If E and F are two Banach spaces of the same dimension, I use

$$d(E,F) = \inf\{\|T\| \, \|T^{-1}\| \ : \ T \text{ is an isomorphism from } E \text{ onto } F.\}$$

Some well-known properties of these notions are (cf. [5]):
If E is an n-dimensional Banach space, then

$$\lambda(E) \leq d(E, \ell_\infty^{(n)}). \tag{1}$$

If F is another n-dimensional Banach space, then

$$\lambda(E) \leq \lambda(F) d(E,F). \tag{2}$$

S. P. Singh (ed.), Approximation Theory, Spline Functions and Applications, 471–476.

If $E \subset X$ and X is $C(K)$, $L_\infty(\mu)$ or ℓ_∞ then

$$\lambda(E) = \lambda(E, X). \tag{3}$$

2 Projection Constant Does Not Interpolate

Example 1. For every p with $\infty > p > 2$, there exists a sequence of n-dimensional algebraic spaces E^n such that

$$\lambda(E^n_\infty, L_\infty) = 1; \quad \lambda(E^n_2, L_2) = 1$$

$$\lambda(E^n_p, L_p) \geq cn^{(1/2-1/p)}.$$

Proof. We first pick functions $\ell_1, \ldots, \ell_n \subset L_p[0,1]$ such that for $F^n := \text{span}\,\{\ell_j\}_{j=1}^n$ we have

$$\lambda(F^n_p, L_p) \geq \frac{1}{2} n^{(1/2-1/p)}. \tag{1}$$

This is possible (cf. [1]) for any $p \neq 2$. For arbitrary $\delta > 0$, we pick functions $\psi_1, \ldots, \psi_n \in C[1,2]$ such that $0 \leq \psi_j(t) \leq 1$, $\max_t \psi_j(t) = 1$, meas supp $\psi_j < \delta$; supp $\psi_j \cap$ supp $\psi_i = \phi$.

Finally, we pick $\eta > 0$ so that

$$\eta \|\Sigma \alpha_j \ell_j\|_\infty \leq \max |\alpha_j|,$$

for all $\alpha_1, \ldots, \alpha_n \in \mathbf{R}$.

We now consider functions $g_j(t) \in L_p[0,2]$

$$g_j(t) = \begin{cases} \eta \ell_j(t) & \text{if } t \in [0,1] \\ \psi_j(t) & \text{if } t \in (1,2]. \end{cases}$$

Let $E^n = \text{span}\,\{g_j\,;\, j = 1, \ldots, n\}$. Clearly

$$\sup\{|\Sigma \alpha_j g_j(t)|\,;\, t \in [0,2]\} = \max |\alpha_j|$$

and hence $d(E^n_\infty, \ell^n_\infty) = 1$. Thus $\lambda(E^n_\infty, L_\infty) = 1$. Since L_2 is a Hilbert space we have $\lambda(E^n_2, L_2) = 1$. It remains to demonstrate (2.1). Let

$$\tilde{g}_j(t) := \begin{cases} \eta \ell_j(t) & \text{if } t \in [0,1] \\ 0 & \text{if } t \in (1,2] \end{cases}$$

and let

$$\check{E}^n = \text{span}\,\{\tilde{g}_j\,;\, j = 1, \ldots, n\}.$$

Clearly $\lambda(\tilde{E}^n, L_p) \geq \frac{1}{2}n^{(1/2-1/p)}$. But

$$\|g_j - \tilde{g}_j\|_{L_p[0,2]} \leq \delta.$$

It now follows (cf. [4]) that for sufficiently small δ,

$$\lambda(E_p^n, L_p) \geq \frac{1}{4}n^{(1/2-1/p)}.$$

■

Remark. One can not replace ∞ and 2 in Example 1 by other "p". Indeed if $p \neq 2$, then $\lambda(E_p^n, L_p) = 1$ implies (cf. [3]) that E^n has a basis consisting of functions with disjoint support and hence

$$\lambda(E_p^n, L_p) = 1 \quad \text{for all} \quad p.$$

Yet it is probably possible to find a sequence E^n such that for some $\infty < p_1 < p < p_2 \leq 1$,

$$\lambda(E_p^n, L_{p_1}) = \mathcal{O}(1); \quad \lambda(E_{p_2}^n, L_{p_2}) = \mathcal{O}(1); \quad \lambda(E_p^n, L_p) \to \infty.$$

3 Minimal vs. Orthogonal Projections

It seems to be the case with spline functions as well as others (cf. [2]) that the existence of projections in $C_{[0,1]}$ of small norm implies that the norm of the "orthogonal projection" (i.e. the orthogonal projection in $L_2[0,1]$) onto the same space has a small norm as well.

It follows from Example 1 that this is not the case in general. Furthermore,

Example 2. There exists a sequence of subspaces $E^n \subset L_\infty[0,1]$ such that

$$\lambda(E_\infty^n) = 1$$

and

$$\|Q_n\|_{L_\infty \to L_\infty} \geq c \cdot n,$$

where Q_n is the L_2-orthogonal projections from $L_2[0,1]$ onto E_2^n.

Proof. Pick $\varepsilon > 0$ so that $\varepsilon < 1/n$. Consider functions

$$f_k(t) = \begin{cases} 1 & \text{if } t \in [\varepsilon(k-1), \varepsilon k) \\ 0 & \text{if } t \in [\varepsilon(j-1), \varepsilon j]\ ;\ j \neq k \\ \dfrac{1}{n} & \text{if } t \in [\varepsilon n, 1] \end{cases}$$

where $j,k = 1,\dots,n$. Clearly $\|\Sigma\alpha_k f_k\|_{L_\infty} = \max|\alpha_k|$ for all $\alpha_1,\dots,\alpha_n \in \mathbf{R}$ and hence for $E^n = \operatorname{span}\{f_1,\dots,f_n\}$ we have

$$\lambda(E^n_\infty, L_\infty) = 1.$$

Let $\alpha = \alpha(\varepsilon) = \frac{1-\varepsilon n}{n^2}$. We introduce functions

$$h_1 := f_1, \quad h_k = f_1 + \cdots + f_{k-1} - \left(\frac{\varepsilon}{\alpha} + (k-1)\right) f_k.$$

The functions $h_k \in \operatorname{span}\{f_1,\dots,f_k\} \subset E^n$ and we claim that the functions h_k form an orthogonal basis in E^n_2. To show this we need to show that

$$\langle h_k, f_j\rangle := \int_0^1 h_k(t) f_j(t)dt = 0 \quad \text{for} \quad j = 1,\dots,k-1.$$

$$\begin{aligned}\text{Indeed} \quad \langle f_j, h_k\rangle &= \sum_{\substack{m=1\\ m\neq k}}^{k-1} \langle f_j, f_m\rangle + \langle f_j, f_j\rangle - \left(\frac{\varepsilon}{\alpha} + (k-1)\right)\langle f_j, h_k\rangle \\ &= (k-z)\alpha + \varepsilon + \alpha - \left(\frac{\varepsilon}{\alpha} + (k-1)\right)\alpha = 0.\end{aligned}$$

Observe that

$$h_k(t) = \begin{cases} 1 & \text{if } t \in [0, \varepsilon(k-1)) \\ -\left(\frac{\varepsilon}{\alpha} + (k-1)\right) & \text{if } t \in [\varepsilon(k-1), \varepsilon l) \\ 0 & \text{if } t \in [\varepsilon k, \varepsilon n] \\ -\frac{\varepsilon}{\alpha} & \text{if } t \in (\varepsilon n, 1]. \end{cases}$$

Hence

$$\delta_k = \langle h_k, h_k\rangle = \varepsilon\left[(k-1) + \left(\frac{\varepsilon}{\alpha} + (k-1)\right)^2 + \frac{\varepsilon^2}{\alpha}\right].$$

We now form the projection

$$Q_n = \sum_{k=1}^n \frac{1}{\delta_k} h_k \otimes h_k$$

and estimate its norm.

$$\begin{aligned}\|Q_n\|_{L_\infty \to L_\infty} &= \sup_s \int \left|\sum \frac{1}{\delta_k} h_k(s) h_k(t)\right| dt \\ &\geq \sup_{s\in[0,\varepsilon)} \int_\varepsilon^1 \left|\sum \frac{1}{\delta_k} h_k(s) h_k(t)\right| dt \\ &= \left(\sum_{k=1}^n \frac{\varepsilon}{\alpha\delta_k}\right)[1-\varepsilon n] = \frac{\varepsilon(1-\varepsilon n)}{\alpha} \sum_{k=1}^n \frac{1}{\delta_k} \\ &= \varepsilon n^2 \cdot \sup \frac{1}{\delta_k} \geq c\varepsilon n^2 \frac{1}{\varepsilon} \sum_{k=1}^n \frac{1}{k^2}.\end{aligned}$$

■

Remark. If $\lambda(E^n_\infty, L_\infty) = 1$ then there exists an interpolating projection of norm 1 onto E^n_∞ and hence it is easy to construct a weighted orthogonal projection of norm $1 + \varepsilon$. It would be interesting to know if this is true in general. Let E^n be such that $\lambda(E^n_\infty, L_\infty) = \mathcal{O}(1)$. Does it imply that there exist positive measures μ_n such that $\|Q_n(\mu_n)\|_{L_\infty \to L_\infty} = \mathcal{O}(1)$, where $Q_n(\mu_n)$ is the orthogonal projections onto E^n in the Hilbert space $L_2(\mu_n)$?

4 Projections Onto the Direct Sums of Subspaces

In the case of blending projections there seems to be a relationship between the projection constants of subspaces and the projection constant of the direct sum of these subspaces.

Example 3. There exist sequences of subspaces E^n, $H^n \subset L_\infty[0,2]$ such that $E^n \cap H^n = \{0\}$;

$$\lambda(E^n_\infty, L_\infty) = \lambda(H^n_\infty, L_\infty) = 1 \quad \text{and}$$

$$\lambda((E^n \oplus H^n)_\infty, L_\infty) \geq \frac{1}{2}\sqrt{n}.$$

Proof. As in Example 1, we pick functions $\varphi_1, \ldots, \varphi_n \in L_\infty(1,2)$ such that

$$\lambda(\text{span}\,(\varphi_1, \ldots, \varphi_n)) \geq \frac{1}{2}\sqrt{n}.$$

Next we pick $\psi_1, \ldots, \psi_n$ to be a picked partition of unity on $[0,1]$ so that

$$\|\Sigma\alpha_j\psi_j\|_{L_\infty[0,1]} = \max|\alpha_j|,$$

for all $\alpha_1, \ldots, \alpha_n \in \mathbf{R}$. Finally choose $\eta \geq 0$ so small that $\eta\|\Sigma\alpha_j\varphi_j\|_{L_\infty[1,2]} \leq \max|\alpha_j|$ for all $\alpha_1, \ldots, \alpha_n \in \mathbf{R}$. Now let

$$g_j(t) = \begin{cases} \psi_j(t) & \text{if } t \in [0,1) \\ \varphi_j(t) & \text{if } t \in [1,2]. \end{cases}$$

Let

$$\tilde{g}_j(t) = \begin{cases} \psi_j(t) & \text{if } t \in [0,1) \\ 0 & \text{if } t \in [1,2] \end{cases}$$

and set $E^n = \text{span}\,\{g_j,\ j = 1, \ldots, n\}$; $F^n = \text{span}\,\{\tilde{g}_j\ ;\ j = 1, \ldots, n\}$. Clearly $E^n \cap F^n = \{0\}$; $\lambda(E^n) = \lambda(F^n) = 1$. Consider the operator $Q_n : E^n \oplus F^n \to E^n \oplus F^n$ acting as follows:

$$(Q_n f)(t) = \begin{cases} f(t) & \text{if } t \in [1,2] \\ 0 & \text{if } t \in [0,1). \end{cases}$$

It is easy to see that Q_n is a projection from $E^n \oplus F^n$ onto the span $\tilde{\varphi}_1, \ldots, \tilde{\varphi}_n\}$ where

$$\tilde{\varphi}_n(t) = \begin{cases} \varphi_n(t) & \text{if } t \in [1,2] \\ 0 & \text{if } t \in [0,1). \end{cases}$$

Clearly $\|Q_n\|_{(E^n \oplus H^n)_\infty} = 1$ and $\lambda(\text{span}\,\{\tilde{\varphi}_j\}) \geq \frac{1}{2}\sqrt{n}$ since span $\{\tilde{\varphi}_j\})_\infty$ is isometric to (span $\varphi_j)_\infty$. Now let P_n be an arbitrary projection from $L_\infty(0,2)$ onto $(E^n \oplus H^n)_\infty$. Then $Q_n P_n$ is a projection from $L_\infty(0,2)$ onto (span $\{\tilde{\varphi}_j\})_\infty$ and we have

$$\frac{1}{2}\sqrt{n} \leq \|Q_n P_n\| \leq \|Q_n\| \|P_n\| \leq \|P_n\|.$$

■

References

[1] B. Beauzamy. *Introduction to Banach Spaces and their Geometry,* North-Holland, 1985.

[2] C. de Boor. *A Bound on the L_∞-Norm of L_2-Approximation*, Mathematics of Computation, **30**(1976), 765-771.

[3] J. Lindenstrauss and L. Tsafriri, *Classical Banach Spaces* I, Springer-Verlag, 1977.

[4] J. Marti, *Introduction to the Theory of Bases*, Springer-Verlag, 1969.

[5] A. Pietsch. *Operator Ideals*, North-Holland, 1980.

SUBJECT INDEX